VERTEBRATE REPRODUCTION

VERTEBRATE REPRODUCTION

E. W. Jameson, Jr.

Department of Zoology
University of California, Davis
Davis, California

WILEY

A WILEY-INTERSCIENCE PUBLICATION

JOHN WILEY & SONS

New York / Chichester / Brisbane / Toronto / Singapore

Library of Congress Cataloging in Publication Data

Jameson, E. W. (Everett Williams), 1921–
  Vertebrate reproduction / E. W. Jameson, Jr.
    p. cm.
  "A Wiley-Interscience publication."
  Includes index.
  ISBN 0-471-62635-X
  1. Vertebrates—Reproduction. I. Title.
QP251.J27 1988 88-5891
596′.016—dc19 CIP

Printed in the United States of America

10 9 8 7 6 5 4 3 2 1

To my wife, Sumiko

PREFACE

The principles of evolution and ecology are predicated on the mechanics of reproduction. The environmental forces that ultimately form evolutionary pathways act through an elaborate complex of hormonal, behavioral, and developmental components. These successive events begin at the earliest point—sex determination and embryological differentiation—and continue through courtship and mating.

Students have access to several excellent volumes on reproductive physiology. These texts constitute an essential basis for the study of vertebrate reproduction. It is no shortcoming of these books that they report research primarily on laboratory and domestic species of mammals. For students who wish to explore the reproductive adaptations of wild vertebrates, this volume attempts to integrate internal rhythms of individuals and populations to seasonal and irregular environmental changes.

The diversity of reproductive patterns among both dissimilar and closely related taxa is impressive. This diversity reflects not only evolutionary adaptations to long-term environmental changes but also sometimes great flexibility to short-term fluctuations in the environment. This should caution the student who is prone to place all reproductive modes into a small number of general categories.

E. W. JAMESON, JR.

ACKNOWLEDGMENTS

It has been my good fortune to receive assistance from many generous friends and colleagues, without whose help I could not have assembled this material. Prominent among these friends are N. R. Adams, R. E. Ballinger, M. S. Barkley, R. L. Baldwin, A. T. Cowie, C. Desjardins, V. de Vlaming, L. C. Drickamer, C. A. Erickson, D. S. Farner, B. Gondos, H. J. Grier, R. Grey, L. J. Guillette, Jr., M. W. Hardisty, A. Hedrick, G. Högstedt, F. J. Karsch, L. B. Keith, N. Kline, J. Kenagy, S. J. Legan, W. L. Lidicker, Jr., P. G. Link, J. Loring, R. T. Reiter, W. R. Rice, G. W. Salt, A. M. Shapiro, R. M. Sharpe, R. Shine, M. Simon, N. Stimson, N. E. Stacey, J. Stamps, H. W. J. Stroband, P. F. Terranova, C. A. Toft, P. Q. Tomich, T. A. Uchida, M. H. Wake, R. A. Wallace, K. E. F. Watt, R. J. Wootton, and I. Zucker. To all these generous people I tender my thanks. I also wish to thank E. Kiang of John Wiley & Sons, Inc., for her assistance throughout the editorial and production phases.

E. W. JAMESON, JR.

CONTENTS

PART I

SEXUAL DEVELOPMENT AND GONADAL ACTIVITY

Many aspects of vertebrate reproductive behavior are determined during fetal development. Not only is the sex established at the time of fertilization, but modifications of sexual roles can be altered by embryonic position. Very early—prepubertal—social environments also alter the rate of sexual development. Social interactions also stimulate adult behavior, from gonadal development to courtship and mating. An appreciation of these physiological phenomena is essential to understanding vertebrate reproduction in the field.

1

GONADAL DEVELOPMENT AND SEX DETERMINATION

1.1 EARLY GONADAL DEVELOPMENT

A brief description of the development of the gonads is helpful in understanding their activity as well as some aspects of sex determination, hermaphroditism, and growth. In the vertebrate embryo, early gonadal development begins subsequent to the establishment of major features of the body plan. These processes have been investigated most intensively in laboratory mammals and some domestic species, and the literature has been reviewed by Hardisty (1978), Serra (1983), and Zuckerman and Baker (1977).

1.1.2 Migration of Primordial Germ Cells

Primordial germ cells (PGCs) enter the embryo after it has acquired a longitudinal axis and a cephalic end; prior to their migration, PGCs occupy various extraembryonic sites. PGCs are distinguished at an early stage by their large size (which varies among different taxa) and, in most nonmammalian vertebrates, by their large reserves of lipids, glycogen, and/or yolk. Before and during migration, mammalian PGCs can be recognized by their high alkaline phosphatase activity. The energy reserves are largely depleted by the time the PGCs have arrived at the genital, or gonadal, ridge, on the dorsal lining of the coelomic cavity. Details of this migration are known for very few vertebrates. In almost all cases (salamanders being an exception), PGCs are derived from endoderm and originally occupy an extraembryonic position in a region known as the germinal crescent.

In lampreys PGCs move through the endoderm of the yolk sac and migrate dorsad via the mesentery to the genital ridge. In several unrelated species of fish, PGCs migrate from the gut endoderm dorsally through the mesentery and then laterally to the genital ridges. In other fish species the route appears to be through somatic mesoderm or even myomeres. Hardisty (1978) pointed out that these apparent differences in migration path may be due to the stage in which PGCs are found and that the different routes are not necessarily mutually exclusive. Active amoeboid movement is known for some taxa, and passive passage, in the circulatory system, is also important. In some species the migration may be prolonged. In the golden mullet (*Mugil auratus*), PGCs enter the gonadal ridges when the total embryo length is 9 to 10 cm, at about eight months of age, which is much later than in some related species (Bruslé and Bruslé, 1978).

The origin and movement of PGCs have been explored experimentally in several anurans. In early cleavage stages of species of *Rana* and *Xenopus*, the removal or exposure to ultraviolet radiation of critical areas in the vegetal pole is followed by a reduction or absence of germ cells in the genital ridges. This effect is reduced as cleavage proceeds (Hardisty, 1978). From their original extraembryonic site, anuran PGCs move by amoeboid activity, appearing later in the dorsal mesentery and subsequently in the genital ridges, which are already well developed. Amphibian gonads lie attached to the median ventral margins of the kidneys.

There appear to be basic differences in the origin of PGCs of salamanders (Caudata). Although the very early embryonic position of PGCs has not been determined in these amphibians, destruction of endoderm does not result in sterility or in the failure of PGCs to arrive at the genital ridges, as in anurans. Instead, PGCs of salamanders appear to arise from totipotent cells in the animal pole (Nieuwkoop and Sutasurya, 1976). This proposal advocates the contention that the Amphibia are diphyletic (Schmalhausen, 1968), which is supported by fundamental differences in fertilization, courtship, and geographic distribution.

Migration routes are similar in birds and reptiles. In the early development of the chick (18 hours of incubation), PGCs occupy an extraembryonic position in the germinal crescent, anterolateral to the embryo. In reptiles PGCs are found in a similar position (Fig. 1.1). By 28 hours in the chick, PGCs have entered the circulatory system, where they occur abundantly until 50 to 55 hours. PGCs leave the vascular system through lacunae in capillary walls and migrate by amoeboid movement through the dorsal mesentery to the genital ridges, in both birds and reptiles (Clawson and Domm, 1963; Hubert, 1968).

In mammals, after migration from the yolk sac to the genital ridges, PGCs settle at the genital ridges. Migration is completed by day 10 in the laboratory mouse, by days 16 to 21 in dogs, and by days 35 to 40 in the human embryo. Gonadal development commences promptly; but at the time PGCs enter the tissues that will form the gonads, there is no differentiation into ovarian or testicular tissue.

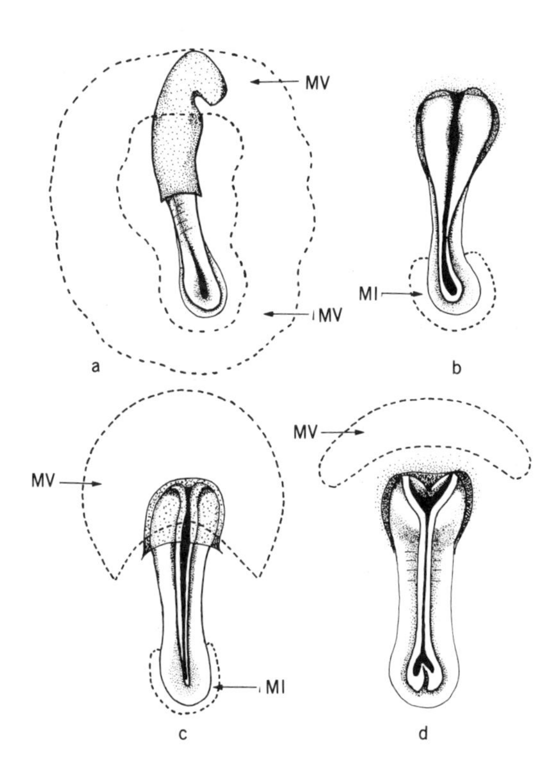

Figure 1.1 Premigratory position of germ cells of several reptiles. Germinal distribution shown by dashed lines. MV, vascular migration; MI, tissue migration. (a) *Anguis fragilis*, (b) *Lacerta vivipara*, (c) *Sphenodon punctatus* and (d) *Vipera aspis*. (From Hubert, 1976, courtesy of *Anée Biologie*.)

1.2 SEXUAL DIFFERENTIATION OF THE GONAD

As they reach the genital ridges, PGCs become enveloped by or enter the germinal epithelium of the gonad. This epithelium folds inward and forms the cortical sex cords. While the gonad is in its early, bipotential condition, the central (medullary) part can develop into a testes and the marginal (cortical) area an ovary (Witschi, 1962). In males the medullary tissue invades the cortical layer. In females (of mammals) the gonad may remain more or less undifferentiated or the cortex may show further development. A well-developed cortex is present in the ovary of the domestic rabbit, for example, but not in the laboratory mouse.

In the early differentiation of ovarian tissue, cortical tissue proliferates at a greater rate than does the medulla, which either regresses or remains in an embryonic, undifferentiated state. Major aspects of gonadal differentiation include the development of seminiferous tubules and Sertoli cells in males and ovarian follicles and stroma in females. Sex cords become Sertoli cells in males and granulosa cells in females. Normally the Wolffian ducts become the vas deferens and seminal vesicles, and the Müllerian ducts become the uterus and oviducts. In caecilians Müllerian ducts persist in the males, and the terminal part becomes glandular (see Chapter 4, Section 4.2). If the embryonic gonad is destined to become testicular, the medulla develops at the expense of the cortex. Thus the embryonic gonad has the potential to develop into either an ovary or a testis, and in some taxa gonadal plasticity persists throughout much or all of the adult life.

The gonad of early larval lampreys is undifferentiated. After metamorphosis gonadal tissue of only one sex develops. In hagfish, gonadal differentiation occurs along a longitudinal axis: the developing gonad is ovarian at the anterior end and testicular at the caudal end. Both functional hermaphrodites and gonochoristic (unisexual) individuals exist.

Some species of teleosts are either sequentially or simultaneously hermaphroditic. There first develops either an undifferentiated ovotestis, with both testicular and ovarian cells scattered more or less evenly, or a condition where male and female tissues occupy separate parts of the same gonad. On the other hand, in some gonochoristic species (in which the sexes are separate), the sex of the individual is established at an early age. The gonadal plasticity that provides for hermaphroditism in fish is confined to oviparous species, since live-bearing fish have well-developed secondary sexual characteristics that provide for internal fertilization and internal development of young.

In the synchronous hermaphroditic fish *Serranus hepatus*, ovarian tissue in the ovotestis differentiates prior to testicular tissue. In very young individuals (1.2–2.7 cm, or at 2–3 months of age), oogonia can be identified and testicular tissue is absent. In fish from 3 to 4 cm, the ovotestis is 1.5 cm long and there is testicular tissue devoid of germ cells. In addition, oocytes are in meiotic prophase. Spermatogonia appear in the testicular section in fish 4.5 cm long. As PGCs differentiate, the cytoplast develops endoplasmic reticulum and Golgi apparatus (Buslé, 1983).

The following sections on early gonadal growth relate primarily to mammalian development.

1.2.1 Early Ovarian Growth

After they arrive at the genital ridges, the PGCs cease amoeboid activity and become more spherical. They continue to divide mitotically and are henceforth known as oogonia. They lie within the sex cords and are connected by cytoplasmic bridges, which are remnants of incomplete cytokinesis. As the cords, which have developed from the splanchnic epithelium, penetrate the medulla, the rete ovarii, derived from the mesonephros, extend from the medulla to the cortex (Fig. 1.2). There is no steroidogenic activity in the mammalian embryonic ovary.

1.2.2 Early Testicular Growth

The mesonephros develops prior to the time at which the testis begins to differentiate. The genital ridges lie adjacent to the mesonephros at about the time the PGCs move to the dorsal margin of the coelomic cavity. The PGCs settle on the surface of the genital ridges and are carried inward with invaginations of the epithelium. As previously mentioned, these invaginations form the cortical (or epithelial) sex cords and become seminiferous tubules. Initially the sex cords are solid and at first grow in length, but not in width. Eventually they do grow in width and develop a lumen, at which point they become seminiferous tubules. From the surrounding mesenchyme,

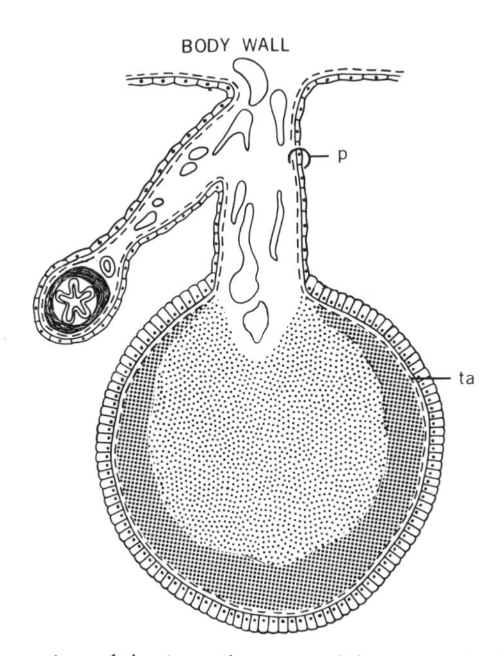

Figure 1.2 Cross section of the juvenile ovary of the mountain beaver (*Aplodontia rufa*), showing the differentiated medulla (open stipple) and the cortex (closed stipple); p, perotineum; ta, tunica albuginea. (From Mossman and Duke, 1973, courtesy of the University of Wisconsin Press.)

Leydig cells differentiate and they synthesize and secrete androgens. Additionally the embryonic testis secretes a nonsteroidal substance that suppresses development of the Müllerian ducts. Gametogenesis does not occur in the fetal testis, as it does in the fetal ovary, but is postponed until the onset of sexual maturity.

Anteriorly the mesonephric tubules become the rete testis and, within the testis, connect to the free, unattached ends of the seminiferous tubules. Posteriorly the rete testis leads to the epididymis. These tubular changes occur rather early in testicular formation and are apparently induced by androgens from the surrounding Leydig cells.

Leydig cells, which have developed from the testicular connective tissue, commence to produce testosterone in the first trimester of development, or at about 70 days in the human (Wilson et al., 1981). It is not clear whether embryonic steroidogenesis is initiated by embryonic pituitary tissue or by placental gonadotropins.

Sertoli cells gradually develop during the period between birth and sexual maturity. In the testicular cords of the fetal testis are *light* and *dark* cells, which are precursors of Sertoli cells. These cells divide mitotically prior to sexual maturity, but their proliferation ceases when spermatogenesis begins. It has been suggested that the dark cells stimulate mitotic divisions and the light cells promote meiosis (Wartenberg, 1983).

1.2.3 Masculinization and Defeminization of the Brain

Early in the development of the individual, tissues become differentially sensitive to stimulation. As growth proceeds, tissues may become refractory to stimulation, and inhibitors may suppress mitosis, perhaps by negative feedback. Negative feedback is familiar in ovarian-hypothalamic-hypophyseal function. One hypothesis proposes that a kind of negative feedback within a tissue results from the formation of suppressants called chalones, which diffuse through an old part of a tissue to a germinative layer.

The nature of adult sexual behavior—male or female—is determined during embryonic life and displayed later, during the adult activities of courtship and mating. The basic sexual orientation is shown by the homogametic sex: that is, female in mammals and male in birds. Steroids secreted by the gonad of the heterogametic sex during embryogenesis alter the basic sexual orientation.

In embryonic mammalian development, testosterone induces development of steroid receptors in the brain (or masculinization, as well as defeminization, of the brain). This codes the brain for both later male behavior and loss of female behavior. Although the embryonic ovary is not steroidogenic, the embryonic testis is active and accounts for serum androgens. Without androgens during this responsive, critical period, the mammalian brain remains unchanged and feminine. The level of androgens circulating in developing females (Challis et al., 1977) appears to be insufficient to alter later brain orientation.

It has been suggested that estradiol, not testosterone, accounts for masculinization and defeminization and that this estradiol results from intracellular aromatization of testosterone (MacLusky and Naftolin, 1981). Although maximal hypothalamic sensitivity seems to occur in fetal life, the critical period for masculinization and defeminization varies among different mammals. There is clearly prenatal sensitivity in most species, and the sensitivity extends past birth in some. In the ferret (*Mustela furo*), coital masculinization may result partly from a surge in testosterone several hours after birth, suggesting that prenatal modification of brain cells by estrogen renders them responsive to postnatal testosterone and that there is a synergistic action of these two steroids. In the ferret, coital masculinization, which begins prior to birth, may continue up to two weeks or more after birth (Baum, 1987).

Embryos of both sexes, however, cannot avoid estrogens that pass from the parent through the placenta. The hypothalamus is unexposed to these exogenous estrogens by an estrogen-binding protein from the embryonic liver. Endogenous androgens are not bound by this material (Raynaud et al., 1971).

Embryonic mammals may be affected by testosterone from an adjacent embryo, and it is well established that intrauterine position in mice alters the genetically determined sexual development of adjacent female embryos (vom Saal, 1983). Although the pathway by which steroids diffuse from a male fetus to a female fetus is not apparent, there is clear evidence for such movement. In polytocous mammals, the position of female embryos can be indicated as 2M (between 2 males), 1M (between a male and a female), and 0M (between 2 females) (Fig. 1.3). Females

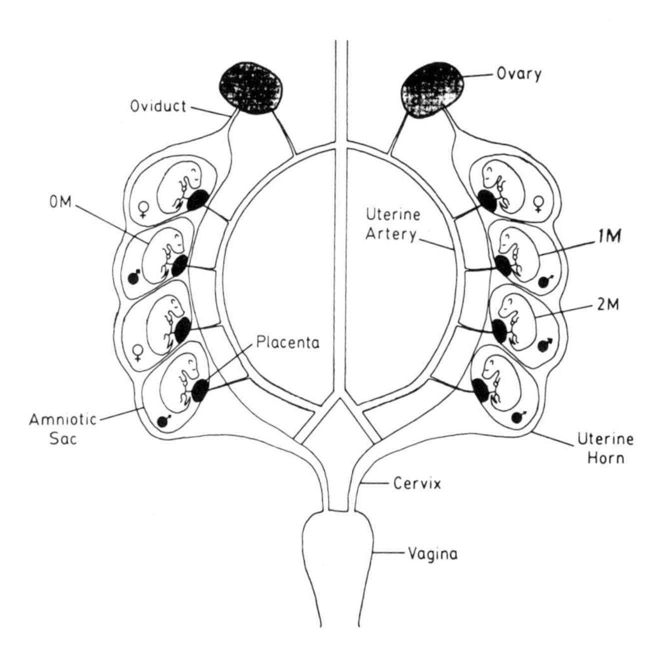

Figure 1.3 A hypothetical arrangement of mouse embryos, with random distribution of sexes, determined by Caesarean delivery at term. "M" indicates the number of male embryos lying next to a given embryo: 1M, lying next to one male embryo, 2M, male embryos on each side, and 0M, female embryos on each side. (From vom Saal, 1983, courtesy of Springer-Verlag.)

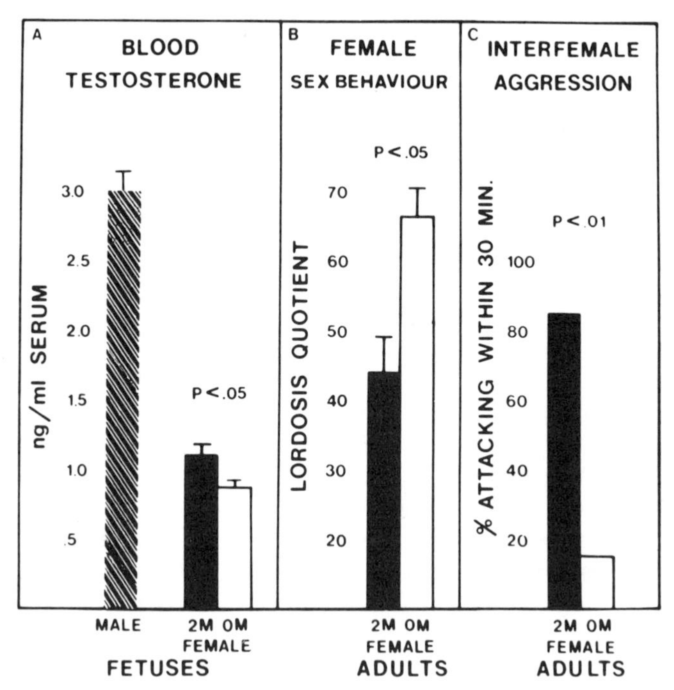

Figure 1.4 (*a*) mean (± SEM) blood titers of testosterone for males and for 0M and 2M (5 pools; 25 embryos per pool) on day 17 of gestation. (*b*) mean (± SEM) lordosis quotient: (Number of lordoses/number of mounts) × 100 for 2M females and 0M female mice (10/group). All females were ovariectomized at 90 days of age and tested for sexual receptivity when 150 days old. Forty-eight hours before testing, females were injected with estradiol benzoate (5 μg) and 4 hours before testing with 200 μg progesterone. A 0M and 2M female were then matched for age and weight, and placed with a stud 1M male for 30 minutes. (*c*) the percentage of 2M and 0M female mice that showed aggression (tail rattling, chasing, and biting) during a 30-minute test. Twenty-eight pairs of matched pairs of 2M and 0M females were observed for aggression when in diestrous phase of estrous cycle. In eight pairs there was no aggression. Results based on the 20 pairs in which there was aggression. Significant levels for (*a*) and (*b*), *t*-test; and (*c*) chi-square comparisons. (From vom Saals, 1983, courtesy of Springer-Verlag.)

placed between two males in a single uterine horn have more amniotic and serum testosterone than do female embryos placed between any other combination of embryos. This intrauterine position is reflected in their adult morphology and physiology (vom Saal et al., 1981). There are also significant differences in adult behavior (Fig. 1.4).

In birds, in which the female is heterogametic, ovarian hormones alter brain function. Embryologically female galliform birds become demasculinized in development. Thus the nervous system is initially bisexual, and its orientation is developmentally determined by ovarian secretions during embryogenesis. Early gonadal steroids (estradiol 17-β) nevertheless appear to account for the masculinizing of the song system in the female zebra finch. Subsequently, such a masculinized female can be induced to sing by the administration of either testosterone or 5α-dihydrotestosterone (Gurney and Konishi, 1980).

In embryonic Japanese quail, testosterone propionate or estradiol benzoate demasculinizes males so that their behavior is feminine. In the female Japanese quail, blocking of estrogen on day 9 of incubation results in masculinization (Adkins, 1976). When estradiol benzoate was injected into laying hen quail (10, 20, 30, 40, or 50 g) for two to three weeks, some of the steroid apparently moved into the yolks of eggs laid a week later, and some males from these eggs showed evidence of copulatory deficiency (Adkins-Regan, 1981). Conceivably, wild females ingesting phytoestrogens as they forage could pass on such steroids via the yolk to their young.

Centers of sexual activity in some mammals exist in the preoptic anterior hypothalamus area (for males) and the ventromedial nuclear area (for females). Thus, with respect to sexual and coital behaviors, the steroid receptor areas appear to be separate for the two sexes. This sort of differentiation is also known to occur in birds, but there is a distinct neural circuitry for song in passerine birds (Balthazart and Schumacher, 1983).

1.3 SEX DETERMINATION

The sex of an individual is usually established very early in gonadal development in most vertebrates. Hermaphroditic fish and some amphibians constitute exceptions, for in some of these species sex reversals may normally occur after most of the somatic growth, or even after sexual maturity. Gender is frequently predicated by combinations of sex chromosomes, but in some ectotherms the direction of sexual development is also affected by environmental factors.

In fish, amphibians, and reptiles the developing gonads remain bipotential, capable of becoming either ovaries or testes, well into advanced stages of embryological development and, in some frogs and turtles, into the first year after hatching. In these species the medullary tissue becomes testicular and the cortical tissue is ovarian, but eventually one develops at the expense of the other.

1.3.1 Hormonal Influences on Embryonic Gonads

Experimentally, exogenous hormones may modify the direction of gonadal differentiation in birds and mammals but will not completely determine the destiny of the embryonic gonad (McCarrey and Abbott, 1979). These situations would not occur in nature.

In commercial fish culture it is desirable to maximize the frequency of female offspring. The sex ratio can be manipulated in two ways. All-female populations of Atlantic salmon and rainbow trout have been obtained by feeding them 17-β-estradiol shortly after their first meal. Also, by feeding the fish dietary androgen, genetic females (XX) become functional males, but they are phenotypically distinguishable from genetic males (XY). When the XX "males" are crossed with normal genetic females, the offspring are all female (Johnstone and Youngson, 1982). Similarly, the grass carp female becomes a functional male following the insertion of a capsule releasing methyltestosterone into the abdominal cavity. The offspring from mating XX males to normal females will produce only normal females (Shelton, 1982).

In the medaka (*Oryzias latipes*), a freshwater teleost, however, exogenous estrogens and androgens administered early in embryogenesis do result in functional sex changes (Yamamoto, 1958). The guppy (*Poecilia reticulatus*), when treated with androgens (methyltestosterone, administered orally to the gravid female) during embryonic development, develop into functional males with spermatogenesis (Takahashi, 1975).

The effect of exogenous steroids on reptiles, which lack a larval stage and which have slight or no absorption of fluids through the skin, is weaker than in amphibians. Exogenous hormones can, however, influence sexual development in turtles. Estradiol benzoate and testosterone propionate injected into the eggs of the snapping turtle (*Chelydra serpentina*) and the painted turtle (*Chrysemys picta*) both cause most or all of the young to develop as females (Gutzke and Bull, 1986).

1.3.2 The Relationship of Sex Chromosomes to Sex Determination

For a long time much of the information on sex chromosomes of lower vertebrates was based on genetic evidence (Mittwoch, 1967), for there is little apparent difference between the sex chromosomes in fish, amphibians, and reptiles. Heteromorphic (het) sex chromosomes are now known for some species of teleost fish, amphibians, and reptiles. In several kinds of the lungless salamanders (Plethodontidae), het sex chromosomes clearly characterize males. In anurans het sex chromosomes are known in males of the genera *Rana* and *Hyla*, but in females in *Xenopus* and *Bufo* (Becak, 1983). Although visually dissimilar het sex chromosomes are less common in fish, amphibians, and reptiles than in birds and mammals, that does not necessarily preclude a genetic sex determination for ectotherms. It may be more realistic to view environmental influences, especially temperature, as modifying the role of genetic sex determination.

Many reptiles appear to be homomorphic. Among those lizards in which het sex chromosomes are known, males are heterogametic in about half of the species and females are clearly heteromorphic in the other half. Male het sex chromosomes have been found in chelonians (King, 1977), and sex chromosomes are homomorphic in crocodilians.

In birds the females are heterogametic, and in most mammals the male is the heterogametic sex. Some exceptions to this general rule are now known for mammals. Although the inheritance of sex chromosomes provides clues to the determination of sex, critical chromosomal aspects lie in the specific mechanisms by which features of one sex are promoted while those of the other sex are suppressed.

1.3.3 The H-Y Antigen

The X chromosome in mammals bears many genes that affect somatic development and do not relate to sexual development. The homologue in males, the Y chromosome, lacks alleles to many of the genes on the X chromosome. The Y chromosome has been associated with a histocompatibility (H-Y) antigen that accounts for the rejection of male skin grafts by females, and this H-Y antigen refers to antigens on mammalian male cells that cause formation of antibodies by females.

The basic pattern of feminine development in mammals seems to be modified by the Y chromosome. The medulla (of the gonad) becomes enlarged and testicular under the influence of the H-Y antigen. The H-Y antigen seems to control differentiation in the heterogametic sex: that is, it promotes testicular development in mammals and ovarian development in birds.

From very early stages of cell division until the formation of the blastocyst, the H-Y antigen of the male is active, and in its absence the mammalian gonad becomes an ovary (Haseltine and Ohno, 1981). Thus the genetic sex, established at fertilization (in mammals), is followed by gonadal differentiation. The H-Y antigen has been assumed to determine maleness in mammals and femaleness in birds. However, the occurrence of phenotypic male mice (XX) with well-developed testes but no H-Y antigen seems to indicate that it has another function. These particular phenotypic XX male mice were sterile, suggesting that the H-Y antigen may be responsible for spermatogenesis (McLaren et al., 1984).

Because the H-Y antigen is usually associated with the heterogametic sex, it had been assumed to be determined by a gene on the Y chromosome (Wachtel et al., 1975). It has been suggested, however, that the gene for the H-Y antigen may be autosomal and that there is a suppressor gene on the X chromosome, with a gene on the Y chromosome that is antagonistic to it (Wolf et al., 1980). In humans the testis-determining factor is located on the short arm of the Y chromosome, and the gene for the H-Y antigen is on the long arm or centromeric region of the Y chromosome (Simpson et al., 1987). In at least some nonmammals, sex steroids appear to induce the H-Y antigen (Zaborski, 1982).

1.3.4 Variations in Distribution of Mammalian Sex Chromosomes

Although the H-Y antigen appears to be on an autosomal locus, the Y chromosome is nevertheless very closely tied to masculine development in mammals. It is interesting therefore to review some examples of mammals in which sex determination violates the usual XX female and XY male pattern.

In marsupials, as in eutherian mammals, males are XY and females XX. In some species, however, supernumerary sex chromosomes occur: there are X, X_1, and X_2 chromosomes and there are $X_1Y_1Y_2$ males and $X_1X_1X_2X_2$ females (Hayman and Rofe, 1977). Among the spiny anteaters (*Tachyglossus aculeatus* and *Zaglossus bruijnii*), the female has two pairs of chromosomes of which the male has only one homologue. These chromosomes are designated X_1 and X_2, and a female is $X_1X_1X_2X_2$

and a male X_1X_2Y. Thus both sexes of the spiny anteaters have het sex chromosomes. In the remaining monotreme, the duckbill (*Ornithorhynchus anatinus*), the male/female sex chromosome arrangement is the more usual XY/XX (Murtagh and Sharman, 1977).

The wood lemming (*Myopus schistocolor*), a vole of northern Eurasia, has XY chromosomes, but a gene on one X chromosome of an XX female can suppress the gene for testis formation on the Y chromosome. These XY females, moreover, can produce both XX and XY females but not XY males (Fredga et al., 1976). Thus most wood lemmings are female. In the wood lemming the H-Y antigen may occur in females of both XY and XX genotypes (Wiberg et al., 1982). A similar mechanism of sex determination has been described for the lemming (*Dichrostonyx torquatus*) by Beneson (1983). The effectiveness of the H-Y antigen may result from an absence of an H-Y receptor. That is, testicular development depends not only on an adequate dose of the H-Y antigen but also on the specific gonad receptor.

Another variation is seen in the mole rat (*Ellobius lutescens*), a vole (Arvicolidae) of the Old World. The mole rat has XO sex chromosomes that are visually identical in both sexes. In this species both XX and OO combinations are lethal (Ohno, 1977). A similar situation seems to occur in the spiny rat (*Tokudaia osimensis*), a murid rodent known only to exist on the island of Amami Oshima of southern Japan (Fredga, 1983). Various other ratlike rodents exhibit similar departures from the familiar XX/XY female/male mammalian pattern.

In the creeping vole (*Microtus oregoni*) of western North America, females are XO and males are XY. In spermatogenesis, nondisjuction produces XXY and OY germ lines, and only the OY survive. Females develop when ova are fertilized by O sperm, which lack a sex chromosome. When XO chromosomes are differentiated into oogonia in the embryonic development of the female, nondisjunction of the X chromosome results in XX oogonia, and such females are homogametic (Ohno et al., 1966).

In mice there may be a sex reversal factor on the distal end of the X chromosome as a result of a translocation from the Y chromosome (Singh and Jones, 1982). This could result in XX individuals that are phenotypically male. Other variations on the mammalian Y chromosome, as well as on autosomes, are transferable to the X chromosome and affect phenotypic sex (Simpson, 1982).

In humans there is an occasional departure from the heterogametic XY condition of the male. Males with XX sex chromosomes are estimated to occur with a frequency of about 1 in 20,000 (Burgoyne, 1984). This condition may result from a transfer of some of the Y chromosome to the paternal X chromosome. The unequal size of the two X chromosomes supports the proposed heterogeneity of the XX chromosome (Evans et al., 1979). These XX males are also H-Y positive.

1.3.5 Environmental Influences on Sex Determination

In nature, sex ratios of some species of vertebrates show a marked departure from the 50:50 male:female ratio at birth reported for most species. These differences, sometimes correlated with habitat and sometimes with season, have stimulated researchers to incubate eggs at various temperatures and photocycles. It is now well established that seasonal and environmental variations in sex ratio are determined by temperature in the early embryonic stages of some fish, amphibians, and reptiles. In one species of fish, variations in day length favor the development of one sex over the other.

In fish the influence of temperature in the determination of sex is known for several species. In the hermaphroditic cyprinodont *Rivulus marmoratus*, of the Gulf of Mexico, the ovarian part of the ovotestis develops at the expense of the testicular element at low or moderate temperatures, and maleness is favored at high temperatures and short day lengths (Harrington, 1971).

Sex determination in the Atlantic silverside (*Menidia menidia*, an estuarine fish of the Atlantic coast of North America, is under the control of both environmental temperature and genotype. Young hatched from eggs spawned in July develop mostly into females, but from August to September males prevail among the hatchlings. Under controlled laboratory conditions eggs incubated under cold fluctuating temperatures (11–19°C) yielded mostly females, whereas those kept under warm fluctuating temperatures (17–25°C) produced mostly males. A genetic influence was apparent, however, because sex ratios at both temperature ranges varied with spawn from individual females (Conover and Kynard, 1981). The effect of temperature, moreover, is more pronounced at lower latitudes (Fig. 1.5).

In a number of frogs, toads, and salamanders the larvae and young adults contain only ovarian tissue, with vestiges of the medulla. About a year later the genotypic males develop testes (Foote, 1964). For example, in Taiwan all young of the bullfrog (*Rana catesbeiana*) are female at metamorphosis, and testicular tissue subsequently develops in genetic males (Hsu and Liang, 1970).

Temperature is a well-established determinant of sex in some reptiles. Among oviparous squamates the low end of the viable temperature range, about 27°C, favors the production of females, and males predominate when eggs are kept at near 31°C. Among chelonians the opposite response is seen (Bull, 1980). In the alligator (*Alligator mississipiensis*), males emerge from eggs incubated at 34°C, whereas incubation temperatures below 30°C result in broods entirely of females (Ferguson and Joanen, 1982).

One might ask whether genetic sex determination and temperature sex determination are mutually exclusive. It can be argued that in a temperature-sensitive species with het sex chromosomes, there will be XY females and therefore a 25% chance for the occurrence of YY (lethal) genotypes (Bull, 1980). This rationale assumes, however, that with genetic sex determination there must be markedly heteromorphic sex chromosomes, with the smaller sex chromosome bearing very few gene loci. It is possible that only slightly het sex chromosomes may meet the requirements of genetic sex determination. If the sex chromosomes are nearly homomorphic, which seems to be the case in many ectotherms, the two modes of sex determination would not seem to be mutually exclusive. Perhaps the genetic control of sex determination became the more influential mode with the development of endothermy, under which marked variations are less likely to occur, and that the role of temperature preceded the marked heteromorphic sex chromosomes of birds and mammals.

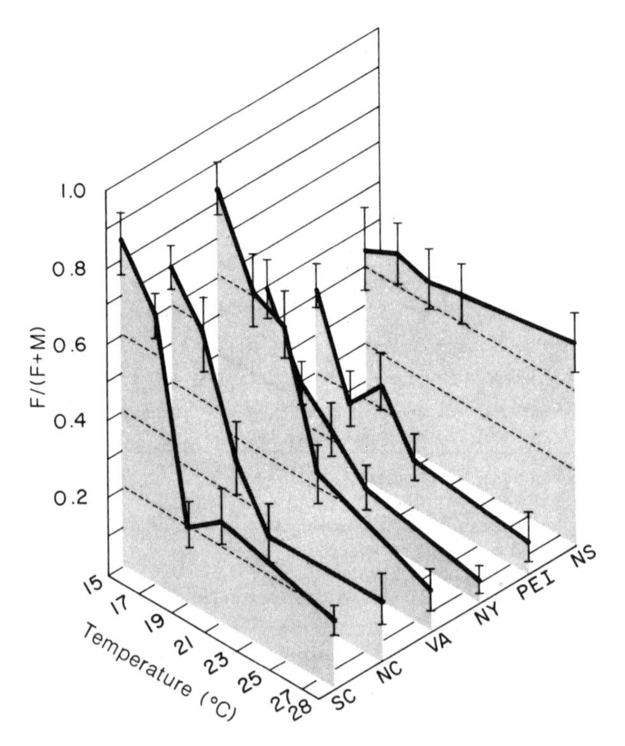

Figure 1.5 The relationship between sex ratio [F/(F + M)] and temperature in laboratory-reared silversides (*Menidia menidia*) from materials collected from South Carolina, North Carolina, Virginia, New York, Prince Edward Island, and Nova Scotia. Vertical lines indicate 95% binomial confidence limits. (From Conover and Heins, 1987, reprinted by permission from *Nature*, vol. 326, Fig. 1, p. 496, © 1987, Macmillan Magazines Limited.)

1.4 PARTHENOGENESIS

Some all-female species of vertebrates transmit an apparently unchanging genome from one generation to another, with or without the aid of males. This is accomplished by the production of diploid ova, probably through a suppression of cytokinesis in meiosis (Uzzell, 1970). Parthenogenesis has been reported, with some variation, in teleost fish, salamanders, and reptiles and is by far the most common among reptiles.

1.4.1 Fish

Although most fish are gonochoristic, hermaphroditic and parthenogenetic species exist. Heteromorphic chromosomes have been found, and possibly some sex-determining genes are located on autosomes (Price, 1984).

All-female populations of fish are known in the Poeciliidae, in southern Texas and northeastern Mexico. The well-known *Poecilia formosa* is a hybrid of two sympatric species. *P. sphenops* and *P. latipinna* of southern Texas and adjacent Mexico (Hubbs and Hubbs, 1932). Subsequent attempts to reconstruct this hybridization in the laboratory have failed, and one of the presumed parents (*P. formosa*) is now known to consist of several species (Turner et al., 1980). *P. formosa* mates with a male of one of the two gonochoristic species, and sperm initiates cleavage, but the sperm pronucleus does not unite with the egg nucleus, which is diploid. This type of parthenogenesis is called gynogenesis.

A variant on this pattern is hybridogenesis, in which the diploid egg nucleus does unite with the haploid sperm pronucleus to produce a triploid female. The male genome is discarded in oogenesis, so that subsequent ova contain only the two genomes of the female. Because of the heterogeneity of the paternal contribution, such a hybridogenetic population is phenotypically more variable and presumably more adaptive. This pattern is found in several all-female species of *Poeciliopsis* of Sonora, Mexico (Moore et al., 1970).

1.4.2 Amphibians

There are triploid all-female populations of the salamander genus *Ambystoma* in parts of northeastern United States and adjacent areas of Canada (Uzzell, 1964; Uzzell and Goldblatt, 1967). Some of these populations are derived from *A. laterale* to the north, *A. texanum* in Ohio (Sessions, 1982), and *A. jeffersonianum* in the northeastern United States (Kraus, 1985). The occurrence of all-female species from the hybridization of *A. laterale* and *A. texanum* resulted from the removal of forests and the tilling of land no more than 160 years ago (Kraus, 1985). Both diploid and triploid populations exist, reproducing by gynogenesis and hybridogenesis. A diploid population on an island in Lake Erie exists without any other species of *Ambystoma* and is clearly parthenogenetic. Surprisingly, no other parthenogenetic amphibians are known. In these salamanders the parental chromosomal condition ($3n$ or $2n$) is preserved by the suppression of cytokinesis in the last mitotic division prior to meiosis (Uzzell and Goldblatt, 1967).

1.4.3 Reptiles

Parthenogenesis has been reported for six families of lizards and one family of snake. Some of these families are of hybrid origin and some appear to have arisen by some other mechanism. In southwestern United States and parts of Mexico, interspecific hybridization has produced a number of parthenogenetic species of whiptail lizards (*Cnemidophorus*). When their distribution is carefully plotted on maps, their occurrence is clearly clustered about river valleys (Fig. 1.6), which are subject to sporadic flash floods. Flooding may have been critical in the development of these all-female species, for they occur in every major drainage system in southwestern United States (Cuellar, 1977). In one of these species (*Cnemidophorus*

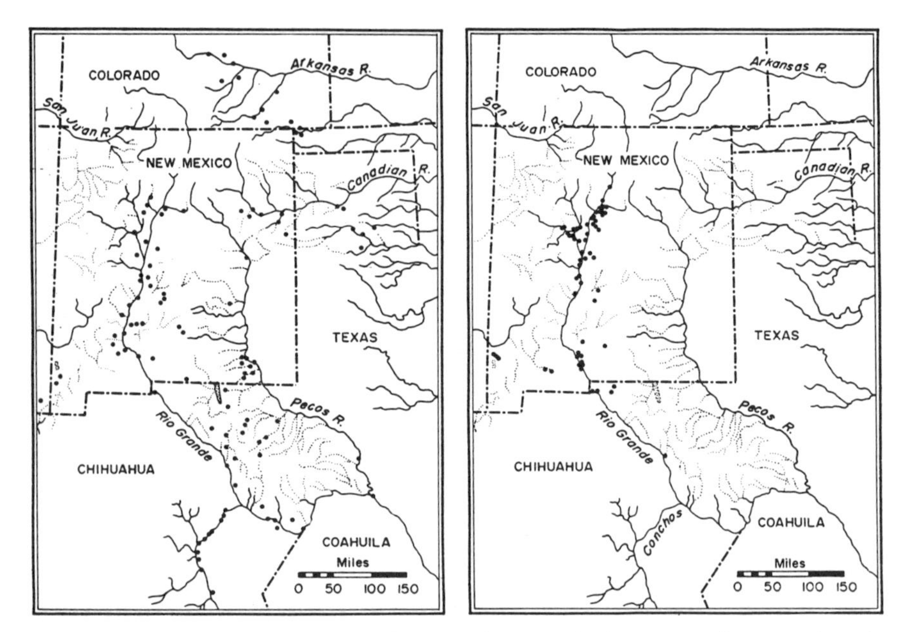

Figure 1.6 *Left*: The riparian distribution of *Cnemidophorus tesselatus*. *Right*: The riparian distribution of *Cnemidophorus neomexicanus*. Both are parthenogenetic species of teiid lizards. (From Cuellar, 1977, *Science*, vol. 197, figs. 3 and 4, pp. 837–843, © 1977 by the AAAS.)

uniparens) there is no cytokinesis in the mitotic division just before meiosis (Cuellar, 1970). Parthenogenesis in the Old World genus *Lacerta* is associated with glaciated elevations in the Causcasus, along the headwaters of major river systems, with the bisexual forms at lower elevations (Darevsky, 1966). The antiquity of parthenogenesis in these two lizard genera is unknown but is presumably post-Pleistocene.

Among parthenogenetic reptiles, both diploid and triploid species exist. Most of these species apparently arose as a result of hybridization. The triploid *Hemidactylus vietnamensis*, which may be the form sometimes referred to as *H. garnotii*, appears not to have evolved from hybridization (Darevsky et al., 1984).

1.4.4 Birds

Among domestic fowl, fertile eggs are sometimes found from isolated hens. Because the female is heterogametic (WY) and the YY combination is lethal, all offspring are males (WW). This is known to occur in chickens and the domestic turkey. If this occurs in wild birds, it might explain some examples of apparent mating outside of a pair or polygynous group.

Parthenogenesis does not normally occur in mammals.

1.5 HERMAPHRODITISM

Hermaphroditism is the occurrence of both male and female gonads within one individual. In vertebrates this condition is most common among fish, where it is expressed in several different forms, but it is rare among other vertebrates. Hermaphroditism does not occur in amphibians, except for the incipient ovary, or Bidder's organ, in male toads (Bufonidae). As we noted in Section 1.2, the teleost gonad is amphipotential in the early life of many species. Hermaphroditism in fish is restricted to species with external fertilization (Warner, 1975), and most hermaphroditic species are marine. In some species the gonad regularly develops as an ovary and, at a species-specific age, transforms to a testis. Sometimes accompanying these changes are changes in secondary sexual characteristics as well. Presumably sex reversals occur at an age and stage when gonadal tissues, germ cells, and associated ducts are sensitive and responsive to a shift in balance of estrogens and androgens. This stage is governed by genetically programmed changes modified by social and environmental factors.

When the individual is initially a functional female and subsequently becomes a functional male, the hermaphroditism is referred to as *protogynous*. The hermaphroditism in the reverse transformation, when the individual is first a male, is called *protandrous*. Protogynous species are more common, but protandry is known to occur in eight families of teleosts (Shen et al., 1979). Less common are species that are simultaneously male and female, or *synchronous* hermaphrodites; but these species are rarely self-fertile. In some protogynous fish some genetic males become functional females, but other males, called primary males, develop directly into functional males. This situation is referred to as *diandry* or *biandry* and is known to occur in both the Labridae and the Scaridae (Reinboth, 1967). Primary males are recognized by the absence of ovarian tissue or ovarian ducts.

In the eel-like freshwater teleost *Monopterus albus*, sex reversal occurs as a normal event in growth (Fig. 1.7). This protogynous hermaphrodite is female until about 30 cm in length (or 24 months of age). At one year of age *M. albus* are sexually mature females with ripe ova and no male germinal tissue. At 30 months testicular growth develops at intervals along connective tissue of the gonadal capsule (Liem, 1963). They then gradually pass through an intersex stage, between 28.0 and 45.9 cm, but sperm and ova are not known to occur simultaneously. Most individuals more than 42 cm in length (or 36 months of age) are males.

1.5.1 Environmental Conditions Inducing Sex Reversals

In teleost fish, density can favor the preponderance of one sex over the other. In most hermaphroditic species, crowding results in the transformation of females to males. Neither the determining mechanism nor the advantage of a greater proportion of males under environmental duress are understood (Chan and Yeung, 1983).

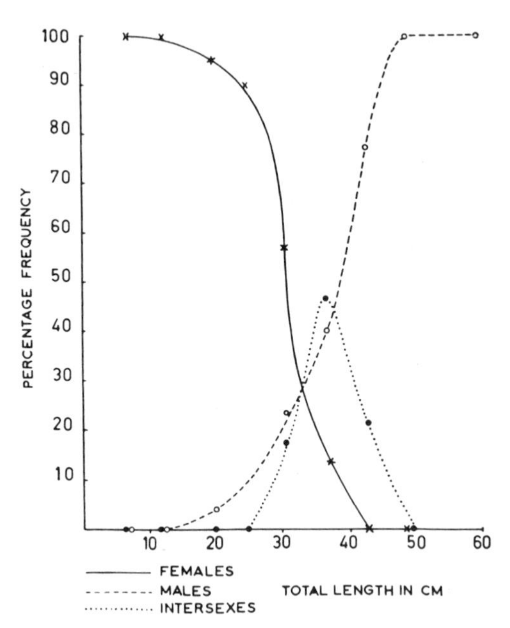

Figure 1.7 The percentage frequency of males, females, and intersexes relative to total length in the synbranchiform teleost, *Monopterus albus*, from field samples. (Liem, 1963, courtesy of the American Society of Ichthyologists and Herpetologists.)

1.5.2 Socially Induced Sex Reversals

Hermaphroditism is known in many species in at least 14 families of marine teleosts (Warner, 1984). Some species are female but transform to males under social stimuli, and not as a natural consequence of age or size. In several of these families, sometimes referred to as *coral reef fish*, a small group of females breeds with a single male. The male's disappearance (or removal at the hand of an experimentor) is a stimulus for a large female to transform into a male. Therefore the size at which sex reversal occurs varies with the local situation, although in such species all males are derived from large, functional females. A minimal number of females is required to induce one of them to become a male, and this number seems to vary with the species (Shapiro, 1984). Such socially induced sex reversal precludes the occurrence of a group of mature females without a male.

1.5.3 Evolutionary Significance of Sex Reversals

There would seem to be several apparent advantages to change from one sex to another, either in the entire population or only in one or a few individuals. If every individual is potentially female, there is a doubling of productivity. Some benthic hermaphrodites appear to be synchronous hermaphrodites with scattered populations (Mead et al., 1964), and each individual is a potential mate for any other it might meet. Synchronous hermaphroditism, in which all siblings are of the same sex, precludes inbreeding. Ghiselin (1969) suggested that sex changes take place to enable reproductive efficiency to correlate with sex and size, with male efficiency being greater at larger sizes in those species in which small males breed infrequently and reproductive activity of males increases with size. On the other hand, if mating is random and not dominated by the largest males, it then becomes advantageous for an individual to become a female when larger, inasmuch as the fecundity of a female generally increases with size. These hypotheses have been discussed at length by Warner (1975, 1978), Charnov (1982), and others. These three potential advantages, reflecting the diversity of hermaphroditism, do not necessarily apply equally to all hermaphroditic species.

1.6 SUMMARY

Germinal tissue in the gonads is derived from primordial germ cells (PGCs). These cells remain separate from somatic elements and outside of the embryo until the embryo's central axis is determined. PGCs then migrate from their extraembryonic position to the genital ridge, on the dorsal wall of the coelomic cavity. In fish and anurans migration is through the gut, by amoeboid activity, and then dorsad through the supporting mesentery. In birds and reptiles PGCs occupy an early site in the germinal crescent, anterolateral to the head, and move at least partly via the circulatory system, eventually leaving the capillaries and moving up the dorsal mesentery. In mammals PGCs enter through the hind gut and then via the mesentery, avoiding the circulatory system. Pseudopodia are lost when the PGCs enter the genital ridge.

During mammalian embryonic life, circulating androgens cause sexual differentiation in some brain cells that affect sexual behavior in the adult. It is believed that androgens become converted to estradiol in some brain cells, but embryonic ovaries are not steroidogenic and the female brain remains unmodified.

The direction of sexual development can be controlled environmentally by two sorts of influences: genetic sex determination or temperature sex determination, or a combination of the two. A gene for the histocompatibility, or H-Y, antigen, on the long arm or centromeric region of the Y chromosome, accounts for the rejection of male skin grafts on otherwise genetically identical females and accounts for at least some aspects of maleness. The H-Y antigen locus is autosomal and may be suppressed by a gene on the Y chromosome that is antagonistic to the suppressor on the X chromosome.

The usual mammalian XX:YY female:male pattern is sometimes altered, so that XX males and XY females may exist. There are several possible means by which this can occur.

In some fish, amphibians, and reptiles the temperature of the early embryo can determine the direction of sexual development. In some species, sex determination results from a combination of genetic and thermal influences.

Some species of vertebrates consist only of females. Various mechanisms exist for producing offspring with two or more genomes. In the teleost family Poeciliidae, a diploid egg is stimulated to cleavage following mating with the male of another species. Cleavage results from the presence of a sperm, which does not join with the egg nucleus. In some poeciliids the sperm of a sympatric species not only stimulates cleavage but unites with the egg genomes to create a triploid individual. In these examples the male genome is discarded in oogenesis. Some species of salamanders (*Ambystoma*) are either diploid or triploid and arise from the same mechanisms as noted for the Poeciliidae.

Six families of lizards and one family of snakes are parthenogenetic, and males are unknown. Unlike the examples of poeciliids and ambystomid salamanders, exogenous influence is not needed to stimulate cleavage of the diploid egg. Both diploid and triploid species of parthenogenetic species of reptiles are known. Many of them have arisen from natural hybridization.

Parthenogenetic strains have been found in some domestic fowl. Because in birds the female is heterogametic (WY), all males are homogametic (WW), and the YY combination is lethal, products of parthenogenetic reproduction are all males (WW).

Sex reversals are normal in the lives of many fish but are confined to species in which fertilization is external. The sex reversal may depend entirely on age, the change in gender occurring at a certain size. In other species all individuals develop into females. As they become sexually mature, they are dominated by a single male. The removal or disappearance of the male stimulates the largest female to become a male.

BIBLIOGRAPHY

Austin CR and Edwards RG (Eds) (1981). *Mechanisms of Sex Differentiation in Animals and Man*. Academic, London, New York.

Armstrong CN and Marshall AJ (Eds) (1964). *Intersexuality in Vertebrates Including Man*. Academic, London.

Clark WC (1973). The ecological implications of parthenogenesis. Pp 103–113. In *Perspectives in Aphid Biology* (Ed: Lowe AD). Bulletin 2, The Entomological Society of New Zealand, Auckland.

Clarke I (1982). Prenatal sexual development. Pp 101–147. In *Oxford Reviews of Reproductive Biology*, vol 4 (Ed: Finn CA). Oxford Univ, Oxford.

Ghiselin MT (1969). The evolution of hermaphroditism among animals. *Quarterly Review of Biology*, **44**: 189–208.

Hardisty MW (1978). Primordial germ cells and the vertebrate line. Pp 1–45. In *The Vertebrate Ovary* (Ed: Jones RE). Plenum, New York.

Price DJ (1984). Genetics of sex determination in fishes—a brief review. Pp 77–89. In *Fish Reproduction: Strategies and Tactics* (Eds: Potts GW and Wootton RJ). Academic, London.

Reinboth R (Ed) (1975). *Intersexuality in the Animal Kingdom*. Springer-Verlag, Berlin.

2

REPRODUCTIVE HORMONES

2.1 THE FUNCTIONS OF HORMONES

The effect of a hormone is a property of both the hormone and its target tissue(s). The relative similarity of a hormone among different classes of organism does not necessarily imply a commonality of responses. The broadly occurring pituitary hormone prolactin, for example, has a wide variety of responses, including mammogenesis and lactogenesis in mammals, broad patch formation in birds, dermal pigmentation in ectotherms, and water balance and parental care in fish. Many hormones not only stimulate or suppress the activity of the target organ but additionally have some rather general effects. There is a general increase in motor activity with a rise in gonadal steroids, and in laboratory rats, for example, exploratory activity increases during estrus (Martin and Battig, 1980). High levels of androgens seem to account for behavioral activity that is sexual in only an ancillary way. In both birds and mammals an increase in persistence toward a goal is positively correlated with serum androgen. In chicks, injections of testosterone elevate the persistence with which they seek a preferred food (Andrew and Rogers, 1972). Vasopressin has the general effect of elevating blood pressure.

In target tissues are cells that contain surface receptors, or binding sites. Presumably the sensitivity of a given tissue is proportional to the abundance of binding sites. Movement of a steroid through the lipophilic cell membrane poses no problem, and concentrations of steroids rapidly accumulate within cells of target tissues. Protein hormones, such as a gonadotropin, contact a receptor on the outer surface of a cell membrane. This may be followed by an adenylyl system, which in turn induces formation of cAMP. Cyclic AMP appears to mediate activities such as ovarian steroid synthesis. Internalization of the protein-hormone receptor complex occurs for many hormones, presumably for intracellular metabolic inactivation. It has been suggested that for some protein hormones, internalization may lead to further stimulation of the target cell by mechanisms that remain to be defined but may involve binding to intracellular organelles and/or the nucleus.

Within the cell the steroid may form a steroid-receptor complex within the cytoplasm or the nucleus. Following attachment to an acceptor site on chromatin (Fig. 2.1), the complex may initiate a specific target organ response in addition to accelerating RNA synthesis and cellular metabolism. Also, the presence of estrogen within cells of the uterus, for example, is followed by an increase in both estrogen and progesterone receptors, which are maximal prior to ovulation. After ovulation and the postovulatory increase in progesterone, receptors of both types decrease (Stormshak, 1979).

Steroid hormones closely resemble one another and are produced primarily by the gonads and by the adrenal cortex. Some leave their source as prehormones to be transformed to the active state elsewhere, usually at the target organ.

Prostaglandins comprise a large number of hormonelike lipids that are produced in many tissues. Because they are quickly metabolized in the lungs, they are short lived. They are usually synthesized from prostanoic acid near their target site, and their effects tend to be local. They were first found in seminal fluid and were erroneously believed to emanate solely from the prostate gland, and hence they were called prostaglandins. Prostaglandins have a broad spectrum of functions, and some are important in reproduction. Some prostaglandins function in parturition and also in the production of proteins within the placenta. More than 10 different prostaglandins are distributed in human semen, but their significance is not known. Prostaglandins also occur in the maternaal segment of the placenta and in the uterus, oviduct, and mature follicle. Generally they stimulate smooth muscle contractility and blood flow. They may also modulate steroidogenesis and shorten luteal life span in nonpregnant mammals.

A special group of hormones, catecholamines, are also neurohormones, which transmit nervous impulses across synapses or between a nerve and an effector organ.

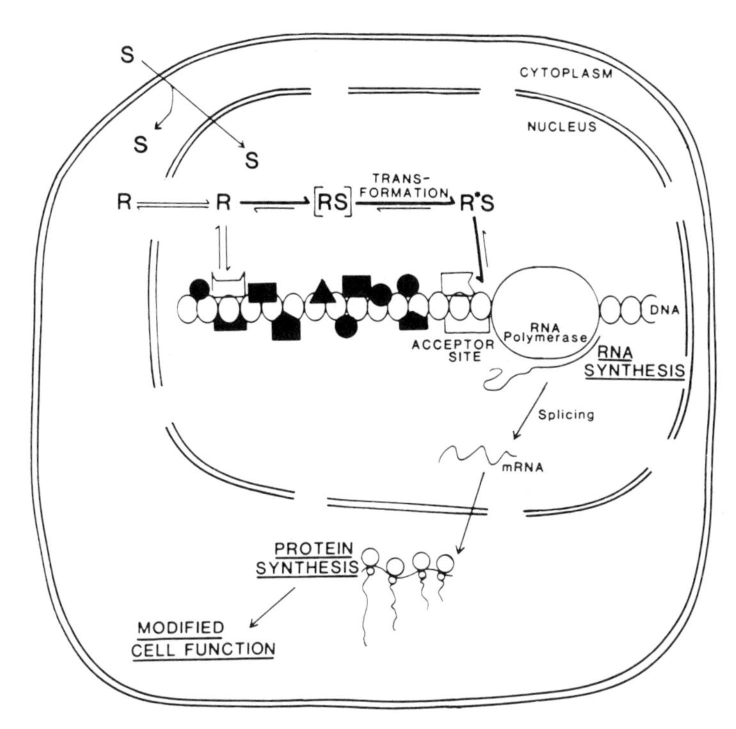

Figure 2.1 A model of steroid-receptor interaction and induction of cellular response. S, steroid; R, receptor. (From Walters, 1985, courtesy of The Endocrine Society, 1985, p. 532.)

The main catecholamines ae epinephrine, norepinephrine, and dopamine. They influence reproductive structures either through the nervous system or by altering the metabolic rate of certain tissues and/or blood flow. Smooth muscle of tubular parts of the reproductive system of both sexes is innervated by the sympathetic nervous system. There are synchronous fluctuations of norepinephrine and gonadal steroids in the human oviduct (Dujone et al., 1976), and their local occurrence (histochemically determined) parallels that of adrenergic nerves. The role(s) of catecholamines in normal reproduction is unclear and remains controversial.

2.2 THE HYPOTHALAMUS AND ITS HORMONES

The gonads do not operate in isolation from conditions in the rest of the body but are controlled, in part at least, by hormonal messages from both the hypothalamus and the anterior pituitary. Gonadotropins from the anterior pituitary may stimulate metabolism, growth, and vascularization of the gonads in addition to stimulating steroidogenesis. In turn, the output of gonadotropins is affected by varying levels of steroids released by the gonads. In addition, hormones from the hypothalamus not only stimulate the synthesis and release of gonadotropins but may also inhibit secretions from the anterior pituitary. In mammals, and to a lesser degree in some other vertebrates, the hypothalamic-pituitary-gonadal axis is influenced by the pineal gland. This chapter outlines the functions and interrelationships of the hormones controlling gonadal activity.

2.2.1 Structure of the Hypothalamus

The hypothalamus is a loosely defined area in the ventral region of the diencephalon. A posteroventrad projection is usually referred to as the posterior pituitary. Two hypothalamic regions, the supraoptic and the paraventricular nuclei, contain large secretory neurons that store their products (oxytocin and vasopressin) in axon terminals in the adjacent posterior pituitary, from which they are released into the vascular system. The hypothalamus of tetrapod vertebrates has a portal system, a plexus of capillaries, that carries hormones to the anterior pituitary.

The median eminence of the avian hypothalamus contains functionally distinct anterior and posterior parts, which are provided with separate portal connections to rostral and caudal lobes of the anterior pituitary. Within the hypothalamus are numerous centers that affect such diverse activities as water balance, thermoregulation, appetite, and reproductive activity. These various discrete functions are mediated by different areas and some paired ganglia or nuclei. Their connections provide a pathway for responses to both exogenous and endogenous stimuli, including those factors that modify or control reproduction.

The hypothalamus of mammals not only is richly vascularized but contains nerves that connect to various parts of the brain. In contrast, the hypothalamus of the hagfish (*Myxine*) is poorly developed, but it does include a portal system leading to the anterior pituitary. The role of the hypothalamus in elasmobranchs is equivocal, inasmuch as the ventral lobe of the pituitary receives blood from the internal carotid artery and there is no portal connection from the hypothalamus to the anterior pituitary (Dodd, 1975). The hypothalamus of the dogfish (*Scyliorhinus canicula*) nevertheless secretes a substance that releases gonadotropin (Jenkins and Dodd, 1980). In teleost fish the hypothalamus lacks a portal connection to the pituitary, but there are neurosecretory fibers extending from the nucleus lateral tuberis of the hypothalamus to glandular cells in the anterior pituitary. These neurosecretory fibers may secrete gonadotropin-releasing hormone (GnRH) and regulate gonadotropin hormone (GtH) synthesis and release (Peter and Crim, 1979; Peter, 1982). The hypothalamic-pars distalis portal system is found in the living coelacanth (*Latimeria chalumnae*), in which the portal system reaches the cerebral lobe of the pars distalis but not the rostral lobe (such a separation of the pars distalis does not occur in other bony fish). In *Latimeria* the rostral lobe receives a branch of the internal carotid, resembling the condition found in elasmobranchs (Lagios, 1975).

2.2.2 Hormones of the Hypothalamus

A major role of the hypothalamus is the synthesis of GnRH, which induces gonadotropin production and release by the anterior pituitary. In bony fish, however, gonadotropin release is stimulated by a different hypothalamic secretion (Donaldson, 1973).

GnRH causes a tonic and low-level release of both LH and follicle-stimulating hormone (FSH). Fluctuating sensitivity of the anterior pituitary and pulsatile GnRH release may account for temporal variations in amounts of LH and FSH release. Preovulatory increased secretion of estrogen from the ovary is followed by a substantial LH release. In addition to stimulating the anterior pituitary, GnRH may have inhibitory influences on the gonads. GnRH, or a GnRH-like peptide made in the ovary, also affects the ovary directly, depressing the activity of both granulosa and theca cells (Sharpe, 1982). It blocks estradiol secretion by granulosa cells and also androgen synthesis by theca cells. In cultured granulosa cells, GnRH suppresses production of cAMP, which is necessary for steroidogenesis (Knecht et al., 1983a, 1983b; Jones and Hseuh, 1983).

In teleost fish the anterior preoptic area of the brain secretes a gonadotropin-release-inhibiting factor (GRIF), which may suppress the release of gonadotropins (Stacey, 1984). Studies on the goldfish and the rainbow trout indicate dopamine as an inhibitor of gonadotropin release from the teleost pituitary (Peter, 1983).

Norepinephrine (or noradrenaline) effects the release of LH and FSH and can thus stimulate ovulation. Norepinephrine in the anterior hypothalamus fluctuates during the estrous cycle, in synchrony with the phasic secretions of gonadotropins. Serotonin, on the other hand, blocks the discharge of GnRH, inhibiting the release of LH. Dopamine, the third catecholamine from this region, blocks the release of prolactin and sometimes LH (Kordon, 1978). A GnRH-associated peptide from the hypothalamus is an inhibitor of prolactin secretion and also a stimulator of the release of gonadotropins in rat pituitary tissue in vivo (Nikolics et al., 1985).

2.3 THE ANTERIOR PITUITARY

The pituitary gland, or hypophysis, is a composite organ consisting of an adenohypophysis (formed from Rathke's pouch, a dorsal outpocketing from tissue that also gives rise to the roof of the mouth) and a neurohypophysis (derived from the infundibular process of the diencephalon of the brain floor). The neurohypophysis includes a neural lobe, which joins to the caudalmost part of the adenohypophysis; the neural lobe constitutes the posterior lobe (or pars nervosa) of the pituitary gland. In most vertebrates the apex of the adenohypophysis becomes differentiated as the anterior lobe (or pars distalis). The remainder of the adenohypophysis, which joins the posterior lobe, becomes the intermediate lobe (or pars intermedia). In birds there are rostral and caudal lobes of the adenohypophysis, and both regions secrete gonadotropins. As pointed out previously, the separate lobes of the avian adenohypophysis are joined to the hypothalamus by separate portal systems, and this separation may have functional significance (Follett, 1984). The intermediate lobe is absent in birds and some mammals (e.g., the beaver, armadillo, cetaceans, and probably others). In species lacking the intermediate lobe, a thin layer of nonglandular tissue separates the posterior lobe from the anterior lobe.

The adenohypophysis of the hagfish is not discretely divided and is distinct from the neurohypophysis. The pituitary has a disputed role in gonadal activity in these cyclostomes, but most students find some reduction of gonadal size after hypophysectomy. The pituitary of elasmobranchs consists of a united pars intermedia and pars nervosa, and together they constitute the *neurointermediate lobe*, whereas the pars distalis is divided into three lobes (rostral, median, and ventral). The ventral lobe of elasmobranchs appears to be the most important in reproduction. It develops a narrow stalk and produces a glycoprotein gonadotropin. Removal of the ventral lobe is followed by gonadal regression.

2.3.1 Hormones of the Anterior Pituitary

Seven protein hormones are synthesized in the anterior pituitary of both sexes. Those of mammals are the best known, and most nonmammals have similar, though perhaps not identical, hormones. Most directly concerned with reproduction are the two gonadotropins—FSH and LH. Prolactin, in addition to having a broad spectrum of functions, affects several important aspects of reproduction, which vary with the taxa concerned. Two other hormones, thyroid-stimulating hormone (TSH) and growth hormone occur widely among the vertebrate classes. The anterior pituitary synthesizes a peptide hormone, adrenocorticotropic hormone (ACTH), which increases the production of steroid hormones from the adrenal cortex. This portion of the pituitary also produces β-lipotropin, from which β-endorphin is derived. Except for prolactin, these hormones are released on specific hormonal signals (releasing hormones) from the hypothalamus. The release of prolactin is suppressed by dopamine from the hypothalamus.

2.3.2 Gonadotropins

Because the roles of gonadotropins have been most carefully studied in mammals, their functions in these vertebrates will be treated first. Two gonadotropins, FSH and LH, control gonadal activity. They are produced by both sexes and exert distinctive effects on the gonads of each (see also Chapter 4, The Testis, Section 4.3.1). In the female, FSH stimulates the growth and development of the follicle in the early preovulatory phase and initiates steroidogenesis. LH increases vascularization and continues the follicular growth and the synthesis of ovarian steroid hormones. The increased ovarian blood flow facilitates the future passage of the ovarian hormones, estrogen and progesterone, into the general circulation. LH also promotes the conversion of cholesterol into progesterone and estrogen.

Macropod marsupials possess both LH and FSH, and their anterior pituitary is regulated by a negative feedback from steroid hormones similar to the negative feedback of eutherians (Tyndale-Biscoe and Evans, 1980). Follicular growth and ovulation in the tammar wallaby (*Macropus eugenii*) are dependent upon pituitary secretions, and both processes cease after hypophysectomy (Hearn, 1975).

Not all gonadotropins come from the anterior pituitary. The placenta of some mammals produce a glycoprotein called chorionic gonadotropin, which may be similar in its role to both LH and FSH. One of these placental gonadotropin, pregnant mare serum gonadotropin (PMSG), which is often designated as equine chorionic gonadotropin, is commonly used in experimental studies. Human chorionic gonadotropin (hCG), a glycoprotein, has primarily LH activity: it is unable to induce ovulation but does maintain the corpus luteum.

The pituitary exerts only a minor influence on gametogenesis in lampreys, and hypophysectomy merely slows the rate of gonadal metabolism. Moreover, the hypothalamus of lampreys apparently does not control the slight gonadotropic influence of their pituitary (Dodd et al., 1960; Dodd, 1975). Following the removal of the pituitary in *Lampetra fluviatilis*, however, there may be some follicular growth but no ovulation (Larsen, 1974). Apparently there is only incipient pituitary control of reproductive activity in these primitive vertebrates.

The four distinct and separate lobes of the pituitary of elasmobranchs all produce gonadotropins. Gonadotropins from the ventral lobe of the dogfish (*Squalus acanthias*) seem to control steroidogenesis (Dodd et al., 1960), and maximal secretion of gonadotropins is recorded during the breeding season (Dodd, 1975). It is not known if the ventral lobe of the dogfish produces either FSH, LH, or perhaps a different gonadotropin, but the dogfish (*Scyliorhinus canicula*) produces a gonadotropin (GtH) that has properties of LH (Sumpter et al., 1978).

Teleostean fish are presumed to have two gonadotropins, together designated as GtH, but initially there was thought to be only one hormone. These gonadotropins have distinct effects: one is vitellogenic and the other maturational (Idler and Ng, 1983). There are two types of secretory cells in the anterior pituitary of fish, and the current consensus is that GtH exists in two forms (Stacey, 1984). In fish, hypophysectomy is followed by gonadal regression (Fontaine, 1975). The release of GtH in teleosts may result not only from the stimulus of GnRH but also from the withdrawal of GRIF.

In amphibians, reptiles, and birds two gonadotropins occur, but they may not be identical to mammalian gonadotropins. In birds LH stimulates the production of estrogen, progesterone, and testosterone, and FSH promotes the growth of the follicle. Prolactin, acting synergistically with FSH and LH, may also stimulate follicular growth. As the follicle increases in size, estrogen synthesis and release increase, and the largest follicles, those nearing ovulation, begin to release more progesterone and less estrogen. Testosterone and progesterone together stimulate activity of the avian oviduct.

Among the different classes of vertebrates, the testis shows some variation in its responses to LH and FSH. In amphibians and mammals LH is of paramount importance in stimulating the production of androgens, although FSH and prolactin have a synergistic stimulatory effect. In vitro preovulatory and postovulatory tissue of the painted turtle (*Chrysemys picta*) is steroidogenic when exposed to either mammalian or avian LH. Estrogen predominates from preovulatory tissue and progesterone from postovulatory incubates (Callard et al., 1976). The reptilian testis is far more responsive to mammalian FSH, as measured by androgen output, than it is to mammalian LH, when each gonadotropin is administered separately. Although the potency of gonadotropins in reptiles (squamates, crocodilians, and chelonians) varies with the source of the hormone, reptilian and avian pituitaries stimulate both steroidogenesis and spermatogensis in the lizard *Anolis*. Regarding the specificity of testicular response to gonadotropins, reptiles differ more from amphibians, birds, and mammals than these three classes differ among each other (Licht, 1979). Chronic exposure to ovine FSH stimulates testicular growth and spermatogenesis in the anole, and similar, but less pronounced, effects are produced by LH. In vivo FSH tends to have a longer half-life than does LH in lizards, perhaps accounting for the difference in the observed effects in these reptiles (Licht, 1979).

Gonadotropins from all ectotherms (fish, amphibians, and reptiles) resemble, but are not identical to, mammalian LH and FSH, and gonadal response varies from one species to another. In all these vertebrates gonadal response to gonadotropins is greatly influenced by temperature (Licht, 1979). Two gonadotropins from the amphibian anterior pituitary are comparable to FSH and LH of both reptiles and mammals (Licht and Papkoff, 1974). Experimental administration of mammalian LH and FSH to hypophysectomized frogs indicates that steroidogenesis and spermiogenesis are stimulated by LH and that spermatogenesis is effected by FSH. This pattern seems to hold for most amphibians (Lofts, 1974).

2.3.3 Prolactin

Prolactin and growth hormone are secreted predominately from different parts of the anterior pituitary. Prolactin occurs in all jawed vertebrates, and the prolactin-producing cell has been found in all vertebrates examined except agnaths. Prolactin, growth hormone, and placental lactogen have structural similarities, suggesting a common evolutionary history. Human prolactin and growth hormone have some 40% of their amino acid sequences in common (Malcolm, 1981).

Prolactin has numerous known and postulated functions, more than are known for any other hormone, but its similarity to growth hormone may account, in part at least, for the versatility of prolactin. Binding sites for prolactin are known for several diverse tissues. Prolactin seems to have a variety of functions in teleost fish: it is well known to enhance osmoregulatory ability and promote lipogenesis. In some groups (Cichlidae) it promotes parental care. In salamanders (Salamandridae) prolactin is reported to promote the movement to water prior to spawning (Mazzi, 1970). In birds prolactin not only promotes parental care (e.g., broodiness) but in pigeons and doves (Columbidae) it also stimulates cells lining the crop to synthesize and release "pigeon milk."

In mammals prolactin is important not only in the growth of mammary tissue but also in the production of milk. It acts synergistically with both estrogen and progesterone from the ovary as well as with ACTH and growth hormone from the anterior pituitary (see Lactation). In lactating rabbits, for example, lactogenesis is interrupted by hypophysectomy but is restored by injections of either sheep prolactin or human growth hormone (Cowie et al., 1980).

Prolactin is under inhibitory control, and its release is prevented by dopamine, the primary prolactin-inhibiting factor, produced in the hypothalamus. Prolactin is released only when dopamine is prevented from entering the portal system to the anterior pituitary. The destruction of this portal system is followed by a rise in secretion of prolactin. At least some natural release of prolactin may result from the action of a prolactin-releasing factor, which suppresses the action of dopamine.

Estrogen is known to inhibit the action of dopamine on prolactin release (Cowie et al., 1980) and thus stimulate the secretion of prolactin.

Apart from endogenous influences on prolactin secretion, levels of this hormone may be affected by the photocycle. In some examples these changes are clearly correlated with gonadal changes, which also follow day length. A number of species of teleost fish have circadian rhythms of circulating prolactin (Matty, 1978). Long days are known to stimulate the release of prolactin in several kinds of birds, for example, the migratory, or Japanese, quail (*Coturnix coturnix*), and peaks of serum prolactin coincide with gonadal activity. Cattle also have a phase-sensitive period, and prolactin release is increased by both an increase in day length (from 8L:16D to 16L:8D) and exposure to skeletal day lengths (6L:8D:2L:8D) (Tucker, 1981). Prolactin levels also exhibit marked seasonal changes in domestic cattle under a natural photocycle (Fig. 2.2). Prolactin release in sheep (Bosc et al., 1982; Lincoln et al., 1982) and also in the golden hamster are similarly under control of the photocycle (Steger et al., 1983).

2.3.4 Prolactin and Gonadal Activity

Prolactin has varying roles in gonadal activity. In the domestic rabbit and in some other domestic and laboratory mammals, prolactin promotes the maintenance of the corpus luteum. Prolactin may inhibit ovulation in the postpartum human female, either via the hypothalamus or possibly by direct effect on the ovary (Hamada et al., 1980). It is well established that prolactin, when elevated by suckling young, suppresses ovarian activity and that this suppression reflects the frequency and duration of suckling (McNeilly et al., 1982). Prolactin may also be released as a reflex from nervous stimulation of coitus, with the accumulation of prolactin in

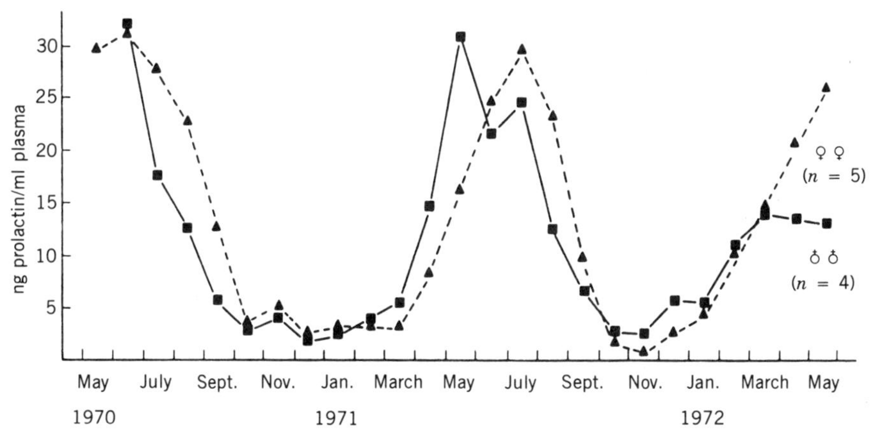

Figure 2.2 Seasonal variations in basal prolactin levels in nonlactating male and female cattle, from birth to puberty. (From Karg and Schams, 1974, courtesy of the *Journal of Reproduction and Fertility*.)

follicular fluid, where it enhances the secretion of progesterone from granulosa cells (McNatty et al., 1974). It is also released after coition in the male, but with no known significance, although it does induce LH receptors in a number of mammals.

Experimentally, prolactin has an antigonadal effect in many birds and reptiles, apparently depressing LH and FSH. But whether prolactin is stimulatory or suppressive depends upon the time of day it is administered. There is a circadian rhythm in the sensitivity to exogenous prolactin, in addition to its natural rhythmical release (Meier, 1975). Prolactin is progonadal in some rodents and some birds (see Chapter 4, The Testis, Section 4.3.2).

2.4 THE POSTERIOR PITUITARY

As previously pointed out, the hypothalamus contains large secretory neurons that project into the posterior pituitary, where they release their hormones. The hormone oxytocin plays a major role in uterine contractions and milk letdown or milk ejection, as well as in maternal behavior (at least in some mammals). The release of hormones from the posterior pituitary is under neural control.

2.5 GONADAL HORMONES

The steroidogenic cells of the vertebrate ovary are rather similar among the several classes of vertebrates, which is also true for the testis (see Chapter 4, Section 4.3). Gonadal organization may differ from one class to another or within a class, but the cells responsible for steroidal synthesis remain the same (Fig. 2.3).

Most gonadal hormones are steroids, ultimately derived from cholesterol, and they are chemically similar. Progesterone, evolved from cholesterol, can be converted to androgens and estrogens, and androgens in both sexes can be converted to estrogens. In some mammals the placenta synthesizes progesterone, but in others the bulk of progesterone is produced by the corpus luteum.

Progesterone, androgens, and estrogens are also synthesized in the adrenal cortex. These hormones not only have clear-cut physiological effects but are also partly responsible for behavioral patterns in both sexes. In young (sexually immature) vertebrates the adrenal cortex is presumably the major source of sexual steroids.

2.5.1 Estrogen

In most vertebrates estrogen is formed and secreted by granulosa cells of the mature follicle, but it may be formed, in lesser amounts, by the corpus luteum. It has been suggested that estrogen synthesis results from the cooperation of theca interna cells and granulosa cells (Fig. 2.4). The placenta becomes a major source of estrogen in some mammals, and the estrogen appears in bound form in the urine of pregnant mammals of many taxa. Estrogens may also be synthesized by the blastocyst and, in small amounts, by the mammary gland.

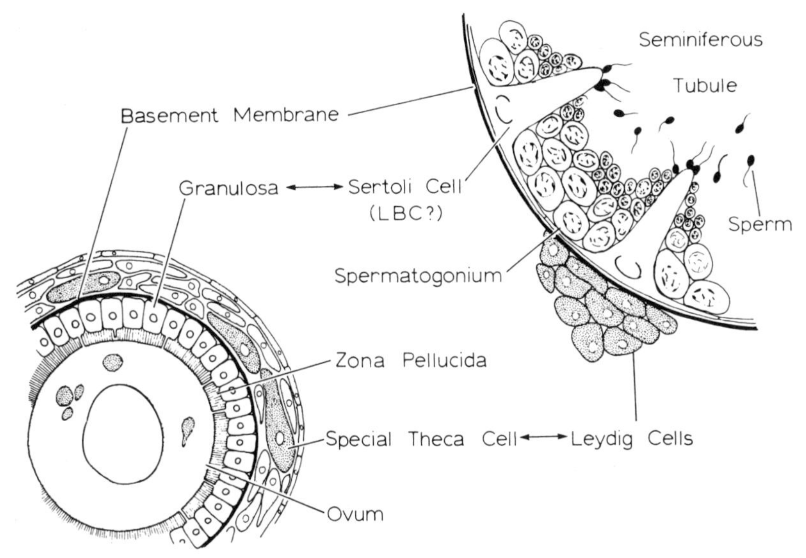

Figure 2.3 Steroidogenic cells in teleost gonads. (From Hoar and Nagahama, 1978, courtesy of Reproduction, Nutrition, Development, I.N.R.A.)

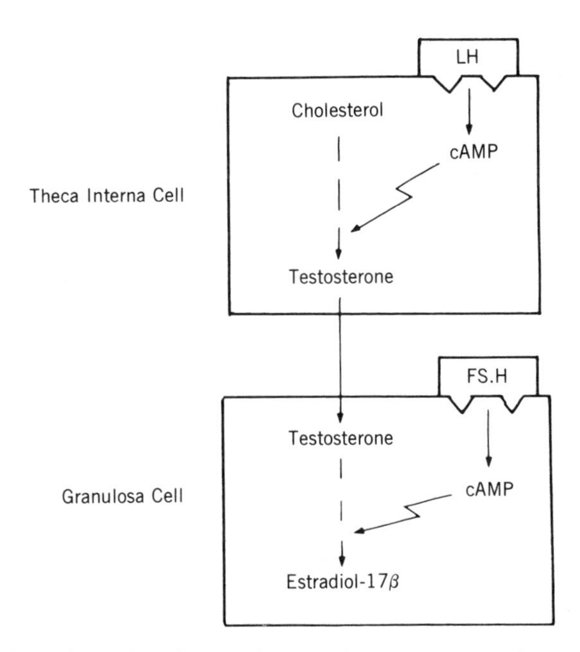

Figure 2.4 A hypothetical pathway of LH and FSH action in the regulation of the synthesis of estrogen in follicles. (From Armstrong DT, Goff AK and Dorrington JH, Regulation of follicular estrogen biosynthesis, 1979, courtesy of Raven Press.)

The source of ovarian steroids in bony fish is not completely agreed upon. In the guppy (*Poecilia reticulata*) steroid synthesis appears to occur in the granulosa cells (Lambert, 1970). In the goldfish (*Carassius auratus*) and several species of Pacific salmon (*Oncorhynchus kisutch* and *O. gorbuscha*), theca cells appear to synthesize steroids (Hoar and Nagahana, 1978). Both theca and granulosa cells appear to produce steroids in the preovulatory follicle of the live-bearing bufonid toad (*Nectophrynoides occidentalis*), and after ovulation corpora lutea continue to show steroidogenic activity (Xavier et al., 1970).

Target organs of estrogen include mammary tissue and the uterus, oviduct, and hypothalamus. A major function of estrogen is the rapid development of such structures as uterine epithelium and mammary tissue through the stimulation of protein synthesis. In humans estrogen promotes fat deposition on the hips and growth of underarm and pubic hair. In these processes estrogen usually acts together with one or more other hormones. Estrogen has immediate effects on target organs, where it may persist for up to two hours (in ovariectomized rats), but is quickly metabolized elsewhere in the body (e.g., the liver and blood) (Heap and Illingworth, 1977).

2.5.2 Progesterone

Progesterone is secreted by the granulosa cells in the preovulatory follicle, and the secretion is increased after ovulation by the corpus luteum. Additionally, the placenta in many mammals produces large amounts of progesterone.

Numerous functions are attributed to progesterone. Serum levels increase slightly prior to ovulation and stimulate the growth of the oviduct, uterus, and immediately adjacent tissue in nonmammalian vertebrates. Following ovulation, progesterone serum levels increase manyfold, except in oviparous vertebrates. Progesterone is the hormone associated with pregnancy in placental mammals, but it also occurs in marsupials, monotremes, and in many other vertebrates as well. In addition to its effect on the uterus, progesterone aids in stimulating mammary growth. Progesterone also depresses or blocks gonadotropin release from the anterior pituitary. It elevates body temperature and may account, in part at least, for increased body temperature in pregnant bats upon emergence from hibernation.

Postovulatory levels of progesterone may persist for some time in live-bearing ectotherms. In the viviparous perches (*Hysterocarpus traski* and *Cymatogaster aggregata*), serum progesterone is low and probably has no role in maintaining pregnancy (de Vlaming et al., 1983). Progesterone in at least some anurans appears at ovulation, but in oviparous species there seems to be no functional corpus luteum (Lofts, 1974). In live-bearing amphibians, however, the corpora lutea persist, and there are high levels of progesterone until birth of the young. Progesterone is characteristic of viviparous reptiles and may remain in the circulation until parturition (Callard and Ho, 1980; Callard and Lance, 1977; Highfill and Mead, 1975), but its role is not clear. In turtles there is a brief preovulatory rise in progesterone, much as there is in birds.

Progesterone has been detected in the serum of many wild birds and probably occurs seasonally in all birds with the nearing of ovulation. In the domestic fowl, estrogen declines and progesterone sharply rises with the onset of laying. During follicular maturation of the domestic hen, there is a continuous rise in progesterone and a concurrent decline in the secretion of estradiol (Bahr et al., 1983). No real corpus luteum is formed from the postovulatory follicle of birds, and discrete behavioral effects of progesterone are generally not known (Farner et al., 1983).

The sources of progesterone during pregnancy among different groups of mammals vary significantly. In the domestic rabbit, the goat, and the pig, preservation of the placenta is completely dependent upon progesterone from the corpus luteum. On the other hand, in some species the placenta gradually assumes the production of progesterone. This is true of the horse, in which the endometrial cups (of embryonic origin) of the placenta synthesize progesterone. In some bats the corpus luteum may regress in early pregnancy (e.g., the African *Nycteris luteola*; Matthews, 1941) or in later pregnancy (e.g., the Australian *Eptesicus regulus*; Kitchener and Halse, 1978). (See also Section 3.2.1.)

2.5.3 Androgens

Androgens are synthesized by the testes of all, and the ovaries of quite a few, if not most, vertebrates. Androgens are also the major product of testicular steroidogenesis and are synthesized by Leydig cells of the testicular interstitial tissue. In the ovary, androgens are produced by theca cells in response to LH stimulation in laboratory rats, and such androgens stimulate both endometrial and myometrial hypertrophy (Armstrong et al., 1979). Androgens may be converted to estrogens in target organs.

There are discrete cycles of testosterone in some female reptiles, and peaks may occur with or preceding peaks of estradiol. The significance of appreciable amounts of testosterone in females may relate to the conversion of testosterone to estradiol (Callard et al., 1978).

Androgens in female birds account for the slight development of male traits, such as the comb in a hen. In some species, such as phalaropes (Phalaropodidae), normally high androgen levels in females produce male behavioral traits, such as courtship, and also brightly colored plumage, which is usually characteristic of only males. Androgens may also function synergistically with other ovarian hormones. Small amounts of androgens are circulating in both sexes and can be detected in the very young. In puppies, for example, behaviorally ineffective levels of testosterone occur in circulation, but in amounts far below those of a sexually mature male (Hart and Ludewig, 1979).

In adult males androgens (mostly testosterone) induce activity of the Sertoli cells and the seminiferous tubules. Androgens also account for male behavior, including territoriality and courtship, and for the development of secondary sexual characteristics. Androgen implants increased calling and aggressiveness in territorial cock red grouse (*Lagopus lagopus scoticus*), and estrogen implants in territorial cocks resulted in a decline in song, loss of territory, and desertion of mates (Watson and Parr, 1981).

Both male and female anoles (*Anolis carolinensis*) exhibit male sexual behavior when treated with dihydrotestosterone. Injections or silastic implants of testosterone induce receptivity in females, presumably being converted (aromatized) to estrogen in the brain (Adkins and Schlesinger, 1979).

2.5.4 Relaxin

Relaxin has long been known as a hormone that softens cartilaginous connections in the pelvic girdle. It is a polypeptide produced in the ovaries and especially in the corpus luteum. It is found not only in mammals but also in the sand shark (*Odontaspis taurus*), a viviparous shark that continues to ovulate throughout pregnancy (see Chapter 8, Live-Bearing Ectotherms, Section 8.2). It is also produced by the placenta of at least some mammals (rabbits and horses). It is most abundant in the luteal phase of the sow, increasing during pregnancy and declining after parturition (Fig. 2.5). By softening the birth canal and relaxing the cervix, relaxin facilitates parturition. Additionally, relaxin inhibits spontaneous uterine contractions, perhaps

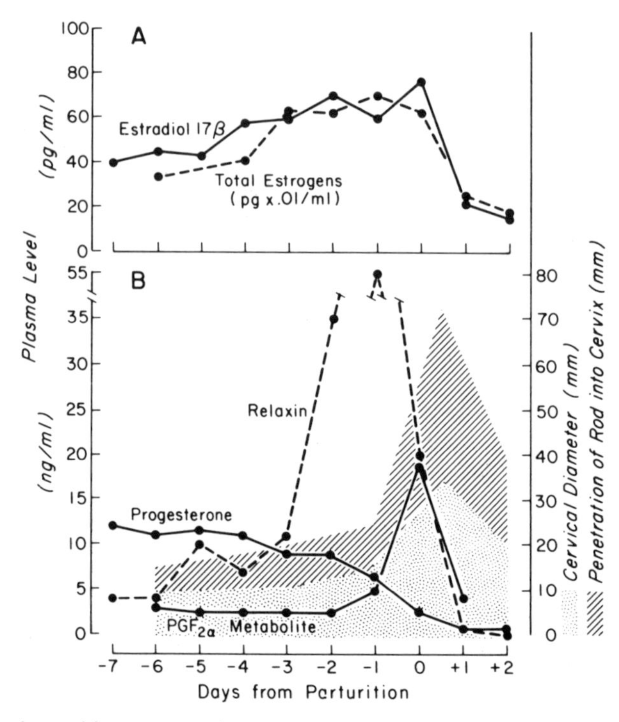

Figure 2.5 The sudden rise in relaxin one to three days before parturition in the sow, together with levels of (*a*) estrogens, (*b*) progesterone, $PGF_{2\alpha}$ metabolites, and cervical dilation. (From Sherwood OD and Downing SJ, 1983, The chemistry and physiology of relaxin, courtesy of Raven Press.)

preserving the placenta and fetuses after progesterone has declined, until this suppression is overridden by high levels of oxytocin and prostaglandins (Sherwood and Downing, 1983). Relaxin inhibits the release of oxytocin (Dayanithi et al., 1987).

2.5.5 Oxytocin

As previously pointed out, this hormone, a peptide, is synthesized by neurosecretory cells in the hypothalamus and released from the posterior pituitary. It is also produced by the ovary (Wathes and Swann, 1982) and may play an important role in the ovulatory cycle (Heap, 1983). In addition, it is synthesized by the placenta. Levels of circulating oxytocin and progesterone parallel each other after ovulation, when the corpus luteum is active, and elevated concentrations of oxytocin have been found to occur in the corpus luteum of the cow (Wathes and Swann, 1982). In sheep, concentrations of oxytocin are greater in the ovarian vein than in the ovarian arterial circulation, and the secretion of ovarian oxytocin is probably stimulated by $PGF_{2\alpha}$ from the uterus (Flint and Sheldrick, 1982). The function of ovarian oxytocin is not clearly understood, but it may inhibit the synthesis of progesterone and thereby hasten the demise of the corpus luteum. Possible roles of gonadal oxytocin are reviewed by Wathes (1984).

A major function of oxytocin is the stimulation of smooth muscle contraction: oxytocin causes myofibrin contraction in the lactating breast tissue, inducing milk letdown, and also initiates uterine contractions at the time of parturition. The presence of a fetus in the birth canal causes additional oxytocin release, an example of positive feedback. Oxytocin may be released during intercourse, also stimulating milk letdown in a lactating woman.

Oxytocin also occurs in males. In the male rat and rabbit, oxytocin increases after ejaculation and is followed by sexual refractoriness (Stoneham et al., 1985).

2.5.6 Inhibin

It has long been postulated that a gonadal hormone limits the production of FSH. A polypeptide had long ago been detected, but its exact nature was elusive. Its presence is apparent when castration is followed by a rise in FSH. Although it is well accepted that inhibin, from seminiferous tubules in testes and granulosa cells in the ovarian follicle, limits pituitary release of FSH (Main et al., 1979; Steinberger and Steinberger, 1976), the nature of inhibin has been difficult to define. When cultured in vitro, Sertoli cells of the rat secrete a nonsteroidal substance that suppresses the synthesis and release of FSH (Francimont et al., 1981). This action, which describes the role of inhibin, clearly affects the anterior pituitary and perhaps also the hypothalamus (Steinberger, 1981).

Analysis of inhibin from porcine follicular fluid indicates its occurrence as two subunits, α and β. There are also dimers of the β subunit, and either subunit β_A or β_B can combine with subunit α and inhibit secretion of FSH (Vale et al., 1986; Ling et al., 1986). The two subunits of β, moreover, may combine with each other and stimulate the production and release of pituitary FSH (Fig. 2.6). This subunit combination is called activin (Ling et al., 1986). Neither inhibin or activin affects LH or prolactin release.

2.5.7 Feedback

Not only is the ovary sensitive to LH and FSH from the anterior pituitary, but both the hypothalamus and the pituitary respond to varying levels of the steroid hormones secreted by the ovary. This feedback relationship accounts for fluctuations in secretory activity of the hypothalamus, the anterior pituitary, and the ovaries and also for the cyclicity of ovulation in polyestrous species. This phenomenon is best known for mammals but probably occurs in all vertebrates. Castration of the rainbow trout (*Salmo gairdnerii*) is followed by a five-fold increase in GtH, which in these individuals is reduced by pituitary implants of 11-ketotestosterone (Billard, 1978). There may also be a positive feedback in sexually developing teleost fish (e.g., salmonids and eels), which may promote sexual maturation. In juvenile trout testosterone has been shown to directly promote maturation of gonadotrophs (Gielen et al., 1982).

As already pointed out, when circulating levels of estrogens are low, levels of serum LH and FSH are higher. The release of gonadotropins follows stimulation of GnRH from the hypothalamus, through its portal circulation, to the anterior pituitary. The rise in gonadotropins stimulates the activity in one or more of the larger follicles, causing an increase in plasma estrogens.

The rise in circulating levels of estrogens depresses FSH release by increasing the pituitary threshold to GnRH. This is the *negative feedback*. While the circulating estrogen suppresses the release of FSH, there is a continued but depressed release of LH. Inasmuch as the movement of GnRH into the anterior pituitary continues, LH synthesis and accumulation also continues. With the gradual increase in the release of estrogens, however, a point is reached at which estrogen from the large follicle(s) reaches a stimulatory level. This increase of estrogens (especially estradiol), together with the buildup of LH in the anterior pituitary, seems to account for the sudden release of LH that is associated with ovulation. This is a *positive feedback* of estradiol. Thus the preovulatory surge of LH does not require additional GnRH.

With the decline of estrogen, as in a sterile ovulatory cycle, the negative feedback disappears. This is followed by a subsequent rise in circulating LH and FSH, indicating a lowering of the threshold of sensitivity of the pituitary to GnRH.

Progesterone has a clear negative feedback effect on the release of LH. In the ovariectomized ewe, Silastic capsules that released levels of progesterone simulating the midluteal phase produced a significant decline in LH release. But the reduction was not as great as in the midluteal phase of the normal estrous cycle, suggesting that estrogen also plays a role in the reduction of the tonic LH secretion (Karsch et al., 1977). Prior to the preovulatory surge of LH in the ewe, there is a rise in estradiol and, together with the rise in progesterone, a marked decline in LH (Hauger et al., 1977), also suggesting a negative feedback effect of progesterone (Fig. 2.7). Postovulatory levels of progesterone in the ewe allow the release of only low levels

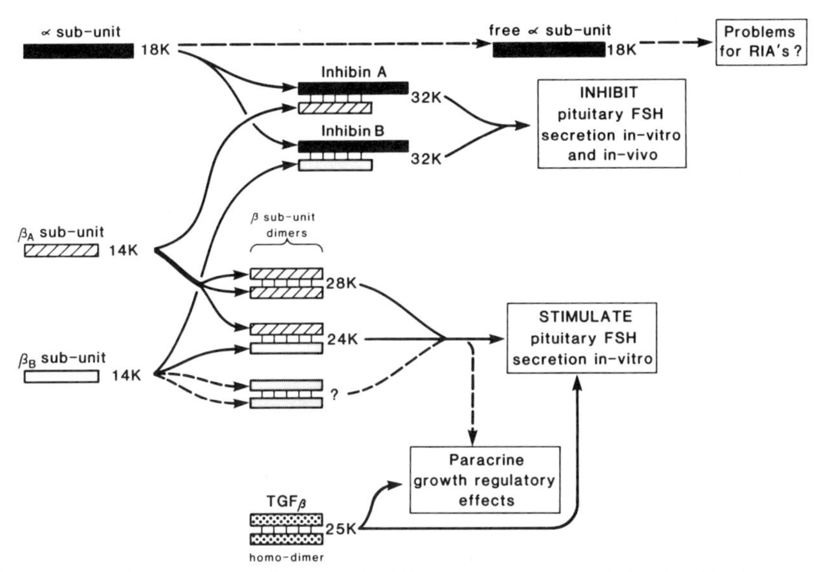

Figure 2.6 The α and β subunits of inhibin (from porcine follicular fluid) and their combination either to inhibit or to stimulate FSH activity, with indicated relative molecular masses. Solid lines show proven pathways and broken lines possible pathways. (From Tsonis CG and Sharpe RM, 1986, Dual gonadal control of follicle-stimulating hormone, *Nature*, vol. 321, p. 724, Fig. 1, courtesy of Macmillan Journals Limited.)

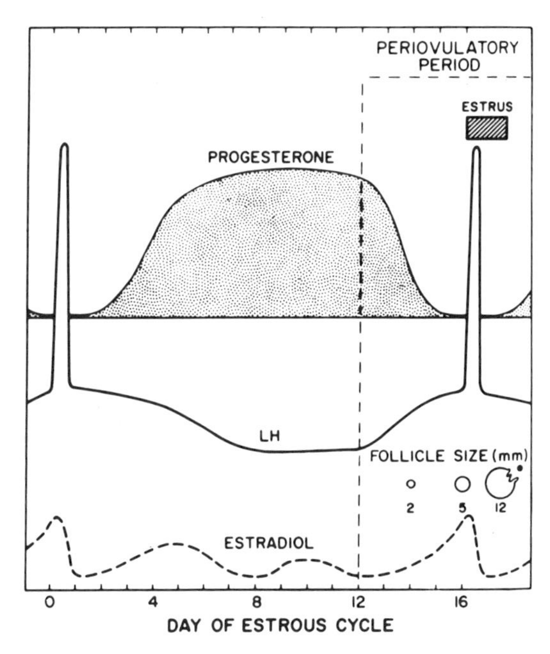

Figure 2.7 Fluctuations of LH, estradiol, and progesterone in the 16-day ovulatory cycle of the ewe, showing the sharp rise of LH with the disappearance of serum progesterone. (From Legan and Karsch, 1979, courtesy of *Biology of Reproduction*.)

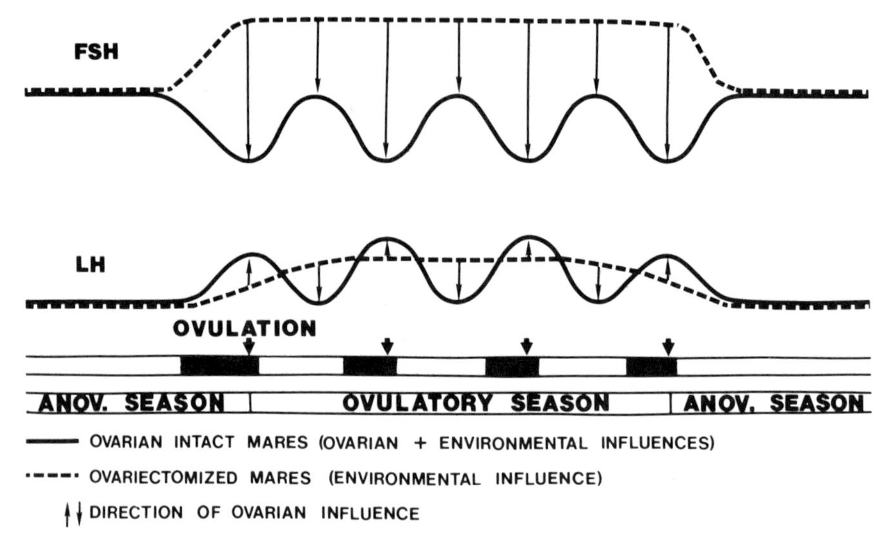

Figure 2.8 The combined effects of the ovarian and photoperiodic influences on circulating concentrations of LH and FSH in mares. (From Ginther, 1979, courtesy of Allanheld, Osmun & Company.)

of LH, and the gradual decline of the corpus luteum may contribute to the preovulatory surge of LH. This effect is shown by the surgical removal of the corpus luteum, which is followed by premature estrus and ovulation. It is also shown by the absence of estrus and ovulation after a lutectomy followed by a progesterone implant (Karsch et al., 1978).

There is apparently no negative feedback for progesterone alone, but when it occurs together with estrogens, it does increase the negative feedback of estrogens on FSH. In mammals such as some primates, the corpus luteum produces both progesterone and estrogen, which in such species prevent high levels of FSH. In some other mammals the corpus luteum secretes only progesterone, and new follicles continue to develop immediately after ovulation. In the domestic fowl, progesterone has a positive feedback on both the hypothalamus and the anterior pituitary, stimulating the release of LH. This modification in birds enables the rapid development of a full clutch of eggs, with the release of only a single egg at one time.

The negative feedback relationship is also shown by ovariectomy. In the absence of gonadal steroids and inhibin, LH and FSH increase in the general circulation but still reflect the stimulation of long days (Fig. 2.8).

2.6 SUMMARY

Hormones are chemical messengers that are released by one specific gland into the circulatory system. Although hormones are carried throughout the body, they induce a precise effect in certain organs or tissues. Having receptors for one or more hormones, these target organs or tissues are sensitive to these hormones and passive to others. The reproductive system is controlled by hormones that function among the hypothalamus, the anterior and posterior pituitary glands, and the gonads. This system of organs is sometimes referred to as the hypothalamic-pituitary-gonadal axis.

The hypothalamus is sensitive to both exogenous and endogenous signals, which can be either neural or hormonal. The hypothalamus is a major regulator of reproductive events, but in some mammals the pineal gland influences the hypothalamus. By releasing hormones that either stimulate or depress specific parts of the anterior pituitary gland, the hypothalamus affects the gonads. Gonadotropin-releasing hormone (GnRH), from the hypothalamus, is transported via a portal system directly to the anterior pituitary in tetrapod vertebrates and birds and induces the release of gonadotropins—follicle-stimulating hormone (FSH) and luteinizing hormone (LH)—into the systemic circulation. In elasmobranchs hypothalamic hormones are released into the internal carotid artery. In teleost fish, neurosecretory cells extend directly from the hypothalamus into the anterior pituitary. In teleost fish there are two gonadotropins, together referred to as gonadotropin hormone (GtH). Prolactin, from the anterior pituitary, and thyroxin also affect gonadal activity.

Several steroid hormones are synthesized by the gonads. Estrogen is produced by the ovary and, to a lesser extent, by active mammary tissue. Progesterone is formed by the ovary and most conspicuouslsy by the corpora lutea in the postovulatory ovary of most live-bearing vertebrates. Progesterone is also produced by the placenta of many mammals. Androgens are produced in great amounts by the testis, but they are also synthesized by the ovary, where they are mostly converted to estrogens. Steroids are also synthesized by the adrenal cortex.

Steroids play major roles in gametogenesis, gestation in live-bearing vertebrates, and oviposition in egg-laying forms. Many different secondary sexual characteristics are induced by gonadal steroids, and these hormones also have profound effects on such behaviors as courtship and parental care. Steroids also affect the activities of the hypothalamus and the anterior pituitary.

Relaxin, a polypeptide from the ovary, softens the pelvic girdle prior to parturition and may also inhibit uterine contractions.

Oxytocin, a peptide from the posterior pituitary and also from the ovary, functions in milk letdown and sometimes promotes parental care.

Inhibin, a polypeptide from the gonads, limits the secretion of FSH.

Feedback, either positive or negative, is the effect of a hormone on an endocrine gland whose secretion stimulates the secretion of that hormone. An important feedback is the negative effect of estrogen from the ovary, which depresses FSH release from the anterior pituitary.

BIBLIOGRAPHY

Adler NT (Ed) (1981). *Neuroendocrinology of Reproduction: Physiology and Behavior*. Plenum, New York, London.

Asdell SA (1964). *Patterns of Mammalian Reproduction*. Cornell Univ, Ithaca.

Austin CR and Short RV (1979). *Reproduction in Mammals*. Book 7. *Mechanisms of Hormone Action*. Cambridge Univ, Cambridge.

Bentley PJ (1976). *Comparative Vertebrate Endocrinology*. Cambridge Univ, London.

Beyer C (Ed) (1979). *Endocrine Control of Sexual Behavior*. Raven, New York.

Bhatnagar AS (Ed) (1983). *The Anterior Pituitary Gland*. Raven, New York.

Brenner RM and West NB (1975). Hormonal regulation of the reproductive tract in female mammals. *Annual Review of Physiology*, **75**: 273–302.

Donaldson EM (1973). Reproductive endocrinology of fishes. *American Zoologist*, **13**: 909–927.

Epple A and Stetson MH (Eds) (1980). *Avian Endocrinology*. Academic, New York.

Farner DS (Ed) (1973). *Breeding Biology of Birds*. National Academy of Sciences, Washington.

Follett BK (1984). Birds. Pp 283–350. *Marshall's Physiology of Reproduction, vol 1. Reproductive Cycles of Vertebrates*, 4th ed. (Ed: Lamming GE). Churchill Livingstone, Edinburgh.

Holmes RL and Ball JN (1974). *The Pituitary Gland. A Comparative Account*. Cambridge Univ, London, New York.

Karim SMM (Ed) (1975). *Prostaglandins and Reproduction. Advances in Prostaglandin Research*. Univ Park, Baltimore.

Lofts B (1972). Reproduction. Pp 107–218. *Physiology of the Amphibia* (Ed: Lofts B). Academic, New York, London.

Lofts B, Follett BK, and Murton RK (1970). Temporal changes in the pituitary-gonadal axis. *Memoirs of the Society of Endocrinology*, **18**: 545–575.

Martini L, Motta M, and Fraschini F (Eds) (1970). *The Hypothalamus*. Academic, London.

Nalbandov AV (1976). *Reproductive Physiology of Mammals and Birds*. Freeman, San Francisco.

Norris DO and Jones RE (Eds) (1987). *Hormones and Reproduction in Fishes, Amphibians, and Reptiles*. Plenum, New York.

Saint Girons H (1970). The pituitary gland. Pp 135–199. *Biology of the Amphibia 3, Morphology C* (Eds: Gans C and Parsons TS). Academic, New York.

van Tienhoven A (1983). *Reproductive Physiology of Vertebrates*, 2nd ed. Saunders, Philadelphia.

3

THE OVARY

3.1 OVULATORY CYCLES AND OVARIAN STRUCTURE

In all vertebrates the ovary produces ova and synthesizes hormones, but the basic ovarian plan has several morphological variants. In all developing ovaries there are oogonia, oocytes, and, as the follicle develops, theca cells and granulosa cells. Oogonia do not usually occur in the postnatal ovaries of elasmobranchs, birds, and mammals, but some exceptions do exist (see Section 3.1.1.1). Connective tissue develops into theca cells, and interstitial tissue may synthesize hormones in addition to those produced by the follicle and the corpus luteum. Stromal tissue lies between the follicles and it contains both nerve endings and circulatory vessels.

Cyclostomes have little or no stromal tissue, and the single ovary in these vertebrates tends to be saclike. The ovary of teleost fish is either a solid structure that sheds ova to the peritoneal cavity (called gymnovarian) or, in live-bearing and other fish, a hollow and usually median structure that releases ova into its own lumen (called cystovarian). Amphibians and reptiles have very little stroma and few or no interstitial cells. The amphibian ovary, like that of some teleosts, is hollow with inner ovigerous folds, formed from embryonic medullary cells, and a lining of germinal cells, formed from primordial germ cells (PGCs). Stromal elements are similarly limited in birds and monotremes.

The mammalian ovary has a rich vascular system, which must be adequate not only for providing nutrients and oxygen but also for conducting the business of endocrine communication. In some mammals there is a close association of ovarian and uterine vascular systems, suggesting direct hormonal communication between these two structures. Although vascular relationships of a given animal do not change with age, the amount of blood moving through it does change with the reproductive state. The ovary has sympathetic and parasympathetic innervation.

In vertebrates the ovaries are basically paired. In cyclostomes the genital ridges fuse to form a median ovary. In many teleost fish, especially the viviparous groups, the ovaries fuse to form a single median structure. In teleosts the ovaries are hollow and are characterized by an extensively folded ovigerous lining. Genuine ovarian asymmetry is seen in the hagfish (Myxinidae) and some elasmobranchs (species of *Pristiophorus*, *Mustelus*, and *Zugaena*), in which only the right ovary is functional.

Although PGCs occur symmetrically in birds, gonadal asymmetry develops after PGCs arrive at the genital ridges. In most birds the right ovary does not develop but does persist in a vestigial condition. In some raptorial birds both ovaries develop, but only the left ovary functions. Paired ovaries have been reported to occur, at least sporadically, in pigeons, ducks, and parrots. Perhaps the space required to contain the relatively large avian egg favors the development of only one half of the reproductive tract of female birds and would select against the retention of the paired tract (McCarrey and Abbott, 1979).

The ovaries are paired in most mammals, but a few species typically have only a single ovary developed and functional. The platypus (*Ornithorhynchus anatinus*) has paired ovaries, but only the left ovary and oviduct function; both ovaries are functional in the spiny anteaters (*Tachyglossus aculeatus* and *Zaglossus bruijinii*). Variations are the rule in bats, and in many species only a single functional ovary occurs. In primates (and in some other mammals, for example, the porcupine), which normally produce only a single ovum in a single ovulatory cycle, estrogen is secreted almost solely by the ovary containing the developing follicle, with little or none secreted by the inactive ovary.

3.1.1 Oogenesis

The growth and proliferation of oogonia and their development to the oocyte stage constitute oogenesis. A narrower concept considers oogenesis to be the growth of oogonia to the onset of the first meiotic division, the point at which oogonia become primary oocytes. Within this period both mitotic division and degeneration (atresia) of oogonia take place. These changes occur early in ovarian development in lampreys, elasmobranchs, birds, and mammals but are seasonal events in other vertebrates. They constitute the prefollicular development of ova. Although oogenesis and folliculogenesis are sequential aspects of a single extended process—the development of ova—it is convenient to separate them in discussion.

3.1.1.1 Oogonia and Oocytes. In the embryonic ovary, germ cells divide mitotically, and mitosis in the ovaries of elasmobranchs, birds, and most mammals ceases at birth. Intercellular bridges characterize these cells, and such bridges are almost always absent in PGCs (Fig. 3.1). Some oogonia continue to divide mitotically after others have initiated the first meiotic division and produced primary oocytes. The two processes are thus not entirely mutually exclusive, and oogonia may persist in the postnatal ovary of a few mammals.

In almost all mammals the primary oocyte remains arrested in meiotic prophase I for a prolonged period, and it does not resume growth and division until shortly before ovulation. It has been postulated that meiosis is suppressed by secretion of an oocyte maturation inhibitor from granulosa cells (Lindner et al., 1983).

In some mammals, such as voles (Arvicolidae), hamsters (Cricetidae), rabbits (Leporidae), weasels (Mustelidae), and lemurs (Lemuridae), oocyte formation takes place in the first few days or weeks of postnatal life. In marsupials, which are born in a state comparable to a eutherian fetus, oogenesis proceeds postpartum. In the tammar wallaby (*Macropus eugenii*), for example, oogonia form after birth, and the first meiotic division forming primary oocytes occurs at 24 to 30 days of age (Alcorn and Robinson, 1983).

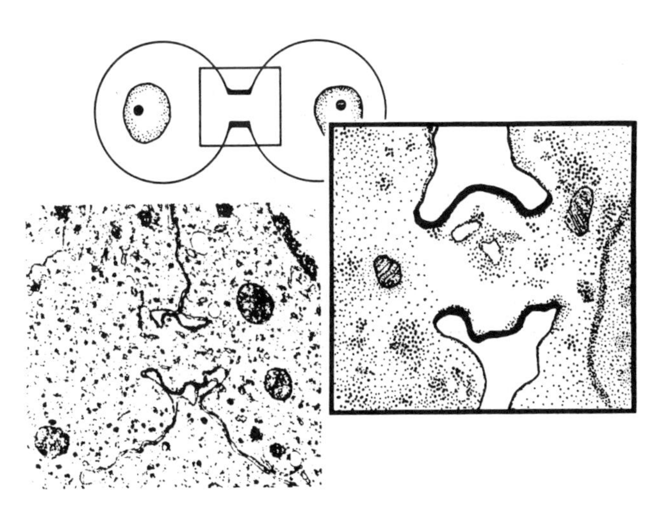

Figure 3.1 Intercellular bridges connecting adjacent oogonia and oocytes. These bridges allow cytoplasmic exchange between cells and are known to occur in a broad spectrum of vertebrates, including many mammals. (From Gondos, 1978, courtesy of Plenum Publishing Corporation.)

3.1.1.1.1 Fish. Oogonia divide actively in the larval stage of cyclostomes. In the early embryonic ovary of bony fish, extensive oogonial mitosis occurs. Proliferation is followed by the formation of clusters, or *nests*, of oogonia. The ovary of adult teleosts usually has a supply of oocytes. In seasonally breeding bony fish, oogonia usually divide after spawning, so that these oocytes mature as a synchronized population of germ cells. Oogenesis is irregular and more or less continuous in tropical species, which have a prolonged breeding season. In any case a new generation of primary oocytes follows the release of eggs, and several successive generations of oogonia may occur annually. Oocyte maturation is stimulated by a gonadotropin (or gonadotropins), apparently together with one or more steroids from the ovary and the adrenal cortex (Truscott et al., 1978). Progestogens and 11-deoxycorticosteroids induce final maturation of the teleost oocyte in vivo (Goetz, 1983).

3.1.1.1.2 Amphibians. In urodeles and anurans oogonia form in the larval ovary. After each seasonal ovulation in the adult, waves of oogonia move near the surface. Mitotically dividing oogonia occur in clusters, and such synchrony is perhaps assisted by intercellular bridges. Similarly, the annual formation of oocytes is synchronized, a necessary pattern in species releasing large numbers of eggs within a brief period. Synchronous maturation of oocytes in the toad (*Bufo bufo*) may be achieved by the suppressive action of growing oocytes on those that have not yet begun growth, resulting in a large number of ova of rather uniform size (Jørgensen et al., 1970; Jørgensen et al., 1978). In tropical amphibians several generations of oocytes may exist at any given time, and ovulation may occur two or several times a year. Development is usually arrested at prophase I, as in most vertebrates, to resume shortly before ovulation.

In mitosis of anuran oogenesis the cytoplasmic divisions may not occur at all, so that instead of the cells being held together by cytoplasmic bridges there are multinucleate oocytes. In the bell toad (*Ascaphus truei*), after the final three oogonial mitoses, each oocyte has eight nuclei; in late oogenesis, maturation eliminates all but one (Macgregor and Kezer, 1970). In some species of marsupial hylid frogs of the Neotropics, apparently the same process produces oocytes that may have up to 2000 nuclei (del Pino and Humphries, 1978).

3.1.1.1.3 Reptiles. In reptiles, in contrast to fish and amphibians, there are fewer oogonia of uniform size at any given time. Mitosis begins in the embryonic ovary, and oogonia persist in the adult in groups called germinal beds. Mitotic division within a single bed is synchronized. Meiosis to prophase I occurs in the adult. Oogenesis in reptiles occurs at a distinct point in the annual reproductive cycle, but this point varies with each group. In turtles maturation of oocytes follows oviposition but is not completed until the following spring. In freshwater species oogenesis may be annual, and there may be several sets of follicles in those species that lay two or more clutches a year (Moll, 1979).

3.1.1.1.4 Birds. Most studies concern the domestic fowl and domestic strains of the Japanese, or migratory quail (*Coturnix coturnix*). Mitotic divisions of oogonia occur in the embryonic ovary of birds and cease at hatching. There are relatively few intercellular bridges, perhaps accounting for only limited synchrony in avian oogenesis. Clusters of developing oogonia occupy only a small part of the ovary at any one time, and these groups tend to be in different stages of development. At the time of hatching, avian oocytes are mostly in meiotic prophase I. Follicles are continuously lost to atresia.

In teleost fish, amphibians, and reptiles, meiosis II is completed after fertilization. In the event no sperm are present to stimulate the completion of the second meiotic division, as in some parthenogenetic species, the chromosomes of the nucleus of the secondary oocyte together with those of the polar body contain the full (diploid) complement, and fatherless offspring may develop. In birds and most mammals meiosis is completed by the time of ovulation.

3.1.2 Folliculogenesis

Folliculogenesis has been most carefully followed in mammals and is reviewed by Midgely and Sadler (1979), Peters (1978, 1979), Williams and Hodgen (1983), and others; for nonmammalian vertebrates see Tokarz (1978). The following initial comments refer to mammalian folliculogenesis, and some differences in nonmammalian vertebrates will be noted.

At birth each oocyte is surrounded by a single layer of undifferentiated spindle-shaped cells, which lie in cell nests closely associated with the cords and rete ovarii. These cells constitute a population of nonproliferating follicles. Initial growth is shown by an increase in the size of the oocyte and in the number of surrounding granulosa cells. Continuation of the development of primary follicles occurs throughout the active life of the individual, but may be seasonal and seems to be independent of gonadotropins or sexual steroids.

The gradually enlarging oocyte moves toward the surface of the ovary. The adjacent granulosa cells are characterized by concentrations of Golgi, mitochondria, rough endoplasmic reticulum, and lipids. Granulosa cells originate in the vicinity of the cords and may be products of the invaginated epithelial covering. Granulosa cells produce steroids. They also secrete a fluid, which at least partly surrounds the oocyte; this fluid-filled region is the antrum. An acellular membrane, the zona pellucida, develops around the oocyte.

One or more follicles mature before ovulation. In most mammals folliculogenesis includes the formation of the antrum and the synthesis and release of estrogen, progesterone, and androgen; it includes an infusion of yolk in most nonmammals. The antrum is absent in the oocyte of the platypus, this space being filled with yolk. The antrum is very small is some, if not most, shrews (Soricidae).

In the developing follicle an outer layer of theca cells surrounds the inner layer of granulosa cells. These layers are targets of luteinizing hormone (LH), follicle-stimulating hormone (FSH), and prolactin. FSH and estradiol increase LH receptors on granulosa cells, and estradiol may induce growth in the number of granulosa cells in some mammals. During follicular development the granulosa cells proliferate, and some form a clump, the cumulus oophorus, which surrounds the oocyte and connects it to the wall of the follicle. With continued enlargement, the cumulus separates from the follicle wall, which then becomes thin in the region adjacent to the surface of the ovary.

Theca cells encircle the oocyte in its initial growth. Theca cells differentiate into two layers, the theca interna and theca externa. Both of these layers are provided with capillaries, which fail to extend to the granulosa cells. Theca cells are sensitive to LH, and granulosa cells respond to FSH. Theca cells produce androgens, which stimulate the formation of LH receptors in the granulosa cells; additionally, the androgens may provide for the synthesis of estradiol by the granulosa cells. Thecal androgens also induce the formation of prolactin receptors in the granulosa cells, providing for the synthesis of progestin by the mature follicle (Birnbaumer and Kirchick, 1983). Luteinization of the granulosa cells commences prior to ovulation.

The primary oocyte now resumes meiosis, and the nucleus, still diploid, begins to assume a position near the edge of the cell. The completion of meiosis I generally occurs only hours before ovulation. In this condition the secondary oocyte leaves the ovary and enters the upper oviduct. The second meiotic division proceeds to metaphase, where the oocyte remains until the sperm enters the egg. Penetration of the zona pellucida and plasma membrane of the ovum stimulates the completion of the second meiotic division, releasing a second polar body. The pronulcei of sperm and egg then unite to form the zygote, restoring the diploid complement of chromosomes.

In many mammals one or more oocytes are released at ovulation, but in some species it is the primary oocyte. In some insectivores sperm may enter the ovary and penetrate the unovulated oocyte; this has been observed in the short-tailed shrew (*Blarina brevicauda*) (Pearson, 1944).

Maturation (resumption of meiosis) of oocytes of fish has been reviewed by Goetz (1983). From a large number of in vitro studies on a variety of teleosts, 11-deoxycorticosteroids and progestogens appear to be the most powerful steroids in effecting final oocyte maturation. The direct effect of estrogens in this process appears to be much less important. In seasonal ovarian recrudescence of the rainbow trout (*Salmo gairdneri*), there are initial (preovulatory) high pulses of gonadotropin hormone (GtH) release, which seem to regulate estradiol synthesis and oogenesis (Zohar et al., 1982). In amphibians and reptiles stimulation of the ovary, for both oogenesis and steroidogenesis, is induced by pituitary gonadotropins, similar to the hormonal stimulation described for mammals. Oogonia proliferate mitotically, in a synchronous fashion, so that a large number initiate meiosis and reach prophase I simultaneously at about the time of ovulation.

Teleosts that spawn several times a year have oocytes at various stages of development. The ovaries of semelparous fish (those that spawn only once and die) have oocytes all in the same stage of development, in contrast to iteroparous species (those that seasonally spawn once in successive years), which have two distinct generations of oocytes.

In nonmammals, as in mammals, there is a zona pellucida composed of mucopolysaccharides. Exchange of nutrients and other materials occurs through the zona pellucida, but it forms a barrier to the entrance of sperm. A zona pellucida surrounds the developing follicle of elasmobranchs; as the follicle matures, this membrane is called the zona radiata. The teleost follicle is similarly surrounded by a zona pellucida (Fig. 2.4), which forms adjacent to the granulosa cells and may be a product of them (Wourms and Sheldon, 1976). Surrounding the maturing anuran oocyte are layers of cells apparently homologous to theca and granulosa cells of other vertebrates. During the infusion of vitellogenin, a vitelline membrane develops about the oocyte and within the cellular layers. This membrane is called the vitelline envelope (or zona radiata) and is probably homologous to the zona pellucida of other vertebrates (Grey et al., 1974).

3.1.2.1 Yolking. In most nonmammalian vertebrates the egg is characterized by a substantial amount of nutrient material known as *yolk*, which is composed mostly of lipids and proteins. Yolk volume is reduced in many viviparous nonmammalian vertebrates and is virtually absent in a few (see Chapter 8, Livebearing Ectotherms, Sections 8.3.4 and 8.5.2.1). Physically, the addition of yolk is the most conspicuous aspect of folliculogenesis in nonmammalian vertebrates. Yolk is most abundant in those species in which (1) the condition of young at birth is advanced and (2) the young receive little or no parental care. Yolk is least abundant in those species in which (1) young are born or hatched in a relatively undeveloped or larval condition and (2) the young receive parental care from the mother either before or after birth, including in utero parental nourishment. The sudden growth seen in the developing ova of birds and most ectotherms comes from an infusion of the yolk precursor, vitellogenin. In all vertebrates vitellogenin comes from the liver, and its release into the circulatory system and acquisition by the follicle(s) are both under hormonal control.

Vitellogenesis in lampreys includes the release of vitellogenin, as in other oviparous vetebrates (Wallace et al., 1966). In the Pacific hagfish (*Eptatretus stouti*), estrogen induces vitellogenesis (Yu et al., 1981). In the lamprey (*Lampetra fluviatilis*), implantation of estradiol (0.5 mg/animal/month) was followed by an increase of serum calcium and ovarian mass, apparently as a result of an induction of yolking (Pickering, 1976). Normally, vitellogenesis begins before lampreys ascend rivers on their anadromous migrations. In the spotted dogfish (*Scyliorhinus canicula*) there is an annual cyclicity of the thyroid gland, and the changes are more pronounced in the female. Thyroidectomy causes a failure of vitellogenesis in this species (Lewis and Dodd, 1974; Dodd, 1983).

Initial vitellogenesis of amphibians includes a great increase of cytoplasm. In temperate latitude species this event typically follows the laying of the previous generation of eggs and precedes hibernation.

In amphibians yolking is induced by estrogens, which cause the release of fat from fat bodies, and this fat is carried in the blood to the liver. Gonadotropin administration induces the release of yolk proteins from the liver, but not in ovariectomized females (Nicholls et al., 1968). After the release of vitellogenin from the liver, the incorporation of these yolk proteins from plasma into oocytes is mediated by gonadotropins (Wallace and Jared, 1969). The appearance of vitellogenin in the blood is concurrent with an increase in serum calcium, and when released from the liver, vitellogenin binds free calcium ions. This situation exists in the clawed frog (*Xenopus laevis*), in which circulating vitellogenin was found following treatment with human chorionic gonadotropin and estrogen. This response also occurs in vitellogenic females and estrogen-treated males (Wallace, 1978). Vitellogenins have been identified in a broad spectrum of oviparous vertebrates.

In amphibians yolk consists mostly of protein, with lesser amounts of lipids and glycogen. The protein is mostly incorporated into yolk platelets, which have a delicate crystalline structure. The release of yolk precursors from the liver is induced by estrogen, and the uptake of vitellogenin by oocytes is stimulated by FSH in many diverse kinds of oviparous vertebrates. Growth hormone may promote or assist the estrogen-induced release of vitellogenin from the liver in some iguanid lizards (Callard et al., 1972; Callard and Lance, 1977).

In both hypophysectomized and intact tiger salamanders (*Ambystoma tigrinum*), in vitro and in vivo administration of prolactin enhances maturation and ovulation prior to the normal reproductive season, and thyroxine blocks the effect of prolactin (Norris and Duvall, 1981). Ovulation in the toad (*Bufo bufo*) is stimulated by immersion in water and not by amplexus (Heusser, 1963).

In an in vivo experiment with ovariectomized lizards (*Lacerta vivipara*), only estradiol stimulated vitellogenin synthesis, and there was no response from progesterone or androgens (Gavaud, 1986).

The anamniotic egg of amphibians contains a large yolk sac over which small blood vessels not only carry food and oxygen to the developing embryo but remove metabolic wastes. It is essential that the surface be in contact with water or moist air so that gas exchange can be rapid. In the amniotic egg of birds and egg-laying reptiles, oviposition is on land. Birds always have rigid-shelled eggs, and oviparous reptiles have eggs with either rigid or leathery shells. In both reptiles and birds these shells contain pores and allow the passage of gases and some fluids. The rather coriaceous shell of many reptilian eggs allows for a substantial increase in egg dimensions and mass from absorption of water. Hard-shelled eggs are characteristic of many species of turtles, crocodilians, and geckos. The adaptive significance of eggshell texture and thickness is equivocal.

The avian egg contains adaptations that allow for its survival on land. In addition to the yolk, there is an abundance of albumen, which is entirely enclosed in an egg membrane and eggshell. Both the yolk and albumen consist of proteins and lipids. Enveloping the yolk and extending the long axis of the egg are fibrous strands of albumen, the chalazae. In both birds and reptiles the shell, which contains calcium carbonate, is deposited by the shell gland, the homologue of the mammalian uterus. Calcium is extracted from the circulation, which draws on calcium intake from the intestine; on a calcium-free diet, calcium can be mobilized from bone. Calcium moves across the shell gland by active transport under hormonal stimulus, possibly as a result of estrogen secreted after ovulation (Eastin and Spaziani, 1978).

3.1.3 Atresia

Of the multitude of oogonia present in the neonatal ovary of all vertebrates, very few reach the secondary oocyte stage. Throughout the various stages of follicle formation, atresia accounts for a mass disappearance of potential ova. Although atresia is a continuous process, it is appreciably greater prior to sexual maturity of the female than it is subsequently. Atresia frequently involves pairs of follicles, perhaps as a consequence of intercellular bridges. In more sexually mature birds and mammals, atresia attacks larger, more nearly mature follicles. In the older follicles, atresia is accompanied by a loss in blood supply. In mammals necrosis attacks the older oocytes: while the theca cells hypertrophy and become surrounded by lipid droplets, the oocyte is last. Following atresia, phagocytic leucocytes and granulosa cells fill the cavity of the follicle.

If not ruptured in ovulation, mature follicles normally become atretic, but they may survive in the ovary of hibernating bats. In the autumn several antral follicles develop in the right ovary of the greater horseshoe bat (*Rhinolophus ferrum-equinum*), and all but one are atretic by spring. The surviving oocyte seems to be maintained by an abundant supply of lipids and carbohydrates from the cumulus cells (Oh et al., 1985b). In the southern elephant seal (*Mirounga leonina*) the blastocyst lies free in the uterus for several months, during which time folliculogenesis results in a series of immature follicles wherein the theca interna become luteinized and then atretic. Later, after implantation, a second wave of follicular activity occurs, and these follicles also degenerate.

Atresia occurs in teleosts as in higher vertebrates. Oogonia and oocytes are phagocytized by granulosa cells, which then become corpora atretica, or "corpora lutea." A steroidogenic role has sometimes been attributed to them, but there is little evidence for this (Nagahama, 1983). In most oviparous teleosts a postovulatory corpus luteum–like structure does not develop. In amphibians and reptiles atresia occurs both before and after folliculogenesis.

Although there may seem to be a great waste of oocytes, the larger atretic follicles produce (or contain) appreciable amounts of steroids. In a mature mammalian ovary the large atretic follicle, now without an oocyte, will sometimes become luteinized and produce substantial amounts of progesterone, comparable to that produced by the corpus luteum (Byskov, 1979).

3.1.4 Ovulation in Mammals

The process by which one or more primary oocytes mature and are released from the ovary constitutes the ovulatory cycle. As part of this recurring event, there is a cyclical ebb and flow of circulating hormones from the hypothalamus, pituitary, gonads, and some other structures. These hormones not only induce follicular and uterine growth and development but also produce behavioral changes that announce the female's pending fertility to nearby, and sometimes distant, males of the same species. Most mammals have one or several cycles annually, but a few have a single annual cycle and some have only one in a lifetime.

The period of folliculogenesis leading up to ovulation is called the follicular phase, and the period from ovulation to the demise of the corpus luteum is termed the luteal phase. The length of the luteal phase varies considerably among various taxa, generally being longer in live-bearing forms. In marsupials the luteal phase is of the same duration as pregnancy, which is greatly abbreviated.

The physical changes that effect the development of an ovum are accompanied by rather pronounced hormonal changes, which are well known for some mammals. These hormonal changes include the role of the hypothalamus in controlling the activity of the anterior pituitary as well as the feedback mechanisms from ovarian steroids. Additionally exogenous signals affect the ovulatory cycle, including photoperiod, temperature, food, and courtship, and they are discussed in subsequent chapters. Here we shall review the major hormonal changes along with their endogenous cyclical causes and effects.

As already noted the initial endogenous stimulus of the cycle is the gonadotropin-releasing hormone (GnRH) from the hypothalamus. During the period of follicular growth, there is a low and rather constant (tonic) release of GnRH, which induces a steady, low-level release of both LH and FSH. Initially estrogen, through its negative feedback effects, suppresses the release of both GnRH and FSH; but as estrogen secretion increases, there is a continued, but low-level, synthesis and release of LH. In the preovulatory ovary the level of FSH rises, probably because of an increase in GnRH. This rise is followed by an increase in follicular size (due to an increase in follicular fluid) and an increase in estrogen secretion. This production of estrogen is apparently a synergistic response to both FSH and LH.

Estrogen finally reaches a stimulatory level, and there is then a sudden and substantial rise in the release of LH. This is followed by the discharge of one or more ova from one or more follicles on the surface of the ovary. This well-documented pattern is seen in the human female (Fig. 3.2) and is rather typical for

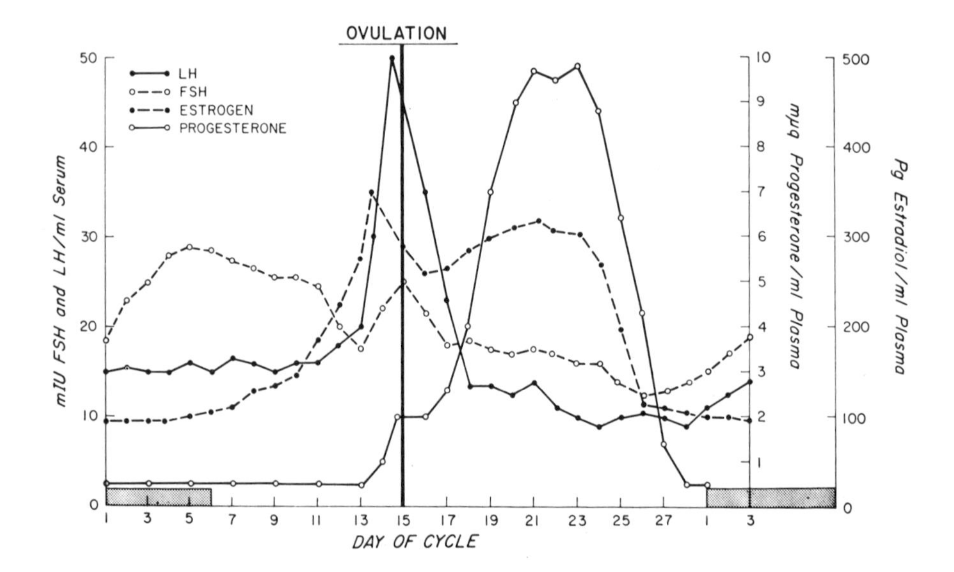

Figure 3.2 Levels of ovarian steroids and gonadotropins throughout the human menstrual cycle. Levels of LH (●——●), FSH (○-------○), estrogen (●——●), and progesterone (○-------○). (From Taymor et al., 1972, courtesy of C. V. Mosby Company.)

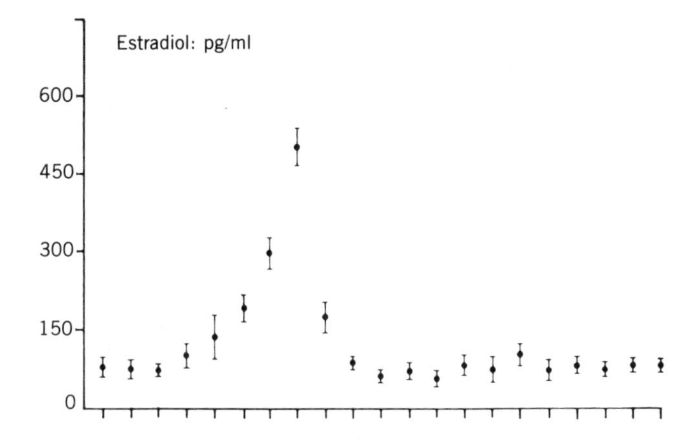

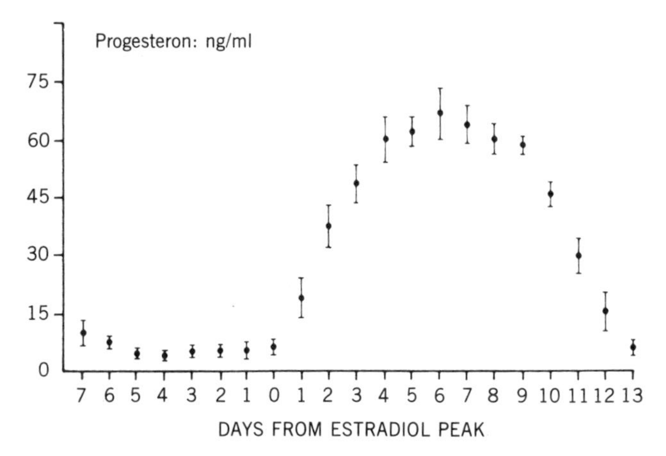

Figure 3.3 The mean (± SEM) levels of serum estradiol (top) and progesterone (bottom) from nine capuchin monkeys (*Cebus apella*) through an entire menstrual cycle. (From Nagle et al., 1979, courtesy of *Biology of Reproduction*.)

most vertebrates. In most primates the level of serum estradiol appears to remain steady through the menstrual cycle, with a brief rise just before ovulation (Fig. 3.3).

Prior to ovulation serum progesterone increases in placental mammals, birds, and many other vertebrates. This steroid emanates from the preovulatory follicle and/or the adrenal cortex, most of it probably coming from the follicle. However, serum progesterone rises before there is an increase from the ovarian vein, and the adrenal cortex may be a source (Feder, 1981). The action of progesterone is enhanced when the target tissue has been previously sensitized by estrogens; such synergistic response requires that many of the target organs of the two hormones be the same. Although estrogen and progesterone usually have distinct and separate peaks—

estrogen being preovulatory and progesterone being postovulatory—their occurrences broadly overlap in the ovulatory cycles of many species.

In many, perhaps most, mammals lactation delays or prevents ovulation, apparently through elevated levels of prolactin. Physiological levels of prolactin inhibit FSH-induced synthesis of estrogen in cultured granulosa cells (Dorrington and Gore-Langton, 1981).

Gonadotropins that appear in one cycle affect the timing and growth of follicles in the subsequent cycle. FSH release during proestrus and estrus of rats and hamsters stimulates the development of follicles for the following cycle. Experimentally inhibited FSH release in the hamster results in a reduced number of follicles at the next estrus (Chappel and Selker, 1979). Additionally, in the hamster the LH surge prior to ovulation not only stimulates preantral follicles that will mature at the next proestrus but may be the synchronizing signal for the theca cells to synthesize progesterone. After the LH surge, the preantral follicles synthesize androstenedione and estradiol (Terranova et al., 1983).

Prostaglandins have at least two roles in the regulation of the ovarian cycle. Following gonadotropin release, prostaglandin $F_{2\alpha}$ ($PGF_{2\alpha}$) increases in the mature mammalian follicle, probably from granulosa cells and perhaps from stimulation by LH. Prostaglandins stimulate the synthesis of both cAMP and steroids. They also cause contraction of both intestinal and uterine tissue during menses and may promote the extrusion of the ovum from the follicle.

In the theca of several groups of vertebrates are elements resembling smooth muscle, and presumably their contraction is followed by ovulation (DiDio et al., 1980; Goetz, 1983; Larsen et al., 1977). There are microfilaments in theca cells resembling smooth muscle cells, and they presumably account for the follicular contraction at the time of ovulation (Goetz, 1983). Levels of $PGF_{2\alpha}$ drop immediately after ovulation. The role of prostaglandins in ovulation is suggested by a failure of ovulation following administration of prostaglandin inhibitors (aspirin or indomethacin) and by a reversal of this effect by injections of prostaglandins (Behrman, 1979).

Experimentally PGEs and PGFs induce labor and terminate pregnancy in several laboratory mammals (Labhsetwar, 1975). Also, both PGEs and PGFs have a luteolytic effect on the corpus luteum and seem to bring about its demise at the close of the luteal phase of the cycle. Prostaglandins are released from the uterus of the ewe and carried directly through the uterine vein, which lies appressed to the ovarian artery (Fig. 3.4). This arrangement may allow a countercurrent exchange of $PGF_{2\alpha}$ to the ipsilateral ovary. Prostaglandins uncouple LH from adenylcyclase, resulting in cAMP production and thus preventing LH from acting on the corpus luteum. Prostaglandins from the uterus cause diminished regression of the corpus luteum in several well-studied mammals, including the guinea pig, ewe, cow, and mare (Ainsworth et al., 1979).

3.1.5 Ovulation in Nonmammalian Vertebrates

Ovulation in fish is associated with a rise in GtH (Fig. 3.5). Among several salmonid fish, increased release of GtH occurs with gonadal recrudescence, and a conspicuous

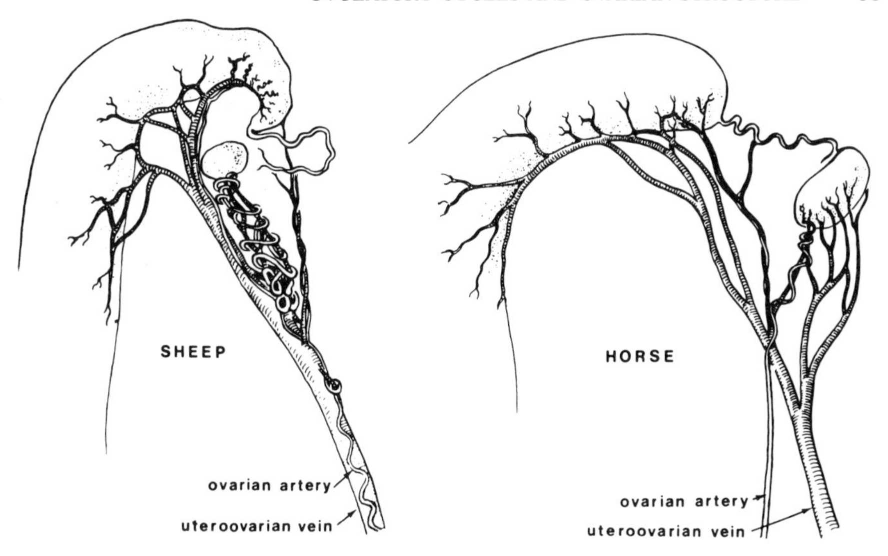

Figure 3.4 The uterine and ovarian vascular systems of a sheep and a horse. In the sheep, which has an ipsilateral communication from the uterus to the ovary, the ovarian artery is closely applied to the uteroovarian vein. There is little or no uteroovarian communication in the horse. (From Ginther, 1974, courtesy of the *Journal of Animal Science*.)

rise is seen at the time of ovulation and spermiation (Peter and Crim, 1979; Stacey, 1984). GtH secretions in at least some salmonid fish rise with an increase in day length (Crim et al., 1976; Breton and Billard, 1977). In the carp (*Cyprinus carpio*), GtH increases in the summer, the season of ovarian growth, but is low in the spring, at the time of spawning (Bieniarz et al., 1978). In some bony fish, mammalian LH induces ovulation (Jones, 1978; Goetz, 1983). In some teleosts ovulation occurs together with a rise in corticosteroids from both ovarian and extraovarian sources, perhaps the interrenal tissue (Dodd, 1975).

In vivo administration of prostaglandins E_1, E_2, and $F_{2\alpha}$ were all followed by ovulation in female goldfish in which natural synthesis of prostaglandins had been blocked by indomethacin (Stacey and Goetz, 1982). $PGF_{2\alpha}$ is generally the most effective inducer of ovulation and may account for the follicular constriction in at least some teleosts. PGF is known to increase about the time of ovulation in a number of fish species. Prostaglandins also account for spawning behavior (see Chapter 9, Parental Care).

The presence of eggs in the oviduct of a female goldfish has been shown to be the stimulus for spawning behavior. Removal of eggs by stripping will stop the spawning behavior, but spawning behavior resumes when the eggs are injected back into the ovarian lumen (Stacey and Liley, 1974). This behavior is under the control of prostaglandin; nonovulated females will continue spawning behavior following prostaglandin injection (Stacey and Goetz, 1982). Prostaglandins have

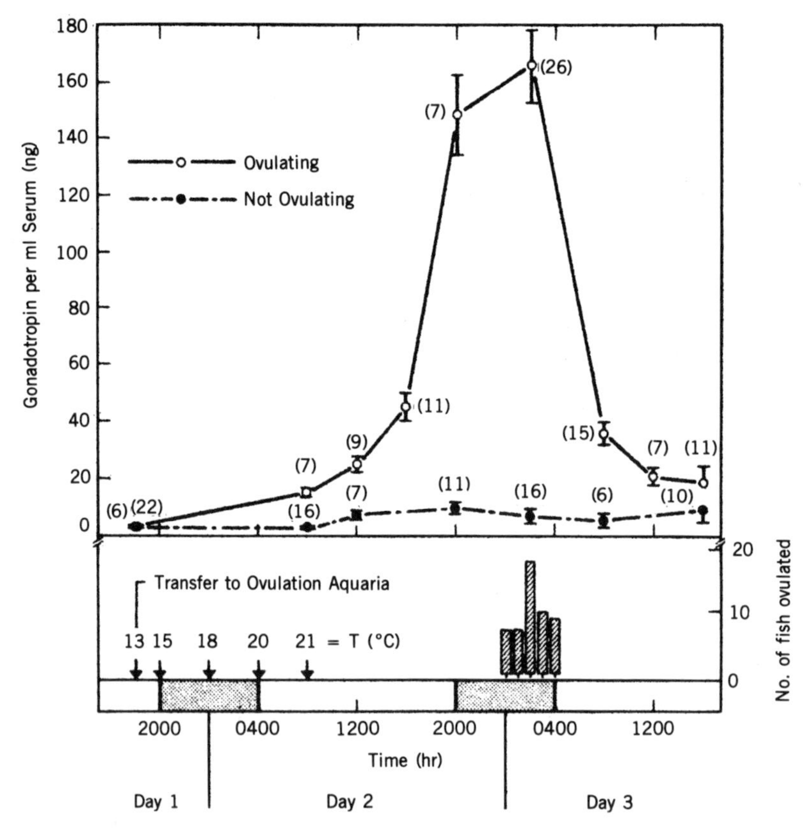

Figure 3.5 Times of ovulation and serum gonadotropin (GtH) (mean ± SE) of sexually mature goldfish kept under 16L-8D photocycle and warmed from 13° to 21°C. Number of fish ovulating per each hour is shown in the lower graph. GtH measured in serum from ovulatory fish sampled at time of ovulation (0000–0400 on day 3) were pooled and expressed as a mean at 0200 on day 3. (From Stacey et al., 1979, courtesy of *General and Comparative Endocrinology*.)

been implicated in spawning behavior of several unrelated species of fish. Experimentally, spawning behavior can be suppressed by indomethacin, prostaglandin inhibitor, and restored subsequently by prostaglandin injection (Stacey, 1983).

There is some variation in the pathway mature ova follow when released from the ovary. In cyclostomes gametes are released directly into the coelomic cavity. In elasmobranchs the eggs are shed into slender oviducts; the lower oviducts are expanded and secrete a shell or, in live-bearing forms, house embryos. In teleost fish, ovulation varies: eggs may be released directly into the coelomic cavity (gymnovarian, as in salmonid fish) or into the ovarian lumen (cystovarian), in which case an oviduct is connected to the ovarian capsule. Unless fertilization is internal, as in some cystovarian species, oviposition soon follows ovulation (in several hours or days). The daily photocycle determines the time of day of both ovulation and oviposition of many species (Stacey and Goetz, 1982; Stacey, 1984).

In teleosts a large number of eggs are ovulated within a short period, and they have a short life span if not soon laid and fertilized. It is essential, therefore, that ovulation is cued not only by endogenous factors but also by the presence of a male and, in oviparous species, an appropriate site for oviposition (Stacey, 1984).

Ovulation in terrestrial ectotherms (amphibians and reptiles) seems to be controlled by gonadotropins. Experimentally, anurans with mature ova will ovulate following the administration of LH (Lofts, 1974). Maturation of anuran ova normally follows a rise in gonadotropins. Gonadal steroids appear to constitute the direct cause of ovulation, for ovulation occurs as a result of the administration of steroids, especially progesterone, the release of which is normally induced by a rise in gonadotropins (Lofts, 1984). Amphibian ovaries are gymnovarian, as in most teleosts. Ova leave the ovaries and enter the coelomic cavity; they enter the oviducts at the ciliated funnel-like ostia.

Ovulation in *Bufo* may be the result of gonadotropin release (Schuetz, 1972). Progesterone induces the final meiotic division in vitro (Thornton and Evennett, 1969). As the egg passes through the amphibian oviduct, a series of capsules enclose the ovum. It is enclosed by a vitelline membrane, which at this time becomes free of follicle cells. These capsules provide physical support for the developing embryo and probably control passage of materials to and from the embryo. The exterior capsule is usually not sufficiently hard to provide adequate physical protection against small predators, but the contents of the capsules of some salamanders (Salamandridae) and anurans (*Bufo* spp.) are sufficiently toxic to deter predators.

Ovulation has been carefully studied in several kinds of domestic fowl and the canary, which seem to exhibit a general pattern (Follett and Davies, 1979). Ovulation follows a peak in LH release. A series of follicles mature until a clutch is completed. Ovulation occurs during specified periods in the female's daily activity cycle. The space during which the LH surge may occur is called the *open period* and lasts about eight hours in the domestic fowl. It begins at the end of the daylight period and is apparently part of an entrained circadian rhythm. In the turkey hen there is a distinct surge of both LH and FSH prior to ovulation (Fig. 3.6). Both gonadotropins seem to be essential to ovulation, but the separate role of each is not clear (Mashaly et al., 1976).

There are high levels of LH in the immature fowl, but they drop conspicuously as follicles develop in the mature hen. High levels of LH induce synthesis of estrogens and serum estrogens, which in turn depress the release of LH. The preovulatory surge of LH results primarily from secretions of progesterone, but experimentally the effect of progesterone requires priming by previous levels of estrogen (Sharp, 1981). Through positive feedback the preovulatory release of progesterone induces the LH surge in birds. During follicular maturation and with the approach of ovulation in the domestic (white Leghorn) hen, steroidogenesis by the granulosa cells increases while activity of the theca cells declines (Bahr et al., 1983).

The posterior lobe of the avian pituitary contains arginine vasotocin, a hormone to which the avian uterus is very sensitive (Munsick et al., 1960). Circulating arginine vasotocin increases immediately prior to ovulation, and this hormone may constitute a stimulus for oviposition (Sturkie, 1976).

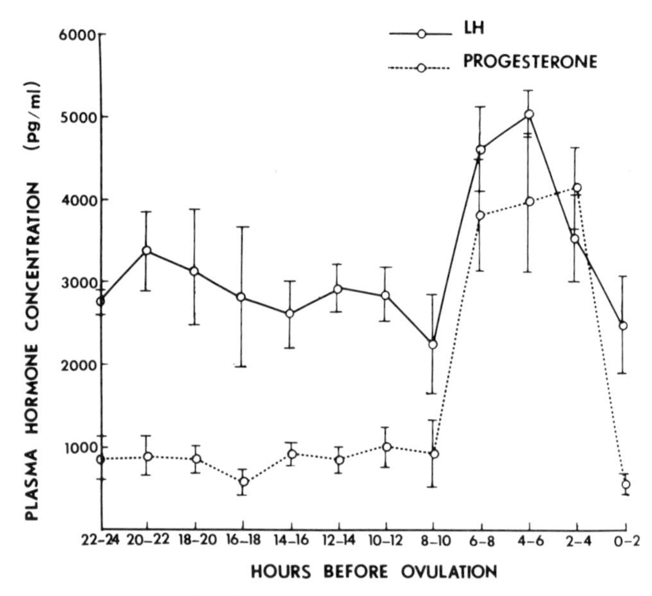

Figure 3.6 Progesterone and LH levels in the plasma of turkey hens during the ovulatory cycle. (From Mashaly et al., 1976, courtesy of *Poultry Science*.)

The avian ovulatory cycle differs from that of mammals in (1) the serial ovulation of single ova, until a clutch is formed, and (2) the large amount of yolk laid down with the ova during vitellogenesis. Following ovulation, the egg remains in the oviduct for some hours, during which time the egg is fertilized and provided with a shell and increases in mass. In many birds one egg is laid every day until the clutch is complete, and until the last egg is laid, oviposition is shortly followed by ovulation. The presence of an egg in the oviduct postpones the next ovulation.

Prostaglandins appear to have a role in regulating oviposition in birds. PGF increases fourfold in the preovulatory follicle of the domestic fowl 4 to 6 hours before the estimated time of ovulation, and this level increases about a hundredfold 24 hours later. Although indomethacin, which depresses levels of prostaglandins, does not prevent ovulation, it does delay laying up to 48 hours (Day and Nalbandov, 1977; Hertelendy and Biellier, 1978). Prostaglandins cause an increase in electrical activity of uterine muscle, the intensity varying with the dosage. This suggests the mechanism of their role in oviposition (Shimada and Asai, 1979).

3.1.6 Induced Ovulation

Some vertebrates ovulate only after, and as a consequence of, coition: these species are called induced ovulators. Induced ovulation is characteristic of the domestic rabbit, the ferret, the domestic cat, the vole, and many others. The short-tailed shrew (*Blarina brevicauda*), an abundant insectivore of eastern North America, is an induced ovulator, as are probably many other soricid shrews. Ovulation is also induced in the house, or musk, shrew (*Suncus murinus*) (Hasler and Nalbandov, 1978). Although this phenomenon has been reported mostly for mammals, it may be widespread in other vertebrates with internal fertilization. It has been noted in the dogfish (*Mustelus canis*) (Hisaw and Abramowitz, 1939) and may well occur in other elasmobranchs. Although distinct in concept from spontaneous ovulation, the differences become somewhat blurred in reality.

The human female, for example, usually ovulates at midcycle, but there is evidence for ovulation induced by coition. Ova and corpora lutea may be present during any phase of the cycle, and conception may occur at any time, including during menstruation. Evidence for induced ovulation is derived from instances of conception following rape in women with definite knowledge of the last menstruation, and induced ovulation could reasonably result from the intensity of intercourse during rape (Clark and Zarrow, 1971). Also, the occasional failure of the rhythm method for birth control might be a manifestation of induced ovulation at a time other than the point at which ovulation would be expected to be spontaneous.

In induced ovulators stimulation of the cervix is followed by an increased release of LH, as a consequence of GnRH release from the hypothalamus. The hypothalamus apparently receives a neural communication from impulses initiated during cervical stimulation. Coition not only stimulates the pelvic nerve, thereby initiating the release of GnRH, but also induces a release of prolactin from the anterior pituitary. This prolactin is responsible for full development of the corpus luteum (Smith et al., 1975). Copulation in the vole (*Microtus agrestis*) causes a large release of GnRH, which in turn stimulates a surge in LH (Milligan, 1981; Versi et al., 1982). In the domestic cat, coitus is promptly followed by a rise in LH, and ovulation occurs about one day later (Shille et al., 1983).

Among species of voles (Arvicolidae) known to be induced ovulators, the ovulatory cycle is irregular. In *Microtus townsendii*, for example, estrous cycles begin at intervals of from two to eight days, and most vaginal smears indicate estrus or diestrus (MacFarlane and Taylor, 1982). Prolonged estrus in a species with induced ovulation enhances the likelihood of a fertile mating; this is one mechanism for promoting a rapid population increase.

Species with short life spans, high reproductive potential, and early sexual maturity tend to have induced ovulation. Although ovulation is usually induced by coition, olfactory signals sometimes have the same effect. A substantial body of evidence shows that odors of males may stimulate ovulation in females of the same species (see Chapter 6, Courtship and Mating, Section 6.2.2).

3.1.7 Unilateral Ovulation

In mammals ovarian activity and implantation are commonly random, with more or less equal activity on the left and right sides, but in some groups asymmetry exists. A departure from the norm is seen in (1) the release of an ovum always from the same side and (2) implantation always on one side only, which may not be ipsilateral to the active ovary.

In most bats ovulation and implantation appear to be random, but in some well-studied species there is a clear unilateral pattern. In at least two species of horseshoe bats (*Rhinolophus ferrumequinum* and *R. hipposideros*), only the right ovary releases ova, and implantation is always in the right uterine horn (Matthews, 1937). A similar pattern is seen in the free-tailed bat (*Tadarida brasiliensis*) of the New World (Sherman, 1937). Follicles do not develop in the left ovary, but the development of interstitial tissue suggests that steroidogenesis occurs in this "inactive" ovary (Jerrett, 1979). Among vesper bats (*Myotis* sp. and *Eptesicus* spp.) ovulation is random, but implantation is always in the right uterine horn (Baker and Bird, 1936; Guthrie, 1933). In the Indian vampire bat (*Megaderma lyra*) both ovulation and implantation are confined to the left side (Ramakrishna, 1951). A unique situation is seen in the bent-winged bats (*Miniopterus*) of the Old World. In at least three species (*M. schreibersii, M. australis*, and *M. natalensis*) only the left ovary ovulates, but the zygote invariably implants in the right uterine horn (Matthews, 1941; Richardson, 1977). Although the California leaf-nosed bat (*Macrotus californicus*) ovulates from the right ovary, removal of the right ovary is followed by activation of the left ovary (Bleier and Ehteshami, 1981).

In the mountain viscacha (*Lagidium peruanum*), a South American chinchillid rodent, ovulation is invariably from the right ovary and implantation always follows in the right horn. However, removal of the right ovary is followed by ovulation from the left ovary and implantation in either horn (Pearson, 1949). In the rat such compensatory growth is follicular, so that the enlarged ovary releases a greater number of ova (Peppler and Greenwald, 1970). Presumably the elimination of one ovary reduces inhibin and estrogen, allowing an increase in FSH, which causes the increased growth of the remaining ovary.

This asymmetry of the reproductive tract almost always appears to be associated with litters of one. Perhaps suppression of one half of the uterus allows greater development of the opposite half. Such asymmetry, however, is clearly lacking in those primates that usually have single births.

In some groups of mammals that characteristically have a single young, ovaries are alternatively ovulatory. This pattern is seen in seals, sea lions, porcupines, and many primates, but it is not asymmetrical. Alternating ovarian activity is also seen in some reptiles. The gecko (*Lepidodactylus lugubris*) on Oahu, Hawaiian Islands, lays a clutch of two eggs, and each ovary functions alternately (Jones et al., 1978).

3.1.8 Estrus or Heat

Cyclical changes in hormonal levels not only prepare the female for the production of a mature ovum and its release into the oviduct but also induce important behavioral changes. In all vertebrates the female becomes receptive to the sexual advances of a male, a condition known in mammals as estrus or *heat*. In almost all vertebrates certain behavioral signals, and frequently scents, announce the receptivity of the female. Estrus occurs shortly before ovulation and mating times so that the introduction of sperm is synchronized with ovulation. Notable exceptions are those species in which sperm is stored in the reproductive tract of the female. In these groups (some bats, many reptiles, some amphibians, and some fish), ovulation may follow mating by days, months, or even years.

In addition to olfactory advertising by a female mammal in estrus, some species may tend to move about more, and thereby increase the likelihood of an encounter with a male. In captives of two species of kangaroo rats (*Dipodomys microps* and *D. merriami*), which are spontaneous ovulators, there is intense locomotor activity from just before to just after estrus (Wilson et al., 1985).

Although most nonprimates mate only at estrus, the Asiatic house shrew (*Suncus murinus*) appears not to experience estrus, and mating seemingly occurs only at the will of an aggressive male (Dryden, 1969; Hasler and Nalbandov, 1978). In this shrew the uterus and vagina seem not to depend upon estrogens. There is no uterine and vaginal atrophy after ovariectomy, and sexually inexperienced shrews will mate after ovariectomy (Dryden and Anderson, 1977).

Behavioral aspects of estrus precede physiological aspects and enhance the likelihood of sperm meeting a recently ovulated ovum, or coition may induce ovulation. Behavioral estrus may be somewhat prolonged: in the domestic dog, heat may persist for more than a week after ovulation, while estrogen levels remain high (Concannon et al., 1975). In many mammals tactile stimulation of the genital region of the female at a point in proestrus may induce lordosis, an arching of the back indicating receptivity to the male. Postures vary among different groups of vertebrates, but the message is the same.

In several mammals the estrous cycle may be interrupted by a period of relative inactivity, or anestrus, at which time the ovary regresses to a small portion of its active mass. If fertilization occurs, estrous cycles cease. If there is a sterile mating or no mating, however, another follicular phase and ovulation may follow immediately or with only a brief delay. There may also be a postpartum estrus, at which time fertile mating may occur. A succession of estrous cycles is called a polyestrous pattern. This is characteristic of many species of small mammals. The polyestrous pattern and postpartum ovulation both enhance reproductive capacity. The occurrence of a single annual ovulatory cycle is called monestrous.

Silent heat is a phenomenon observed in sheep and some other mammals prior to the reproductive season. In silent heat there may be one or more ovulatory cycles in which ovulation occurs without overt behavioral signs. This presumably results from a lower threshold for ovulation than for the external manifestations of heat in response to gonadotropins (Robinson, 1951). Some wild mice may have one or more ovulatory cycles, with the formation of corpora lutea, at the beginning of the reproductive season while the vagina remains closed. Normally the vagina opens prior to ovulation and closes after ovulation.

3.1.9 Menstruation and Menopause

The ovulatory cycle in primates is not profoundly unlike that of other vertebrates. However, in Old World primates, including humans, and a few other mammals, there is an appreciable buildup of the endometrium, the uterine lining. This is seen also in the elephant shrew (*Elephantulus* spp.) of Africa. With the decline of serum progesterone toward the end of the cycle, the endometrium loses its hormonal support. The thickened endometrium then becomes loosened and sloughs off, with a concomitant loss of blood, the menstrual flow.

In the human female, menarche, the onset of menstruation, usually starts at about 12 years of age and continues, with interruptions for pregnancies, until sometime between the ages of 48 to 55. Menarche occurs earlier in cultured societies today than in the nineteenth century, presumably as a result of improved nutrition. Following the release of progesterone from the corpus luteum, body temperature rises above the normal 37°C for some two to three days, which may be taken as evidence that ovulation has occurred (Döring, 1963).

Adolescent sterility is well established for humans. Among aboriginal tribes in many parts of the world, sexual intercourse commences at menarche, or sometimes before, but pregnancy is usually rare until four to six years later. This phenomenon has been observed by numerous anthropologists and is clearly not attributable to either abortions or any sort of birth control. Although estrogen fluctuations can produce secondary sexual characteristics as well as menstruation, ovulation does not occur without the midcycle surge of LH together with estrogen stimulation. Adolescent sterility is the occurrence of estrogen cycles (and menses) in the absence of adequate progesterone and without LH; it has been found to occur in most mammals in which it has been explored. Anovular cycles occur with the greatest frequency in the early years and gradually become minimal between 26 and 40 years, after which they increase somewhat (Fig. 3.7).

Menarche in the rhesus monkey (*Macaca mulatta*) is seen at about three years of age and, as in the human, it may not be accompanied by ovulation. In initial

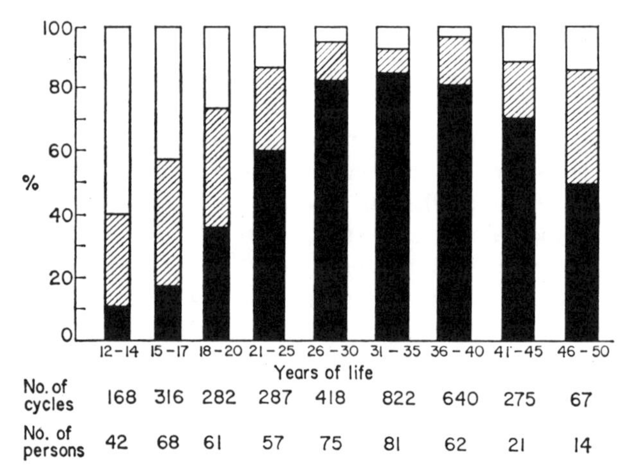

Figure 3.7 Secular changes in ovulation in the human female. Black areas indicate incidence of normal ovulation; shaded areas show shortened (luteal) phase, and white areas show anovular menstrual cycles. (From Doring, 1969, courtesy of the *Journal of Reproduction and Fertility*.)

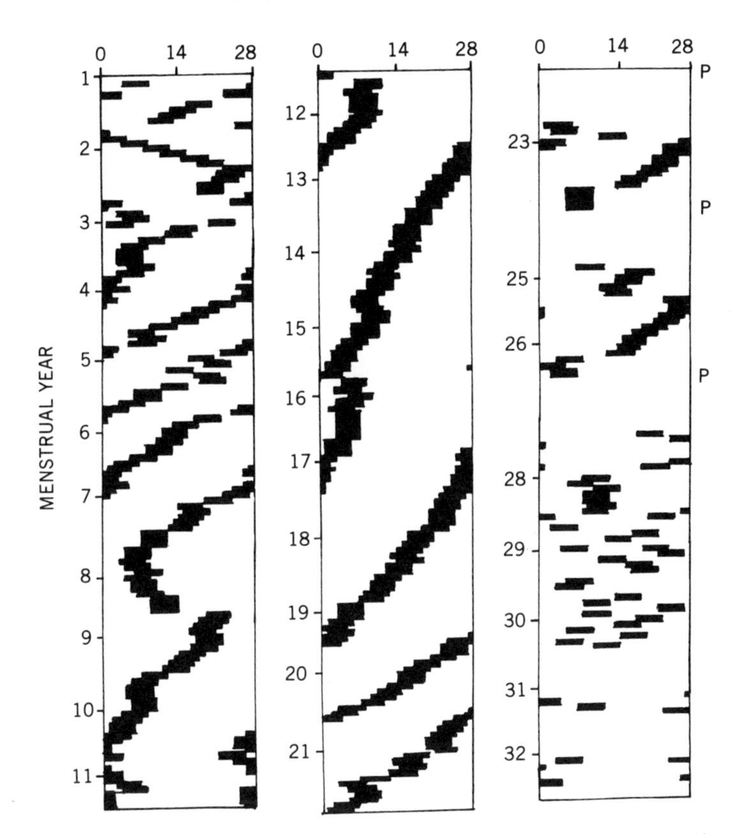

Figure 3.8 A history of menstrual activity of one human female from the time of menarche (year 1) until menopause. In the right column, P indicates pregnancies. (From Reilly and Binkley, *Psychoneuroendocrinology*, vol. 1, The menstrual rhythm, 1981, courtesy of Pergamon Press.)

menstrual cycles there is only a brief luteal phase with little progesterone secretion. The hypothalamus appears not to respond to injections of estradiol, and this failure may account for the prepubertal, premenarchial condition (Dierschke et al., 1974). Maturity in the rhesus monkey is eliminated by destruction of the median basal hypothalamus, abolishing gonadotropin secretion by the anterior pituitary; but menstrual cycles are restored by treatment with GnRH in both adult and immature females. This indicates that in immature females sexual maturity is not dependent upon performance by either ovaries or the anterior pituitary (Wildt et al., 1980). Current thought is that final maturation of hypothalamic control over gonadotropin release does not occur in the female until puberty, unlike the male, where the hypothalamic-pituitary-gonadal axis matures within a short period after birth.

Toward the end of fertility in the human female, menstrual cycles uusually become less frequent than the "normal" 28 days (Fig. 3.8), with the final menstrual flow being called menopause. The gradual decline in ovarian function is called the climacteric. Menopause is sometimes also considered to be the gradual decline in ovarian activity, resulting from a decrease in sensitivity to gonadotropins. The onset of this transition from an active to an inactive ovary most commonly occurs in the late forties, but in some individuals ovarian activity continues until age 55 and occasionally until the late sixties (Wharton, 1967). Conception and healthy fetal development rarely occur beyond the age of 55. Conception after menopause in the human is very unusual, but there are examples of ovulation and conception after cessation of menstruation (Watson, 1944; Wharton, 1967).

After menopause the ovary is shrunken and reduced in mass in contrast to an active premenopausal ovary. There may be many normal primordial follicles after the cessation of regular menstrual flows, but they lack the capacity to be stimulated by high levels of gonadotropins. Estrogen synthesis declines gradually to very low levels, during which time androgen production remains constant; and the conversion of some of this androgen accounts for the presence of low levels of serum estrogen (El-Badrawi and Hafez, 1980).

Menopause is a known, definite phenomenon in the human female and at least some other Old World primates. In the macaque (*Macaca mulatta*), menopause occurs after 25 years of age, and at 30 years of age females are considered post-menopausal (van Wagenen, 1972). After the cessation of menstrual bleeding in the rhesus monkey, there are persistent high serum levels of gonadotropins and low levels of ovarian steroids (Hodgen et al., 1977). Also, the African elephant (*Loxodonta africana*) experiences a decline in reproductive activity with age and a cessation of reproduction after about 50 years (Laws et al., 1975).

Age-related reproductive changes increase from 10 to 12 months in the laboratory rat, at which time the frequency of irregular cycling increases (Fig. 3.9). As female rats grow older, uterine concentration of estrogen receptors steadily decreases (Hseuh et al., 1979). Subsequently, there is a gradual increase in frequency of anovulatory

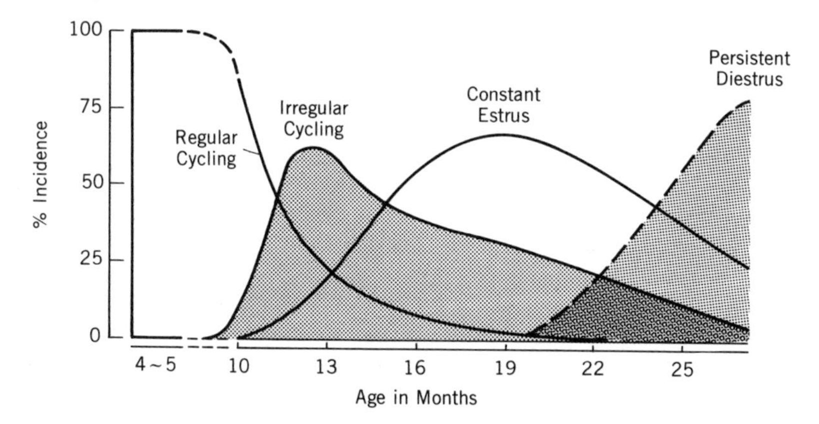

Figure 3.9 Secular changes in the regularity of the ovulatory cycle of laboratory rats, between 9 and 26 months of age. (From Lu et al., 1979, courtesy of *Biology of Reproduction*.)

female rats, with reduced production of ovarian steroids and also diminishing cyclical releases of gonadotropins and prolactin (Lu et al., 1979).

Senescence and a decline in ovarian activity have been noted in such fish as the guppy (*Poecilia reticulata*), the molly (*P. formosa*), and the fighting fish (*Betta splendens*), as well as some marine fish. There is a reduction in vitellogenesis and an increase in atresia (Woodhead, 1979).

3.2 THE CORPUS LUTEUM

The earliest phylogenetic appearance of the corpus luteum is that of a corpus luteum–like structure in an ascidian (*Ciona intestinalis*). In this urochordate the expended follicles regress, but the unruptured (atretic) follicles become luteinized and have been called "preovulatory corpora lutea." They are not known to be secretory.

In many vertebrates ovulation is followed by luteinization of the remnant tissue in the collapsed follicle, and this constitutes the corpus luteum. The occurrence is highly variable among nonmammalian vertebrates. It is usually absent or short-lived in oviparous forms. Prior to ovulation both FSH and estrogens seem to increase the number and sensitivity of LH binding sites. Following ovulation, with the concurrent surge of LH, a blood supply reaches the granulosa cells in the ruptured follicle, and luteinization of this tissue induces the development of the corpus luteum (or corpora lutea). The granulosa cells are receptive to a host of steroids, peptides (LH and prolactin), and prostaglandins, which are assumed to induce steroidogenic enzymes (Hseuh et al., 1983).

Luteinization actually begins prior to ovulation in most species. In the event of pregnancy, prolactin levels are high and help maintain the corpus luteum in certain mammals. In a sterile cycle, prolactin remains low and the corpus luteum degenerates rather quickly. Prolactin prolongs the functional life span of the corpus luteum by increasing LH binding sites on granulosa cells. In some mammals, such as the guinea pig, the corpus luteum survives without prolactin.

The corpus luteum appears as a conspicuous yellow protrusion on the surface of the ovary in most viviparous vertebrates. It produces progesterone, and its development among various taxa may perhaps vary with their need for this steroid. Corpora atretica and postovulatory corpora lutea develop in both oviparous and viviparous elasmobranchs, but their function is not clear. The development of secretory corpora lutea in sharks has been suggested (Chieffi, 1962), and corpora lutea of the viviparous torpedo (*Torpedo marmorata*) are reported to be steroidogenic (Te Winkle, 1972).

Several teleost fish produce luteinized corpora atretica from unovulated follicles. They occur in the long-jawed goby (*Gillichthys mirabilis*), but they are not known to be secretory. The postovulatory follicles in this teleost are quickly resorbed and do not form corpora lutea (de Vlaming, 1972).

Among oviparous amphibians the ruptured follicle contains granulosa cells and some lipids, but there is little evidence of secretory activity. In several live-bearing species (both ovoviviparous and viviparous), functional corpora lutea form and secrete progesterone until the birth of young. In the viviparous toad (*Nectrophrynoides occidentalis*) of Africa, for example, the corpora lutea remain throughout pregnancy and secrete progesterone (Xavier, 1977). Corpora lutea persist in both egg-laying and live-bearing caecilians. The function of the corpus luteum in oviparous species is equivocal, but there could be some intrauterine development prior to oviposition (Wake, 1977).

The role of the corpus luteum in reptiles is not clearly understood. Corpora lutea invariably follow ovulation in reptiles, but the duration is brief in egg-laying species. In two oviparous lizards (*Crotophytus collaris* and *Eumeces obsoletus*) progesterone synthesis is greatest during luteogenesis and is positively correlated with atresia (Fox and Guillette, 1987). Luteal regression occurs approximately at the time of oviposition in many lizards, and the decline in gonadal steroids may cause oviductal contractions associated with oviposition (Jones and Guillette, 1982). In viviparous species it persists and presumably accounts for a high level of progesterone, whereas there is an ovulatory peak in oviparous reptiles (Fig. 3.10). In the New World natracine snakes, all of which are viviparous, the corpus luteum may persist throughout pregnancy and produce progesterone. In the garter snake (*Thamnophis elegans*), luteectomy performed in the first trimester does not terminate pregnancy, although parturition is delayed and about half the young have placentae with yolk still in the yolk sac (Highfill and Mead, 1975b). In the DeKay's snake (*Storeria dekayi*), a

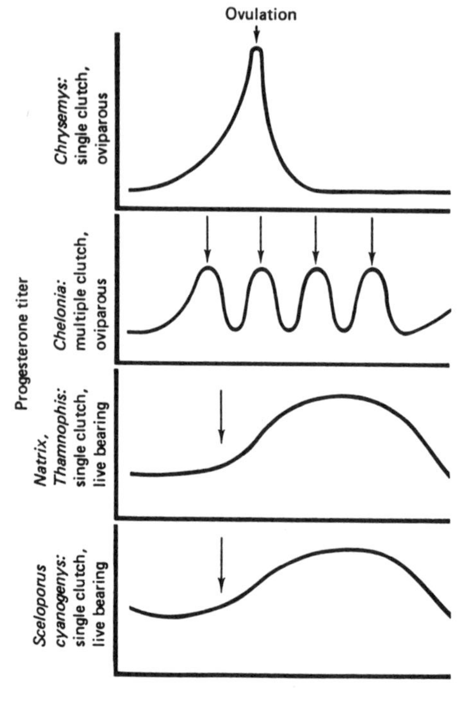

Figure 3.10 Plasma progesterone titers during the annual reproductive cycle of several reptiles. The two upper patterns represent turtles, showing an ovulatory peak of progesterone, and the lower two patterns show the postovulatory peak of progesterone in some live-bearing snakes and lizards. (From Callard and Ho, 1980, courtesy of S. Karger.)

small natracine of eastern North America, the corpus luteum contains 21-hydrolase, an enzyme implicated in the synthesis and secretion of progesterone (Colombo and Yaron, 1976).

Unlike the corpus luteum of eutherian mammals, that of the tammar wallaby (*Macropus eugenii*) continues to grow after hypophysectomy and stimulates uterine growth in the absence of the hypophysis. Removal of the corpus luteum in the pregnant tammar wallaby before day 17 interferes with successful parturition. Presumably both progesterone and relaxin from the corpus luteum are essential for a healthy birth in this marsupial (Young and Renfree, 1979).

Large corpora lutea develop in monotremes (*Ornithorhynchus anatinus* and *Tachyglossus aculeatus*) and persist until the eggs are laid. In an individual with near-term eggs in the oviduct, very high levels of progesterone were noted (Griffiths, 1978).

3.2.1 Preservation of the Corpus Luteum

The corpus luteum attains its maximal role only if the ovum is fertilized, and if fertilization does not occur, progesterone secretion is less than maximal. The life of the corpus luteum following fertilization varies among different taxa, but the corpus luteum of pregnancy is essentially similar to that of a sterile cycle, except that it persists longer and may become larger in a fertile cycle. In the guinea pig the corpus luteum forms in sterile cycles and may persist much longer than in the rat. This may account for the long ovulatory cycle (16–18 days) of the guinea pig. Also, in the domestic dog the corpora lutea remain functional in the absence of mating, and progesterone release is substantial up to 80 days after estrus (Concannon et al., 1975). In a number of laboratory rodents, however, there is a small amount of postovulatory progesterone for about one day, and such mammals are called *incomplete cyclers* (Yoshinaga, 1978). In these mammals the corpus luteum becomes a major source of progesterone only after an ovum is fertilized.

The corpus luteum can be maintained by several substances: (1) prolactin, from the anterior pituitary, secreted as a result of coition; (2) LH from the anterior pituitary, also secreted as a result of stimulation of the cervix; (3) estrogen from the blastocyst; and (4) a placental gonadotropin, such as pregnant mare serum gonadotropin (or PMSG), from the fetal cells of the placenta (Hamilton et al., 1973). In the ferret (*Mustelus putorius*), prolactin is necessary to maintain the corpus luteum and its production of progesterone during the first half of pregnancy. Subsequent progesterone levels decline until parturition, but mammary development suggests continued secretion of prolactin. Seemingly the corpus luteum of the ferret becomes decreasingly responsive to prolactin (Murphy, 1979).

Among its several effects, progesterone in some domestic and laboratory mammals stimulates the uterus to secrete $PGF_{2\alpha}$ in the absence of fertilization. This material is luteolytic, leading to the decline of the corpus luteum (Hearn et al., 1977; Henderson et al., 1983). Oxytocin from the corpus luteum of the ewe may stimulate release of $PGF_{2\alpha}$, which, in turn, by a positive feedback may induce release of luteal oxytocin. This cycle would amplify the luteolytic effect, causing the demise of the corpus luteum (Flint and Sheldrick, 1983). In contrast to this relationship of uterine prostaglandins and the corpus luteum in some domestic mammals, uterine prostaglandins do not cause a regression of the corpus luteum in primates, including the human. In primates ovarian $PGF_{2\alpha}$ blocks LH binding by the corpus luteum, and although $PGF_{2\alpha}$ does depress serum levels of progesterone, it does not induce regression of the corpus luteum. Functions of the corpus luteum vary among different taxa of mammals. In many mustelid carnivores and in some kangaroos, for example, the corpus luteum seems to regulate the embryonic diapause of the blastocyst, probably via photoperiodic cues (in the mustelids) or via nervous cues from the nipples or photographic cues (in the kangaroos). In either case, resumption of luteal activity seems to be related to implantation of the blastocyst. (For further comments on the preservation of the corpus luteum, see Section 7.2.)

3.2.2 Accessory Corpora Lutea

Some species of mammals have more, sometimes many more, corpora lutea than fertilized ova. These corpora lutea are usually postovulatory structures, but some may result from unovulated follicles (corpora atretica). Some may be secretory, but for the most part their role is unknown.

In the early pregnancy of the horse, there is a single corpus luteum. This corpus luteum secretes progesterone until implantation, after which PMSG is secreted by the placenta. PMSG stimulates the ovary to further ovulation, producing a series of ovulations, and the subsequently formed corpora lutea continue to secrete progesterone throughout pregnancy (Cole et al, 1931; Yoshinaga, 1978).

Accessory corpora lutea have been noticed in carefully observed laboratory populations of the Australian hopping mouse (*Notomys alexis*). These structures are typical corpora lutea that result from postfertilization ovulations and occur in both pregnant and lactating females (Breed, 1981). Similarly, the Canadian porcupine (*Erithizon dorsatum*) usually has a single young. A single follicle ruptures and becomes luteinized, and the remaining structures become atretic and luteinized (Mossman and Judas, 1949). The African elephant (*Loxodonta africana*) ovulates after conception, and there may be several very long-lived corpora lutea in each ovary. In three species of elephant shrews (*Elephantulus myurus, E. capensis*, and *Macroscelides proboscideus*), each ovary releases about 60 eggs at one time, with the subsequent development of 60 corpora lutea in each ovary. The total lutein tissue in these species, however, seems no greater than in *Elephantulus intufi*, which releases one or two ova and forms one or two very large corpora lutea (van der Horst, 1944).

3.3 SUMMARY

The vertebrate ovary is penetrated by capillaries that provide for the transfer of oxygen, nutrients, hormones, sometimes vitellogenin, and metabolic wastes.

The mitotic division and development of oogonia to produce primary oocytes occurs in the embryonic ovaries of lampreys, elasmobranchs, birds, and mammals, but it is seasonal and may be annual or sporadic in fish. Early oogonia may be connected by intercellular bridges in many vertebrates, the result of incomplete separation of the cell membranes. In some amphibians there is nuclear division without total cell division, resulting in multinucleate oocytes.

As the diploid oocytes enlarge, they move to the ovarian margin, where they remain in meiotic prophase I until after ovulation. Granulosa cells surround the oocyte and produce steroids. An acellular membrane, the zona pellucida, surrounds each oocyte. This membrane in nonmammals is sometimes referred to as the zona radiata (elasmobranchs) and vitelline envelope (amphibians). In most mammals granulosa cells secrete a fluid that surrounds the oocytes, and in most nonmammals yolk adjoins the oocyte. A layer of theca cells grows about the granulosa cells, and both layers are targets for LH, FSH, and prolactin. Theca cells synthesize androgens, which provide for the production of estradiol by the granulosa cells. Prior to ovulation, luteinization of the granulosa cells provides for the synthesis of progestin.

Meiosis I is completed shortly before ovulation. The second meiotic division begins shortly after ovulation but is completed only after sperm penetrate the zona pellucida. Fertilization follows, restoring the diploid number of chromosomes. In nonmammals estrogen induces the release of vitellogenin (the precursor of yolk) from the liver. Incorporation of yolk proteins from the plasma is mediated by gonadotropins.

Many oocytes become atretic and die during meiosis I, after which leukocytes and granulosa cells occupy the follicular space. Atresia also afflicts mature follicles that are not ovulated. Even though they do not contribute to a possible continuation of the genotype, atretic follicles synthesize substantial amounts of gonadal steroids.

A sudden rise in LH is seen just prior to ovulation. Fine contractile microfibrils, resembling smooth muscle, run through the theca and contract at ovulation, apparently ejecting the ovum. At the time of ovulation in fish, prostaglandins ($PGF_{2\alpha}$) appear to induce follicular constriction. Gonadotropins are also associated with ovulation in amphibians and reptiles.

In several diverse groups of mammals and perhaps in some elasmobranchs, ovulation follows as a result of coition. The stimulus of the cervix induces GnRH release, which stimulates the discharge of LH and prolactin. This is called induced ovulation and provides for a prolongation of the receptive state of a female, thus enhancing the likelihood of a fertile mating.

At about the time of ovulation in many vertebrates, ovarian hormones induce behavior that signals the receptivity of the female to sexual attention from males. Frequently these changes include odors that both attract and arouse sexually active males.

In the human, Old World primates, and a few other mammals, endometrial growth increases during the ovulatory cycle. This growth is lost sometime after ovulation, with the decline of supporting steroids from the ovary. In the event of a fertile mating, this lining is preserved for implantation of the blastocyst. In the human female endometrial loss is called menstruation or menses. It occurs approximately monthly from the onset, at about 12 years of age, to menopause, which occurs from the late forties until as late as 55 in some women.

After the release of the ovum in mammals and in viviparous nonmammals, the tissue remaining in the follicle becomes luteinized and is termed the corpus luteum. In the corpus luteum the granulosa cells continue to secrete progesterone. The corpus luteum may be absent in many oviparous species and does not occur in birds. Progesterone from the corpus luteum plays many roles during pregnancy.

BIBLIOGRAPHY

Goetz FW (1983). Hormonal control of oocyte final maturation and ovulation in fishes. *Fish Physiology*, **IXB**: 117–170.

Greenwald GS and Terranova PF (Eds) (1983). *Factors Regulating Ovarian Function*. Raven, New York.

Greep RO (Ed) (1983). *Reproductive Physiology IV. International Review of Physiology* 27. Univ Park, Baltimore.

Jones RE (Ed) (1979). *The Vertebrate Ovary*. Plenum, New York.

Midgley AR and Sadler WA (Eds) (1979). *Ovarian Follicular Development and Function*. Raven, New York.

Milligan SR (1982) Induced ovulation in mammals. Pp 1–46. In *Oxford Reviews of Reproductive Biology*, vol 4 (Ed: Finn CA). Oxford Univ, Oxford.

Mossman HW and Duke KL (1973). *Comparative Morphology of the Mammalian Ovary*. Univ Wisconsin, Madison.

Peters H and McNatty KP (1980). *The Ovary. A Correlation of Structure and Function in Mammals*. Univ California, Berkeley, Los Angeles.

Rankin JC, Pitcher TJ, and Duggan R (1983). *Control Processes in Fish Physiology*. Croom Helm, London, Canberra.

Romanoff AL and Romanoff AJ (1949). *The Avian Egg*. Wiley, New York.

Serra GB (Ed) (1983). *The Ovary. Comprehensive Endocrinology*. Raven, New York.

Spies HG and Chappel SC (1984). Mammals: Nonhuman primates. Pp 659–712. In *Marshall's Physiology of Reproduction*, vol 1. *Reproductive Cycles of Vertebrates* (Ed: Lamming GE). Churchill Livingstone, Edinburgh.

Sunndararaj BI (1981). *Reproductive Physiology of Fishes*. UN Development Program FAO UN, Rome.

Wallace RA (1985). Vitellogenesis and oocyte growth in nonmammalian vertebrates. In *Developmental Biology*, vol 1 (Ed: Browder LW). Plenum, New York.

Wiegand MD (1982). Vitellogenesis in fishes. Pp 136–146. In *Proc Internat Symp Reprod Physiol Fish* (Eds: Richter CJJ and Goos HJT). Central Agricultural Publications and Documents, Wageningen.

Zuckerman L and Weir B (1977). *The Ovary*, 2nd ed., 3 vols. Academic, New York.

4

THE TESTIS

4.1 TESTICULAR FORM

Chapter 2 outlines the hormonal activities that underlie testicular functions in general, including spermatogenesis. This chapter discusses testicular activity with emphasis on steroidogenesis and formation of male gametes. Testicular metabolism, like that of the ovary, is both stimulated and controlled by several endogenous factors. The roles of exogenous factors are treated in Chapters 15 and 16.

In the more primitive vertebrates the testis is elongate and sometimes lobed, but in most reptiles, birds, and mammals it is ovoid. It lies closely associated with the kidney, adopting its excretory ducts for the discharge of germ cells and seminal fluids. Genital ducts are absent in cyclostomes, in which gametes are released into the body cavity and thence pass through pores into the urinary ducts, which lead directly to the cloaca. In primitive form (e.g., holostean fish), spermatozoa are released into efferent ducts of the kidney and from there via mesonephric ducts to the cloaca. In amphibians the situation varies among taxa: efferent ducts of the testis lead to the kidney, and the spermatozoa may or may not leave together with urine. In most vertebrates the testes are paired and more or less equal in size. Only a single testis develops in male cyclostomes.

In nonmammalian vertebrates the testes are always situated entirely within the body cavity. In most mammals they lie either just within the ventral surface of the body cavity and covered by skin (the cremaster sac) or in a scrotum, but in a few unrelated groups of mammals the testes lie within the coelomic cavity. This latter heterogeneous assemblage is sometimes referred to as the testicondidae and includes cetaceans, elephants, sirenians, monotremes, and some others. The possible significance of this variation is discussed in Section 4.5.

4.1.1 Mammalian Testis

Because the mammalian testis has received much more study than that of nonmammals, it will be described first. Many features appear to apply to the testis of other vertebrates.

In most mammals the testis is ovoid with a strong superficial covering, the tunica albuginea (an extension of the peritoneum), which is provided with branches from the testicular artery and sympathetic nerve endings from the autonomic nervous system. The testis of many mammals is divided into many sections or lobules, which are separated by septa. Each lobule contains a densely packed system of seminiferous tubules surrounded by interstitial tissue, which contains Leydig cells, circulatory vessels, and nerves.

The seminiferous tubules are convoluted and crowded; they are mostly single, but some are branched. The seminiferous tubules are connected, at each end, to rete tubules (or rete testis), which constitute the beginning of the efferent system. The rete tubules are an anastomosed system that joins the efferent ducts. They connect the seminiferous tubules to the epididymis. Spermatozoa are remodeled (mature) during their passage through the head, body, and tail of the epididymis. Also, spermatozoa may be stored in the cauda epididymis.

The seminiferous tubules contain germ cells and Sertoli cells, the latter of somatic origin. Although germ cells divide throughout the sexually active life of the individual, Sertoli cells cease to do so in early infancy. In cross section the seminiferous tubules consist of an outer layer of myoid cells, which are capable of peristolic constrictions and which move germ cells toward the rete tubules. There is also a basement membrane, which delimits the seminiferous tubules. The Sertoli cells lie at or near the lining of the seminiferous tubules. Germ cells lie embedded in Sertoli cells, joined by complex processes. The close association of the developing germ cells and Sertoli cells suggests a critical role for the latter in spermatogenesis. Sertoli cells provide nutritional and physical support for spermatocysts and spermatids. The Sertoli cells are the main targets of follicle-stimulating hormone (FSH) and testosterone. Leydig cell activity, which is induced by FSH, has testosterone as a major product.

4.1.2 Elasmobranch Testis

In the testes of sharks and rays germinal cells form into closed, hollow follicles (or seminiferous follicles) or ampullae, within which spermatogenesis takes place. These follicles contain Sertoli cells, but the question of the occurrence of Leydig cells is apparently not settled (Dodd and Sumpter, 1984).

4.1.3 Teleost Testis

The testis of teleost fish is rather similar to that described for mammals. It contains seminiferous tubules, which enclose Sertoli cells and germ cells. The tubules are surrounded by interstitial tissue, with Leydig cells, but it lacks the lymphatic system characteristic of mammals. The Sertoli cells in fish, as in other vertebrates, are intimately involved with the division and development of germ cells. Leydig cells are steroidogenic. They appear to be homologous to the Leydig cells of tetrapods and should not be confused with "lobule boundary cells" described by some authors (Grier, 1981). In some taxa the Leydig cells move into the margins of the seminiferous tubules.

Two types of teleost testes can be distinguished on the basis of their tubules. Most bony fish have an *unrestricted* spermatogonial testis, in which the spermatogonia may occur anywhere along the length of the anastomosing, or branching, tubules. This type of testis has no discrete efferent duct system, and sperm enter the lumen (Fig. 4.1). In the Atheriniformes (which includes several live-bearing families) there is a *restricted* spermatogonial testis, in which Sertoli cells form cysts at the distal end of a tubule (Grier, 1981). Germ cells develop synchronously within a cyst and move as a unit toward an efferent duct as development proceeds (Fig. 4.1). In the golden mullet (*Liza aurata*), unlike other teleosts, there is no germinal epithelium. But, together with spermatogonia, there persist primordial germ cells, which develop into spermatogonia (Buslé, 1982).

4.1.4 Amphibian Testis

In anurans the testis is ovoid and paired, and the efferent ducts lead through the sexual segment of the kidneys to the cloaca. In some species of salamanders (Caudata) the testis is lobular, consisting of a cluster of irregularly sized globules or lobes, which open into a median duct. These lobes increase in number with age, but there is apparently no regular correlation of lobe number and age (Tilley, 1977). Although these lobes lie in two series (the paired testes), they sometimes appear to consist of a single cluster. The lobular linings resemble the seminiferous tubules in function and are presumably derived from them.

Caecilian testicular lobes are similar to those of salamanders, but their number does not change (Wake, 1977a).

In males of toads (Bufonidae) there remains an undeveloped ovary, apparently without steroidogenic activity. This has long been known as Bidder's organ and

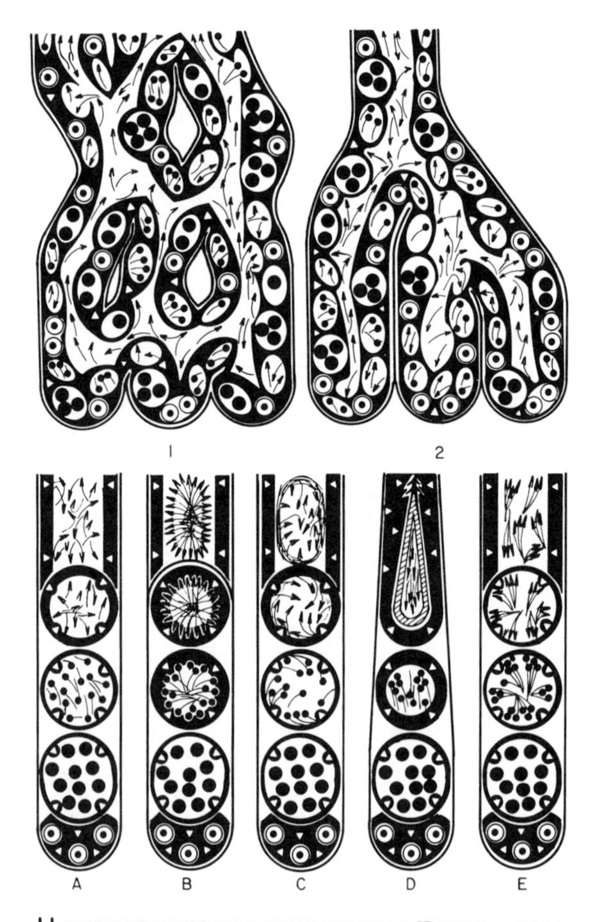

Figure 4.1 Tubule modifications with the teleost testis. The contiguous spermatogonial testis may be either an anastomosing (1) or a branching (2) network of tubules. The restricted spermatogonial testis may be either unmodified (*a*) or with various Sertoli cell–germ cell modifications: Poeciliidae (*b*); Goodeidae (*c*); Horaichthyidae with barbed spermatophores (*d*); or *Anableps dowi* (*e*). (From Grier, 1981, courtesy of the American Society of Zoologists.)

becomes active following castration. In nature there appears to be no function for this structure.

4.1.5 Reptilian Testis

The reptilian testis resembles that of birds and mammals, with densely packed seminiferous tubules between which is a vascular system and interstitial cells. The testes are ovoid in most species but are elongate in snakes. In worm snakes (*Leptotyphlops* spp.) the testis consists of small lobes.

In squamates and the tuatara (*Sphenodon*) the testes tend to be longitudinally displaced, the right lying in front of the left. Sertoli cells occur in reptilian seminiferous tubules. Leydig cells have been reported to occur in many, but not all, species examined (Fox, 1977).

4.1.6 Avian Testis

The avian testis contains a mass of seminiferous tubules and interstitial tissue together with vascular tissue. There are no divisions into lobules. Sperm pass through the avian testis through the epididymis and are stored in a seminal sac, an expansion of the vas deferens. Avian testes are subequal, the left being slightly larger than the right. They may lie close to the air sac system of birds and are perhaps cooled slightly below the bird's core temperature.

4.2 ACCESSORY GLANDS

Several secretory structures discharge fluids as spermatozoa pass to the urethra. Near the junction of the epididymides and the urethra are the paired Cowper's glands and the paired seminal vesicles together with the median prostate gland. These structures and such features as antlers, sexual coloration, and the scrotum develop with the seasonal increase of testosterone and are also maintained by testosterone.

The prostate gland secretes citric acid and many other compounds, and the seminal vesicles contribute fructose. These materials, together with sperm, constitute seminal fluid or semen. Glandular secretions aid in the survival and movement of sperm in the female genital tract.

In squamates spermatozoa pass through the sexual segment of the kidney, which in males develops under the stimulation of testosterone. It has been suggested that secretions from the sexual renal segment may promote survival of spermatozoa (Cuellar et al., 1972). The sexual renal segment may serve the same role in nutrition of spermatozoa as performed by the seminal vesicles in other vertebrates.

Preputial glands, present in both sexes, discharge a secretion into urine. Preputial odors, combined with those of urine, constitute one means of olfactory communication between the sexes.

The chemical composition of accessory gland secretions varies among taxa. The seminal vesicle of the fruit bat (*Pteropus giganteus*) produces fructose, and the prostate and ampullary glands secrete both fructose and citric acid (Rajalakshmi and Prasad, 1970). In the human male the seminal vesicles are the major source of prostaglandins (Gerozissis et al., 1982). The male caecilian has well-developed Müllerian glands. The distal parts are glandular and join the cloaca. They produce both fructose and glucose, as well as acid phosphatase (Wake, 1981), the secretion resembling the seminal fluid of both birds and mammals.

4.3 STEROIDOGENESIS

The synthesis and release of testicular steroids is best known for mammals, and the initial comments on testicular steroidogenesis relate to these vertebrates. Like the ovaries, the testes are stimulated by both luteinizing hormone (LH) and FSH from the anterior pituitary. FSH acts directly on the Sertoli cells, which synthesize tubular proteins and also control tubular calcium metabolism. FSH is also required for complete spermatogenesis.

Steroidogenic activity of the testis begins in the embryo and increases, irregularly, until prepuberty. Embryonic production of androgens may, in fact, exceed that of neonatal and prepubertal stages. Prenatal testosterone accounts not only for the embryonic development of the male genital tract but also for the masculinization of the hypothalamus.

Leydig cells (in the interstitial tissue surrounding the tubules) synthesize steroids, and conversion of steroids, from one compound to another, can occur in Sertoli cells. Most steroidogenesis occurs in the Leydig cells, and they produce the overwhelming amount of the mammalian androgens—testosterone, androstenedione, androstanediol, and dihydrotestosterone. Steroids are synthesized from cholesterol, which is provided through lipoproteins synthesized in the liver or derived from acetate within the testis. Pregnenolone is an intermediate product, and both estrogens and androgens may be end products. Androgens can also be aromatized by fat tissue. There is also a low level of estrogen synthesis within the seminiferous tubules in neonatal mammals, and estrogen may also be converted from testosterone in Sertoli cells. In males estrogen may depress the release of LH and therefore of testosterone.

4.3.1 Control of Steroidogenesis by Gonadotropins

The testis in general and the Leydig cells in particular are under the control of LH. The testicular interstitial cells of mammals contain LH receptors and respond to LH by synthesizing androgens, primarily testosterone. In seasonally reproducing species these two hormones fluctuate together.

Gonadotropin-releasing hormone (GnRH) receptors also occur on Leydig cells, and a material like GnRH, possibly of Sertoli cell origin, has been reported to occur in the seminiferous tubules (Sharpe et al., 1982). This substance from the seminiferous tubules stimulates androgen synthesis and in this manner may adjust androgen synthesis by Leydig cells (Steinberger et al., 1983). Although testosterone circulates throughout the body, most of it passes directly to the seminiferous tubules, close to its site of origin.

During the reproductive season, a mature male mammal may exhibit repeated peaks of testosterone within a 24-hour period. This reflects sequential peaks of LH. Levels of the gonadotropins respond to levels of circulating testosterone and also to levels of estrogen, which depresses FSH and LH.

This outline of mammalian testicular steroidogenesis may apply, in general, to steroidal synthesis in most vertebrates. The control of testicular activity and the site of testicular steroidogenesis are essentially the same in all classes (Callard and Ho, 1980; Grier, 1981).

Leydig cells are present in the testes of elasmobranchs and they synthesize androgens. Sertoli cells (or nurse cells) are also said to be steroidogenic in sharks and rays.

Steroidogenesis in amphibians resembles that outlined for mammals. Leydig cells occupy the region about the seminiferous tubules. Like the germ cells and seminiferous tubules, the Leydig cells may exhibit marked seasonality in species of temperate climates, reflecting exogenous cues to the hypothalamus. Although early literature indicated Sertoli cells as the source of salamander androgens, it is now apparent that within the lobular walls are cells similar to, and probably homologous with, Leydig cells (Lofts, 1974). These cells are sometimes called lobule boundary cells and are active at the onset of spermatogenesis and before the discharge of spermatozoa. Seasonal changes include an accumulation of cholesterol and lipid-rich material after reproduction, and these materials may persist until the onset of spermatogenesis the following year. There is an increase in sensitivity to gonadotropins with a rise in body temperature in ectotherms, and there may also be seasonal changes in gonadal sensitivity to gonadotropins (Muller, 1977).

The production of androgens by the reptilian testis is reported to occur in both Sertoli cells and Leydig cells. Lipids increase in the nonreproductive seasons in the seminiferous tubules and especially in the Sertoli cells. Histochemical evidence and in vitro studies indicate the presence of cholesterol before and after spermatogenic activity (Lofts, 1977). The apparent androgenic activity of Sertoli cells is much less than that of Leydig cells, and the latter appear to produce the major amount of serum androgens (Licht, 1984). The teiid lizard (*Cnemidophorus tigris*) has a vascularized circumtesticular sheath of tissue, beneath the tunica albuginea, that contains Leydig cells (Currie and Taylor, 1970). Androgen production by these Leydig cells is about 10 times that of interstitial Leydig cells (Tsui, 1976). In the racerunner (*Cnemidophorus sexlineatus*) there is a substantial variation in the diameter of the circumtesticular Leydig cells but not in the interstitial Leydig cells (Johnson and Jacob, 1984).

4.3.2 Other Mediators of Testicular Activity

In addition to the stimulatory effects of gonadotropins on the testis, other hormones affect its activity. Testicular steroidogenesis is altered by the presence of serum levels of thyroxin and prolactin, which can be either stimulatory or inhibitory. In the case of thyroxin, the inhibition appears to be mutual. For example, reciprocal fluctuations of thyroxin and testosterone have been observed in the badger (*Meles meles*), and in castrated males a high level of plasma thyroxin is depressed after injections of testosterone (Maurel, 1978).

On the other hand, in several families of lizards thyroxin is essential for complete gonadal function. In many midlatitude species of reptiles circulating thyroxin peaks in the spring and summer, with much lower levels reported for winter (Lynn, 1970).

In birds thyroxin appears to suppress testicular activity. There are reciprocal cycles of thyroxin and testosterone in the domestic (Pekin) duck, and injections of thyroxin depress levels of serum testosterone, although not of plasma LH. Castration, moreover, is followed by a rise in thyroxin. It is suggested that a seasonal rise in thyroxin in this bird depresses reproduction (Gourdji and Tixier-Vidal, 1966).

A similar pattern exists in the occurrence of these two hormones in the Indian weaverbird (*Ploceus phillipinus*) and the spotted munia (*Lonchura punctulata*). In these birds thyroidectomy is followed by an increase in testicular activity (Chandola and Thapliyal, 1978). In the annual reproductive cycle of the green-winged teal (*Anas crecca*), there is also a reciprocal occurrence of testosterone and thyroxin, and injections of the latter are followed by a decline in circulating testosterone, with no effect on LH levels. As in the preceding examples, castration is followed by a rise in thyroxin levels, and thyroidectomy is followed by an increase in testosterone (Assenmacher and Jallageas, 1978).

The presence of the thyroid gland in the myna (*Acridotheres tristis*) is essential, however, for normal testicular responses to changes in the natural photocycle. Perhaps this thyroid function at times of sudden, but short-term, changes in ambient temperature could modify seasonal changes in gonadal activity (Chaturvedi and Thapliyal, 1983).

The role of prolactin in modifying reproductive activity seems to vary among different taxa: it is apparently essential and clearly stimulatory in some mammals and inhibitory in others. An important function is the increase of androgen uptake by accessory reproductive organs such as the prostate, seminal vesicle, and epididymis (Orgebin-Crist and Dijiane, 1979). Prolactin-binding sites occur within testicular tissue in the laboratory rat, and prolactin, in turn, increases LH-binding sites on Leydig cells and may thus indirectly stimulate spermatogenesis (Aragona and Freisen, 1975).

Perhaps the clearest evidence for the support of gonadal activity by prolactin comes from experimental work with the golden hamster (*Mesocricetus auratus*). In this rodent, as in many wild rodents, testicular size declines following exposure to short day lengths (or blinding), and testicular regression is accompanied by a dramatic drop in circulating prolactin. In some mammals, including the hamster, the pineal gland affects serum levels of prolactin (Reiter, 1980). Prolactin stimulates steroidogenesis in interstitial cells, and a decline of prolactin is followed by testicular regression (Reiter, 1977). When hamsters were kept under short day lengths (5L:19D) and treated three times daily with prolactin injections (250 g/day), there followed (1) an increase in testicular LH receptors and (2) an increase in testicular activity (Bartke et al., 1975; Bex et al., 1978). Also, in blinded golden hamsters weights of testes and associated reproductive structures were maintained by subcutaneous injections of ovine prolactin (Matthews et al., 1978). Apparently, normal testicular function in the golden hamster depends upon prolactin, LH, and FSH, and any two are ineffective without the third (Fig. 4.2).

In rams secretion of prolactin and testicular size are usually inversely related to activity. This relationship may change, however, when the photocycle is manipulated, and the relationship may be more coincidental than dependent or causal (Lincoln et al., 1982). Thus the sometimes presumed "antigonadal" effect of prolactin in

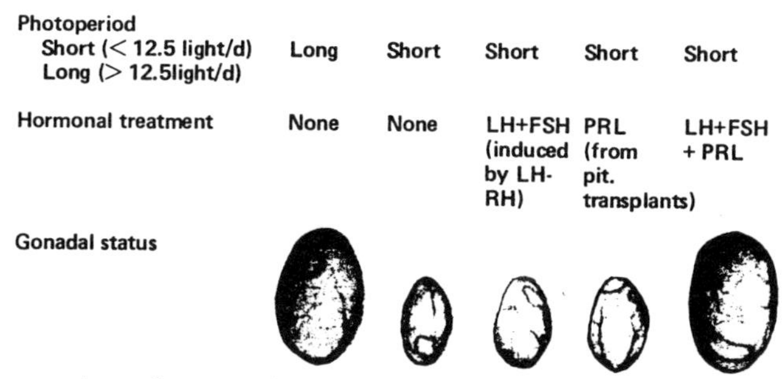

Figure 4.2 The influence of LH, FSH, and prolactin (PRL) on the testis of the Syrian hamster kept under short day lengths. LH-RH is gonadotropin-releasing hormone (GnRH). (Reiter, 1983, courtesy of Plenum Publishing Corporation.)

sheep may result simply from the vernal appearance of prolactin in an autumnal breeding species and therefore be more apparent than real.

Prolactin clearly stimulates testicular activity in some birds. In pigeons and doves prolactin stimulates the crop to synthesize and release pigeon milk, which is the initial food of the young. While feeding young, pigeons may form another clutch of eggs. In some other multibrooded species the brood patch is maintained by prolactin while gonadal activity persists (Lofts, 1970). Annual fluctuations of prolactin parallel testicular enlargement in the domestic duck (Fig. 4.3), suggesting that

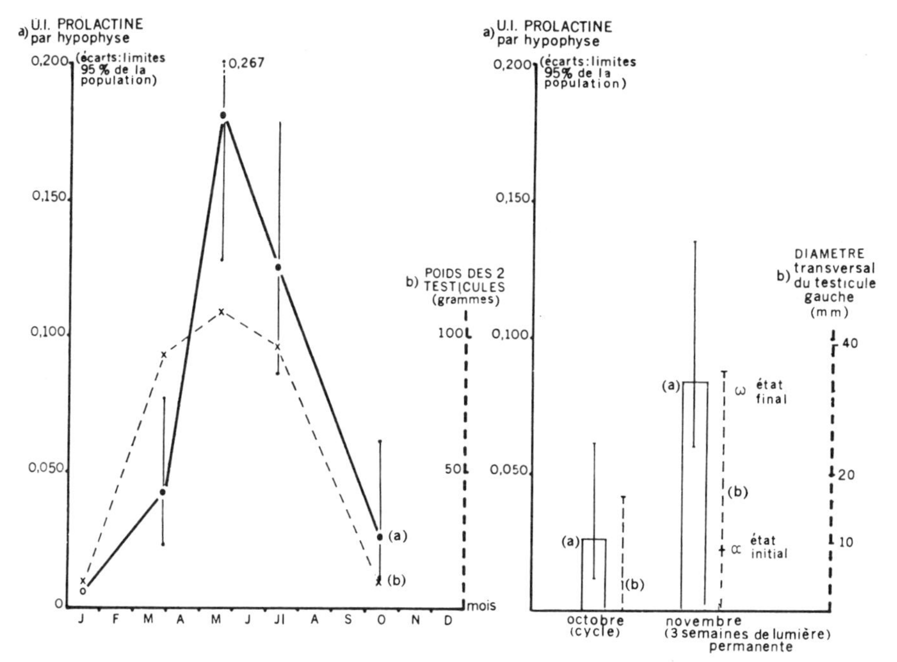

Figure 4.3 *Left*: variations in hypophyseal prolactin and testicular mass in the Pekin duck during the annual photocycle. *Right*: autumnal changes in hypophyseal prolactin and testicular mass after three weeks of photostimulation. (Gourdji and Tixier-Vidal, 1966, courtesy of the Academie des Sciences, Paris.)

prolactin is stimulatory in this bird (Gourdji and Tixier-Vidal, 1966). Following photostimulation of prolactin, an exposure to high temperatures (30°C compared with 18°C) induced an additional rise in prolactin in hen turkeys; it also induced broodiness (Halawani et al., 1984).

4.4 SPERMATOGENESIS

During the sexually inactive season in most seasonally reproducing vertebrates, the testes regress to a small fraction of their active mass and spermatogenesis and steroidogenesis are greatly reduced. This seasonal rhythmicity is under hypothalamic control, induced either directly or indirectly by environmental cues. The change in testicular mass reflects seasonal fluctuations in size and development of the seminiferous tubules, which comprise the bulk of the contents of the testis. There is also a reduction in size (but not in number) of Sertoli cells and Leydig cells. In extreme testicular reduction, which is characteristic of many species of small vertebrates in mid- and high latitudes, spermatogenesis is, in effect, almost at a standstill and rarely proceeds beyond meiosis I (Fig. 4.4). Thus mature spermatozoa are absent during the nonbreeding season. In fully regressed testes of many kinds of bats, mice, squirrels, and other small vertebrates, the cauda epididymides are shrunken and devoid of sperm.

In contrast to the embryonic ovary, gametogenesis does not occur in the embryonic testis. Some species of surfperch (Embiotocidae), however, appear to be an exception. The embryonic testes of *Amphigonopterus aurora* and *Micrometrus minimus* develops prior to birth and contains various stages of developing germ cells, with abundant transforming spermatids and spermatozoa at birth (Hubbs, 1921).

Production of spermatozoa is best known for mammals, to which the following introductory comments refer.

The seminiferous tubules and spermatogenesis are supported by testosterone. This has been shown by hypophysectomy and androgen replacement in rats. Hypophysectomy is normally followed by a regression of the seminiferous tubules, but testosterone may preserve the activity of the tubules in the absence of secretions from the anterior pituitary (Clermont and Harvey, 1967; Steinberger, 1971; Wells, 1942). The close proximity of the natural source of testosterone (the Leydig cells) probably allows continuous infiltration of testosterone. FSH may be needed for uptake of testosterone by Sertoli cells.

Lining the seminiferous tubules are germ cells, sometimes called gonocytes. In the early stages of spermatogenesis, germ cells, or gonocytes, produce spermatogonia. Initially stimulated by testosterone, the spermatogonia undergo several mitotic divisions. This commences in the embryonic testis, and spermatogonia are always in close association with one or more Sertoli cells. As in the mitotic divisions in oogenesis, these cells may remain connected by intercellular bridges. The final mitotic division produces primary spermatocytes. Secondary spermatocytes, still diploid, result from the first meiotic division. The second meiotic division (the reduction division) results in spermatids, still connected by intercellular bridges

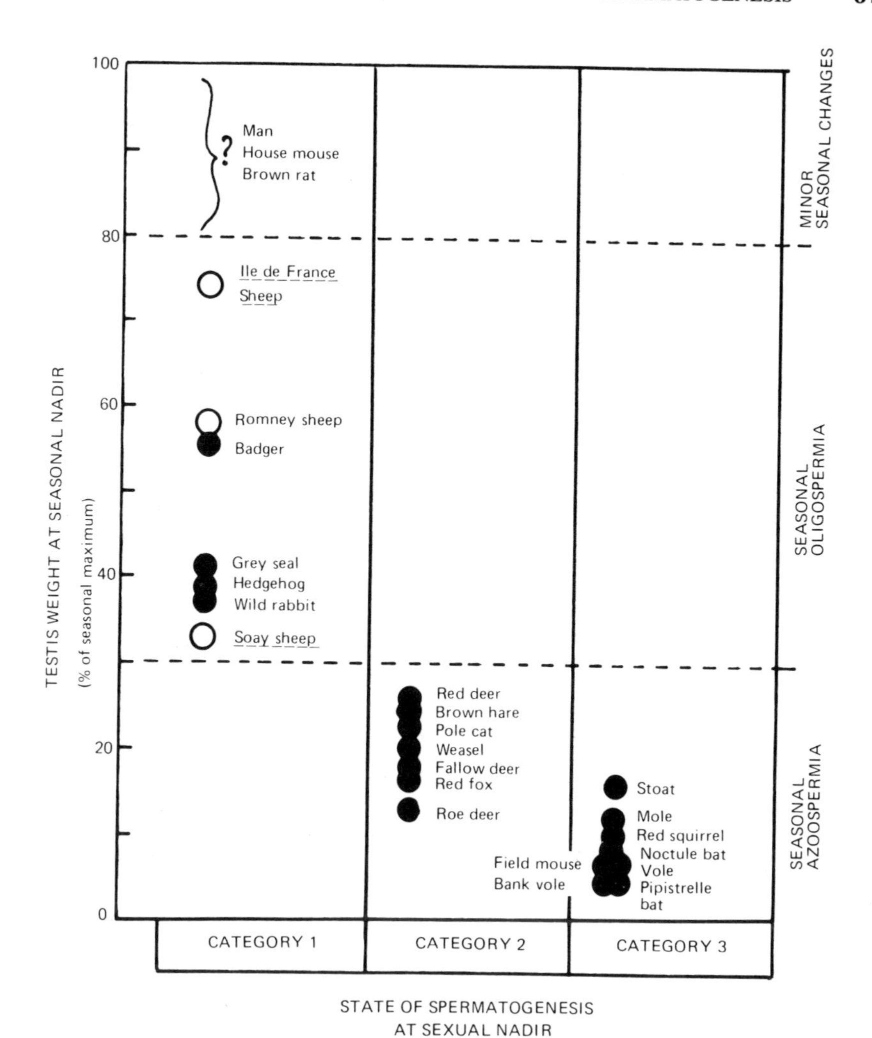

Figure 4.4 The relationship between testicular mass (percent of seasonal maximum) and state of spermatogenesis at the time of maximal seasonal regression. Category 1, with some complete spermatogenesis; Category 2, with some meiosis but no complete spermatogenesis; Category 3, with no meiosis or spermatogenesis. [From Lincoln, 1981, Seasonal aspects of testicular function. In *The Testis* (Burger H and de Kretser D, courtesy of Raven Press.)]

(Fig. 4.5). In contrast to the maturation of germ cells in the fetal ovary, meiosis in the testis is delayed until just prior to the onset of sexual maturity. Androgens from the human fetal testis apparently suppress the release of gonadotropins, which are at levels much lower than in the female fetus (Clements et al., 1976).

In spermatogenesis of elasmobranchs mitotic divisions begin at the anterior end of the testis and proceed caudad, following a migration of the ampullae. Epithelial

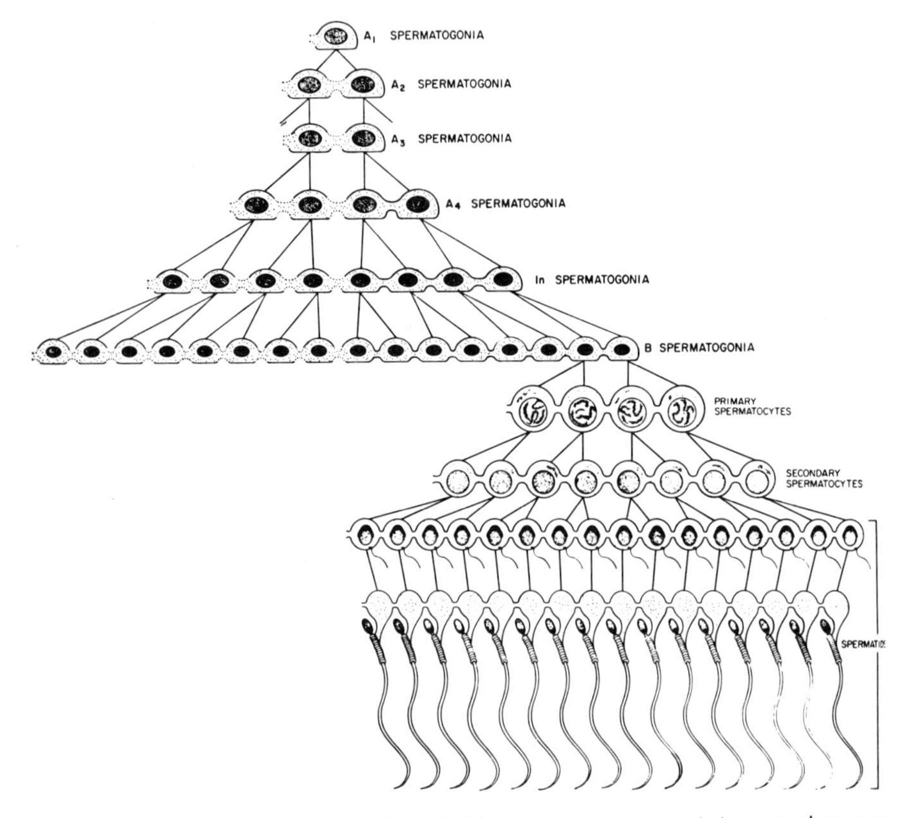

Figure 4.5 The retention of intercellular bridges in spermatogenesis in several generations of cell division. (From Dym and Fawcett, 1971, courtesy of Academic Press.)

cells surround the ampulla as spermatogonia divide, and these epithelial cells (also called nurse cells) are comparable, if not homologous to, Sertoli cells. A Sertoli cell surrounds a spermatogonium and seems to promote further development of the spermatogonium. Spermatocysts form as the ampulla moves toward the testicular margin. Eventually an ampulla will release spermatozoa, and the remaining ampullar elements are resorbed (Dodd and Sumpter, 1984).

In fish and amphibians there are clusters of spermatogonia, among which spermatogenesis is synchronized. These clusters, or nests, are enclosed in seminiferous tubules, which are sometimes called lobes, seminiferous follicles, or spermatocysts. Within these spermatocysts, or follicles, are follicle cells, or nurse cells, which are similar in function (and probably homologous) to Sertoli cells. Frequently these nurse cells are called Sertoli cells, but usage varies among authors.

In the Indian teleost *Horaichthys setnai*, mature germ cells are encased within a barbed spermatophore. This spermatophore is secreted about the mature spermatocysts by Sertoli cells. As the germinal cyst moves from the bulbular anterior end of the testis to a narrowed region of the efferent ducts, the barb-bearing end becomes slender and pointed (Fig. 4.6). After release of the spermatophore, the short efferent ducts, which were formed from Sertoli cells, seem to separate from the underlying tubule basal lamina. The basal lamina preserves tubule integrity within the testis and constitutes the foundation that guides the Sertoli cells in their movement and maturation as they approach the narrow caudal end of the testis (Grier, 1984). (See also Section 4.6.)

In amphibians spermatozoa develop in discrete sections, variously referred to as spermatocysts, germinal cysts, capsules, and lobules, which occupy the seminiferous tubules. Developing spermatogonia are closely associated with follicle cells, which are the precursors of Sertoli cells. The spermatogonia divide mitotically several times, and unlike the Sertoli cells of mammals, the follicle cells also divide. With the onset of meiosis, spermatogonia become spermatocytes, and the follicle cells are referred to as Sertoli cells.

Spermatogenesis is seasonal in most midlatitude amphibians. After mating, which usually takes place in the spring, spermatogenesis produces gametes for the following year. In frogs and toads (Anura) spermatogenesis occupies the entire testis, but in salamanders (Caudata) it proceeds from one region, or lobule, to another in a wave-like sequence, from the caudal tip of the testis forward. In the red-backed salamander (*Plethodon cinereus*), for example, spermatogenesis commences in March (in Maryland), after mating, and the swollen, active region gradually advances until it occupies the anterior part of the testis in autumn (Sayler, 1966).

In many reptiles spermatogenesis and steroidogenesis may not be concurrent, and spermatogenesis may take place either immediately prior to or shortly following mating. In postnuptial spermatogenesis, which characterizes turtles and some snakes,

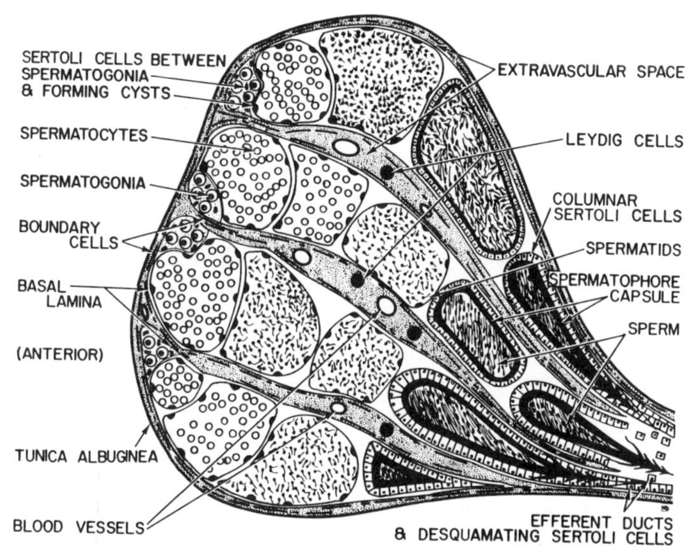

Figure 4.6 A diagrammatic sectional view of the testis of *Horaichthys setnai*. (From Grier, 1984, courtesy of the American Society of Ichthyologists and Herpetologists.)

spermatozoa develop in late summer and are stored in the epididymides; Leydig cell activity and steroidogenesis occur in the spring. In prenuptial spermatogenesis, as in lizards and some snakes, Leydig cell activity and steroidogenesis are greatest during spermiogenesis (Licht, 1984).

Two annual peaks of testicular androgens have been reported for some lizards. One occurs at the time of maximal testicular mass, and a second postnuptial peak has been detected during testicular regression (Licht, 1984).

Avian spermatogenesis is similar to that in mammals. Germ cells reside in convoluted seminiferous tubules. Several generations of spermatogonia divide mitotically, greatly increasing their populations. Spermatocytes divide twice, meiotically, producing spermatids, which subsequently mature into spermatozoa (Lofts, 1978). From the avian testis spermatozoa are released into an epididymis, which leads to a vas deferens, where the gametes may be stored. Sperm storage in the epididymis of birds is usually of short duration, and sperm may not survive there for extended periods. In the Japanese quail (*Coturnix coturnix*), spermatozoa contained by ligatures in the ductus deferens survive for only three days (Clulow and Jones, 1982).

4.4.2 Spermiogenesis

Spermiogenesis is the process by which profound cytological changes in spermatids result in their becoming spermatozoa, following the second meiotic division. This occurs while the germ cell is embedded in the cytoplasm of the Sertoli cells. The spermatid develops a terminal tail and an apical acrosome, and the cell contents, especially the mitochondria, become markedly altered. Clusters of spermatozoa may remain attached to each other and they separate when they leave the wall of the tubule. Most of the cytoplasm of the spermatid is shed, and this "residual body" is phagocytized by Sertoli cells.

In most mammals the clones of attached spermatids mature simultaneously, and an entire generation is released as a unit. Thus there is a conspicuous uniformity of germ cell development in any given section of a seminiferous tubule. Replenishment of spermatogonia in vertebrates occurs continuously, by mitotic division of gonocytes or spermatogonia (in mammals). Unlike oogenesis, which in birds, mammals, and elasmobranchs is completed in the embryonic gonad, the meiotic divisions leading to primary and secondary spermatocytes continue throughout the reproductively active life of the male. Functional maturation of spermatozoa occurs mostly in the body, or corpus, of the epididymis, and the cauda epididymis is the main site of storage. Although testosterone and FSH are both needed for spermatogenesis, FSH is also required for spermiogenesis (Steinberger, 1971).

In seasonally reproducing vertebrates that store sperm, spermiogenesis occurs as spermatids move into the lumen of the tubules and away from the Sertoli cells. As the seminiferous tubules regress and shrink in diameter, the spermatozoa pass into the epididymis, where they remain until mating. In many mammals some sperm remains in the epididymides after the mating period; the sperm may be removed by luminal macrophages or by macrophages in the epididymal membrane.

In mammals spermatogenesis is continuous throughout the year, even in seasonally reproducing species in which the testes regress in the nonreproductive season. FSH is required for efficient sperm production, for in the presence of FSH, atresia declines. Variations in serum FSH determine the rate of germ cell degeneration and, ultimately, the seasonal changes in fertility. In many midlatitude reptiles, there is a lapse of spermatogenesis in the nonreproductive season (Licht, 1984), which probably occurs in many other ectotherms of temperate regions.

4.5 THE TESTICULAR ENVIRONMENT

In ectotherms the testes are internal and have the same temperature as the core. In birds, also, the testes are internal and lie attached to the dorsal wall of the coelom. Passage of air through the air sac system may possibly reduce slightly the testicular temperature of birds. In anesthetized Japanese quail the testicular temperature is 0.6°C cooler than adjacent visceral tissues. This variation is attributed to endothermic reactions of the tissues and not to cooling by air sacs (Jarman and Thompson, 1986).

The body temperature of most birds lies well above 40°C. A mean temperature of passerine birds is 40.6°C, and core temperatures up to 46°C are not uncommon. Apparently heat does not inhibit testicular activity in birds. The rate of avian spermiogenesis is reported to be about 10 times greater than the rate for mammals (Amann, 1970).

In mammals the testes of many species migrate to a position at least partly outside of the abdominal cavity, but in a few taxa (the testicondae or testiconda) the testes remain permanently and entirely within the body. Testiconid mammals include the elephants, hyraxes, cetaceans, and some others. In species with a discrete reproductive season, the testes enlarge and descend into a scrotum during the period of sexual activity and regress and withdraw within the body cavity during the season of sexual inactivity. During mammalian prenatal and early postnatal development, the testes migrate to the region of the future scrotal development. The precise path seems to vary among different orders but may be assisted by pressure brought about from growth of the liver and viscera.

In several species of hyrax (Procaviidae) and the two species of elephants (Elephantidae), for example, there is no scrotum. These mammals experience core and testicular temperatures of nearly 39°C. In the African elephant, core temperatures range from 36.2 to 38.9°C (Sale, 1969; Short et al., 1967; Hanks, 1977), and the hyrax (*Procavia capensis*) has core temperatures from 34.4 to 38.9°C (Millar and Glover, 1973). In both these families high temperatures clearly do not impair testicular function. Mammalian core temperatures vary between 33 and 39°C in most normally active individuals. Higher levels of temperature are similar among species with abdominal testes and scrotal testes, but lower core temperatures occur among species in which the testes are abdominal (Carrick and Setchell, 1977).

In those mammalian groups in which the testes descend into a well-developed scrotum, it is clear that scrotal temperatures are lower than core temperatures. In species with pendulous scrotal testes (many ungulates, primates, and most marsupials), both the size of the scrotum and the distance it descends from the body decrease in response to a decline in ambient temperature. This response is achieved by a contraction of the dartos muscle (tunica dartos), which lies beneath the scrotal skin, causing a folding and wrinkling of the scrotal wall. Contrariwise, in warm ambient temperatures, the scrotum becomes slightly larger and drops to a lower position. These adjustments tend to maintain a fairly constant testicular temperature of 3 to 7°C below the individual's core temperature. In those groups with a well-developed scrotum, the testicular artery and testicular vein lie appressed in the spermatic cord between the abdomen and testes. Presumably this results in a countercurrent exchange, with a reduction of heat entering the testes from the body.

There seems to be no compelling reason to suppose that testicular temperature was originally especially critical within the Mammalia. The need for a low testicular temperature is clearly a characteristic of only those mammalian taxa in which the testes are suspended in a scrotum. On the other hand, circumstantial evidence suggests that the movement of the testes to a cooler site has been an evolutionary tendency to provide a lower temperature for the epididymides.

Experimental evidence indicates that the epididymides require a body temperature lower than that of the core. The redirection of the epididymis to an abdominal position while the ipsalateral testis remained in its original scrotal position revealed that storage of sperm (retained by ligatures) resulted in survival for only 3 to 4 days in rats and 8 to 10 days in rabbits while sperm remained viable in the contralateral and scrotal epididymis (Bedford, 1977).

Indeed, in those species in which the testes lie in a superficial cremaster sac (many insectivores and rodents), the epididymis forms a conspicuous swelling at the caudal part of the testis, exposing it to ambient temperatures. In hystricomorph rodents the testes lie within the abdomen, and the cauda epididymides may lie in postanal or cremaster sacs. Some of these species have been called facultative cryptorchid (Weir, 1974). An extreme example is the natural cryptorchid *Proechimys semispinosus*, a hystricomorph rodent of the mountains of western South America, in which only the epididymides extend into the cremaster sac (Bedford, 1977). The distal site and relatively avascular condition of the epididymis, in addition to its insulation by fat, may account for its temperature being significantly below that of the testes (Brooks, 1973). Also, in a number of mammals with scrotal testes, the scrotum is furred except for the skin covering the epididymides.

An examination of the relative positions of the testis and epididymis also suggests that it is the epididymis that requires a temperature below that of the core (Fig. 4.7). The epididymis is the site of storage of spermatozoa, and experimental evidence indicates that a lower temperature is needed for prolonged survival of spermatozoa. If a lower testicular temperature had been the main cause for the development of the scrotum, the most simple arrangement would be for the testis to occupy the scrotum with the epididymis remaining internal (Fig. 4.7e), a condition that does not occur (Bedford, 1977).

4.6 SPERMATOPHORES

Spermatozoa are released in small packets called spermatophores in several groups of fish, including both elasmobranchs and teleosts, as well as in salamanders (Caudata).

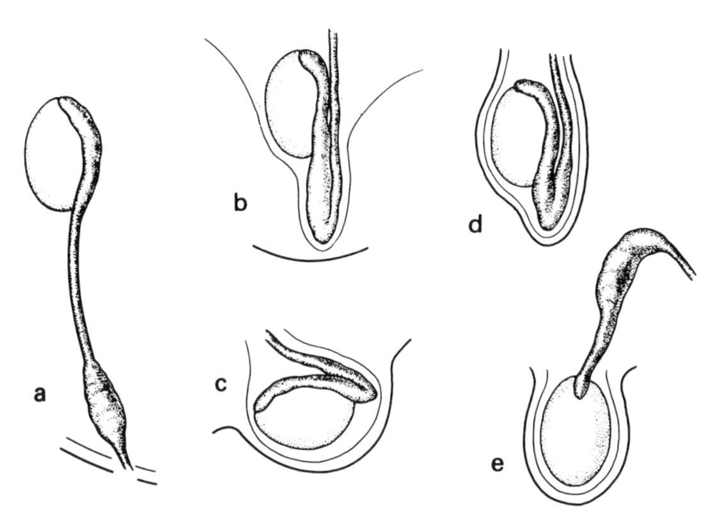

Figure 4.7 Various arrangements of the mammalian testis and the epididymis. The typical testicondid condition as seen (*a*) in elephants and the hyrax; (*b*) the natural cryptorchid condition, in which the cauda epididymis extends into the cremaster sac, as in *Glaucomys*, *Proechimys*, *Suncus*, and *Solenodon*; (*c*) the scrotal epididymis and testis of the horse and the boar; (*d*) the most frequent arrangement of scrotal testis and epididymis; (*e*) and a nonexistent hypothetical condition that would be expected if the scrotum had evolved solely to lower testicular temperature. (From Bedford, 1977, courtesy of the Australian Academy of Sciences.)

The spermatophore is a means of ensuring the efficiency of internal fertilization in an aquatic milieu. See Mann (1984) for a review.

In the rabbitfish, *Callorhycnhus antarcticus*, a holocephalan, the epididymides and seminal vesicles secrete a membranous noncellular covering in which the sperm are packaged. In elasmobranchs, all of which have claspers (modified pelvic fins) that serve as intromittent organs, sperm may also be concentrated in spermatophores. This condition has been described for the basking shark (*Cetorhinus maximus*). The seminal vesicles provide several jellylike envelopes about clusters of sperm, and this spermatophore is literally flushed into the uterus together with a large amount of fluid from the seminal vesicles (Matthews, 1950).

The spermatophores of poeciliid fish (guppies, mollies, and swordtails and their relatives) are weakly encapsulated clusters of sperm. They are placed into the uterus of the female by a modified anal fin. Spermatophores also occur in live-bearing brotulid and embiotocid fish. In the surfperch (*Cymatogaster aggregata*), an embiotocid of the eastern Pacific, sperm are densely packed as they leave the testicular cysts. Spermatophore capsule proteins are synthesized by the cells of the efferent ducts; the spermatophore membranes remain intact while within the seminal fluid but dissolve when they enter the increased pH of the ovarian fluid (Gardner, 1978). In *Horaichthys setnai* (Adrianichthyidae) of India the spermatophores are barbed (see Section 4.4) and do not actually enter the reproductive tract of the female. Instead, the barb implants near the genital orifice of the female, and when in seawater, the

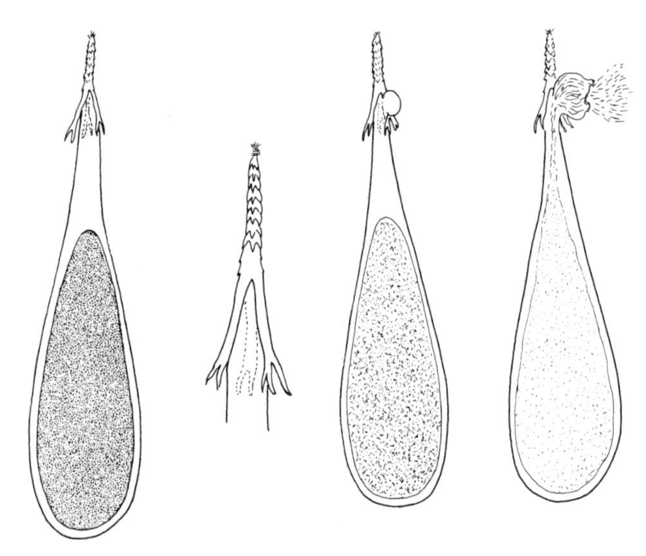

Figure 4.8 Spermatophores of *Horaichthys setnai*. *Left*: in freshly extruded condition, with details of barbed spine. *Right*: the basal swelling of the spermatophore and escape of sperm upon exposure to water. (From Mann, 1984, after Kulkarni, 1940, courtesy of Springer-Verlag.)

spermatophore ruptures and slowly releases spermatozoa (Fig. 4.8). Presumably the slow and directed release of sperm allows some to enter the genital tract of the female (Kulkarni, 1940).

Internal fertilization is the pattern among most salamanders. There is no intromittent organ, but sperm are contained in a spermatophore, which the male deposits on a substrate, such as a rock or submerged piece of vegetation. The spermatophore is subsequently picked up by the cloacal lips of the female. The spermatophore of salamanders consists of a basal stalk, which becomes attached to the substrate, and an apical cap, which contains the spermatozoa (Fig. 4.9). In the plethodontid

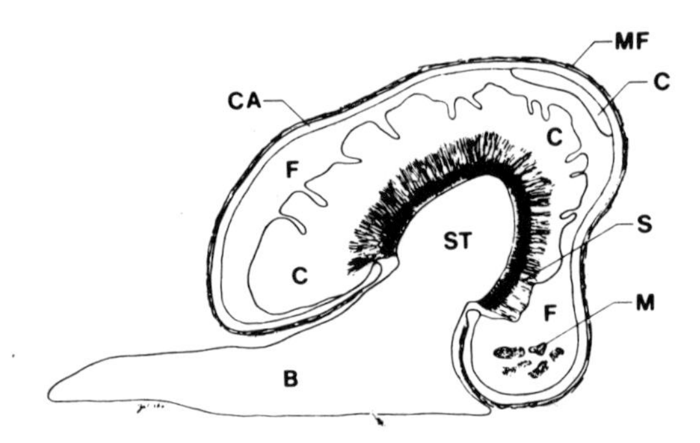

Figure 4.9 Spermatophore of the salamander, *Plethodon jordoni*, through the midsagittal plane: MF, mucous film; CA, capsule; F, fine granular layer; C, coarse granular layer; S, spermatozoa; M, clusters of melanin granules; ST, stalk; B, base. (From Zalisko et al., 1984, courtesy of the American Society of Ichthyologists and Herpetologists.)

salamanders *Eurycea lucifuga* and *E. longicauda*, Kingsbury's and pelvic glands produce the spermatophore cap and enlarge prior to the release of spermatozoa (Williams et al., 1985). The female breaks off the cap, or part of the cap, from the stalk, allowing spermatozoa to escape into her genital tract (Zalisco et al., 1984). For additional discussion of spermatophores, see Section 7.1.4.

4.7 SUMMARY

Paired testes occur in virtually all vertebrates and serve dual functions: hormone synthesis and gamete production. Surrounded by a protective layer, the tunica albuginea, the testis is comprised of a mass of convoluted seminiferous tubules, which are surrounded by interstitial cells, blood vessels, nerve cells, and Leydig cells. Efferent ducts are modified mesonephric tubules and, in elasmobranchs, bony fish, and amphibians, lead through the kidney and thence to the cloaca.

The embryonic testis synthesizes and secretes androgens but does not experience gametogenesis. Apparent exceptions are at least some species of surfperch (Embiotocidae), in which embryonic testes may contain spermatozoa.

The testis responds to LH, FSH, and, in some taxa, also to prolactin and depends upon these hormones for its activities. LH stimulates steroidogenesis by Leydig cells. These cells secrete androgens (mostly testosterone and androstenedione), which, together with FSH, stimulate the seminiferous tubules and their Sertoli cells. Germ cells, in close association with Sertoli cells, divide mitotically as spermatogonia. They later divide meiotically, to mature first into spermatocytes and eventually into spermatids. Functional maturation of germ cells, forming spermatozoa, occurs in the epididymides.

In most vertebrates testicular activity is seasonal. Androgens not only promote activity within the seminiferous tubules but account for seasonal development of secondary sexual characteristics.

In many mammals the testes lie just beneath thin skin, in a cremaster sac, or are suspended in a scrotum. In either position the temperature of the testes is lowered. A lowered temperature may enhance the survival of spermatozoa stored in the epididymides.

In some elasmobranchs, teleost fish, and salamanders spermatozoa are packaged in membranous sacs called spermatophores, which probably aid in internal fertilization.

BIBLIOGRAPHY

Berger H and de Kretser D (Eds) (1981). *The Testis. Comprehensive Endocrinology*. Raven, New York.

Desjardins C (1978). Endocrine regulation of reproductive development and function in the male. *Journal of Animal Science* (Suppl II), **47**: 56–77.

Francimont P and Channing CP (Eds) (1981). *Intragonadal Regulation of Reproduction*. Academic, London, New York.

Gomes WR and Van Demark NL (1970). *The Testis*, vols I–III. Academic, New York, London.

Roosen-Runge EC (1977). *The Process of Spermatogenesis in Animals*. Cambridge Univ, London.

Setchell BP (1978). *The Mammalian Testis*. Cornell Univ, Ithaca.

Sharpe RM (1982). The hormonal regulation of the Leydig cell. Pp 241–312. In *Oxford Reviews of Reproductive Biology*, vol 4 (Ed: Finn CA). Oxford Univ, Oxford.

Sharpe RM (1983). Local control of testicular function. *Quarterly Journal of Experimental Physiology*, **68**: 265–287.

Steinberger E (1971). Hormonal control of spermatogenesis. *Physiological Review*, **51**: 1–22.

Tahka KM (1986). Current aspects of Leydig cell function and its regulation. *Journal of Reproduction and Fertility*, **78**: 367–380.

5

SOME ADAPTIVE MODIFICATIONS OF GROWTH

5.1 THE NATURE OF GROWTH

The purpose of this chapter is not to review in any detail the growth of vertebrates but rather to point out some aspects of growth rates that constitute adaptations to certain life history patterns.

Growth results from cell division and encompasses differentiation, including the development of special cells and tissues. There are separate rates of growth for different tissues, so that structures appear when appropriate to the need of the organism. There are also different rates of regrowth or repair of tissues. These rates are genetically controlled and are characteristic for a given taxon.

As we have seen, the formation of primary oocytes proceeds in the embryonic ovary of cyclostomes, elasmobranchs, birds, and mammals but not in the ovary of other vertebrates. In most vertebrates the growth and repair or replacement of nerve cells is limited to the early life stages, but epidermal cells and erythrocytes are continuously replaced and liver cells are readily replaced. Ligament tissue does not normally repair itself, except under the stimulus of some exogenous material.

The single-celled zygote must change form with the progression of mitosis. In time many kinds of cells differentiate into tissues; some tissues continue to grow throughout life, whereas others are suppressed and lose mitotic potential at birth or hatching. Eventually a final, or adult, form results, after which some tissues are replaced at a constant rate, some periodically, and some not at all. This rate is genetically programmed but does not preclude responses to exogenous factors.

In the absence of mitosis, heterotrophy of existing cells causes growth, as in skeletal muscle. In some structures cell division permits the enlargement of existing units, such as the kidney nephron or alveolus of a lung, but not the increase of such units. These constraints on growth vary from one taxon to another and are more restrictive in birds and mammals than in ectotherms. The mechanism of terminating growth in endotherms is unknown.

5.2 GROWTH IN ECTOTHERMS

All animals have an innate limit to growth. Although growth is said to be indeterminant or without limit in ectotherms, growth does become extremely slow in large individuals and, for all practical purposes, virtually ceases.

Metabolic rate, growth, and reproduction are not entirely separate and independent of each other. Rapid growth requires a high metabolic rate, and the correlation is best seen in fish, which cannot escape their ambient temperature, and some desert-dwelling amphibians, which need to minimize their exposure to possible desiccation. The mosquito fish (*Gambusia affinis*) reared on a uniform diet grows and becomes sexually mature much more rapidly at higher temperatures (Fig. 5.1). Growth in the pike (*Esox lucius*) is correlated with both nutrition and temperature. In a population of pike monitored from 1944 to 1982, growth varied with annual cumulative temperatures but was always slower for males than for females. For young in the years from 1944 to 1959 and from 1966 to 1967, growth in the first four years was correlated with degree days greater than 14°C, and 81% of the variation in mean mass was accounted for by temperature that obtained in the first years of life. In the following 20 years, growth was poorly correlated with temperature. From 1961 to 1967 there were three cool and no warm summers, and growth was slow in contrast to 1968 to 1976, during which period there were six warm summers out of seven and growth was more rapid (Fig. 5.2). Warmer temperatures were correlated with growth, however, only when there was adequate food (Kipling, 1983).

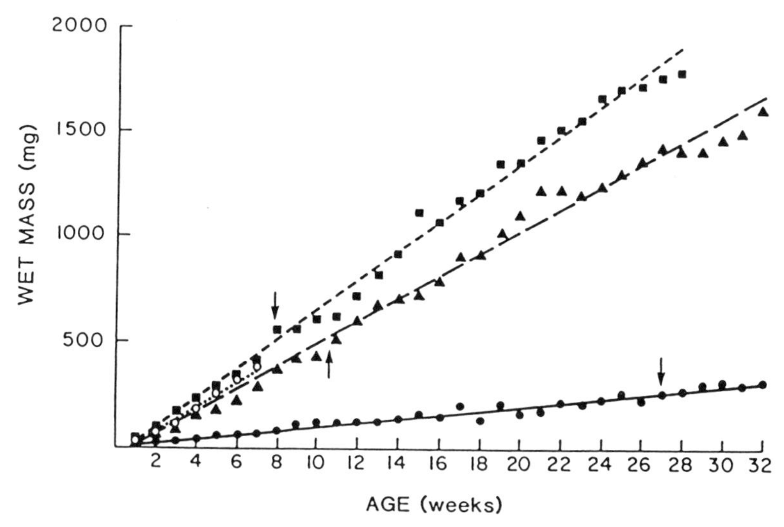

Figure 5.1 Mean wet body mass of the mosquito fish (*Gambusia affinis*) at 20°C (solid circles), 25°C (triangles), and 30°C (squares). The short line of open circles indicates data for diel cycling temperatures (20–30°C, mean 25°C). Arrows indicate mean body mass and age at first reproduction for the first 10 females at each temperature. (From Vondracek et al., 1988, courtesy of *Environmental Biology of Fishes*.)

The relationship between growth rate and environmental seasonality is easily seen in a short-lived species like the stickleback (*Gasterosteus aculeatus*). In upland lakes of Wales this stickleback is an annual species, with the maximal growth rate in the first month after hatching. Growth slows in the autumn and ceases in the winter, with the decline in food and drop in temperature (and metabolism), and resumes in the following spring (Fig. 5.3), prior to reproduction and death (Allen and Wootton, 1982a). The increase of body mass in the laboratory is very sensitive to nutrition, with the most rapid changes being in the liver, which grows quickly under ample food (Allen and Wootton, 1982b).

5.2.1 Saltatory Ontogeny

Differentiation is more sensitive to temperature restraints than growth is, and growth depends upon the stage of differentiation. Certain stages, such as metamorphosis, are especially vulnerable and are passed through rapidly (Stearns, 1976). During such transitional stages, the individual frequently changes habitats and is temporarily exposed to both stressful physical factors and heavy predation.

Balon (1984) emphasized the evolutionary significance of saltatory ontogeny, the synchronized development of several structural systems as the organism passes from one functional state to another. Progression through modes of existence requires the simultaneous perfection of several systems, and such coordinated growth distinguishes saltatory ontogeny from a hypothetical gradual development of all systems

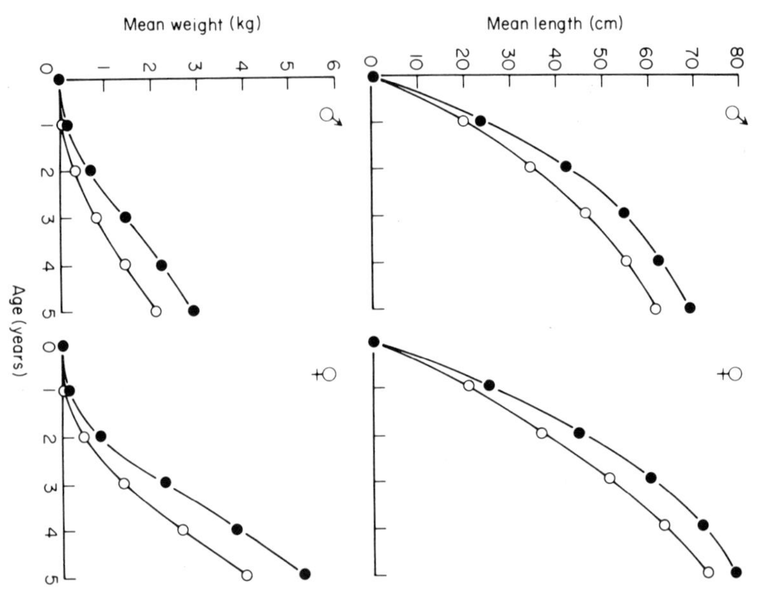

Figure 5.2 Mean lengths (*top*) and weights (*bottom*) for the first five years of a slow-growing (open circles) and a fast-growing (solid circles) year class (or generation) of pike (*Esox lucius*). (From Kipling, 1983, courtesy of the British Ecological Society.)

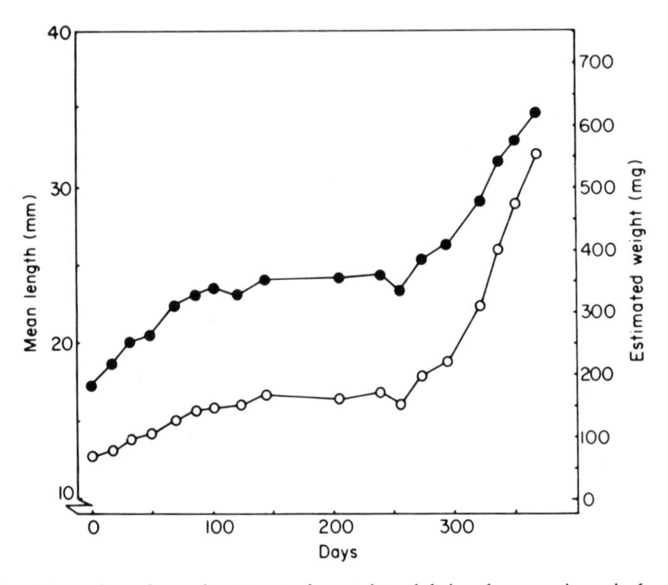

Figure 5.3 Mean length and estimated weight of fish of mean length for 1977 year class of stickleback (*Gasterosteus aculeatus*) in Lake Frongoch (Wales). Day 0 is 5 July 1977). (From Allen and Wootton, 1982a, courtesy of Academic Press.)

at the same rate. With the approach of a new functional state, coordinated growth accelerates, effecting the synchronization of systems development (Fig. 5.4). That is, before a young fish can leave a quiet, resting, and essentially nonfeeding state, it must first develop opercles with motor control, jaw muscles, and fin development, all with skeletal support. These several systems must form together if the individual is to advance to a new level of activity (Fig. 5.5).

5.2.2 Intraspecific Variation in Egg Size

Among many marine fish that lay planktonic eggs, those spawned late in the season are smaller than those produced at the beginning of the season (Fig. 5.6). This may reflect the greater productivity of older and large individuals, for they typically spawn in advance of the less fecund younger and smaller individuals. On the other hand, among a group of North Sea gadoid fish (cod, haddock, whiting, and Norway pout), eggs laid near the end of the spawning season tend to be reduced in mass, both in the sea and in aquaria (Hislop, 1984). This variation could also result from an endogenous control, for egg size and dry mass declined with the season in a captive population of haddock (*Melanogrammus aeglefinus*) maintained on controlled caloric intake (Hislop et al., 1978). Larger eggs result in larger larvae, which

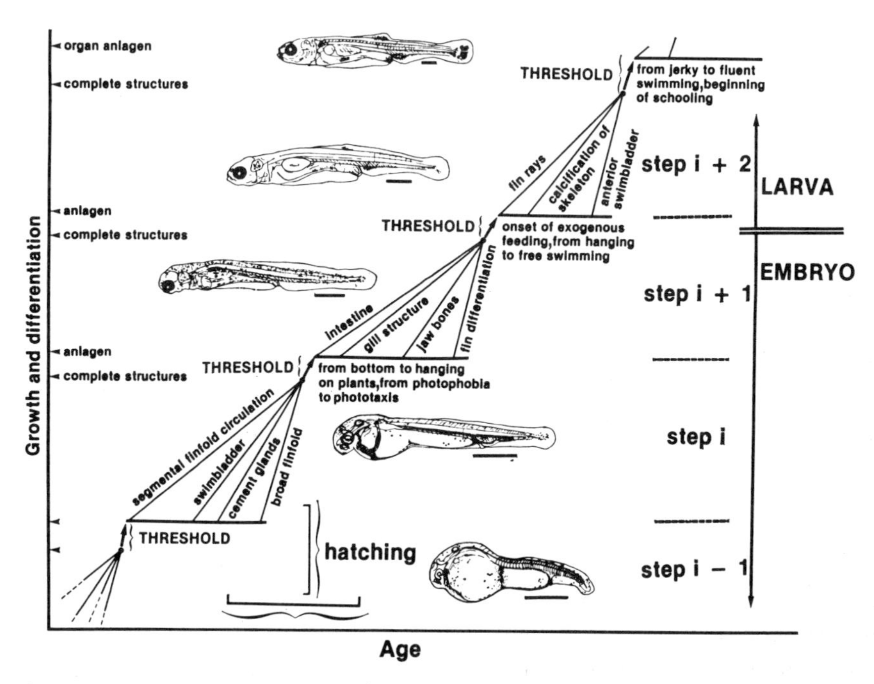

Figure 5.4 Consecutive steps in the growth and differentiation of the zope (*Abramis ballerus*). (From Balon, 1986, courtesy of *Rivista di Biologie*.)

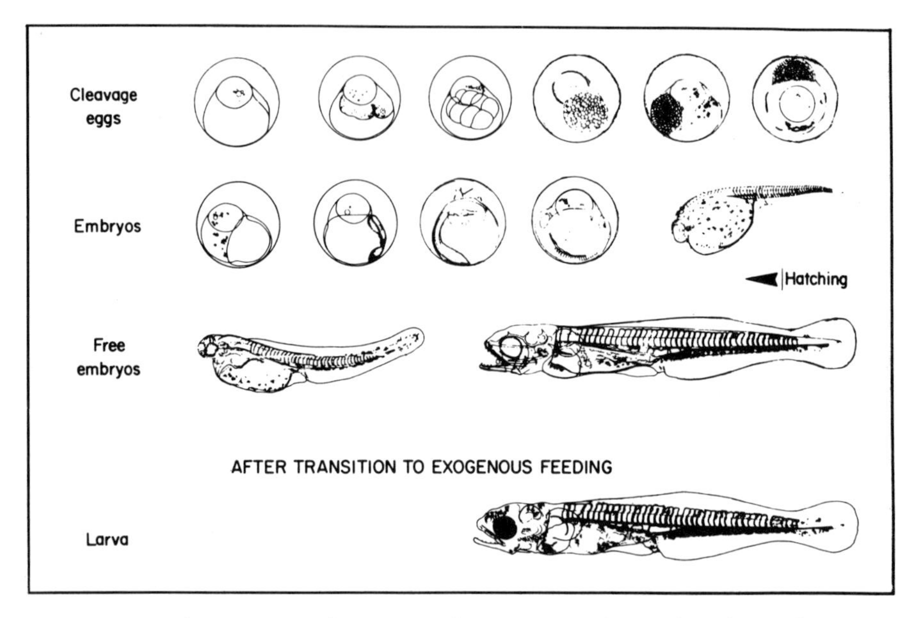

Figure 5.5 Early ontogeny of a nonguarding egg-scattering rock and gravel spawner with pelagic larvae (*Stizostedion vitreum*). (From Balon, 1984, courtesy of Academic Press.)

survive for longer periods without food; and larger eggs are spawned prior to the maximal production of plankton (Bagenal, 1971).

For a given species of pelagic teleost, egg size tends to decrease from winter to early summer—or with an increase in water temperature—and larger eggs produce larger larvae. Presumably, much early mortality occurs during the period when the egg is drifting passively at the surface. Small eggs appear to be more vulnerable to predation than large eggs are, but the more rapid growth of small eggs minimizes their loss. There is a positive correlation between incubation time and egg mass. When incubation time is short, as it is in the warm water of summer, it is advantageous for eggs to be smaller, because their short incubation time (1–2 days) reduces the duration of their vulnerability. In this way egg size relates to incubation time, predation, and seasonality. Larger larvae, from larger eggs, are more suited to feed on the larger food items that prevail in the winter plankton.

5.2.3 Neoteny

Some species of salamanders become sexually mature before metamorphosis and the loss of gills, a condition known as neoteny, paedogenesis, or paedomorphosis. The retention of gills is permanent or facultative and temporary, depending upon the taxon. One view considers neoteny to be precocious sexual development, and another holds that neotenic salamanders are not really precocious in the development of the gonadal system but that they have retained their gills into adulthood. The latter concept seems more logical. Generally neotenic salamanders become reproductively active at approximately the same size and body mass as metamorphosed individuals of the same species. Metamorphosis may occur at various sizes and ages, but there is a minimal body mass at which sexual maturity is attained. Some species of neotenic salamanders are permanently aquatic and never lose their gills. Neoteny does not seem to occur as a normal event in the life cycle of any species of caecilian or anuran.

Neoteny is facultative in some species of salamanders. These species characteristically breed in ponds in arid or semiarid regions, where amphibians risk desiccation in the terrestrial environment. Remaining in the ponds, species such as the Nearctic tiger salamander (*Ambystoma tigrinum*) retain their gills, either throughout their lifetime or until the pond should dry up. Drying of a pond, during a period when rainfall is slight, forces metamorphosis, and such terrestrial salamanders then seek underground burrows.

Obligate neoteny characterizes salamanders that live in large lakes, especially lakes surrounded by arid terrain, and underground waterways. The various species

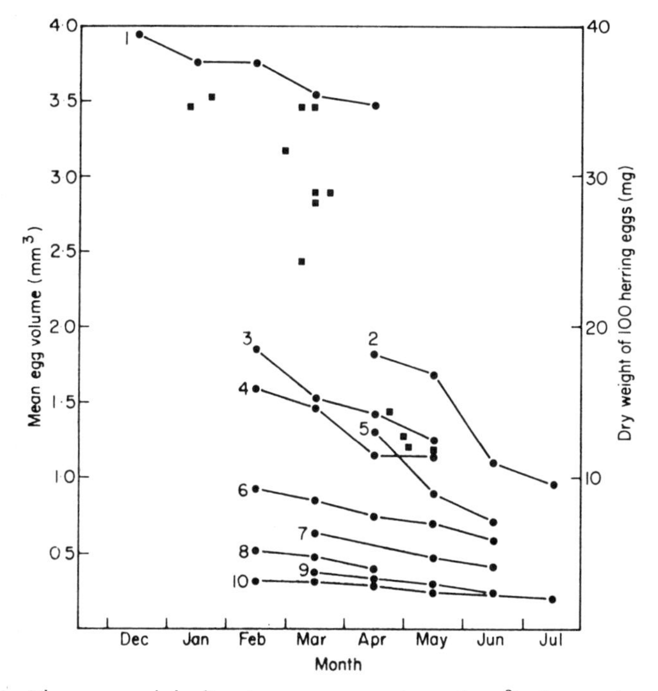

Figure 5.6 The seasonal decline in mean egg volume (mm^3) of several species of fish: 1, *Pleuronectes platessa*; 2, *Trigla gunardus*; 3, *Solea solea*; 6, *Merlangius merlangus*; 7, *Sprattus sprattus*; 8, *Platichthys flesus*; 9, *Rhinonemus cimbrius*; 10, *Limanda limanda*. Solid unconnected squares indicate mean dry weights of herring eggs from different spawning groups. (From Bagenal, 1971, courtesy of the *Journal of Fish Biology*.)

of mud puppies (*Necturus* spp.) of North America dwell in many large lakes and major rivers, and their European relative, the olm (*Proteus anguineus*), lives in streams in caves. Several species of plethodontid salamanders, such as *Haideotriton wallacei*, *Typhlomolge rathbuni*, and *T. robusta*, are obligately neotenic. They occur in subterranean watercourses, and because the food in supply is in water, there is no advantage to their becoming terrestrial.

In nature, as in the laboratory, thyroxin induces metamorphosis in salamanders. Experimentally thyroxin and thyroid-stimulating hormone (TSH) cause the final morphological changes, including the loss of gills, in facultatively neotenic *Ambystoma* (Taurog, 1974). Larvae of *A. tigrinum* pretreated with TSH picked up significantly more labeled iodine than untreated individuals did (Norris and Platt, 1973). Those species that are obligate neotenic forms do not generally respond to treatment with thyroxin.

5.3 EFFECT OF ENDOTHERMY

Ectothermy is perhaps the major constraint on growth rate for fish, amphibians, and reptiles; and growth is consistently faster in birds and mammals (Case, 1978). Endothermy is acquired almost at birth in precocial species but is a gradual process in altricial forms.

In infant mammals, which are nourished by highly divergent qualities of milk, species similar in neonatal size may have very different rates of early growth. The great cost of rearing altricial mammals lies in lactogenesis, a responsibility lying solely with the female. Milk is a major means of thermoregulation in altricial mammals, for it transfers warmth as well as nutrition to the neonate.

5.3.1 Adaptive Growth of Birds

Rate of growth of altricial birds in about twice that of altricial eutherian mammals. Although metabolic rates of adult birds are only slightly higher than those of mammals, the infants of both classes have lower metabolic rates than the adults do. The lower metabolic rates of neonates presumably reflect their heterothermy. Some rapid growth in altricial birds is gained at the with a the delay in the development of endothermy. Growth rates are generally slower in tropical altricial passerine birds than in similar species in mid- or high latitudes (Ricklefs, 1973). In high-latitude species this faster rate would seem to be an adaptation to the need for early preparation for migration.

David Lack (1968) proposed that growth rate in birds is determined by the combined factors of food supply and nestling mortality. Because young in an open nest are exposed to predation, it is to their advantage to grow as rapidly as their parents can provide food. Lack's thesis fits some species but not others. Rapid growth may enhance reproductive potential if it accelerates sexual maturity (Gadgil and Bossert, 1970). Young birds are indeed vulnerable to predation, but it is not apparent that food always modifies the growth rate of nestlings. Growth rates of nestlings of most species are not readily adjustable to food supply, but the demand for food is adjusted through clutch size. The food supply during vitellogenesis and yolking determines the number of eggs to be laid and predicts the availability of food to young (Ricklefs, 1979).

5.3.1.1 Hole Nesting in Birds. Passerine birds that regularly nest in holes enjoy the luxury of slow growth. Most hole-nesting species tend to take a longer time to fledge young than those species nesting in open, unconcealed places do (Lack, 1968). Case (1978) suggested that the rapid growth of small altricial birds is favored by selection, inasmuch as they occupy nests that are exposed to predators, in contrast to the concealed nests of altricial small mammals. This proposal is complicated, however, by other important factors, such as the nature of parental care and duration of heterothermy.

Ricklefs (1973, 1979) discussed patterns of avian growth, and the student should consult his thorough reviews of general posthatching growth of birds. He pointed

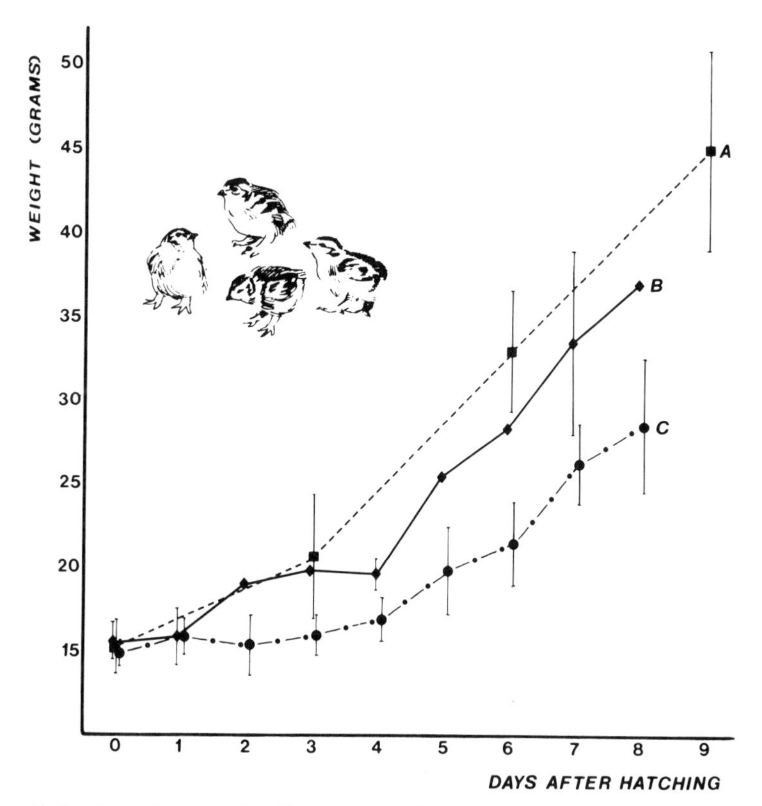

Figure 5.7 Growth rates of willow grouse chicks from small (*b*) and large (c) broods during three cold and wet summers (1975–1977) and large broods (a) during an extremely warm and dry summer (1978), in Trany, northern Norway. Mean weights ± SD. Sample size (number of weights): A = 130, B = 28, and C = 96. At hatching (0 days) chicks in each brood are pooled in one sample. (From Erikstad and Andersen, 1983, courtesy of *Ornis Scadinavica*.)

out that altricial birds have growth rates three to four times those of precocial birds, and precocial birds that forage for their own food grow most slowly. Growth rates are inversely related to adult weight and appear to be as high as physiologically possible (Ricklefs, 1973). This suggestion is compatible with observed high rates of survival in years of food abundance.

Precocial nidifugous species, such as ducks (Anatidae) and quail (Phasianidae), grow more slowly after hatching than altricial birds do, although there are some exceptions (Ricklefs, 1973, 1979). Growth rates may respond to food availability in precocial species, in which the chicks forage largely on their own. Chicks of willow grouse (*Lagopus lagopus lagopus*) grow more rapidly in warm, dry summers; in cold, wet years the growth is not only slower but is especially retarded in larger broods (Fig. 5.7). Larger broods are covered for more prolonged periods by hens and thus have less time for foraging (Erikstad and Andersen, 1983).

Presumably energy drawn from growth is put into thermoregulation and to the rapid development of feathers. These young have a high consumption of insects and other animal food. Young of both puddle and diving ducks, for example, initially subsist largely on midge larvae and other insects.

One of the difficulties in some comparisons is the lack of a distinct dichotomy between the altricial and the precocial states. Species like gulls and hawks are semialtricial, with a dense covering of down, but cannot walk for some time after hatching. Saltatory ontogeny, as discussed by Balon (1984) for fish, also obtains for birds. Both flight and thermoregulation require an adequate covering of feathers as well as flight feathers, and these aspects develop at an early age in precocial species.

5.4 MAMMALIAN SEXUAL MATURITY

In some small mammals, such as voles and other small rodents, there is ovarian activity prior to the opening of the vulva, the female genital orifice. This occurrence is not uncommon in late winter, shortly before the onset of overt reproductive activity, and may possibly indicate the simultaneous influence of both photoperiod and food. In prepubertal rhesus monkeys, ovulation and mensus result from hourly infusions of gonadotropin-releasing hormone (GnRH), and the infantile condition returns when treatment ceases (Wildt et al., 1980).

Long before an individual attains its adult mass, important changes occur in secondary sexual features as well as sexual behavior. The rates of these changes are sometimes clearly adaptive; other times the significance of the rates of change observed in the laboratory is equivocal.

5.4.1 Density-Dependent Growth

In many species growth accelerates following the depletion of populations. This phenomenon is most clearly apparent in species that have been captured over extended periods for commercial use.

In the southern elephant seal (*Mirounga leonina*), growth is slower in unhunted populations than in those that have a long history of exploitation (Carrick et al., 1962). The harp seal (*Pagophilus groenlandicus*) in the northwest Atlantic Ocean experiences a more rapid growth and attainment of sexual maturity in populations reduced by commercial sealing (Sergeant, 1973). In the northern fur seal (*Callorhinus ursinus*) of Alaska, growth and sexual maturity have accelerated under the population-reducing pressure of hunting (Sergeant, 1973). The large whales, which have been drastically reduced in number, today grow more rapidly and reproduce at an earlier age, although at the same body mass, than they did 50 years ago (Gambell, 1973).

Growth rate in the African elephant (*Loxodonta africana*) is inversely proportional to population density. The age of sexual maturity is from 8 to 30 years, perhaps the most variable of any vertebrate (Laws, 1969).

5.4.2 Social Influences

Prior to the attainment of sexual maturity, the development of the young individual is modified by the presence and odors of conspecific adults. Most behavioral factors are clearly either stimulatory or suppressive. Stimulatory factors come from sexually active individuals of either the same or the opposite sex, whereas most suppressive factors emanate from adults of the same sex. This situation provides a potential means by which population structure and density can influence reproductivity.

Odors can play two functional roles: (1) *primer* pheromones affect changes or a succession of changes in developmental rates, such as accelerating or retarding gonadal growth, and (2) *signaling*, or *releaser*, pheromones elicit specific and reversible behavioral responses. Primer pheromones of mice have several distinct effects. The presence of a male accelerates the estrous cycle in female house mice, and females kept together in a group without a male tend to become anestrous; in the case of immature females, the first estrus is delayed. Receptors for male mouse pheromones appear to be located in the vomeronasal organ (Lomas and Keverne, 1982).

The specific roles of primers have been discovered and studied in laboratories but are not well known in nature. Densities of mice in cages far exceed those encountered in the field, and concentrations of odors must be much greater in colonies of captives. The laboratory researcher can manipulate the sex ratios of his or her animals to a degree perhaps never encountered in nature. Also, although one can study the role of castrated males in the laboratory, such an individual will not occur in nature.

Most research on behavioral influences of sexual development is based upon observations of laboratory stocks of the house mouse (*Mus musculus*), and the action of these responses in the field is somewhat equivocal. It is, nevertheless, not unreasonable to speculate on the possible significance of primers in nature, for some natural events would tend to simulate, in a diluted way, the effects seen in captive populations.

5.4.2.1 Stimulatory Effects. Many odors are contained in the urine, and some of their effects are seen in urine removed directly from the urinary bladder. The active ingredient is clearly material in the plasma. This finding does not preclude possible additional odors from such structures as the preputial gland. In certain rodent species odors from the urine of intact males accelerate sexual maturity in females and shorten the ovulatory cycle. When groups of 10 anestrous prairie deer mice are kept together without a male, they show a peak of estrus three days after the introduction of a male, and estrus is accelerated simply from the exposure to urine of an intact male (Bronson and Marsden, 1964). There is a protein, androgen-dependent compondent in male urine that accelerates sexual maturation in female mice (Vandenbergh et al., 1975). This effect of odor of male urine, both on acceleration of female estrous cycles and sexual maturity, is called the Whitten effect. It is caused by airborne odors from male urine and is seen only from the urine of an intact (not castrated) male (Whitten, 1956). The manifestations of the Whitten effect are assumed to travel through an olfactory-hypothalamic-hypophyseal tract, with the release of a gonadotropin being a critical point (Bronson, 1976). The stimulatory effect of the male on sexual maturity and estrus in female mice is produced through an increase in the secretion of follicle-stimulating hormone (FSH) (Bruce, 1966).

The odor that stimulates estrus in mice is specific: urine from male house mice (*Mus musculus*) does not induce estrus in anestrous prairie deer mice (*Peromyscus maniculatus bairdii*), nor does odor from male deer mice affect female house mice (Bronson and Dezell, 1968). Odor from male prairie deer mice, however, does hasten sexual maturity in females of the same species, the effect being confirmed by a greater number of corpora lutea in contrast to that in control females, which had been exposed to urine from castrated males (Teague and Bradley, 1978). The Whitten effect has also been noted in the cactus mouse (*Peromyscus eremicus*) of the Mohave Desert (Skryja, 1978) and in the hopping mouse (*Notomys alexis*), a native murid of central Australia (Breed, 1975).

The pheromonal effect of male house mouse urine on sexual maturity of female mice declines rapidly after castration. Twenty-four hours after the administration of testosterone, the pheromonal effect is restored, and the effect is in proportion to the amount of exogenous androgen administered (Lombardi et al., 1976). The stimulatory effect of the odor of urine from male mice (*Mus musculus*) increases with the sexual experience of the male (Drickamer, 1981).

The same phenomenon occurs in laboratory rats. Young rats reared with an adult male from weaning become sexually mature, mate, and conceive from five to nine days earlier than when reared in isolation from males. Females reared with males experience first estrus in 36.5 ± 0.83 to 39.8 ± 1.04 days, whereas first estrus in isolated females occurs in 45.1 ± 1.62 to 46.9 ± 0.69 days. There was no difference in overall body size between the two groups (Vandenbergh, 1975).

Odor from urine of a pregnant and lactating house mouse prolongs estrus in adult females. Sexual maturity of female wild house mice is accelerated by exposure to the odor of urine from pregnant or lactating adult females. Control females (exposed to the odor of plain water) experienced first estrus at 58.9 ± 1.6 days in contrast to the early initial estrus of mice exposed to urine from gravid females

(45.3 ± 1.9 days) and from lactating females (48.3 ± 1.6 days) (Drickamer, 1986).

An effect, however, is seen in young females (before day 29) exposed to urine odors from a single adult estrous female, such odors accelerating the time of first estrus. This effect is seen in summer, but not in winter, and is greater in urine collected in the middle or near the end of the dark part of the daily photocycle (Drickamer, 1986).

5.4.2.2 Suppressive Effects. The opposite of the Whitten effect occurs when female mice are housed together in the absence of a male. When house mice are kept in groups of four females and no male, estrous cycles tend to become prolonged, because of an extended diestrus (van der Lee and Boot, 1955). This early suppressive effect is apparently caused by an odor in female urine; it is lost by ovariectomy and restored by estradiol implants (Clee et al., 1975).

A maturation-delaying pheromone—an odor that delays sexual maturity—is increasingly potent at greater population densities in the laboratory and quite possibly in the field. This pheromone does not need support from ovarian hormones and is clearly different from the pheromone that affects estrous cycles in adult females. The maturation-delaying pheromone depends on one or more hormones from the adrenals (Drickamer, 1981).

The presence of adult females tends to delay or retard sexual maturity of female mice. In manipulated populations of wild house mice, this effect on young females increases with the density of adult females (Coppola and Vandenbergh, 1987). Experimental removal of adult females in a population of *Clethrionomys rutilus* in southern Yukon was followed by rapid sexual maturity and pregnancy in young females, and no pregnancies occurred in young females in a control area (Gilbert et al., 1986). An experimental removal of all small mammals (by snap trapping) at a peak in a vole cycle resulted in a decline in density of *Clethrionomys glareolus* in southeastern Norway. Concurrently, young females born in the spring became sexually mature and pregnant by late June. Young females on the control plot (untrapped) did not breed in their first summer (Bondrup-Nielsen and Ims, 1986). It is not clear whether this is a specific suppressive effect of the presence of adult (overwintered) females on the control plot or a density-dependent effect.

There is a comparable suppressive effect of adult males on immature males. In the prairie deer mouse this effect is at least partly olfactory. Urine from adult males applied to the nares of young (weanling) males slows development of the seminal vesicles, which does not occur after olfactory bulbectomy (Lawton and Whitsett, 1979).

Obviously much of the research on pheromones has been done with mice. It is not surprising that similar intersexual reactions regulate sexual development in fish, which have a finely developed olfactory sense. Among captive swordtails (*Xiphophorus variegatus*), adult males retard maturation of smaller males, a phenomenon also seen in wild populations. In this way a preponderance of adult males suppresses the increase in the number of adults joining the population (Borowsky, 1978).

5.5 ADAPTIVE GROWTH RATES IN MAMMALS

Relative growth rates in mammals approximately follow the altricial-precocial dichotomy seen for birds, but the pattern is complicated by a diversity of life cycles, including the relative scarcity of migration in mammals as well as the frequency of hibernation in small mammals of midlatitudes.

In some high-latitude mammals, growth is necessarily very rapid, because the young must achieve independence at an early age. This is especially true in the many polar species that migrate from their place of birth in the winter. The harp seal (*Pagophilus groenlandicus*) in the northwest Atlantic Ocean has a body mass of 10.8 kg at birth and a daily gain of about 2.5 kg during a nine-day period of nursing. Of this daily increase, 1.9 kg is deposited in the sculp (skin with attached fat), with the balance going into core growth. After weaning, total body mass ceases to increase and begins to decrease after day 18 (Stewart and Lavigne, 1980). During lactation, the infant elephant seal (*Mirounga anguistrostris*) increases its natal mass by 10.4% daily, while the mother sustains a daily loss of 4.2% per day. Throughout lactation, the mean mass increase of the infant is 55% of the mother's loss. The Weddell seal (*Leptonychotes weddelli*) of the Antarctic weighs some 29 kg at birth and, on a diet of fat-rich milk alone, doubles its weight in two weeks (Lindsey, 1937).

5.5.1 Seasonal Influences

In many species of mammals sexual maturity varies with the season of birth, and in some voles sexual activity may either commence at several weeks of age or be delayed for six months or more. These differences appear to be mediated by both food and day length.

Seasonal growth may reflect the effects of reproductive hormones. Gonadal steroids seasonally induce the growth of epidermal structures of sexual significance: antlers, nuptual tubercles, display plumage, and sexual pigmentation. The seasonal changes in mass of vertebrate gonads result from both cell division and cell growth. The seasonal development of secondary sexual characteristics, such as antlers of a deer or nuptual tubercles of a minnow, also constitute cell growth. Exogenous factors influence both the rate and the degree of growth. Extended light results in gonadal growth in many vertebrates, and food is needed for growth of any animal.

Females of species of *Clethrionomys* are territorial and especially sensitive to crowding; possession of a suitable territory is a prerequisite to reproduction. During a low year in a cycle of *C. glareolus*, when there may be a surplus of breeding territories, young females mature in their first summer. In years of population increase, densities suppress sexual maturation of the young until the following year (Wiger, 1982). In addition, long days (16L:8D) accelerate ovarian and pituitary growth as well as increase body mass in young female voles (*Microtus agrestis*), in contrast to voles kept on short days (8L:16D). The presence of an intact adult male complicates the situation by increasing the long-day influence on ovarian and pituitary growth but not on body mass (Spears and Clarke, 1986).

A very common pattern of sexual maturity in wild rodents and lagomorphs is early reproduction in individuals born in the spring, but for those born in late summer sexual activity is delayed until the following spring. The cause or causes of this pattern remain unknown. Both prompt sexual maturity in the spring and dormant sexuality in summer could result from both photoperiod and diet.

In England females of the wild rabbit (*Oryctolagus cuniculus*) born early in the year mate in the summer, whereas females born in the summer do not attain sexual maturity until the following year (Brambell, 1944). Similarly, in the hare (*Lepus europaeus*) young males born before May become sexually mature in the year of their birth, but those born from May to July mature the following year (Lincoln and MacKinnon, 1976).

In Japan the same phenomenon is known for the red-backed vole (*Clethrionomys rufocanus*) in Hokkaido (Maeda, 1963) and for the wood mouse (*Apodemus speciosus*) in Honshu (Tatsukawa and Murakami, 1976). Reviewing studies on wild populations of the bank vole (*Clethrionomys glareolus*), Alibhai and Gipps (1985) noted that young born early in the breeding season breed that same season, but individuals born in summer become sexually mature the following year. In the South African gerbil (*Tatera brantsi*), females born in autumn (before the beginning of June) begin to ovulate at about three months of age, whereas those born from June to October become sexually mature in the following January, at about eight months of age. Why young females do not breed from October until the end of January is not known (Measroch, 1955).

Among those larger mammals that mate in autumn and give birth in late winter or spring, sexual maturity and initial breeding may occur at some 16 to 18 months of age. Some deer (Cervidae), for example, are known to be stimulated by short day lengths and mate in autumn. Why female fawns may not breed in the stimulatory short days in the year of their birth is not clear. Female fawns of the black-tailed deer (*Odocoileus hemionus columbianus*) do not breed in their first year in nature but may do so in captivity. The basis for this difference is not clear, inasmuch as precocious captive fawns may approximate in mass wild fawns of the same age. A high-protein diet of the captives might account for the observed difference in age of first reproduction (Mueller and Sadleir, 1979).

5.5.2 Variations with Food

In birds growth rate varies both from time to time and from year to year within a species and is most closely related to food supply. Growth of young of the long-tailed jaeger or skua (*Stercorarius longicaudus*) varies with the abundance of microtine rodents on which the adults prey (Andersson, 1976). In years of prey abundance early growth is very rapid (Fig. 5.8).

5.5.3 Adaptations to Other Life History Aspects

Many vertebrates have one or more aspects of their life cycle that either allow a slow rate of growth or require a rapid rate of growth, and these diverse strategies tend to be genetically programmed.

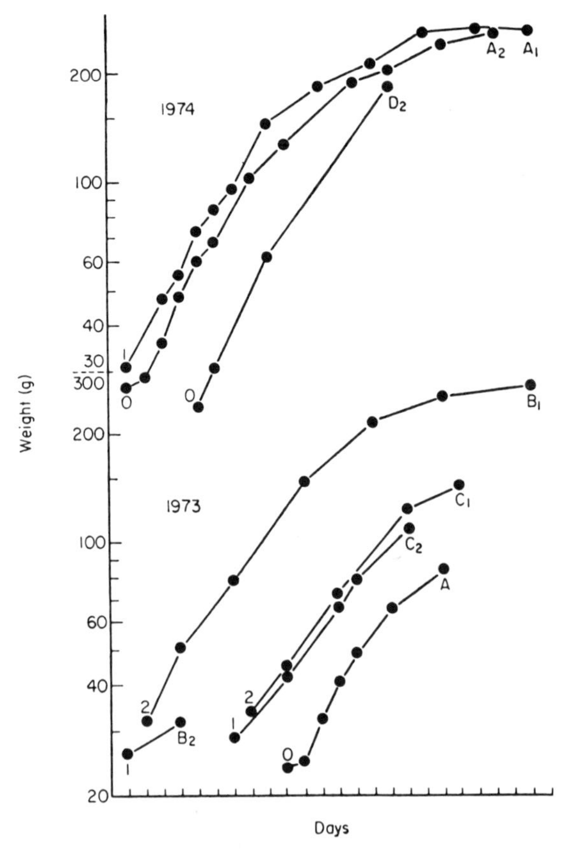

Figure 5.8 Differential growth rates of chicks of the long-tailed skua in 1973 (bottom) and 1974 (top). There was a greater density of rodents in 1974. Data in 1973 from five chicks from three pairs (A, B, and C). Data in 1974 from three chicks from two pairs (A and D). Figures at the bottom of each curve indicate chick age at first weighing. (From Andersen, 1976, courtesy of the *Journal of Animal Ecology*.)

5.5.3.1 Hibernation. Growth rates among hibernating species are more rapid than those seen in nonhibernators of comparable size. Rapid growth tends to compensate for the shorter time available for growth. This is clearly seen in species of chipmunks (*Tamias* spp.): *T. palmeri*, a species with a long period of hibernation, has a more rapid growth rate than species with briefer durations of hibernation (Hirschfeld and Bradley, 1977). Ground squirrels (*Spermophilus* spp.) exhibit the same relationship (Pengelley, 1966). Within a group of four species of western North American ground squirrels are three that hibernate (*Spermophilus lateralis*, *s. tereticaudus*, *S. mohavensis*) and one that remains active all winter (*Ammospermophilus leucurus*); the three hibernators display much more rapid postnatal growth (Fig. 5.9).

Growth rate in the golden-mantled ground squirrel (*Spermophilus lateralis*) in the laboratory is more rapid in animals from areas with shorter annual snow-free summers. In two populations from the same elevation (2650 m), the growth rate is faster in young born of females captured in an area where there is seven months of complete snow cover than in young from females where there is only intermittent snow cover for five and a half months (Fig. 5.10).

5.5.3.2 Geographic Variations. Gestation in two species of tropical African squirrels (*Paraxerus palliatus* and *Funisciurus congicus*) is substantially longer than gestation in midlatitude tree squirrels. Litter size, moreover, is smaller and natal weights are greater in the tropical species than in tree squirrels of the temperate climates. The total weight of a litter is about the same as for midlatitude tree squirrels of comparable size, and the postnatal growth is more rapid than in midlatitude tree squirrels. These aspects of developmental rates, litter size, and gestation are believed to be adjustments to an extended reproductive season (which includes a postpartum estrus), all of which take advantage of an extended period of availability of food (Viljoen and Du Toit, 1985).

The tundra hare (*Lepus othus*) has a more rapid growth rate than those species in more temperate regions. It reaches its adult size (hind foot length and body mass) in about 100 to 105 days (Fig. 5.11). Although the duration of growth is about the same as in more southerly species of hares, *L. othus* attains about twice their size (Anderson and Lent, 1977).

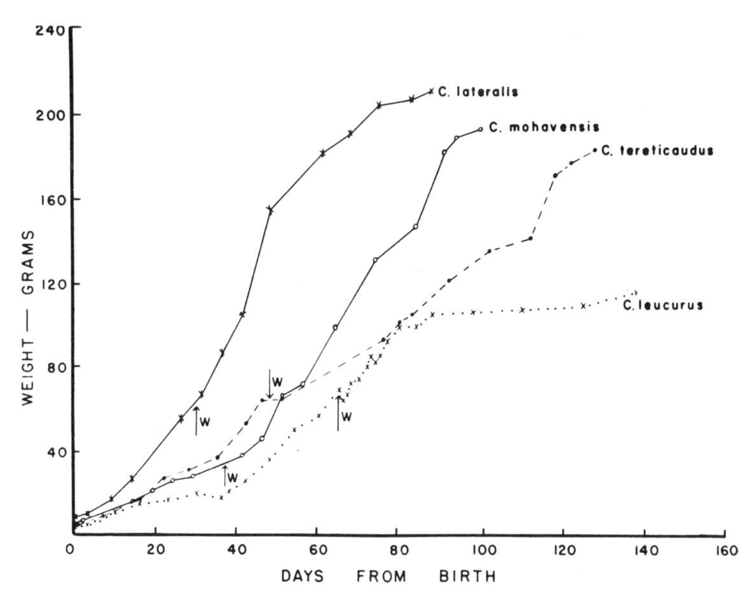

Figure 5.9 Growth rates and weaning times (W) of four species of ground squirrels (genus *Spermophilus* = *Citellus*): *S. lateralis* is a high montane hibernator; *S. mohavensis* and *S. tereticaudus* are desert-dwelling hibernators; *S. leucurus* is a nonhibernating species of the deserts. (From Pengelley, 1966, courtesy of *Growth*.)

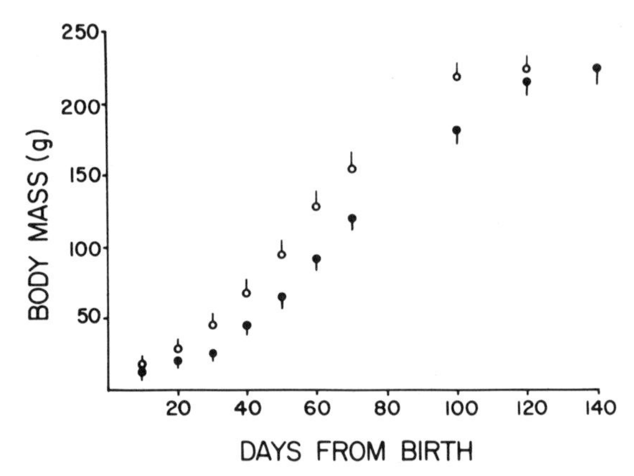

Figure 5.10 Growth rates of captive-born golden-mantled ground squirrels (*Spermophilus lateralis*) from two contrasting habitats at the same elevation in California: solid circles from Bodie (N = 143), a relatively snow-free region; open circles from Castle Peak (N = 59), a region of deep snow. Means ± 2SE. (From Phillips, 1981, courtesy of the *Canadian Journal of Zoology*.)

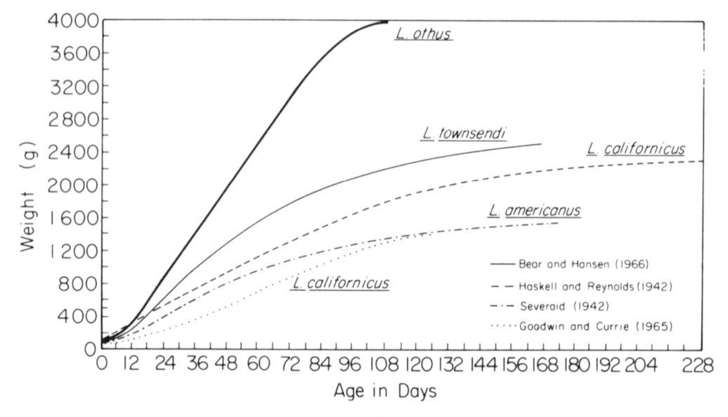

Figure 5.11 Growth of the tundra hare (*Lepus othus*) in contrast to growth of three other species of hares at lower latitudes. (From Anderson and Lent, 1977, courtesy of the *Journal of Mammalogy*.)

5.6 SUMMARY

Vertebrate growth is adjusted to major features of a species' life cycle such as migration, hibernation, and severe changes in seasonal weather. In many vertebrates there is variation in the degree of development prior to hatching or birth. Some species, called *altricial*, are either larval in form at first (as are many fish and amphibians), or relatively weak and defenseless (as are the young of many birds and mammals). The altricial young of birds and mammals are unable to walk and cannot maintain a body temperature much above the ambient. They need to be fed and brooded by one or both parents. In contrast, *precocial* young may be hatched or born in a relatively advanced state, and this allows them to move about and feed for themselves.

The social environment of an infant may influence the speed of growth. Overcrowding usually suppresses growth rates. The sex and reproductive activity of one or more adults may affect the sexual development of the young which are close to it. This is best known in rodents, but seems to occur in some fish as well. The influence is airborne, and comes from urine. Generally a sexually active male accelerates sexual maturity and the time of first estrus in a young female, whereas sexual development is retarded in young females when they are kept only with adult females.

Growth may be more rapid in hibernating mammals than in related species which remain active all winter. Such differential growth rates appear to be genetically determined. Generally slow growth results from less than optimal amounts of food. In fish, growth rates are closely correlated with both food and ambient temperature. Short-term weather changes and accompanying fluctuations in food and temperature account for year-to-year changes in growth rates.

BIBLIOGRAPHY

Blaxter JHS (Ed) (1974). *The Early Life History of Fish*. Springer-Verlag, Berlin, Heidelberg, New York.

Bonner JT (Ed) (1982). *Evolution and Development*. Springer-Verlag, Berlin, Heidelberg, New York.

Case TJ (1978). On the evolution and adaptive significance of postnatal growth rates in the terrestrial vertebrates. *Quarterly Review of Biology*, **53**: 243–282.

Goss RJ (1964). *Adaptive Growth*. Logos, London.

Goss RJ (1974). Aging versus growth. *Perspectives in Biology and Medicine*, **17**: 485–494.

Goss RJ (1978). *The Physiology of Growth*. Academic, New York.

O'Conner RJ (1984). *The Growth and Development of Birds*. Wiley, Chichester, New York.

Ricklefs RE (1969). Preliminary models for growth rates of altricial birds. *Ecology*, **50**: 1031–1039.

Ricklefs RE (1973). Patterns of growth in birds. II. Growth rate and mode of development. *Ibis*, **115**: 177–201.

Ricklefs RE (1979). Adaptive constraint and compromise in avian postnatal development. *Biological Review*, **54**: 269–290.

Summerfelt RC and Hall GE (Eds) (1987). *Age and Growth of Fish*. Iowa State Univ, Ames.

Vandenbergh JG (1987). Regulation of puberty and its consequences on population dynamics of mice. *American Zoologist*, **27**: 891–898.

PART II

PARENTAL INVESTMENT

Much reproductive behavior involves highly specific and genetically programmed intraspecific interactions. Courtship is frequently a mutual affair, with signals from one sex eliciting appropriate responses from the other sex, all of which serve to establish specific and sexual identity and promote the synchronization of gamete release. Exchanges of signals commence at a very early stage, starting with those between the spermatozoa and the oviduct as well as between the ovum and the uterus. There are messages, mostly hormonal and nutritional, between the embryo and its mother, each stimulating responses in the other. Parental care is also a product of interaction between individuals, for while some care behaviors result from hormonal changes within the parent, others are elicited by the young. This can be observed in many vertebrates but is very obvious in lactation.

Individual interaction is studied by both behaviorists and psychologists. Although both groups have similar interests and goals, their methods are sometimes different, and the student should not favor one segment of research to the exclusion of the other.

6

COURTSHIP AND MATING

6.1 MATING SYSTEMS

Vertebrates have developed a surprising array of mating arrangements considering that there are only two sexes. Diversity in the nature of environmental requirements may account in part for varying types of mating assemblages. Some patterns have clear adaptive significance, but the advantages of others remain obscure. Whatever mating patterns obtain in any given species, they constitute competition for mates but may also have become adapted so as to maximize individual and group survival.

Several male and female combinations are possible. Presumably aggregations with several or many individuals of both sexes, such as seen in fish with external fertilization, represent an ancestral condition. Gametes are released into the water over the spawning period so that the union of sperm and ova is random, without regard to the individuals releasing them and without any apparent competition for mates. Refinements of such a generalized reproductive swarm have undoubtedly occurred many times, with a reduction of one sex or the other, or both, so that there is one system or another suitable for any situation that will support life.

6.1.1 Monogamy

The system of one male associating with and remaining with one female for producing and rearing their young is familiar to all Judeo-Christian societies, as are departures from this pattern. Among wild vertebrates monogamous matings are most typical among birds, but even with birds monogamy is sometimes more apparent than real. Monogamy, simple though it may appear, covers a surprising number of different mating arrangements. The monogamous relationship involves a one male/one female relationship. In fish this relationship may be a pair that remains together after spawning for a period of biparental care of the young, or it may simply be two individuals that meet, perhaps repeatedly, to spawn and fertilize eggs without care (Barlow, 1986).

The reader should be forewarned, however, that different authors refer to many sorts of relationships as monogamous (Wickler and Seibt, 1983). Social and reproductive pairs are not necessarily the same. Nonmating and nonreproductive pairs do exist. Many polygynous arrangements may well be monogamous for the female. This at first seems obvious, but it becomes difficult to separate from monogamy which is short lived, and a succession of mates, which constitutes serial monogamy or polygyny (Wittenberger, 1979). The evolutionary biologist must face the hypotheses of fitness and mate choice. Gowaty (1981) further includes within monogamy a system in which males and females occur in equal numbers, with no implication of choice or duration. Such systems may include "pairs" that exercise a substantial amount of infidelity.

Wickler and Seibt (1983) point out that to some extent the purpose determines the concept of monogamy: the sociologist's unit differs from that of the evolutionary biologist, and both are very different from the strictly mathematical concept of the population geneticist.

6.1.2 Lek Systems

The most frequent departure from a monogamous system is seen in a reduction of reproductively active males. This usually results in some competition among males, so that some participate in mating and others do not.

In birds, and sometimes in other vertebrates, males assemble in a concentration to which females come to mate. This arrangement is called a lek system. Breeding males occupy a central position and are surrounded by satellite males, which are usually smaller and younger than the breeding males. Leks must be sufficiently close together so that at least one is available to all the neighboring females. As females enter the lek, they pass by the satellite males, but most of the mating is with the centrally located larger males. Any parental care is provided by only one parent. This system is well known in some kinds of grouse (Tetraonidae) and in snipe but is also seen among fish and amphibians.

In grouse a female probably mates once a season and stores sperm long enough to provide for fertilization for her entire clutch. Her nest is some distance from the lek.

An important aspect of mate choice in lek systems is the size of the lek and number of leks within the home range of a given female, which will determine the number of leks she will visit (Bradbury and Gibson, 1983). In small leks the hierarchy is rigid, and female choice is greatly restricted. Given the possibility of visiting two or more leks, the female then has the option of selecting among leks or among males within leks. It is not clear that one option operates to the exclusion of the other, and among different species, both options probably exist.

The Uganda cob (*Adenota kob*) mate on a breeding ground within which bulls vigorously maintain contiguous territories. Strenuous combat is required to preserve a position on the breeding ground, and there is a constant turover of mature bulls as they move in and out. Not only do males not remain long on the breeding ground but they have a mean life span of some 7½ years in contrast to 12½ years for females (Buechner et al., 1966).

6.1.3 Polygyny

Many situations seem to have favored mating aggregations in which one male mates with several or many females. Sometimes favorable territory is scarce and is dominated by one male. Having been drawn to a favorable, isolated habitat, the females mate with the resident male. Polygyny can be achieved and preserved by (1) engaging in male-to-male combat, (2) defending a resource sought by females, and (3) defending females (Emlen and Oring, 1977). These categories need not be mutually exclusive and are at times ambiguous and difficult to determine, but they remain a useful starting point for investigating polygyny (Ostfeld, 1987).

Perhaps a genuinely polygynous system developed from the lek. Polygyny is probably found among all groups of vertebrates but has been most carefully studied in some fish, birds, and mammals. When these groups of polygynous gatherings are studied with care, they clearly do not compare with the conventional concept of one male/several-to-many females as a reproductive group.

Sea lions (Otariidae) usually assemble in large groups prior to parturition, and cows greatly outnumber bulls. The bulls, however, do not control a specific group of cows but rather defend a territory, and cows move freely through territories of different bulls (Bartholomew and Hoel, 1953). Females of the northern fur seal (*Callorhinus ursinus*) aggregate without regard to the distribution of males and in places not used by bulls for displays and fights. Such loose, temporary aggregations of females are not really dominated by a single bull and are not, strictly speaking, "harems" (Peterson, 1968). Sea lions appear to be loyal to the calving site, to which the cows are prone to return year after year.

It is easy to overemphasize the role of the male in infusing his genotype into the populations, for any given male may not remain long in a position of dominance. In the northern fur seal a bull may hold a territory for a mean of 1.5 breeding seasons, in contrast to a cow who may breed for 5 to 6 seasons (Peterson, 1968).

Polygyny is characteristic of many birds. Commonly the males are distinctively colored and may possess enlarged plumes that are displayed during courtship, such as in many gallinaceous birds. Perhaps the ultimate is reached in the peafowl. Polygyny, however, is not confined to such species, and plural mating with males is rather frequent in species that are not dichromatic and that lack specialized plumes in the male. The socially parasitic indigo birds (*Vidua* spp.) are polygynous. After a male assumes a singing perch, females come successively to mate, up to four females a day (Payne, 1973b).

The male long-billed marsh wren (*Telmatodytes palustris*) may have two mates when the territory is productive enough to allow him to provide food for two broods simultaneously and when he is vigorous enough to keep out intruding males (Verner, 1964). Similarly, the bobolink (*Dolichonys oryzivorus*) may take care of two females and their broods in areas where the food (especially caterpillars and grasshoppers) is abundant (Wittenberger, 1980). The lark bunting (*Calamospiza menalocorys*), a sparrow of central North America, also may be polygynous in habitats where nests and nestlings can be provided with adequate protection from the danger of excess solar radiation (Pleszczynska, 1978). It would appear that these species are facultatively polygynous. The dunnock (*Prunella modularis*), discussed in Section 6.10.1, appears to be facultatively polygynous or polyandrous.

Males of the yellow-rumped cacique (*Cacicus cela*), a Neotropical oriole, court at specific sites. Mating follows at the female's nest site until incubation begins, precluding mating with another male during the female's fertile period. Subsequently, the male returns to his singing site and seeks another female (Robinson, 1986).

The marsh harrier (*Circus aeruginosus*) appears to be monogamous, but a male may mate with and provide food for two females. Bigamous males succeed in fledging almost twice as many young (5.7) as monogamous males do (3.0), and bigamous males make more deliveries of food than monogamous males do (Altenburg et al., 1982). The hen harrier (*Circus cyaneus*) is prone to polygyny, especially after the males are past their first year. In monogamous nests in Orkney, Scotland, hatching success was greater than in polygynous nests, but males with three or four mates had more offspring than monogamous males did (Balfour and Cadbury, 1979).

Several species of shorebirds have developed a particular system designed to maximize the role of both males and females in parental care. On the west coast of Finland, Temminck's stint (*Calidris temmincki*) is successively bigamous. Each female mates with two males in separate territories, and each male mates with two females in his own territory. Males incubate the first clutch, and females incubate the second (Hilden, 1975). The mountain plover (*Eupoda montana*) of North America exhibits similar successive bigamy, differing only in details (Graul, 1973).

Other polygynous species are not tied to a specific site, and polygyny has obviously developed under a variety of environmental circumstances. In western United States, feral horses (*Equus cabellus*) move about in groups dominated by a single breeding stallion. The zebra (*Equus quagga*) is similarly nomadic (Klingel, 1975). Among African antelope the type of mating pattern depends on the permanency of the food supply and the consequent need for one or both sexes to migrate in search of forage. In an area where food is ample through the year, a permanent family group may remain within a single territory. When the forage fluctuates seasonally and both sexes are nomadic, males defend estrous females, but such pair bonds are transitory (Jarman, 1974).

In genuinely aboriginal societies, monogamy is more the exception than the rule: among 185 primitive societies polygynous arrangements exist in more than 85% (Ford and Beach, 1952). The Ono, an aboriginal tribe of Tierra del Fuego, were known to practice a special sort of polygyny that not only staggered generations but increased genetic diversity within the family. In wife-hunting excursions, a marauding band, having located a small group, killed the men and took the women as their wives. A pregnant woman was considered a prize, for if she had a baby girl, the daughter would become a second, but quite unrelated, wife (Bridges, 1951).

6.1.4 Polyandry

Polyandry is known to be common or prevalent in a number of species of birds but, except for human societies, may not occur in mammals. The tropical and subtropical jacanas (Jacanidae) have long been known to be polyandrous. A female mates with a succession of males, and each male sits on a clutch of eggs and rears the young. Each male defends his territory against other males, with assistance from the female (Jenni and Collier, 1972).

Mating in the gray whale (*Escrictius robustus*) has been reported to involve two males and one female, but the relationship of such triads is not clear. In one instance the everted and erect or semierect penis was apparent in each male (Wilson and Behrens, 1982). Cetaceans are among the few mammals to approach heterosexual play venter to venter, and although details of coition were not apparent from a small boat, the orientation of the animals was clear from the position of their flippers.

6.2 THE FUNCTIONS OF COURTSHIP

As food, photoperiod, precipitation, and temperature prepare the individual for reproduction, intersexual communication is necessary to effect the release and union of the mature gametes. This section introduces a few examples to illustrate the mechanisms by which courtship cues facilitate mating and fertilization. It also reviews some hypotheses that may account for observed differences in mating patterns.

6.2.1 Identification of Sex and Species

The two sexes must associate at the proper time, which requires specific and sexual identification and a means of communicating reproductive readiness. Once these conditions have been established, courtship follows. Species that form long-term pair bonds may exhibit a sequence of involved, ritualistic courtship events. The more complex the sequence, the greater is the need for synchrony at the outset. In some vertebrate groups intersexual stimulation continues after mating, and if the sequence does not proceed in the proper order, the reproductive cycle is broken. This phenomenon is seen in the complex, but rigid, breeding cycles of many birds that must defend territories, court, build a nest, mate, lay eggs, incubate, feed, and fledge young.

Significantly different courtship patterns are seen in separated populations of the widespread musk shrew (*Suncus murinus*). In a laboratory study of courtship and copulatory behavior of specimens from Guam and from Madagascar, females from Guam vigorously repelled initial sexual advances, but females from Madagascar seemed to solicit the male's sexual interest. Also, females from Madagascar assumed the lordosis posture when approached or touched by a male. Lordosis did not occur in females from Guam. Prior to intromission, both populations exhibited conspicuous tail movement: females from Guam moved their tails in a horizontal plane in contrast to a vertical movement in females from Madagascar. Because of these behavioral differences, there were difficulties in achieving intromission in attempts at cross-breeding, and young from such matings were both behaviorally and physically abnormal (Hasler et al., 1977). These two geographic groups, both from islands, are characterized by behavioral differences that are usually typical of separate species.

Courtship may not always prevent hybridization. Offspring from a union of two heterospecific individuals is not rare in the laboratory and also occurs in nature. Many examples are known in the Nearctic freshwater sunfish (Centrarchidae), in the live-bearing Poeciliidae, and among many lizards. Large males of the swordtail *Xiphophorus nigrensis* are usually preferred by the female *X. pigmaeus* over conspecific males. Large males of *X. nigrensis*, in contrast to male *X. pigmaeus*, have a long sword and an elaborate courtship, and the visual stimuli of *X. nigrensis* are chosen by female *X. pigmaeus* (Ryan and Wagner, 1987).

6.2.2 Synchronization of Gonadal Cycles

In many vertebrates activity during the preovulatory period is modified courtship. Considered in the broad sense, courtship includes premating signals sent from each sex to the other, so that courtship is a two-way street. Mating in most vertebrates occurs near the time of ovulation, and the female usually requires the stimulus of a courting male before she is sexually receptive. In turn, the male needs some announcement that the female will be attentive to courtship, some signal to draw him to a near-ovulatory female. These patterns are frequently highly specific, presumably having evolved separately, and may today constitute identification signals.

Surprisingly, unisexual populations of parthenogenetic lizards have courtship. Copulatory behavior has been observed in parthenogenetic species of *Cnemidophorus*. Malelike courtship behavior occurs in the postovulatory period, and female behavior is seen in the vitellogenic phase (Moore et al., 1985).

Not only are gamete production, courtship, and mating synchronized between the sexes, but events are so programmed that sequences follow internal rhythms that are adjusted to environmental changes. Gonadal steroids usually govern the onset of courtship and receptivity, but behavioral stimuli from one mate very quickly influence hormonal levels and behavior in the other. The hypothalamus is a major mediator of both endogenous and exogenous stimuli. A grand array of variables enters into this communication system, so that gonadal growth, courtship, and mating occur at such a time that fertilization will occur and chance for survival of the next generation is maximized.

The nature of fertilization influences the degree of synchrony between the sexes. If fertilization is external, courtship ensures a simultaneous release of sperm and ova. This pattern is prevalent in teleost fish and amphibians, in which mating occurs in the water. In a number of both teleosts and amphibians, fertilization is internal. In these species, as well as in reptiles, birds, and mammals, sperm may be stored in the reproductive system of the female, sometimes for extensive periods. If sperm is stored, ovarian activity may not be concurrent with testicular function.

6.3 TYPES OF SIGNALS

Many signals not only bring potential mates together but convey various kinds of information regarding sexual states. These signals may also stimulate, or sometimes depress, the reproductive condition of the other individual. Much of the data on intersexual communication is based on careful measurements made on laboratory animals, especially rodents, large wild and domestic ungulates, and a few teleost fish.

There is a phylogenetic and environmental significance to the types of signals used. Odors must be the oldest means of communication, and the olfactory tract of fish is extremely large. Olfactory abilities are well developed in vertebrates in general, except for most birds, and much courtship information is airborne.

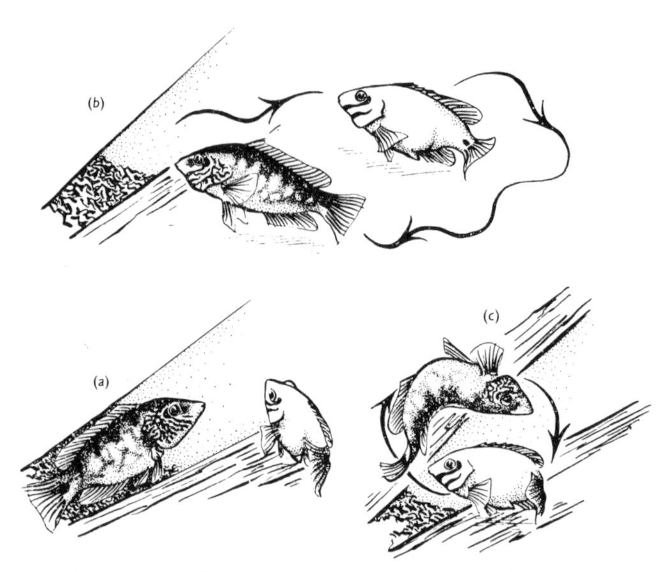

Figure 6.1 (*a*) The male corkwing wrasse (*Crenilabrus melops*) guards and fans eggs laid by a succession of females. (*b*) Following oviposition, the male drives away approaching females from uncovered eggs. (c) After he has covered the last batch of eggs, aggression ceases and courtship ensues. (From Potts, 1974, courtesy of Cambridge University Press.)

Initial male-to-female encounters are sometimes followed by vigorous, apparently hostile, contests, which are similar to male-to-male battles. Walther (1984) suggested that some types of courtship have evolved from, and are extensions of, intermale combat. Such aggressive displays may become courtship displays and vice versa, and a female, in contrast to a competing male, typically responds with various appeasing postures or partial withdrawals. This is illustrated by the marine teleost *Crenilabrus melops*, in which the male guards, separately, each of a sequence of egg clusters laid by a succession of females. The freshly laid eggs are exposed, and the male is aggressive to both males and females. But after he has covered these eggs with algae, aggression toward another approaching female then becomes courtship (Fig. 6.1). After she has deposited her eggs, either she retreats or is chased away by the male (Potts, 1974). These responses by a female mammal amount to courtship just as much as advances by a male: they are mutually stimulating and lead to genital sniffing and eventually to mating. As a male becomes increasingly aroused, the female becomes more receptive.

6.4 VISUAL SIGNALS

Visual signals include both structures, such as plumage of many male birds (and a few female birds) or the colors of fish, and behaviors, such as posturing or the assumption of positions of submission or aggression.

Visual displays are certainly best known for birds, but many teleost fish are sexually dimorphic, and frequently males are distinctively colored. One might imagine that elaborate designs and colors were developed for the sexual arousal of the female, but male-to-male displays are characteristic of some species. The conspicuous extension of the caudal fin of the male swordtail (*Xiphophorus helleri*) serves to identify the sex (Fig. 6.2) and may serve as a stimulatory feature as well.

Visual displays of two sympatric land tortoises (*Geochelone carbonaria* and *G. denticulata*) of South America serve to identify both sex and species on the basis of lateral movement of the head. Smooth movement indicates *G. denticulata*, jerky movement *G. carbonaria*. When two conspecific males meet, the head movement of one elicits the same movement in the other, but males and females from the other species do not respond. If head movements from a given male are not given by the other individual, the challenging male then investigates the cloacal odor and copulates if the other individual is a sexually active female of the same species (Auffenberg, 1966).

Chameleons (Chamaeleontidae) of Africa and India are distinctive not only for their color changes but also for their sometimes bizarre shapes. Males of the chameleon *Chamaeleo hohnelii*, of Kenya, display visually to other males in a rather stylized form of quasicombat. These creatures may occupy a position on a branch for long periods, but a resident male changes color and shape at the approach of an intruding male. A pastel yellow and blue pattern becomes intense in each male, and they expand their throat so as to appear larger. As they advance, each raises his forequar-

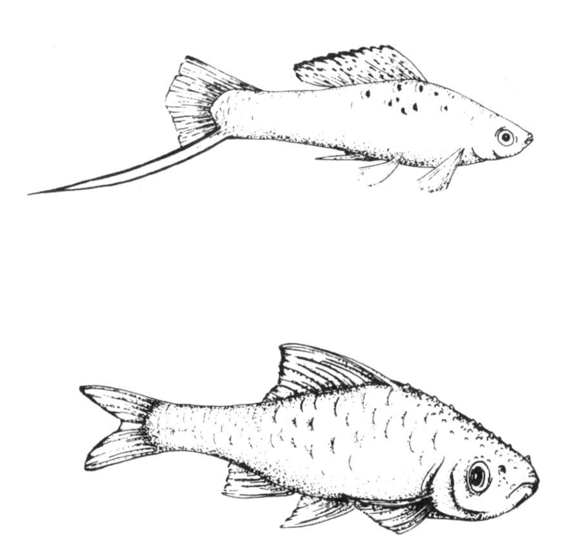

Figure 6.2 *Top*: the extension of the lower lobe of the caudal fin and modification of the anal fin (which is an intromittent organ) distinguish the male swordtail (*Xiphophorus helleri*) from the female. *Bottom*: the male *Notropis venustus*, a minnow (Cyprinidae), is adorned on the head and anterior dorsum with deciduous protrusions called nuptial tubercles or pearl organs, which develop at the time of spawning. (Katherine West.)

Figure 6.3 Courtship displays. *Left*: the inflated throat, or gular pouch, of the lizard genus *Anolis* is accompanied by head movements. *Right*: the tree frog (*Hyla*) moves air from the inflated throat to the lungs and back to produce a call that will attract females. (Katherine West.)

ters, sways from side to side, and attempts, in vain, to bite the flank of the other. The intruding male eventually retreats. If the intruder is a female, she rocks back and forth on one foreleg and the opposite hind leg, apparently a visual cue that she is a potential mate (Bustard, 1965).

The anoles (*Anolis* spp.) of the New World display visually not only by inflating a gular pouch (Fig. 6.3) but also by moving their head. Both serve as specific identifiers as well as courtship maneuvers.

6.5 VOCALIZATIONS

Vocalizations are almost universally employed by vertebrates in intersexual communication. In some groups, such as some rodents, these sounds are ultrasonic—above the range of human perception—and may be more frequent than we realize. When wavelengths lie within the range of human detection, a tyro field zoologist soon learns the distinctive calls and songs of local species. Not only are the sounds characteristic for each species, but different sounds have special meanings within a species. Being controlled by androgens, vocalizations are also seasonal. As with other sorts of signals, vocalizations serve both for specific identification and as a stimulant.

Figure 6.3 (Continued)

Many species of teleost fish create sounds with a broad spectrum of wavelengths and have an excellent sense of sound perception. Some fish noises predominate during the reproductive season and presumably have a role in courtship. However, considering the abundance and diversity of fish and the distinctive sounds that many fish emit, we know relatively little of the behavioral significance of teleost vocalizations (Tavolga et al., 1981).

Also of great potential significance to specific and sexual identification are the electrical currents produced by a number of fish. These electrical "songs" have been recorded and the physiology of their synthesis studied by a number of students, but the role of electricity in the sexual behavior of teleosts is not well understood (Heiligenberg, 1977).

Among the three distinct orders of amphibians, only the anurans are conspicuously vocal. Males of almost all species call from a prospective mating and oviposition site, either in water or on land. The call frequencies are distinctive for each species, and females are drawn to the call of males of the same species. Males move air back and forth between their lungs and air sacs and over their vocal cords (Fig. 6.3).

Reptiles are usually quiet animals and communicate by means other than sound. Geckos and some turtles are notable exceptions.

Bird songs provide a species with a unique vocal identification, and these sounds from one sex, usually the male, serve to elevate gonadotropins in the opposite sex (see Chapter 9, Parental Care). Songs are distinctive not only for a given species but often for an individual as well. Vocalizations also elicit different sorts of behavior, according to their length and modification.

Songs in the wood warbler (*Phylloscopus sibilatrix*) serve both to announce the presence of the male to other males and to attract females. A "long song" advertises

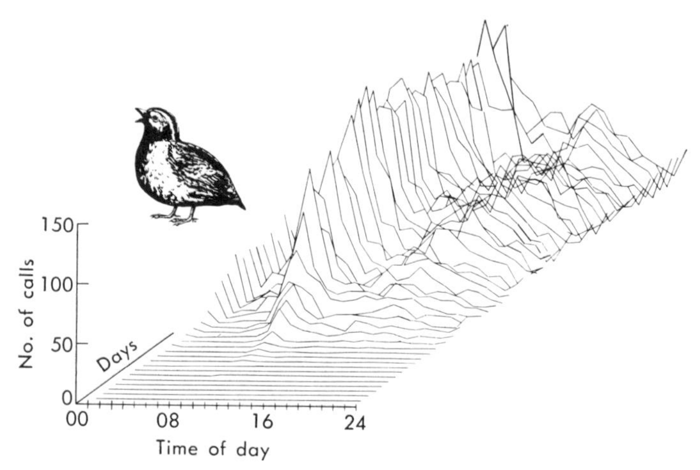

Figure 6.4 The development of crowing in Japanese quail under a 16L:8D photocycle from 22 days of age. Crowing commences at about day 30, with a peak early in the morning and a later smaller peak in the afternoon. (From Wada, 1983, courtesy of the Japan Scientific Societies Press.)

for a female, but vocalization changes to a "short song" following the establishment of a pair bond. Near the periphery of his territory, however, the paired male may sing a long song and attract a second female nesting in an adjacent territory (Temrin, 1986).

Vocalization may be under both exogenous and endogenous control. When growing Japanese quail are kept under long photoperiods (16L:8D), they begin to call at about 30 days of age, at the time circulating androgens begin to increase. Calling ceases after castration but resumes following testosterone replacement by Silastic implants. The pattern of calling, however, is bimodal, with peaks at the first appearance of light and again some eight hours later (Fig. 6.4). Thus calling per se is dependent upon testosterone levels, but there is a circadian rhythmicity to calling in spite of the steady release of testosterone from the implant (Wada, 1983).

As red deer stags collect and defend harems of hinds, stag behavior includes frequent roaring, low vocalizations of about 120 Hz. In captive herds in New Zealand, recorded roars of stags as well as roaring, in addition to pursuit by a vasectomized stag, each advanced the date of estrus of hinds (McComb, 1987).

6.6 PHEROMONES AND CHEMICAL ATTRACTANTS

Olfactory advertising is of special value in solitary species, like the hamster, and also in many mammals that are essentially nocturnal. Although olfactory signals are also important in diurnal mammals, visual cues may prevail in sexual rituals. Among many artiodactyles, which are both diurnal and nocturnal, odors are important in the communication of such matters as territorial boundaries, individual recognition, and alarm, but sexual communication is often visual or vocal. Unlike other signals, odors persist on the substrate, sometimes for very long periods.

Olfactory signals are far more common than we usually suspect, for most animal scents are undetected by humans. Except for birds most vertebrates use odors that transmit messages between sexes and within one sex. This is a world from which we are excluded, but observation and experimentation reveal the importance of olfaction in sexual communication.

6.6.1 Jacobson's Organ or Vomeronasal Organ

Sexually significant odors are frequently carried via nerves from the vomeronasal organ or Jacobson's organ. It first appears in elasmobranchs and Dipnoi, and is in these vertebrates provided with its own ennervation. The nerves go to a part of the brain somewhat separate from the olfactory bulb. This specialized epithelium develops in amphibians and some reptiles (where it is usually called Jacobson's organ) as well as many mammals (where it is known as the vomeronasal organ). It is absent in birds. In amphibians it constitutes a ventral outpocket of the nasal passage; but in reptiles it occurs as an invagination of the roof of the mouth, and in many squamates the tip of the slender forked tongue can be placed in the cavity, and so convey odors to its sensitive epithelium.

In many groups of vertebrates olfactory cues from one individual may stimulate or suppress responses from another, of either the same or opposite sex. In the lungless salamanders (Plethodontidae), a minute ciliated groove (the nasolabial groove) carries fluids from the edge of the upper lip to the nostril and thence to Jacobson's organ. During the courtship season, the nasolabial groove enlarges in some lungless salamanders. In mammals, olfactory cues are overwhelmingly important in communicating sexual information. If mice are rendered anosmic, patterns such as aggression among males and the sexual pursuit of females by males disappear (Ropartz, 1976).

In mammals Jacobson's organ or the vomeronasal organ occurs as an invagination in the roof of the mouth. This olfactory receptor is developed in some insectivores, bats, rodents, and ungulates and perhaps in others. The vomeronasal organ is absent in cetaceans and in Old World primates. In mammals it is fluid filled and is directly connected to the mouth and/or the nasal passages. Some mammals seem able to move fluids in and out of the vomernasal organ. Evidence suggests that it is sensitive to nonvolatile materials, such as those released in urine by many mammals (Ladewig and Hart, 1980).

6.6.2 Teleost Fish

Olfaction is widely employed as a means of communication in fish (Colombo et al., 1982; Stacey, 1983). Many fish live at a depth where there is no light except that produced by fish themselves, and many surface species are primarily nocturnal. Courtship via odors is very clearly documented for the goldfish, *Carassius auratus*, in which a steroid hormone, 17*a*,20*b*-dihydroxy-4-pregnen-3-one (17,20P), induces oocyte maturation. A female goldfish releases 17,20P into water, inducing gonadotropin hormone (GtH) release into the circulation of males within 15 minutes, which, in turn, induces an increase in the volume of milt (Dulka et al., 1987). This hormone is found in other teleosts and may function as a pheromone in many species. Although some courtship in the black bullhead (*Ictalurus melas*) is tactile, there is probably a large olfactory input as well (Wallace, 1967).

6.6.3 Amphibians

Salamanders locate mates primarily through odors, and some of the tactile signals may include chemical elements as well, so that it is difficult to distinguish between the two. When exposed to waters containing odors of either male or female newts (*Notophthalamus viridescens*) in a maze, males prefer to follow pathways with odors of females (Dawley, 1984). In addition, on the basis of odor, males seek out larger females (Verrell, 1985, 1986). In California two newts, *Taricha granulosa* and *T. rivularis*, may occupy the same creek and spawn at the same time. Their courtship maneuvers are similar, but hybrids rarely occur in nature, inasmuch as males are drawn to odors of females of only their own species (Twitty, 1966).

In plethodontid salamanders chemical signals also induce the transfer of the spermatophore. As a final aspect of courtship, the male applies his chin and mental gland to the head of the female so that it contacts her nasolabial grooves and, reversing the routine, places the base of his tail to her chin. The male then deposits a spermatophore, and the female promptly picks up the tip with her cloacal lips (Organ, 1960). Ambystomid salamanders have a somewhat comparable pattern of courtship, which includes similar signals and may be tactile as well as chemical.

Courtship in the Congo eel (*Amphiuma means*), a large salamander of North America, involves a group of females running their snouts over the body of a male and culminates in his depositing a spermatophore directly in the cloaca of one of them (Salthe, 1967).

Because of the great role played by sound in anurans, one would suspect that odors do not contain specific cues in these amphibians. On the other hand, because caecilians are silent and either aquatic or fossorial, odors carry sexual messages for them.

6.6.4 Reptiles

Intraspecific communication in reptiles is frequently olfactory, and the tongue and Jacobson's organ are the receptors. Olfaction is best developed in those species that possess a "forked" tongue, which is a necessary adjunct to Jacobson's organ. This was shown by Noble in 1937 and has been confirmed many times since. For example, sexual identification in the adder (*Viperus berus*) depends upon olfactory signals, and blocking the Jacobson's organ prevents reproductive activity (Andren, 1982). Skinks (Scincidae) have a well-developed olfactory sense and use odor for specific identification. In the laboratory, testosterone-treated male *Eumeces laticeps* expressed greater interest, as measured in frequency and number of tongue flicks, to female odors than to male odors (Cooper and Vitt, 1984).

Pheromones have been studied in garter snakes (*Thamnophis* spp.), in which the sexes have very little color differences. In the Central American *Thamnophis melanogaster*, males court both intact and ovariectomized females when the latter are estrogen treated and released back to the wild. Courting or copulating males were collected and eight days later examined for serum androgens and testicular morphology: both seminiferous tubule diameter and androgens were significantly higher in males that had been exposed to estrogen-treated females than in males that had been collected prior to the release of those females (Garstka and Crews, 1982). The attracting material may be derived from the liver of reproducing (preovulatory) females. Both serum from estrogen-treated females and homogenates of liver from females not treated with estrogen elicited courtship from males when applied to the backs of intact males. The material from the liver homogenate could be vitellogenin, normally released from the liver during yolking (Garstka and Crews, 1981).

Most terrestrial turtles possess a pair of small external glands in the throat area. These glands are called mental glands and are not conspicuous except during the breeding season and especially during courtship. They are found in many species of Emydidae, in the genus *Gopherus* (the sole North American representative of the Testudinidae) and in the Platysternidae (Winokur and Legler, 1975). During courtship of the tortoise *Gopherus polyphemus*, the male everts his mental glands and moves his head in a vertical arc, and females appear to be drawn to an odor that comes from these glands (Auffenberg, 1966).

6.6.5 Birds

Most species of birds appear to have a poorly developed sense of smell, and odors are not known to communicate sexual messages from one individual to another. Nevertheless, birds are a large and diverse group, and a few species are known to have a very good olfactory sense. Therefore one should not rule out the possibility that some species may emit odors of reproductive significance.

6.6.6 Mammals

In mammals scents are frequently carried in urine, and additionally many specialized glands release odors from the skin. Some of these odors circulate freely in the air, and others, less volatile, are applied to various kinds of substrates, such as twigs and rocks. Many mammals urinate freely upon meeting, and female ungulates frequently release large amounts of urine in front of a male. Extensive urination is also characteristic of many hystricomorph rodents, such as porcupines and guinea pigs and their relatives.

6.6.6.1 Proboscideans. In the two living proboscideans males experience a sexually aggressive phase, musth, which is similar to rut in artiodactyls. The African elephant (*Loxodonta africana*) may experience musth at any time of the year, and the duration is individually variable. These periods are recognized by sexual aggression and dominance, a constant dripping of strong-smelling urine, and a distinct green color to the proximal part of the penis (Poole and Moss, 1981). In the Asiatic elephant (*Elephas maximus*), musth is recognized by aggression as well as by temporal swellings, from which there is an obvious discharge (Jainudeen et al., 1972). Females exhibit an interest in the odor from these swellings (Fig. 6.5).

6.6.6.2 Ungulates. When exposed to the odor of a conspecific, especially of the opposite sex, some mammals curl back the upper lips, allowing greater exposure to the anterior palatine foramina, which lead to the vomeronasal organ. This behavior is called *Flehmen* and is expressed by exposure to odors of urine. Its relationship to mating behavior is obscure.

In goats Flehmen seems to move materials into the posterior part of the vomeronasal organ. Sodium fluorescein dye can move from the nasopalatine ducts into the ventral part of the nasal cavity, whether or not there is Flehmen. After Flehmen, however, the dye is usually carried to the posterior part of the vomeronasal organ, where there is sensory epithelium (Ladewig and Hart, 1980). In male goats Flehmen occurs more frequently after exposure to urine of females in diestrus than in estrus, indicating that Flehmen responded to a particular odor and was not the means of detecting it. Flehmen is virtually eliminated by olfactory bulbectomy, showing that the main olfactory system stimulates Flehmen (Ladewig et al., 1980).

Figure 6.5 Female Asian elephant (*left*) sniffing the temporal gland of a male in musth. (From Katherine West, from a photograph by G. McKay in Stoddart, 1980.)

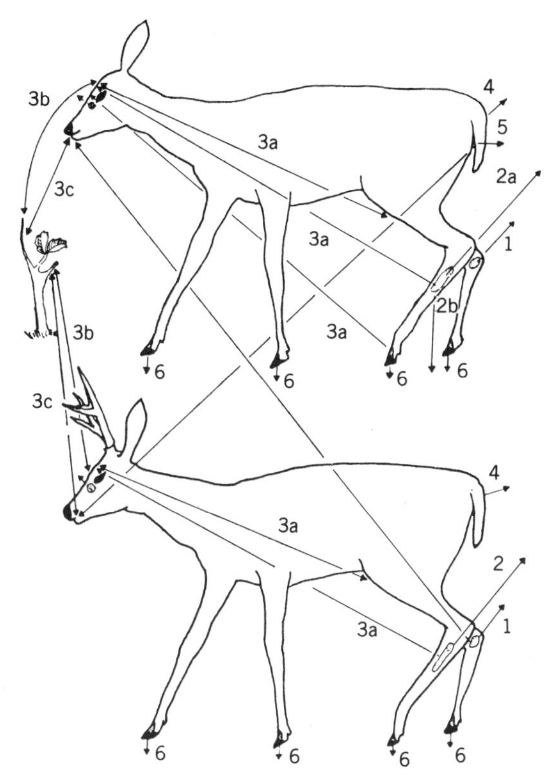

Figure 6.6 Sources of social odors in the black-tailed deer (*Odocoileus hemionus columbianus*). Scents of tarsal origin (1), metatarsal gland (2a), tail (4), and urine (5) are transmitted through the air. While deer are reclining, the metatarsal gland touches the ground (2b). Deer rub their hind legs over the forehead (3a), which is rubbed on dry twigs (3b). Marked twigs may be sniffed and licked (3c). Interdigital glands leave scent on the ground. (From Müller-Schwarze, 1971, courtesy of *Animal Behaviour*.)

6.6.6.2.1 Special Glands. In addition to the distinctive odors of urine, many mammals have dermal glands that may apply directly to special sites in their territories. Many deer have tarsal glands, many antelope have facial scent glands, and some mammals have special odoriferous areas on their arms or belly. On voles and shrews scent glands may be on the flanks. Marking generally serves to identify territorial boundaries and to advertise sexual condition and induce sexual encounters. Thus scents are directed at both sexes.

In addition to marking part of the ground with urine, many artiodactyles urinate slowly with their hind legs held close together so that their inner surfaces become wet. This behavior is known to occur in wild and domestic species and in adults and immatures of both sexes. Presumably the odor so applied serves as individual identification and is not primarily sexual in function.

The black-tailed deer (*Odocoileus hemionus columbianus*) releases distinctive odors from different parts of its body, each odor with its own function (Fig. 6.6).

A tarsal gland scent serves as individual recognition, whereas the metatarsal secretion expresses alarm. The tail and urine release other odors. All of these odors are airborne. In addition, the forehead secretes long-lasting odors, and deer rub their forehead against vegetation to mark territorial boundaries. All of these scents assist in defining the individual and its territory and psychological state (Muller-Schwarze, 1971).

6.6.6.4 *Rodents.* Several well-established physiological responses to olfactory cues occur in mammals and have been most carefully studied in rodents. Such cues have been called *primers*, because they promote a state of reproductive readiness. Primers are frequently carried in urine and depend upon the hormonal state of the individual as well as its social position among its neighbors. The scents may enhance or suppress the endogenous influences produced by hormones (Fig. 6.7). This phenomenon has been most carefully studied in laboratory populations of the house mouse (*Mus musculus*) but is known for other species of small rodents as well. Odors that trigger an immediate response, such as copulation, are called *releasers*.

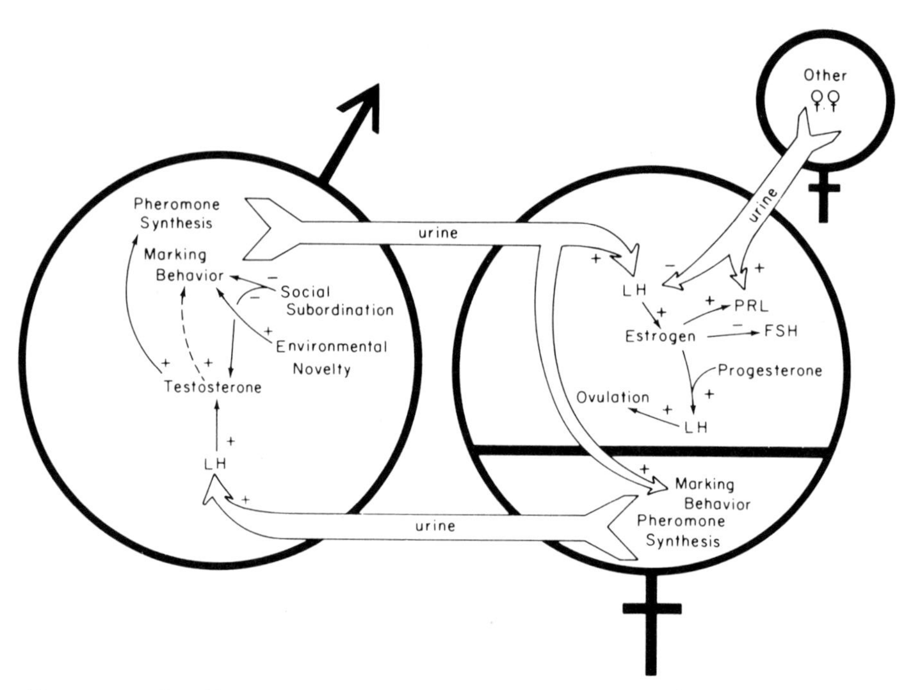

Figure 6.7 The olfactory cueing system of the house mouse (*Mus musculus*), including interactions between adult males and pubertal or adult, nonpregnant females. The primary function of the male's response to female urine is assumed to be an elevation of his pheromonal urinary potency. Solid lines imply day-to-day modulation; dashed lines indicate only chronic support. The separation of hormonal events from marking behavior and pheromonal behavior in females acknowledges the gonadal independence of these latter events. (From Bronson, 1979, courtesy of Elsevier/North-Holland Biomedical Press.)

The odor from the urine of an adult female increases levels of serum testosterone in the male (Maruniak and Bronson, 1976). Presumably this reaction increases the reproductive potential of populations at high densities and so enhances their success in invading new areas (Bronson, 1979). Anovulatory female rats have been found to ovulate when exposed to urine of a male, but this stimulation is reduced if the orifice of the vomeronasal organ is occluded (Johns et al., 1978).

An odor released from the urine of an adult male house mouse (*Mus musculus*) in breeding condition tends to accelerate the sexual maturity of a developing female. This effect is reduced or blocked by litter mates, reducing the likelihood of inbreeding with the male parent or siblings (Bronson, 1979); but this suppression would quickly dissipate after dispersal.

Male hamsters lick vaginal secretions, even when applied to a castrated male. This stimulus is transmitted via the vomeronasal organ (Section 6.6.1), and when its nerve is severed experimentally, males do not proceed to mate (Beauchamp et al., 1976). The stimulus of vaginal secretions in an intact male is followed by a twofold increase in circulating testosterone (Macrides et al., 1974).

6.6.6.5 Primates. Sexual attractants are well known in several primates. The female rhesus monkey (*Macaca mulatta*) produces a male-attracting odor just prior to ovulation (Michael and Bonsall, 1977). Short-chain aliphatic acids, which are rather volatile, are contained in vaginal secretions of at least some primates. However, vaginal secretions contain other constituents as well, and it is not clear which component is the pheromone (Keverne, 1983). Short-chain aliphatic acids obtained from vaginal secretions of rhesus monkeys have been suggested as pheromones. When applied to sex skin of ovariectomized females, they induce mounting and ejaculation of males (Michael et al., 1975). At the time of ovulation, the marmoset (*Sanguinus oedipus*) wipes her labia with the tip of her prehensile tail, which then becomes an object of interest to a male. After being attracted to the vaginal odors and after sniffing the tail of the female, the male proceeds to copulate (Epple, 1975). In primates, including the human female, vaginal secretions support a bacterial flora that, together with gonadal steroids, release volatile aliphatic acids.

Monthly changes occur in human vaginal secretions, but their odors seem to be unrecognized by the human male (Doty et al., 1975). In humans the aliphatic acids attain a peak in the late follicular phase, but this peak does not occur in women taking oral contraceptives (Michael et al., 1975). A variety of interesting tests designed to reveal human ability to distinguish body odors of men and women suggests that discrimination is based more on intensity of odor than on any distinctive sexual quality to odor. Individual differences do exist and are detectable, but they are influenced by such personal habits as diet and shaving of axillary hair (Doty, 1976).

One interesting test involved 16 male and 13 female college students who wore identical T-shirts for 24 hours and did not use soap or deodorants for that period. Of these students, 13 males and 9 females correctly identified the odor from the shirts as (1) their own, (2) worn by a male, or (3) worn by a female. These students described male odors as musky and female odors as sweet (Russell, 1976).

Androstenol ($5a$-androst-16-en-3α-ol) is a steroid synthesized in the human testis, and it occurs in the urine of both men and women. It is also secreted by axillary glands. It has been found in underarm sweat of adult males and can be detected at low levels by humans (Brooksbank et al., 1974). Its role in humans is equivocal. It is a sexual pheromone in the domestic boar, in which it is concentrated in the saliva and induces mating behavior in the sow. It is also synthesized by truffles (*Tuber melanosporum*), in which the concentration may be twice that in the boar's saliva (Claus et al., 1981). Most likely the presence of androstenol in truffles accounts for the enthusiasm with which sows seek them.

6.7 MARKING

Marking by both adult males and females transmits information regarding their sexual state. The transmission of olfactory signals in urine is enhanced by frequent release of small amounts of urine. Such polyuria is typical of many mammals during the reproductive period. Males usually release urine frequently and in small amounts, and this behavior is especially characteristic of socially dominant individuals. Not surprisingly, the urinary bladder of a dominant male is more likely to be nearly empty than that of a younger, less dominant male (Bronson, 1976). Much of the scent, however, comes from the preputial gland. Marking is very common in both wide-ranging mammals, such as ungulates and large carnivores, and very small mammals, such as mice and shrews.

6.7.1 Territorial Marking

Odors placed on vegetation define the individual's territory. The odors are placed not only on the periphery but throughout the territory as well. Odors of an adult male serve both to repel subordinate males and to attract sexually receptive females. Marking by female mice is increased both in a new environment and in the presence of a male.

Sexual messages in the hamster are largely olfactory, and females apply secretions from the vaginal orifice. Two glands that open near the vagina discharge a fluid, and other secretions may come from the vaginal lining. Apparently urine is not involved in marking by female hamsters. Female hamsters mark actively in the follicular phase of the ovulatory cycle but do not mark at estrus. The female hamster applies vaginal scent by partly everting her genitalia against the substrate (Johnston, 1975). The odor is strongly attractive to males, even sexually inexperienced males, and the secretion of a female in estrus applied to a castrated male induces an intact male to attempt coition (Johnston, 1977).

6.8 TACTILE SIGNALS

In some vertebrates touch appears to have sexual significance, quite apart from the other types of signals to which touch may be ancillary. Because it is sometimes difficult to separate tactile messages from chemicals that they might transfer, the significance of touch is not always apparent.

The use of touch in courtship is quite apparent in some vertebrates but has not been given major emphasis in research. In amphibians a substantial amount of courtship includes touch, but this behavior may also function to transmit chemical signals. Tactile signals are paramount in some turtles and snakes, but the use of contact appears in the courtship of many sorts of vertebrates. Male minnows (Cyprinidae), for example, develop hard, rough protuberances called nuptial tubercles or pearl organs on the surface that contacts the female during courtship (Fig. 6.2), and these nuptial tubercles seems to stimulate the female to release eggs.

In colubrid snakes courtship and copulation appear to result from three distinct phases: (1) tactile-chase, (2) tactile-alignment, and (3) intromission. The sequence is fixed and characteristic of a given species, reducing the likelihood of interspecific matings (Gillingham, 1979). As pointed out in Section 6.6, courtship in snakes includes chemical stimulants as well as tactile activities.

In *Python molurus* (an Asiatic member of the boa family), tactile activity is less conspicuous in courtship, but there is an apparently essential use of the spurs of the male. In the tactile-chase phase in this species, the extremely mobile spurs not only stimulate the dorsum of the female but also probe beneath some scales. When his spurs are close to the vent of the female, the raising of an anal scale is followed by the elevation of her tail, which allows apposition of the cloacal areas, and intromission follows. In the Boidae the spurs are essential to courtship and mating (Gillingham and Chambers, 1982). The tactile pattern seems to be similar in the rosy boa (*Lichanura roseofusca*) of North America (Kurfess, 1967).

Among four species of kinosternid turtles, courtship is mostly tactile, after sexual identification has been made from odor. After an individual is determined to be a female, there is a courtship lasting from a few seconds to three minutes, during which time the male nudged and bit the female about the head and neck (Mahmoud, 1967).

6.9 SEXUAL SELECTION AND THE PROMOTION OF FITNESS

Mate selection has been a topic of major interest to zoologists, especially behaviorists. This concept was introduced by Darwin and has been enthusiastically pursued ever since. For many years it has been widely believed that the decision of with whom to mate is made solely by the female. A corollary of this concept, moreover, is that the female is likely to select a dominant or somehow superior male, which is tantamount to selecting a superior genotype. The definition of "dominance" is not simple and straightforward and is not necessarily correlated with social behavior, including sexual performance (Rowell, 1974). Detailed observations introduce genuine doubt as to the universal applicability of the concept of mate choice for a superior genotype, although there is good evidence that a female may select a superior mate in some species. This concept was doubted by Huxley in 1938 and remains in question today.

Several proposed pathways might explain mechanisms by which sexual selection could favor superior genotypes. One, male-male combat, especially in examples when the victorious male increases his access to females, would tend to exclude less physically vigorous males from mating. Two, females are attracted to a possibly neutral secondary sexual characteristic of the male that may be somehow associated with separate, but desirable, "good genes." Three, in the so-called runaway selection, a female may choose a male for a particular trait, attractive but again perhaps deleterious. If her sons bear this trait and succeed in attracting more mates, then they will increase her fitness by producing more offspring than a less "attractive" male might.

Most discussions of mate choice assume that selection is solely the prerogative of the female. A male can maximize his reproductive potential by being less discriminating than females are. In a one-to-one monogamous relationship, it is to the male's advantage to mate with the most superior female, but increased frequency of infidelity by females tends to weaken a male's reason to be selective. If females become temporarily more numerous than males in a monogamous species, then males might select females; but this situation is probably infrequent.

6.9.1 Mate Selection

As pointed out by Halliday (1983), the likelihood of a female selecting a mate on the basis of genetically determined criteria is fraught with two difficulties: the recognition of genotype and the heritability of a trait. There seem to be no clearly demonstrated examples of females (1) being able to recognize a more attractive genotype and (2) selecting a superior genotype as a mate (Parker, 1983).

What are the criteria for predicting that a given male will father offspring of superior fitness? One can most conveniently measure size, age, or social dominance over other males, but do these criteria increase the likelihood of greater survival of offspring or render the sons more attractive as future mates? Among females dispersing after mating in a lek, it would appear impossible to measure breeding success of the males. If a female entering a lek does, in fact, select a specific male, does she choose one who will produce male progeny that will become socially dominant or does she prefer one who will father offspring of maximal viability (Bradbury and Gibson, 1983)? Is there clear evidence that the two characteristics are linked?

There are examples in which females exhibit no apparent selection. Female anoles (*Anolis carolinensis*) associate randomly with males in two-thirds of the cases, with no preference dependent upon body size or dominance (Andrews, 1985).

In the process of mate selection, the criterion that forms the basis of attraction, and possibly selection, may or may not indicate social dominance or physical superiority. It has become traditional to assume that the female has considerable freedom of choice in mate selection, and much research has been analyzed with female choice in mind. In monogamous species with 50:50 sex ratios, however, random selection must be more common than in polygynous species. Sometimes female selection may be influenced by the behavior of other females. In herds of ponies in the New Forest (United Kingdom), dominant mares may threaten subordinate mares and prevent their mating with the herd stallion (Tyler, 1972).

An indirect selection of genotype might result from selection of size. Size can reflect age and therefore competitiveness, but does not necessarily indicate genotype (Halliday, 1983). At least some of the phenotypic variation that forms the basis for mate selection is derived from differences in age, and survival to sexual maturity is, in itself, suggestive of genotypic superiority. Survival to sexual maturity involves selection that is imposed by exogenous factors. In examples of competing males, however, the older male that loses a contest does not necessarily have a genotype inferior to that of the victor. Among herds of feral horses in the Great Basin (United States), successfully competing stallions, those that evict another stallion from their herd of mares, come from an age group of from five to seven years, and the evicted stallion is invariably older (Berger, 1986). Both the winner and the loser have demonstrated adaptiveness and dominance by virtue of having survived food shortages, severe winter weather, and predators and also by virtue of having held together a herd of mares. The vanquished cannot logically be assigned to an inferior genotype.

6.9.1.1 Selection in Leks. Students of lek systems have sought after specific cues on which a female will make her choice (Bradbury and Gibson, 1983). As a female enters a lek, she is continuously exposed to courting males. Assuming lek species resemble other vertebrates with respect to courtship responses, courtship by all males in the lek, individually and collectively, will induce a rise of serum luteinizing hormone (LH) in the female, to the point at which choice may become less critical. This effect does not preclude choice but does reduce the possible significance of distinguishing specific cues from different males.

In many groups of vertebrates males assemble in leks and display to each other, prior to mating with females that come to the lek for the sole purpose of copulation. There is no unequivocal evidence that females choose a male, but there is evidence to suggest that males compete among each other. Male-male display in lekking in the great snipe (*Capella media*) appears to be an assessment of body condition (Avery and Sherwood, 1982). Much of the male-male behavior involves intimidation and fighting, and the oldest males copulate most frequently (Lemnell, 1978).

In reality, mating within a lek appears to be nonrandom, suggesting that females might be making a choice (Bradbury and Gibson, 1983). A major factor in the issue of female choice, however, may be the level of sexual motivation required for mating, and the level of arousal may well result from the courtship of two or more males (Halliday, 1983). This behavior probably occurs in lek systems, so that a female becomes increasingly stimulated by successive courtship from several males (which presumably causes a rise in LH levels in the female). This increasing arousal would cause her to eventually reach a receptive state, at which point mating might occur more or less at random upon reaching the inner group of dominant males.

Among anurans some species occupy and call from a mating site for a prolonged time, apparently competing with other males for the exclusive use of a superior spawning area. Noncompeting (noncalling) male anurans may approach calling males and try to clasp moving females. Other species breed briefly, and the males move actively in search of unattached females. The difference is, in part at least, a response to densities, and calling prevails at low densities whereas searching predominates at high densities (Wells, 1977). Female tree frogs (*Hyla versicolor*) select males on the basis of the duration of their calls, which may indicate their investment in courtship (Klump and Gerhardt, 1987). Males of the Australian leptodactylid frog *Uperoleia rugosa* assemble in a lek and compete for spawning sites. Larger males succeed in combat. Females appear to select the larger males on the basis of their lower call (Robertson, 1986).

Some anurans apparently select a mate, for the association of a female with a male is nonrandom (Licht, 1976; Wilbur et al., 1978). The criteria, however, are not clear. Lack of randomness, in itself, does not necessarily indicate female choice. Male-male combat extends to interference in mating, tending to reduce the effectiveness of a possible selection made by a female but not to eliminate female choice altogether (Diamond, 1981). Mate choice in lek systems is clearly an unsettled issue.

Temperate latitude toads appear to be prolonged breeders and may call for several days. Nonrandom mating in Fowler's toads (*Bufo woodhousei fowleri*) appears to be predicated on size, which is reflected in the call; but there is no correlation between the sizes of males and females in amplexus. Lower ambient temperature alters the calls of Fowler's toads so that they sound larger than they actually are, and it has been suggested that large toads seek colder parts of their habitat so as to enhance their attractiveness to females (Fairchild, 1981). In *Bufo bufo* the larger males are the first to enter a pond for spawning but are less likely to arrive paired than are smaller males that reach the water subsequently. Small males may obtain mates by clasping females on land. Larger males enjoy greater breeding success, apparently from their ability to displace smaller males (Loman and Madsen, 1986).

The toad *Bufo typhonius*, of the Canal Zone, is an explosive breeder, and spawning occupies less than one day. Males actively pursue females and usually mate only once. There is no evidence of selection of males by females (Wells, 1979).

It is interesting to consider the positive correlation between dominance and fitness (the number of offspring surviving to sexual maturity) in industrialized human societies. Except for a period between 1935 and 1960, the correlation has been negative. Social dominance (in terms of education, accumulated wealth, and intellectual level) is correlated with a decreasing number of offspring (Vining, 1986). The availability of dependable means of contraception enables humans to separate reproduction from heterosexual play, and a woman today can select her own degree of fertility, an option not available to other vertebrates. In contrast, from sixteenth to eighteenth century England, when upper-class women tended to turn over newborn infants to wet nurses, conception sometimes closely followed birth, and women of wealthy families not uncommonly had 20 or more children (Fildes, 1986). This phenomenon can also be considered as cultural (and "unnatural") and not applicable to wild populations. Thus, applying concepts of sociobiology to human reproductive rates appears to be very difficult.

Status or rank among nonhuman primates may not correlate well with reproductive success (Bernstein, 1976). Monogamy characterizes many species of primates, but faithfulness is not always as absolute as it appears to be. In the rhesus monkey (*Macaca mulatta*), sexual advances and general aggressiveness are most apparent in dominant males, but there is some sexual activity among subordinate males. It has been observed that copulation of mature females with subordinate males is less obvious than it is with dominant males. Therefore it is possible to overemphasize the reproductive role of the more easily observed and conspicuous dominant male (Drickamer, 1974).

It has been suggested that in primates mate selection is based on factors in addition to favoring a dominant male. Dominance among female gelada baboons (*Theropithicus gelada*) is correlated with reproductive success. Apparently harassment of subordinate females during their estrus either disrupts ovulation or induces very early abortion. Similarly, among captive populations of marmoset monkeys (Callithricidae), the dominant female produces young, and socially inferior females may mate but do not become pregnant. This phenomenon has been observed in *Sanguinus fusicollis*, *S. oedipus*, and *Callithrix jacchus*. It has been suggested that odors from the dominant female suppress reproduction in subordinate females and that these odors are a mixture of urine and vaginal secretions (Epple, 1975). Heterosexual associations of both captive and wild baboons (*Papio doguera* and *P. cynocephalus*) increase during the midcycle. Although large juvenile males have fewer sexual interactions than adult males do, most successful mountings are by young males (Rowell, 1974).

6.9.2 Territory Selection

In view of the possible existence of alternatives to the selection of an attractive genotype in the process of mate choice, one should consider the female's choice of an optimal habitat. If habitat selection is of primary importance to both sexes, then the male mate is chosen with the territory, with no separate choice of a mate. Males often contest each other for a territory, with the largest or most robust of mature males succeeding, and for serving the females that enter their domain. Also, in birds in which courtship includes the presentation of food to the female, females may prefer those males that present the most food. If such offerings are predictive, such a male could prove to be a better provider for offspring, perhaps because his territory is rich in resources.

Among a diversity of vertebrates females clearly tend to mate with males that hold the more desirable territories (Halliday, 1983). This preference is tantamount to selecting a desirable territory. Female bullfrogs (*Rana catesbeiana*), for example, seem to select males on the basis of a suitable water temperature for the development of the eggs (Howard, 1978). Female pied flycatchers (*Ficedula hypoleuca*) select a mating territory and not characteristics (size, color, age, or song) of the male defending that territory (Alatalo et al., 1986). These situations clearly constitute a direct selection of habitat, but there is subsequently an acceptance of one of the resident males.

At variance with these observations are the results of experiments with the long-tailed widowbird (*Euplectes progne*). By artificially altering the length of the tail in a community of this species, Andersson (1982) demonstrated that females selected males with the longest tails, even though short-tailed males and long-tailed males were equally effective in maintaining their territories (Fig. 6.8). There is no indication of tail length being associated with greater fitness.

Males of the various species of Darwin's finches (Geospizinae) maintain territories in which they sing and build nests, and they court females as they move about prior to nesting. Courtship consists of vocal and visual signals and may end in a flight about the territory. Females visit one or several territories before pair bonds are established. Usually there does not appear to be any obvious selection, but after a severe drought, males of *Geospiza fortis* greatly outnumbered females, and selection then appeared to be on the basis of body size, dark color, and territory size (Grant, 1986). In such situations, in which males are more common than females, there could easily be an element of female selection by males.

6.9.3 Male-Male Competition

In many species there is vigorous combat among males either immediately prior to the mating period or throughout the year. Marked sexual dimorphism is often displayed in species with strong intrasexual strife, especially in mating systems in which a single male obtains an assemblage of females with whom he alone mates. Dimorphism in polygynous groups is not universal, however. In horses (Equidae) a stallion may, by himself, herd a group of mares, but except for age differences, there is no apparent sexual difference in size or conformation. Sexual dimorphism is also seen in species that are apparently or definitely nonpolygynous. Males of

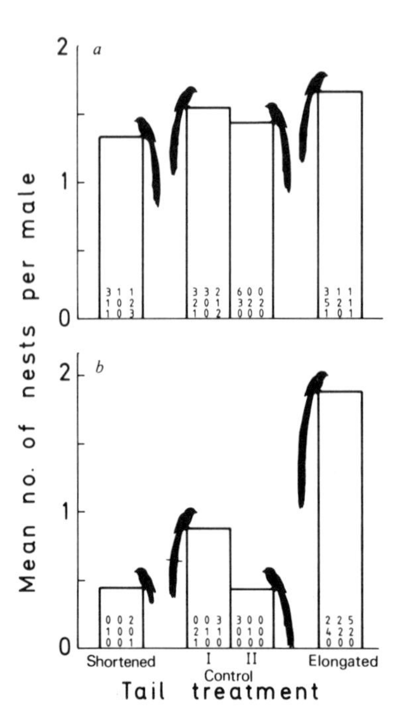

Figure 6.8 The effect of artificially altered tail length in long-tailed widowbirds: (*a*) Mean number of active nests for nine males of each of four categories before experiment; the number of nests are in the bottom of the bars. (*b*) Numbers of new active nests after treatment. See text for discussion. (From Andersson, 1982, reprinted from *Nature*, vol. 299, p. 819, © 1982, Macmillan Journals Limited.)

Figure 6.9 "Boxing" by hares (*Lepus capensis*) during the breeding season has given rise to the myth of the "mad March hare." This has been considered to be male-male combat or a rebuffing of a male by a female. (Katherine West, from a photograph by Holley and Greenwood, 1984, Reprinted by permission from *Nature*, vol. 292, p. 515, © 1984, Macmillan Journals Limited.)

bears and weasels, for example, are considerably larger than females, and although they may not always be monogamous, they are clearly not polygynous.

Some of the best known examples of sexual dimorphism are sea lions (Otariidae) and the elephant seals (*Mirounga* spp.), pinnipeds in which a single male rules a rookery to which females come to have their young and to mate. Among boreal sea lions, bulls remain in cold arctic waters during the winter, while the cows and young migrate to warmer waters. The much greater size of the bulls endows them with a thermoregulatory advantage. Males of pinnipeds frequently range farther and capture larger prey than the smaller females do (Sergeant, 1973). Among large herbivores differences in feeding and selection of food accompany sexual dimorphism (Jewell, 1977).

The hare (*Lepus capensis*) displays one of the more interesting intramale combats. Males establish dominance through "boxing," in which two males face each other and indulge in crude fisticuffs (Fig. 6.9). This behavior has been observed not only between two males (Flux, 1987) but also when a female rebuffs advances of an unwanted male (Holley and Greenwood, 1984).

Although it is easy to imagine that eons of intramale combat would eventually result in a larger male genotype, this would not follow if an increase in size were deleterious. It might well be deleterious in such mammals as horses and antelopes, which must move frequently in search of fresh forage, for a large size would demand more food to sustain it. Other sorts of dimorphism, such as elaborate and colored plumage in male birds and voice in anurans, may function in intramale competition as much as in courtship.

6.9.4 Forced Copulations

Among many different vertebrates are examples of mating with an unwilling female. This phenomenon has sometimes been called "rape," a term to which many people object both on emotional grounds and because of implications that may not apply to wild vertebrates. It is usually committed by a male that does not regularly associate with the female, and it is unaccompanied by precopulatory courtship. Among feral horses forced copulations sometimes result when a sexually experienced stallion encounters mares unprotected by their resident herd stallion. Such matings follow pursuit and biting by the male, but without courtship such as sniffing the genital area of the female, and are with mares not in estrus. In several herds of feral horses in the Great Basin of North America, 14 females (37%) were unable to escape the efforts of alien stallions (Berger, 1986).

Forced copulations in captive mallards (*Anas platyrhynchos*) occur early in the morning, as a female leaves her nest after having laid an egg, and such matings result in a high percentage of fertilizations (Cheng et al., 1983).

6.10 FIDELITY AND INFIDELITY

Mating arrangements among wild vertebrates may, in reality, differ from the generally acknowledged and superficially apparent patterns. The vigorous defense of a strongly territorial male does not preclude one or more females from straying and mating with neighboring males. This phenomenon is known to occur not only in such highly polygynous species as the Priblof fur seal but also in such seemingly monogamous species as squirrels and small passerine birds. Avian infidelity is reviewed by Ford (1983).

A great many kinds of vertebrates seem to be monogamous, but meticulous field observations and experiments indicate that fidelity to territory is frequently greater than loyalty to a mate. Among a number of carefully monitored birds and mammals, apparently monogamous pairs clearly regard copulation as an activity guided by convenience rather than family ties. There are numerous examples illustrating diversity of mating under the guise of monogamy.

6.10.1 Infidelity

If one presumes a superior genotype to constitute the basis for mate selection, one must then account for the apparently appreciable degree of infidelity to the selected mate, for such infidelity would remove any advantage of presumed genotypic superiority.

Some snakes, such as the copperhead (*Agkistrodon contortrix*), in which there is vernal ovulation, mate in both spring and fall, and multiple paternity may occur

(Schuett and Gillingham, 1986). Not only may the adder (*Vipera berus*) mate more than once a season, so that a single brood may have more than one male parent, but multiple paternity in captives exposed to only one male indicates that stored sperm may become active together with freshly contributed sperm (Stille et al., 1986).

The dunnock (*Prunella modularis*), a small European sparrow, is apparently monogamous, but in reality pair bonds may involve more than two individuals. Members of a given pair, however, have separate territories, and the territory of a female may not be congruous with that of her mate and may overlap with territories of one or more neighboring males. By the same token, her mate's territory may extend into those of females of nearby pairs. Copulation may occur with male-female encounters, and this seemingly monogamous bird may, in fact, be polygynous, polyandrous, or polygynandrous (with several males and several females breeding among themselves) in addition to being monogamous. These variations depend upon the size and form of the territories, which, in turn, seem to be determined by the abundance of food (Davis and Lundberg, 1984; Davies, 1985).

Polyandrous females of the dunnock enjoy the attentions of more than one male. They also achieve greater breeding success because more than one male provides for the young and fledglings. Polygynandrous groups are the least successful (Davies, 1986). In polygynous groups males achieve the greatest success (Davies and Houston, 1986).

DNA analysis of four mated pairs of the lesser snow goose (*Anser caerulescens caerulescens*), a presumably monogamous species, and their offspring indicated that in three of these families some of the goslings were the product of either mating outside the pair or intraspecific parasitism (Quinn et al., 1987).

Protein differences, revealed by electrophoretic techniques, may indicate multiple parentage in monogamous animals. In the bluebird (*Sialia sialis*) at least 5% of males and 15% of females were shown to have reared young not their own (Gowaty and Karlin, 1984). Perhaps the most ingenious technique designed to detect infidelity was a vasectomy performed on male redwings (*Agelaius phoeniceus*), whose mates subsequently laid fertile eggs (Bray et al., 1975).

Multiple paternity in birds seems to be most frequent where there are contiguous home ranges. If a male moves to an adjacent territory, he cannot simultaneously prevent intrusions into his own. Males of the pied flycatcher (*Ficedula hypoleuca*) and the collared flycatcher (*F. albicollis*) established a territory and a pair bond and subsequently set up a second territory some distance away and seek a second mate. The temporary absence of the male from the primary territory presents an opportunity for an outside male to enter and mate with the primary female. Multiple paternity in these flycatchers results from a female's mating with more than one male before a full clutch is laid; this occurs in about 24% of their broods (Alatalo et al., 1984).

The male rook (*Corvus frugilegus*), in spite of being more or less monogamous, will mate with females outside his territory. Young females mate with such promiscuous males during the period in which her eggs are being fertilized (Roskaft, 1983).

An increasing number of well-established examples of infidelity are being discovered among vertebrates. External fertilization makes infidelity in most fish and amphibians rather difficult, but mating with subordinate males does occur.

6.10.2 Sexual Deceptions

In some species males resemble females and therefore are not regarded as competitors by "normal" males. They are usually small and lack the distinctive form and color of large and well-developed adult males. This allows them an opportunity to escape the attention of a courting male and therefore to participate in mating without courting.

Among some sequentially hermaphroditic fish are genotypic, or primary, males, which are small. They seem to escape the attention of the dominant male and are able to enter the territories of dominant males and participate in mating. This activity of *sneaking* enables primary males to participate successfully in mating (Warner, 1975). A similar phenomenon is seen in the bluegill sunfish (*Lepomis macrochirus*). Small, inconspicuous males resemble females in size and color. Remaining near a courting pair, such a small male can sneak in and release milt just as the female releases eggs. The smaller males are sexually mature at two years, but the larger parental males (which remain to guard the eggs) mature at seven years (Fig. 6.10), although the growth rates of the two do not differ. These small males may constitute

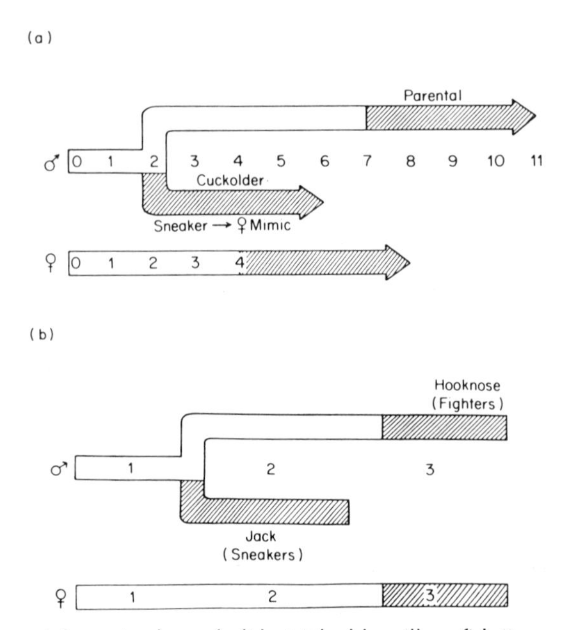

Figure 6.10 Sexual deception by male fish: (a) the bluegill sunfish (*Lepomis macrochirus*), a seemingly monogamous species, may have two males mating with a single female. The responsible and defending (or parental) male is from 7 to 11 years of age (hatched), but sneakers, from two to six years of age, resemble sexually mature females and fertilize some eggs, which are then protected by the parental male. (*b*) the coho salmon (*Oncorhynchus kisutch*) also has males of two sizes. The smaller jacks manage to fertilize some eggs. In each species there is only one spawning female. (From Gross, 1984, courtesy of Academic Press.)

80% of the reproductively active males and fertilize some 14% of the eggs (Gross and Charnov, 1980). The dominant male subsequently remains to fan and guard the eggs, some of which have been fertilized by the sneaking male (Dominey, 1980 and 1981; Gross and Charnov, 1980).

Unlike the bluegill sunfish, the coho salmon (*Oncorhynchus kisutch*) does not guard its nest, but the role of a population of sneaker males is similar. In the salmon, dull-colored precocial males, or *jacks*, ascend rivers at one year of age and at a small size, in contrast to the mature *hooknose* males, which are much larger and more brightly colored and ascend at three years of age (Fig. 6.10). The jaw hypertrophy of the hooknose males is a fighting structure, and some combat occurs among these larger individuals. Males crowd about a female as she constructs her nest. Hooknose males maintain their positions by fighting, whereas the jacks sneak about them and manage to release sperm without having entered combat.

This strategy of precocial maturity may constitute a successful alternative to reproduction by full-sized or normal males (Gross, 1984).

Deception in the red-sided garter snake (*Thamnophis sirtalis parietalis*) is somewhat more refined. The deceptive males, referred to as *she-males*, exude a female pheromone and are courted by normal males. In a field trial, however, the she-males proved to be more successful in mating than the normal males were (Mason and Crews, 1985).

In all these examples the dominant male appears to have been duped into tolerating a bona fide competitor because of the competitor's similarity to a female.

6.10.3 Relationship of Infidelity to Mate Selection

The preceding examples are a few of the many published accounts of irregularities in mating patterns. The purpose is not to emphasize deviations from apparent mating patterns but simply to point out that designated mating systems seem to have validity only as generalities. Any of the mating systems can be disturbed by mating outside the system, thus eroding the concept of a genetic significance to mate selection.

6.11 SUMMARY

Leks, the temporary breeding aggregations of males to which females enter to mate, are known in teleost fish, anurans, birds, and mammals. Leks among these various taxa are not quite the same. Among fish and amphibians oviposition occurs within the lek, a necessary requirement with external fertilization. In birds and mammals oviposition or birth may occur elsewhere. In spite of these differences, the role of leks in courtship is very similar.

Not only do premating activities establish or confirm the specific and sexual identity of individuals, but the exchange of various kinds of signals also tends to communicate sexual readiness and to synchronize the reproductive cycles of both males and females. Courtship signals are species specific and assist in avoiding interspecific matings.

Signals are appropriate to the morphology, habitat, and receptors of each species. Visual signals must occur in good light in order to be received and are logically used more commonly by many teleosts, reptiles, and birds than by the usually nocturnal amphibians and mammals. Sound is frequently employed, for it is efficient at all times in air as well as water.

Odors may have been the earliest means of communication, for the olfactory system of fish is very well developed. As a means of sexual communication, odor is important in all vertebrates but birds. Not only do odors identify species and sex, but the fragrance of one sex may elicit hormonal changes in the opposite sex. The sexual use of odors is best known in mammals. Special scents serve specific functions and may be intersexual or intrasexual. Odors may either be slowly released from the animal's body into the air; placed directly on a substrate, such as a twig or a rock; or carried in urine. Placed on a substrate, odors function not only as intersexual communication but also as territorial markers.

Tactile communication is important in many vertebrates, but probably less so in birds. Because tactile signals can be observed but not easily manipulated experimentally, they are perhaps less well understood than other signals.

Courtship may constitute a means of natural selection by allowing an individual to promote its fitness through choosing a superior individual of the opposite sex. This concept is usually predicated on the choice of a male by a female and implies that the female not only can recognize a superior genotype but actually selects a superior genotype. Most of the evidence remains equivocal.

There are many examples in which male contests reduce the number of breeding males to the most vigorous, and selection is apparently made among males. A victorious male may occupy a superior mating site, to which a female may be attracted. This may be the case in a polygynous system, and the female may select a site in which to lay her eggs or produce her young. The male continues to defend the territory that the female has apparently selected.

Among the various types of mating systems there exists an unknown, but apparently substantial, amount of infidelity. This infidelity would tend to weaken the selection argument for female choice of a mate. Infidelity is known to occur widely among vertebrates, including species with external fertilization as well as those with internal fertilization.

BIBLIOGRAPHY

Bateson P (Ed) (1983). *Mate Choice*. Cambridge Univ, Cambridge.

Beyer C (Ed) (1979). *Endocrine Control of Sexual Behavior*. Raven, New York.

Crews D and Moore MC (1986). Evolution of mechanisms controlling mating behavior. *Science*, **231**: 121–125.

Doty RL (Ed) (1976). *Mammalian Olfaction, Reproductive Processes, and Behavior*. Academic, New York.

Ford NL (1983). Variation in mate fidelity in monogamous birds. *Current Ornithology*, **1**: 329–356. Plenum, New York.

Halliday T (1980). *Sexual Strategy*. Univ of Chicago, Chicago.

Huxley JS (1938). The present standing of the theory of sexual selection. Pp 11–42. In *Evolution—Essays on Aspects of Evolutionary Biology* (Ed: de Beer GR). Oxford Univ, London.

Jewell PA (1977). The evolution of mating systems in mammals. Pp 29–32. In *Reproduction and Evolution* (Eds: Calaby JH and Tyndale-Biscoe CH). Australian Academy of Science, Canberra.

Reese ES and Lighter FJ (Eds) (1978). *Contrasts in Behavior*. Wiley, New York.

Stoddart DM (1980). *The Ecology of Vertebrate Olfaction*. Chapman and Hall, London.

Vandenbergh JG (1983). *Pheromones and Reproduction in Mammals*. Academic, New York.

Walther FR (1984). *Communication and Expression in Hoofed Mammals*. Indiana Univ, Bloomington, Indiana.

7

FERTILIZATION AND EMBRYONIC NOURISHMENT

7.1 FUNCTION AND ROLE OF FERTILIZATION

The exchange of sexual signals during courtship ensures the synchrony of mating and ovulation in most species of vertebrates. In some species ovulation is induced by coition. Additionally, in some species mating precedes ovulation by days, months, or even years, during which time sperm is stored in the reproductive tract of the female.

7.1.1 The Oviduct and Fertilization

The uterus, or the two uterine horns, together with the oviducts develops from the Müllerian ducts. The point at which the oviducts join the uterus is referred to as the uterotubal junction (UTJ), above which is the isthmus. Above the isthmus the oviduct broadens and is called the ampulla. The fimbriated, funnel-like opening, which receives ova from the ovary, is the infundibulum. The entire length of the oviduct consists of blood vessels, connective and glandular tissues, and smooth muscle and may be lined with ciliated epithelium (Gaddum-Rosse and Blandau, 1976). As pointed out in the following sections, specializations for the storage of sperm develop in the oviduct of some vertebrates. The oviducal glands of mammals secrete, into the lumen, materials that promote survival of gametes and zygotes.

Preparation of the uterus for implantation occurs during the normal ovulatory cycle in many mammals, but in some mammals courtship and mating are required for full endometrial development. In some placental mammals coition appears to play a critical role in implantation and initial formation of the placenta. Mating in the laboratory rat consists of a series of several (as many as 10) intromissions, with ejaculation only on the final mounting. These intromissions stimulate, in the female, two daily surges in prolactin release, one diurnal and one nocturnal. These surges of prolactin maintain the corpus luteum, and progesterone from the corpus luteum is needed for a proper uterine environment for implantation. These prolactin surges cease about five days after implantation, by which time the corpus luteum and pregnancy are maintained by hormone secretions from the placenta (Yogev and Terkel, 1978). Similar events lead to two daily surges in the laboratory mouse, and these surges appear to be terminated by the placenta at midgestation (Barkley et al., 1978). Prolactin and rat placental lactogen maintain luteinizing hormone (LH) receptors in luteal tissue and account for the luteal production of progesterone during pregnancy (Gibori and Richards, 1978). The luteotropic role of LH appears to result from the luteotropic effect of estrogen, which is aromatized from testosterone in the corpus luteum (Gibori et al., 1978).

After the ovum is discharged from the follicle, it enters the infundibulum. At the other end of the reproductive tract, sperm travel from the vagina toward the oviduct. Although some of this movement is the result of sperm motility, sperm are also carried by ciliary movement of the epithelial cells and muscular contractions of the uterus and the oviduct in birds and mammals. In the rabbit oviduct, sperm transport is effected mostly by cilia (Halbert et al., 1976). Lipids (phospholipids and cholesterol) occur in mammalian sperm and constitute their main source of energy. Transuterine movement of sperm is promoted by repeated intromissions in the laboratory rat. There is an immediate uterine contraction during intromission and ejaculation, and another delayed series of contractions begins five minutes after ejaculation and continues for 30 minutes to several hours (Toner and Adler, 1986).

As spermatozoa approach the ovum, the latter is slowly carried by ciliary movement (of the oviduct) toward the sperm. Sperm usually meet ova in the upper part of the oviduct, often in the ampulla. Actually only a very small part of the ejaculate ever reaches the upper oviduct.

7.1.2 Capacitation

In mammals, as the ovum enters the ampulla, it is surrounded by a zona pellucida, a protective membrane, which is covered by a loose, "fluffy" layer called cumulus cells. Sperm must penetrate these layers to reach the ovum. Epididymal sperm are not physiologically prepared to fertilize an ovum, and final changes, collectively called capacitation, are performed by the oviduct or parts of the ovum. Initially, there is the elimination of a decapacitation factor, a chemical factor on the surface of epididymal sperm, which interferes with the sperm's ability to fertilize ova. Second, the acrosome reaction is an essential part of capacitation and requires one or more enzymes, especially acrosomal proteinase. Freshly ejaculated semen contains acrosomal proteinase. Freshly ejaculated semen contains acrosomal proteinase inhibitors, which are removed by secretions of the uterine tract. Vesiculation occurs between the plasma and outer acrosomal membranes. Consequently the acrosome develops pores through which lytic materials of the acrosome are released. These materials assist in the sperm's penetration through the protective layers about the ovum.

In the domestic rabbit, capacitation is completed in the oviduct and requires both estrogen and progesterone (Bedford, 1973). Spermatozoa from the epididymides of the golden hamster can be capacitated in vitro with oviducal fluid or fluid from mature ovarian follicles (Yanagimachi, 1973). Capacitation in the uterus is equivocal.

A rather fixed series of events leads to fertilization. In the ampulla, spermatozoa cluster about the freshly ovulated egg. Sperm secrete an enzyme (hyaluronidase) that allows their passage through the cumulus cell layers. Sperm bind to the zona pellucida in mammals or to comparable acellular integuments in lower vertebrates. The attraction of the sperm to receptors on a conspecific zona pellucida may preclude entrance of sperm from a different species (Wassarman, 1987). Sperm penetrate the zona pellucida with the aid of another enzyme, acrosin, released from the acrosome after capacitation.

In New World marsupials sperm are separated when they leave the testis, but they join into pairs while they are in the epididymis. They are joined at their respective acrosomes and remain paired until they reach the oviduct (Temple-Smith and Bedford, 1980). Before fertilization, they part at the acrosome and once again become separate. Sperm are carried by peristaltic movement of the lateral vaginae, presumably stimulated by the high level of estradiol during estrus (Tyndale-Biscoe and Renfree, 1987). Four enzymes (acrosin, arylsulphatase, hyaluronidase, and *N*-acetylhexosaminidase) in the marsupial acrosome may assist the sperm in penetration of the zona.

Following invasion of the zona pellucida by one sperm, the sperm pronucleus joins with that of the ovum, and the zona pellucida (1) loses its attraction for sperm and (2) becomes resistant to penetration by additional sperm. Such a device to protect against polyspermy is called the zona reaction.

Another device is the vitelline block. As an example, the long-fingered bat (*Miniopterus schreibersii fuliginosus*) mates in autumn, and ovulation and fertilization follow shortly thereafter. Only a small percentage of spermatozoa reach the upper oviduct. As spermatozoa pass through the cumulus cell layer, they lose the acrosomal cap and then penetrate the zona pellucida. As many as 100 spermatozoa enter the perivitelline space, and excess sperm are consumed by pseudopodia of the surface blastomeres (Môri and Uchida, 1981).

7.1.3 External Fertilization

The teleost egg is enclosed in an acellular integument, referred to as the chorion or zona pellucida (see Fig. 2.4), which is perforated by an opening, the micropyle. This funnel-shaped orifice has a minute inner opening. The micropyle not only attracts the sperm but permits the entrance of only one, after which the opening is sealed.

In amphibians the egg covering is referred to as the vitelline membrane or chorion, which becomes surrounded by other coatings as it progresses down the oviduct. Sperm penetrate the jelly layers, with the aid of hyaluronidase, and reach the animal pole of the ovum (in anurans), where one pronucleus joins with that of the ovum, and the other sperm disintegrate. Sperm capacitation depends upon contact with the jelly layers, and fertilization does not occur in the presence of an antisera prepared against the jelly (Shivers and James, 1971).

7.1.4 Sperm Storage

In a number of vertebrates male and female gametes mature at different times, and sperm are stored in either the cauda epididymides or the female reproductive tract. When sperm are stored in the reproductive tract of the female, this phenomenon is also called delayed fertilization, but it could just as logically be termed delayed ovulation. Storage of spermatozoa raises several questions relative to their subsequent survival and function: where are they stored, how are they nourished, and what initiates the resumption of activity at the time of ovulation?

Sperm storage has been observed in elasmobranchs, teleost fish, salamanders, some internally fertilizing frogs, many reptiles, a few birds, and several groups of mammals. Because of the diversity of mechanisms of sperm storage and the broad spectrum of taxa within which it is known, one may assume that it has evolved to fill a variety of adaptive functions. Whether storage occurs in the cauda epididymides or in the genital tract of the female, it allows for spermatogenesis to precede oogenesis and also provides for a supply of sperm that can fertilize eggs released over an extended period.

In elasmobranchs sperm are stored in a region of the upper oviduct called the oviducal gland, and fertilization takes place as the ovum moves toward the shell gland (Wourms, 1977). Storage is sometimes prolonged, as in the dogfish (*Scyliorhinus canicula*), in which females have been known to produce fertile eggs after more than two years of isolation from males (Dodd, 1983).

Several families of teleost fish insert sperm into the reproductive tract of the female, sometimes indirectly by means of a spermatophore, and sperm may be preserved within the female for as long as eight months. In poeciliid fish, such as guppies, mollies, and swordtails, sperm are stored in crypts on the surface of the ovary (Tseng, 1972). In the marine live-bearing surfperch (Embiotocidae), sperm, which are stored on the surface of the ovary, are capable of glycolytic metabolism and are presumably supported during storage by sugars in the ovarian lumen (Gardner, 1978). In both of these families fertilization occurs in the lumen of the ovary or within the follicle, and the development of the embryos is intraovarian. Sperm storage will quite likely be discovered in other families of live-bearing teleosts.

Among the Amphibia, fertilization is internal in caecilians, most salamanders, and a few anurans. Sperm has been found in modified areas of the cloaca of only a few species, but storage probably occurs in many others. In the red-spotted newt (*Notopthalmus viridescens*) of eastern North America, spermathecae form from tubular outpocketings of the cloaca of the female. Sperm are apparently nourished by the epithelial lining of the spermathecae, for sperm may survive there for months. At the time of ovulation myoepithelial cells force sperm out of the spermathecae (Dent, 1970). Prolonged sperm storage has been noted in the plethodontid salamander, *Desmognathus fuscus*. In Louisiana courtship and mating precede oviposition by about three months, and viable sperm are present in spermathecae all months of the year. Sperm heads lie appressed to the epithelial lining of the spermathecae, suggesting that they may contribute to nutrition and survival of sperm (Maerynick, 1971). In the bell toad (*Ascaphus truei*), sperm appear to remain viable for up to two years in the lower part of the reproductive tract (Metter, 1964).

The extended storage of spermatozoa is best known in reptiles. The evidence is circumstantial for many species of turtles and squamates, but it has been carefully studied in only a few.

In the garter snake (*Thamnophis sirtalis*), sperm lodge in receptacles in the posterior infundibulum of the oviduct. The receptacles are lined at the entrance with ciliated cells, and the cilia may establish a current against which the sperm move to enter the receptacles. The sperm are maintained by glucose and fructose secreted by walls of the sperm receptacles. Sperm heads lie with the acrosome and sometimes the middle piece appressed to the lining of the receptacles, and it is suggested that at the time of ovulation, stretching of the oviduct wall might eject the sperm (Hoffman and Wimsatt, 1972). Sperm storage seems to characterize many live-bearing iguanid and scincid lizards, for most of these species mate in the spring but ovulate in the autumn.

In several polygynous birds, in which the female mates once a year, the hen receives sufficient sperm to fertilize all of her eggs. This characterizes species that tend to lay large clutches, such as galliform birds and some ducks. Viability of sperm begins to decline after five days, and by day 20 it has declined to less than 10% of the fertility rate of day 5 (Lodge et al., 1971). In the domestic fowl, spermatozoa are temporarily housed in *sperm nests* near the infundibular region of the oviduct or in tubular glands in the uterovaginal junction. These sperm appear to be maintained by the tubule cells. Accumulations of secretory granules in the apices of the tubule cells suggest that cyclic secretions flush sperm from the glands (Burke et al., 1972). Sperm is stored in the uterovaginal junction of turkeys and remains viable for up to seven weeks (Bakst, 1981).

Although sperm may survive several days in the oviducts of placental mammals, this is not generally considered to constitute sperm storage. Sperm may survive some two weeks in the reproductive tract of the domestic dog but for only three days in the human (Gould et al., 1984). Prolonged survival of sperm in the oviducts has been reported for the spiny anteater (*Tachyglossus aculeatus*), resulting in a highly variable duration of gestation (Griffiths, 1978). Also, sperm is known to be stored in the female genital tract of at least two Australian marsupials, *Antechinus stuartii* and *Dasyurus viverrinus* (Tyndale-Biscoe, 1973).

Many species of vesper bats (Vespertilionidae) mate in the autumn, and spermatozoa remain in the reproductive tract of the female until ovulation the following spring, some 130 to 170 days later. Occasionally they may mate again in the spring, but in some species males and females hibernate separately and there is only the autumnal mating. The study of sperm storage in bats has been investigated by many students and reviewed by Racey (1979) and Uchida and Môri (1987). Androgen secretions from brown fat seem to maintain hypertrophy of seminal vesicles in hibernating bats, at a time when the testes are regressing. A nonsaponifiable fraction of brown fat from hibernating little brown bats (*Myotis lucifugus*) induces growth of seminal vesicles in weanling rats equivalent to that of four daily doses of 150 μg of testosterone (Krutzsch and Wells, 1960). Major questions involve the site of storage, mechanisms by which sperm are maintained through the winter, and capacitation the following spring.

In some bats spermatozoa lie against the nonciliated epithelium in the caudal isthmus of the oviduct, and in others they reside in the uterus of the uterotubal junction. After mating in the greater horseshoe bat (*Rhinolophus ferrumequinum nippon*) and in the orange whiskered bat (*Myotis formosus tsuensis*), healthy spermatozoa can be found in the caudal isthmus and uterine spermatozoa are phagocytized. Spermatozoa lie against epithelial cells, which contain cytoplasm that is rich in glycogen. In contrast, sperm may survive in the uterus and/or in the uterotubal junction in *Pipistrellus abramus* (Uchida and Môri, 1987) and in *Myotis lucifugus* (Racey et al., 1987).

Sperm storage in *Pipistrellus abramus* may increase the degree of capacitation. When sperm storage was artificially prolonged, there followed an increase in egg activation and normal development (Môri et al., 1986).

Early summer (October) witnesses seasonal recrudescence of seminiferous tubules in the Cape bat (*Rhinolophus capensis*), at 33° S latitude. Following spermiogenesis from January to April, spermatozoa enter the epididymides (before winter). Unlike most midlatitude bats, the Cape bat mates in the spring, and sperm are stored in the cauda epididymides over the winter (Bernard, 1986). Spermatozoa remain in the cauda epididymides of the Eurasian noctule (*Nyctalus noctula*), a bat that mates in both the autumn and the spring. Although testes regress after spermatogenesis in late summer, sperm remain viable in the cauda epididymides for as long as five months (Racey, 1972).

7.2 EMBRYONIC DIAPAUSE

In many groups of mammals are species in which implantation does not promptly follow entrance of the blastocyst into the uterus. In some groups the blastocyst experiences no growth for months, and in others implantation is delayed for the few final days of lactation of the previous litter. These delays have long been called delayed implantation, but most researchers now refer to such events as embryonic diapause. This is a broader term, for it includes the diapause in eggs of some cyprinodont fish, in freshly laid eggs of some birds, and also in marsupials, which do not actually implant. The following discussion is confined to embryonic diapause in mammals.

Among macropod marsupials the diapausing blastocyst experiences little or no cell division, but in some mammals diapause is a relative matter. In some bats there is slow growth prior to implantation. Embryonic diapause displays a broad variety of physiological patterns in the various taxa in which it is known. Embryonic diapause is perhaps most thoroughly studied in macropod marsupials (wallabies, kangaroos, and allies) and placental carnivores, but several other orders contain diapausing species. A few examples will introduce students to the diversity of diapause in mammals.

7.2.1 Lactational Delay of Implantation

Many polyestrous species mate soon after parturition, and implantation is delayed for a brief period. A rise in prolactin, due to nipple stimulation of the nursing infants, is characteristic of this delay and appears to suppress implantation of the blastocysts of the next litter. The laboratory rat usually has a postpartum estrus within 48 hours after parturition, but subsequently estrus is suppressed by lactation. Lactation depresses levels of gonadotropins and estrogen, and apparently the process of suckling acts independently on the anterior pituitary (via the hypothalamus) to prevent gonadotropin secretion. Levels of circulating prolactin and follicle-stimulating hormone (FSH) in the lactating gravid rat vary inversely: exogenous FSH can induce implantation in the pregnant rat, and prolactin counters the effect of FSH (Raud, 1974). Bromocryptine, a prolactin suppressor, can induce implantation in suckling rats on day 7 (instead of day 11, the normal time in pregnancy of lactation), indicating that postpartum pregnancies are delayed by prolactin secretion during nursing (Flint and Renfree, 1981).

7.2.2 Macropod Marsupials

Most macropod marsupials mate shortly after parturition, so that there is an unimplanted blastocyst in the lumen of the uterus during the extended lactation typical of marsupials. In these species uterine development takes as long as an estrous cycle, and pregnancy is longer than the luteal phase (Tyndale-Biscoe, 1973). At the time of birth, there is also ovulation. Ovulation may take place slightly before or slightly after parturition, and mating may be either prepartum or postpartum. In most macropods mating is postpartum, and in some (e.g., the tammar wallaby) mating may be either prepartum or postpartum. Blastocysts enter diapause, and development ceases for a prolonged period. During normal development of the pouched young, the blastocyst does not grow. Diapause is in the 70 to 100 cell stage, at which point it remains in a cell membrane. There is no mitosis and no change in size throughout diapause (Smith, 1981).

Renewed growth does occur, however, if the pouched young is lost or dies. The experimental removal or natural loss of the pouched young is followed by reactivation of the corpus luteum, which had previously been suppressed by prolactin (Fig. 7.1). Gestation in *Macropus eugenii* is 21 days, and if the pouched young is removed, parturition follows 26.4 days later. Progesterone begins to rise about 22 days before parturition, and the rise occurs 5 or 6 days after removal of the pouched young. The precise roles of the corpus luteum and the uterus in reactivation of the blastocyst are not certain. Resumption of growth of the blastocyst may depend upon secretion of progesterone and its role in the synchrony of endometrial secretions and embryonic growth (Tyndale-Biscoe and Renfree, 1987).

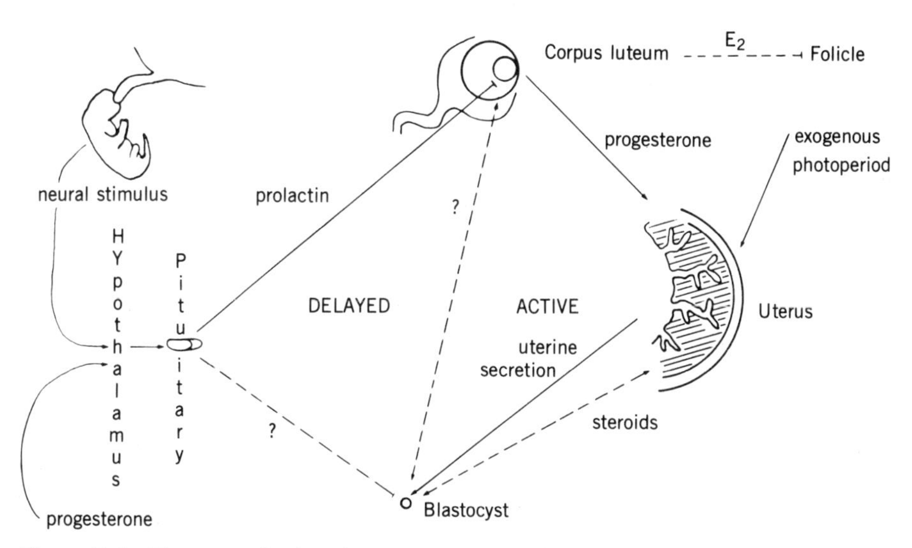

Figure 7.1 The control of embryonic diapause in the tammar wallaby (*Macropus eugenii*). The suckling young induces the release of prolactin, which suppresses secretory activity of the corpus luteum. With the removal or loss of the pouched young, the corpus luteum secretes progesterone, which prepares the uterus for the blastocyst. If removal or loss of suckling young occurs after the summer solstice, decreasing day length, acting through the pineal, suppresses reactivation of the corpus luteum. (From Renfree, 1981, courtesy of the *Journal of Reproduction and Fertility*.)

If the loss or removal of the pouched young occurs after midsummer, as day length is becoming shorter, then diapause persists until the winter solstice, and renewed embryonic development follows the subsequent increase in day length. The seasonal diapause of the tammar is induced by the nocturnal rise of melatonin with the increase of the length of the circadian scotophase (McConnell et al., 1986) (see Chapter 11, Section 11.7). Embryonic diapause appears to be universal among the macropod marsupials.

Three other marsupials also carry unimplanted blastocysts during lactation. Lactating noolbengers (*Tarsipes spencerae*) contain uterine blastocysts (Renfree, 1980). Unimplanted blastocysts may also be in the lactating pygmy possum (*Cercartetus concinnus*) (Clark, 1967), although they may be growing at a slow rate (Tyndale-Biscoe, 1973). In the small dasyurid *Antechinus stuartii*, gestation is prolonged but it includes a period of minimal development; the young are born in the undeveloped state characteristic of marsupials (Fig. 7.2). Final differentiation takes place rapidly but in phases interrupted by developmental arrest (Selwood, 1981).

7.2.3 The Roe Deer

The roe deer (*Capreolus capreolus*), in contrast to other midlatitude deer, mates in July and August, but because the blastocyst is quiescent for four or five months, birth follows mating by some ten months. Cell division and differentiation continue

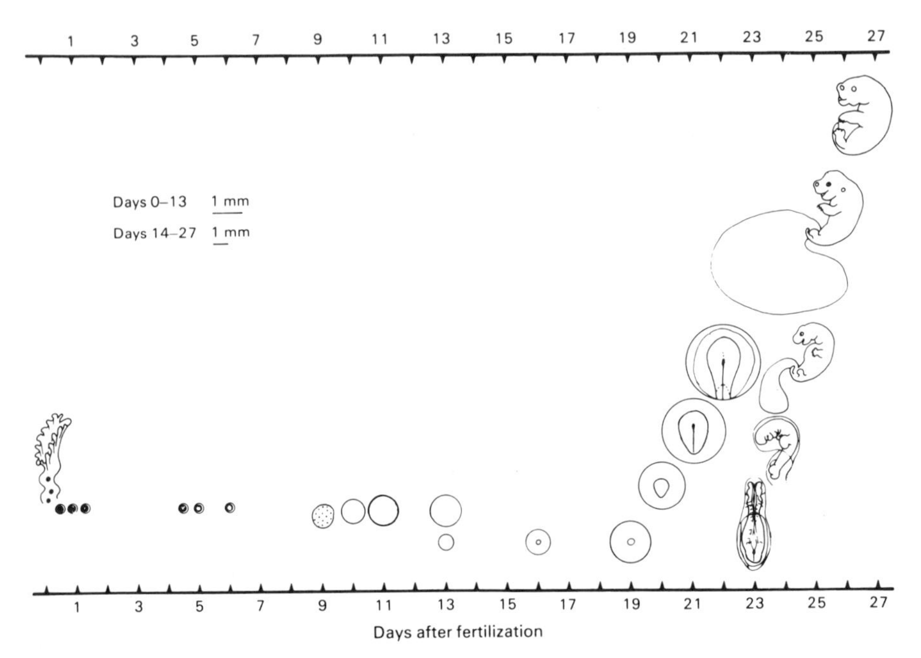

Figure 7.2 Timetable of development from fertilization to birth in the dasyurid marsupial, *Antechinus stuartii*. (From Selwood, 1981, courtesy of CSIRO Editorial and Publishing Unit.)

at a very slow rate until implantation. During diapause, the trophoblast lacks many cytoplasmic organelles. As implantation approaches, organelles appear suddenly, concomitant with sudden elongation of the chorionic sac and perhaps stimulated by the appearance of uterine secretions rich in free amino acids, proteins, carbohydrates, and calcium (Aitken, 1984). During embryonic diapause of the roe deer, the corpus luteum remains well developed, in contrast to that of most species known to exhibit diapause.

7.2.4 Mustelidae

Many species of mustelid carnivores delay implantation for extended durations, but in a large number the blastocysts implant after entrance into the uteri. Many species remain to be studied. The Mustelidae have been among the most carefully studied Eutheria in this respect, but the diversity within the family is surprising, and it is difficult to draw generalities. These carnivores range widely through the tropics to the Arctic, and fossorial, arboreal, and aquatic groups all contain species with diapausing blastocysts. Persistent researchers have clarified aspects of hormonal changes accompanying reactivation of the blastocyst, but exogenous control and adaptive value of embryonic diapause are still not well understood. Several carefully explored species will serve to illustrate embryonic diapause in this family.

Two species of temperate North America, the mink (*Mustela vison*) and the striped skunk (*Mephitis mephitis*), possess the capacity to undergo embryonic diapause but may also implant without delay; thus they seem to represent an intermediate condition. The reproduction of the mink is well known, and variability of its diapause accounts for gestation periods ranging from 40 to 91 days (Enders, 1952). Longer gestations, indicating more prolonged diapause, characterize matings early in the season. Diapause in the mink is reduced when days are long, and experimentally increased day length hastens implantation (Aulerich et al., 1963).

Prior to implantation there is a rise in prolactin, which is luteotropic in the mink (Martinet et al., 1981), and shortly thereafter an increase in serum progesterone. The increase in progesterone precedes implantation in a number of diapausing species, although, experimentally, progesterone alone does not induce implantation. The rise in prolactin in mink appears to follow the increase in day length, as in some other mammals (see Fig. 2.2), and it appears to be the stimulus to reactivation of the corpus luteum, which in turn is associated with implantation. Bromocryptine delays the sequential appearance of prolactin, progesterone, and implantation (Martinet et al., 1981). Second matings in the mink induce a rise in LH and also decrease duration of pregnancy (Møller, 1973), indicating that LH may act together with prolactin in ending diapause. In hypophysized diapausing mink, ovine prolactin induces luteal reactivation and implantation (Murphy et al., 1981). The mechanism for embryonic diapause would seem to resemble that of some other mustelids, and late winter or spring mating would seem to limit the expression of embryonic diapause in the mink.

In many other mustelids there is a prolonged delay before implantation and rapid embryonic growth, and during this period the corpora lutea are also quiescent.

Implantation has been observed to follow activation of the corpus luteum and progesterone secretion. Experimental administration of progesterone, however, does not induce implantation. Ovariectomy is followed by blastocyst death, and ovarian secretions are clearly fundamental to the resumption of blastocyst growth (Mead, 1981).

The spotted skunk (*Spilogale putorius*) has a gestation of 200 days or more in western United States (Mead, 1968a), but in the eastern states, in which populations appear not to experience diapause, gestation lasts from 45 to 55 days (Mead, 1968b). In western populations, concurrent with the increase in daylight from the winter solstice to parturition there is a rise in serum LH and progesterone (Fig. 7.3). While the blastocyst is in diapause, cells of the corpus luteum have spherical nuclei. As implantation approaches, the nuclei enlarge, the Golgi bodies become well developed, and agranular endoplasmic reticulum and lipid droplets increase (Mead, 1981). Daylight appears to be critical in initiating these events, suggesting a role for the pineal gland. When the organism is exposed to unnaturally increased day lengths in midwinter, serum progesterone rises (Mead, 1971). Pinealectomy was not followed by early implantation, but blinding prevents implantation (Mead, 1972). This does suggest a possible role for the pineal gland and melatonin. Melatonin injections or melatonin released from Silastic capsule implants prolongs the duration of the preimplantation period (May and Mead, 1986).

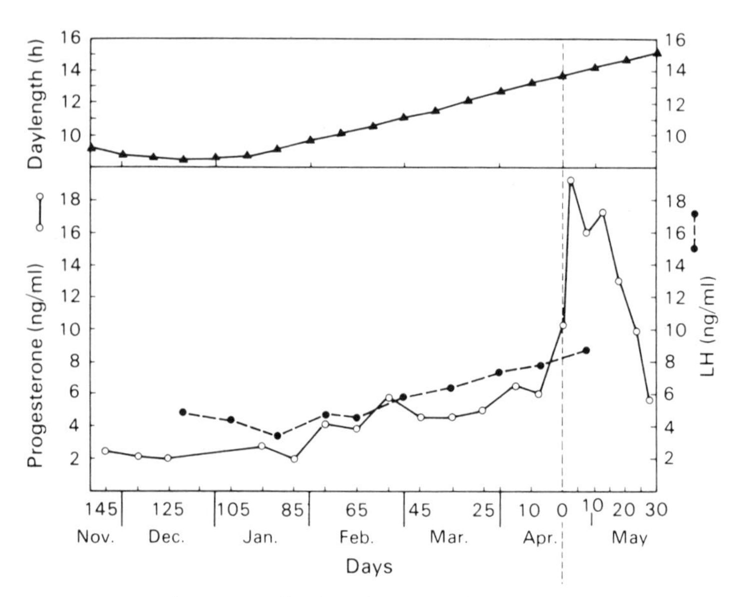

Figure 7.3 Seasonal changes in day length (upper line) and plasma progesterone and LH (lower lines) in pregnant western spotted skunks (*Spilogale putorius*). The vertical broken line indicates implantation (day 0), on 20 April ± 23 days (n = 92 pregnancies). (From Mead, 1981, courtesy of the *Journal of Reproduction and Fertility*.)

The European badger (*Meles meles*), like the mink and the spotted skunk, is sensitive to day length but implants when the dark period is longest. When nights were artificially prolonged in the summer, blastocysts implanted and parturition followed 45 days later, in September, instead of February as under a natural photocycle (Carnovenc and Bonnin, 1981). It is difficult to reconcile implantation under decreasing dark periods, as in macropod marsupials and the spotted skunk, with that of the badger, which implants under long dark periods. It is possible, however, that the hypothalamic-hypophyseal-gonadal axis of the badger becomes refractory to melatonin above certain levels (see Chapter 11, Photoperiodic Responses of Endotherms, Section 11.7.2).

7.2.5 Pinnipeds

Many species of seals (Phocidae) and sea lions (Otariidae) make annual migrations that include a brief association of the sexes at or near the time of parturition. Seals tend to have a short lactation and mate soon after weaning of the young. Sea lions are more likely to have an extended lactation period, during which the pup nurses, and mate shortly after parturition. In either event gestation requires nearly one year, and embryonic diapause, added to the time required for active fetal growth, ensures that the young will be born at the same time every year. This is especially critical in high latitudes, where the growing season is short and when males and females are together for only a brief spell.

Diapause is not total, for mitosis accounts for some growth, but the blastocyst of the northern fur seal (*Callorhinus ursinus*) fails to break out of the zona pellucida until implantation (Daniel, 1981). As in the Mustelidae, progesterone rises with the approach of implantation, and there is also an increase of estradiol and uterine proteins. Nevertheless, as in other diapausing mammals, administration of progesterone and/or estradiol does not alter diapause (Daniel, 1981). Thus, although parturition is apparently precisely scheduled, especially in high latitudes, indicating environmental control, the nature of the control is not apparent.

7.2.6 Microchiroptera

Within the Microchiroptera are three separate mechanisms for separating the time of birth from the time of mating: (1) sperm storage or delayed fertilization, characteristic of many vesper bats (Vespertilionidae); (2) embryonic diapause; and (3) slow, "delayed" embryonic development (Fig. 7.4). Implantation in the bent-winged bat (*Miniopterus schreibersii*) in northeastern New South Wales occurs prior to the completion of hibernation (Wallace, 1978). Delayed implantation characterizes the nonhibernating and seasonal *Rhinolophus lauderi* of northern Nigeria. Mating apparently occurs in autumn, for by late November unimplanted blastocysts are present in the left uterine horn, and they implant before the end of the year (Menzies, 1973).

Blastocyst growth proceeds at a very slow rate in a few species of mammals, causing a postponement of parturition beyond what would be expected for a mammal

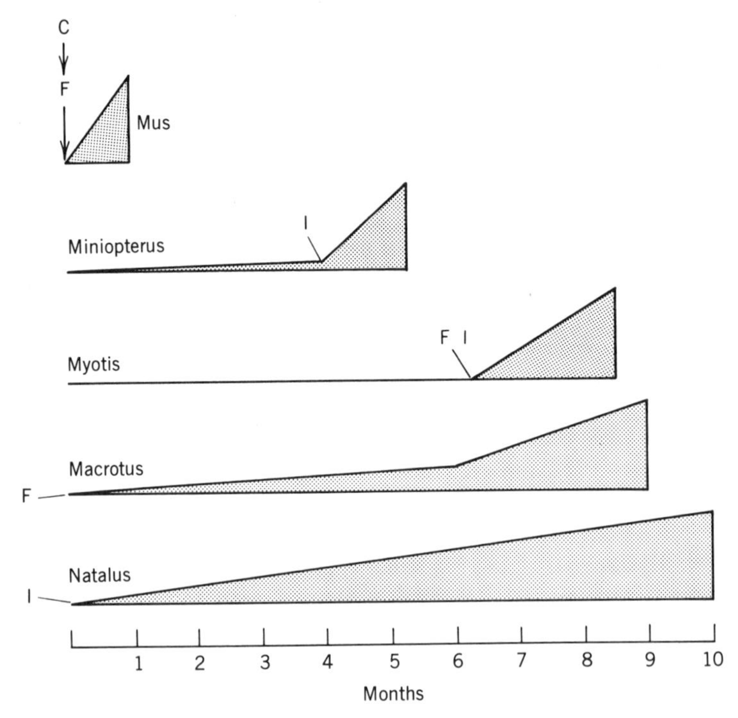

Figure 7.4 Synopsis of various chiropteran placentation schedules in contrast to that of the laboratory mouse: c, copulation; f, fertilization; i, implantation. (From Burns, 1981, courtesy of the *Journal of Reproduction and Fertility*.)

of such a size. In the nonhibernating California leaf-nosed bat (*Macrotus californicus*), fertilization and implantation closely follow mating, and embryonic growth proceeds at a slow, but relatively steady, rate. Normal embryonic development apparently depends upon the ovary, for ovariectomy results in abortion. The rate of development is independent of exogenous temperature, and the control of development is unknown (Burns, 1981). In the long-fingered bat (*Miniopterus schreibersii fuliginosus*), on the other hand, the retarded embryonic development during hibernation seems to be accelerated by an increase in the maternal body temperature at arousal (Kimura and Uchida, 1984; Uchida et al., 1984).

7.2.7 Adaptive Value of Embryonic Diapause

One might ask why so many groups of mammals have modified their reproductive seasonality by the insertion of a prolonged period of embryonic diapause. The advantage is fairly obvious in a few groups, such as the seals (Phocidae) and sea lions (Otariidae): males and females assemble annually, only during the time of parturition, and fetal development does not require a full year. A suspension of embryogenesis prevents unseasonally early births. Macropod marsupials occupy a seasonally arid environment in which they would probably experience difficulty in lactation during much of the year. The pattern of embryonic diapause in kangaroos and wallabies allows births during the early part of the reproductive season, but prevents parturition during periods of environmental hostility.

In some other major taxa embryonic diapause cannot be so easily explained. Why should embryonic diapause characterize so many mustelid carnivores and be absent in other mammals in which active embryogenesis takes approximately the same amount of time? There is no diapause in the Canidae, Sciuridae, or Lagomorpha. Within the Mustelidae, why should populations of the spotted skunk (*Spilogale putorius*) in western United States experience prolonged embryonic diapause, whereas the same species implants promptly in the eastern states and appears also to implant promptly in Mexico and Central America?

The stoat (*Mustela erminea*) experiences early puberty and embryonic diapause, whereas the ecologically similar weasel (*M. nivalis*) has direct development and early sexual maturity and may have two broods in one summer. The potentially greater fecundity of the smaller weasel may provide for a rapid population increase during times of rapid expansion of voles as a staple diet, is a more generalized feeder, and may be less subject to selection for rapid increase (Sandell, 1984). This hypothesis is untested. With its very early puberty, the stoat is also capable of rapid growth in numbers (King, 1984).

One can also consider the enigma of diapause from the aspect of the aberrant seasonality of gonadal recrudescence. Why do the gonadal systems of the eastern and western populations of spotted skunks respond so differently to day length? If gonadal growth could be induced in late summer in the nondiapausing eastern populations, would diapause then follow? The essence of the question then might be the cause of unseasonable gonadal growth, and perhaps not the subsequent event of embryonic diapause. The same question would apply to the roe deer and possibly to some species of bats. For a discussion of this important question, consult Mead (1988), cited at the end of this chapter.

7.3 THE NATURE OF THE PLACENTA

The placenta can more easily be defined by its function than by its structure. In all mammals it exchanges nutrients and oxygen from the material (or sometimes paternal) tissues with metabolic wastes from the fetus. Energy is provided as glucose, increasingly as the fetus nears full term. The placenta includes the various placental analogues, such as hypertrophied gills or pericardial sacs, of ectothermous vertebrates, discussed in the next chapter.

The placenta exchanges respiratory gases much as the lung does: in the placenta, oxygen is dissolved in the plasma and also bound to hemoglobin. The release of oxygen is facilitated by the entrance of carbon dioxide to the maternal plasma, through the Bohr effect, and the Bohr effect occurs in both the fetal and the maternal circulation. In addition, fetal blood has more hemoglobin, and fetal hemoglobin has a greater affinity for oxygen than adult hemoglobin does. In a pregnant human the maternal circulation holds about 15.3 mL/dL of oxygen, but fetal blood may contain 20 to 25 mL/dL of oxygen (Pearson, 1979). Hyperventilation provides for an increased oxygen intake during pregnancy. In the respiratory center of the brain the threshold to carbon dioxide is lowered by progesterone and that of oxygen is elevated by estrogen (Wilbrand et al., 1959). The placenta itself consumes up to 30% of the oxygen that it receives, to support its own metabolism, especially for the synthesis of hormones and for the movement of materials into the fetal circulation against a gradient (Campbell et al., 1966).

The nature of fetal nourishment in ectotherms is discussed in Chapter 8, Section 8.5.2.

7.3.1 Implantation

The blastocyst emerges, or "hatches," from the zona pellucida at about the time it enters the uterus, and the zona may play a role in preventing the blastocyst from joining uterine tissue until it is in the uterus. The process of implantation seems to require action on the part of the blastocyst as well as the uterus. The uterus becomes turgid and fluid filled, and the uterine lining becomes richly and locally vascularized. Hormone synthesis commences early in the life of the trophoblast and may induce receptivity in adjacent regions of the endometrium. In the bent-winged bat (*Miniopterus schreibersii*), materials from the blastocyst induce development of a localized "pocket" on the lining of the uterus, and this grows out to envelop the blastocyst (Fig. 7.5). In several domestic artiodactyls the trophoblast may both promote growth of the uterus and preserve the corpus luteum. After flushing fertilized ova from one horn of mated sows, there was by day 16 a greater arterial flow of the gravid horn, and the corpora lutea on the gravid side released greater amounts of progesterone than those on the nongravid side did (Ford and Christenson, 1979). In the little bulldog

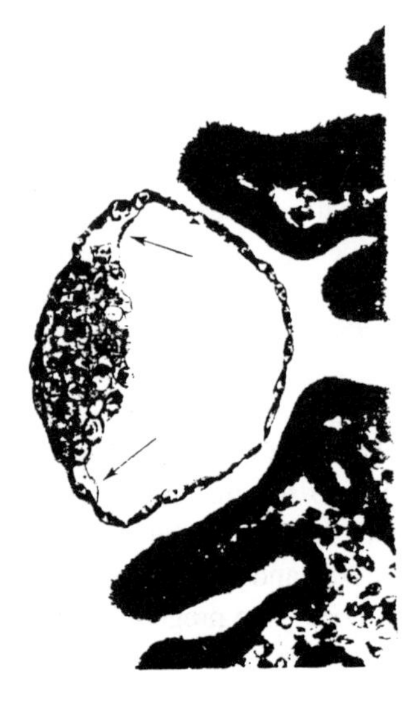

Figure 7.5 The blastocyst of the bent-winged bat (*Miniopterus schreibersii*) showing the growth of uterine lining forming a pocket in which implantation occurs. Ovulation is always from the left ovary and implantation in the right uterine horn. The blastocyst induces the formation of the pocket. Arrows indicate rudimentary endoderm. (From Wallace, 1978, with permission, from the *Journal of Zoology*, vol. 185, © 1978 by The Zoological Society of London.)

bat (*Noctilio albiventris*) a single ovum is released, and the ipsilateral uterine epithelium begins to develop quickly. A utero-ovarian arterial anastomosis may account for ipsilateral communication (Rasweiler, 1978).

The uterus has already been prepared for implantation by estrogen and progesterone from the current ovulatory cycle. The lining, the endometrium, is not only sensitive to hormonal levels but richly supplied with glandular cells itself. As the guinea pig blastocyst is about to enter the lumen of the uterus, there is a rise of uterine blood volume and intraluminal oxygen, both induced by ovarian steroids (Garris and Mitchell, 1979). Not only is the uterus richly vascularized by the time the blastocyst descends from the oviduct, but uterine secretions provide nourishment to the blastocyst prior to implantation. The lack of a rejection reaction between the blastocyst and the endometrium is not well understood, but it presumably depends upon secretions from both tissues.

Implantation in the laboratory rat occurs on day 5 or 6 of pregnancy, at which time there is increased permeability of endometrial capillaries, which may be a response to prostaglandins. Production of prostaglandins peaks on day 5, and indomethacin reduces uterine weight and frequency of implantation, indicating a role of prostaglandins in implantation (Phillips and Poyser, 1981). Gestation is prolonged by the administration of indomethacin, which inhibits the synthesis of prostaglandins. Ovariectomy on day 3, followed by treatment with progesterone and estradiol to promote implantation, did not eliminate the effect of indomethacin. This suggests that the effect of prostaglandins appears to be directly on the endometrium and not on the ovaries (Kennedy, 1977).

7.3.2 Fetal Membranes

In oviparous vertebrates, from some elasmobranchs and teleost fish to birds, the yolk sac is vascularized and not only carries nutrients to the fetus but also exchanges gases. Although yolk is absent in marsupials and placental mammals (and a few other taxa), the yolk sac persists and forms the initial placental line of communication between the fetus and uterus. The allantoic sac collects metabolic wastes from the hind gut. These structures, well studied in the chick embryo, occur in reptiles and mammals but are greatly modified from one taxon to another.

Among several sorts of viviparous vertebrates, placentas are designated according to the source of tissues involved. An essential requirement in nutrition of the eutherian embryo is the participation of a well-vascularized section of the uterus.

As cell division proceeds to about the 100-cell stage in the laboratory rat, the embryo reaches the uterine lumen. By this time the embryo has formed a central cavity and is then called a blastocyst. Part of the cluster of flattened dividing cells on the outer surface forms the ectoderm, or trophoblast, which constitutes most of the blastocyst. The few remaining cells comprise the embryonic disc, or inner cell mass, which is destined to become the embryo per se. From the embryonic disc and beneath the ectoderm extends at thin layer, the endoderm, which envelops the yolk (Fig. 7.6). These layers grow out together and form the blastocyst wall, which becomes known as the yolk sac. By the time the zygote has attained this level of differentiation, the zona pellucida has broken and separates from the blastocyst. At

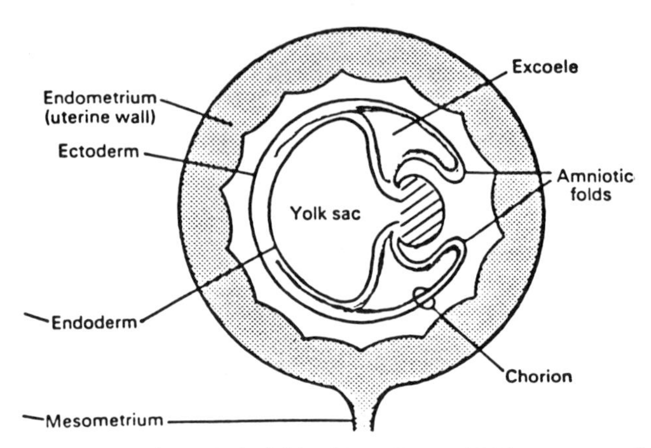

Figure 7.6 Formation of amniotic folds. (From Perry, 1981, courtesy of the *Journal of Reproduction and Fertility*.)

the time the mouse blastocyst implants, approximately three quarters of the cells are directed toward placental formation (Copp, 1979). From this point, the contact of fetal and maternal tissues takes various forms. Wimsatt (1975) has pointed out a great array of differing aspects of implantation in a broad spectrum of mammals.

The yolk sac is the first of the four fetal membranes to develop. While the yolk sac grows downward, away from the embryo, the amnion grows out and over the embryonic disc and encloses the growing fetus, eventually becoming an avascular fluid-filled bag within which the fetus floats (Fig. 7.7). The amnion may be considered a substitute for the aquatic environment in which most anamniotic eggs are laid.

The chorion is composed of the bilaminar sheet of the trophoblast and gradually developing mesoderm and, like the amnion, is avascular. The inner mesodermal lining comes to lie against the yolk sac and develops vascular tissue, forming the yolk sac (or choriovitelline) placenta. The yolk sac provides the vascularization for the chorion. In some taxa (Chiroptera, Insectivora, Lagomorpha, and some others) there is an inversion of the yolk sac layers, without extensive invasion of the mesoderm. This vascular bilaminar tissue lies in contact with the uterine lining.

The yolk sac placenta occurs in marsupials and eutherians, but in eutherians it is later replaced by the chorioallantoic placenta, which becomes the characteristic placenta in most placental mammals. Projecting from the hind gut, the allantois collects nitrogenous metabolic wastes. It lies in the position of the urinary bladder and presumably developed from the bladder with the transition from an amphibian to an early reptile. The allantois is vascularized where it lies in contact with the chorion, and the two structures together form a breathing organ. With the apposition of embryonic membranes and the endometrium, either lamellae or villi develop on the trophoblast surface. These structures are bathed in maternal blood, which escapes from ruptured sections of the maternal circulation. The fetus of the long-fingered bat (*Miniopterus schreibersii fuliginosus*) is initially supported by a single ovoid chorioallantoic placenta, which later disintegrates and is replaced by two smaller irregular accessory placentae (Fig. 7.8).

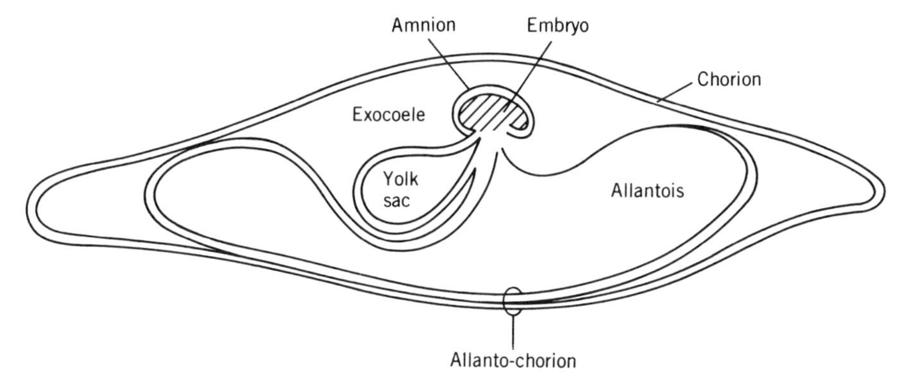

Figure 7.7 Fetal membranes of the pig at 21 days postcoitum (From Perry, 1981, courtesy of the *Journal of Reproduction and Fertility*.)

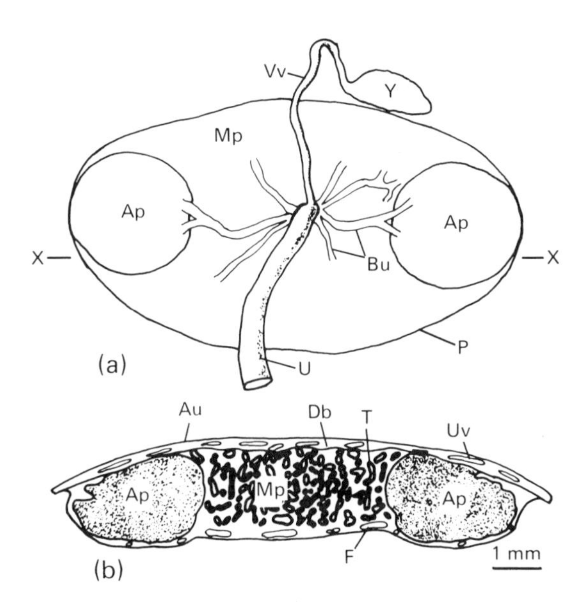

Figure 7.8 Orientation of the main placenta and the two accessory placentae at the transition stage in the long-fingered bat (*Miniopterus schreibersii fuliginosus*). (*a*) Gross appearance viewed from the fetal side; (*b*) Horizontal section at the level indicated by X—X in (a). Ap, accessory placentae; Au, antimesometrial uterine wall; Bu, branch of umbilical blood vessels; Db, decidua basalis; F, fetal blood vessles; Mp, main placenta; P, placental pad; T, trophoblastic tubule; Uv, uterine blood vessels; Vv vitelline blood vessels; U, umbilical cord; Y, yolk sac. (From Kimura and Uchida, 1984, courtesy of the *Journal of Reproduction and Fertility*.)

The yolk sac and the bilaminate chorioallantoic covering are both vascularized and, either singly or together, may serve to exchange nutrients, metabolic wastes, and respiratory gases. In viviparous vertebrates there may be a yolk sac placenta (sometimes with very little yolk) or a chorioallantoic placenta or both. In oviparous amniotes the chorioallantoic membrane is surrounded by a shell, which slows the passage of water but allows the passage of gases, including water vapor. Thus the amniotic egg is a major evolutionary step, allowing terrestrial oviparity as well as viviparity.

The placenta in some mammals includes a decidual tissue, derived from the endometrium. It is deciduate in the sense that it is discarded following expulsion of the fetus.

7.4 PLACENTATION IN MARSUPIALS

In most marsupials the yolk sac persists throughout embryonic life. It becomes vascularized and forms the sole site of exchange between the fetus and the mother, but there is substantial variation from one species to another. The allantois, together with its blood supply, lies within the yolk sac.

A shell membrane surrounds the fertilized ovum of marsupials and remains until approximately the last third of pregnancy, at which point "implantation" can occur. After the rupture of the shell membrane, embryonic development continues, with varying degrees of maternal-embryonic attachment. In the Nearctic opossum (*Didelphis virginiana*), for example, the fetal membranes lie appressed to the uterine wall with no interdigitation. Cells lining the uterus have microvilli, which contact the yolk sac. Yolk sac vascularization transfers nutrients from the maternal circulation.

Among some Australian macropod marsupials, the adjacent membranes may be vascularized but without placental formation beyond the yolk sac. In *Macropus eugenii*, uterine capillaries lie close to the surface, and microvilli of the uterus interdigitate with those of the trophoblast. Microvilli on the surface of the trophoblast indicate active absorption by pinocytosis (Tyndale-Biscoe and Renfree, 1987).

In some other species, such as the koala (*Phascolarctos cinereus*), the allantois makes contact with the chorion. The contact is incipient, without any interdigitation of tissues (Hughes, 1974).

There is generally a sharp division in the mechanism of embryonic nutrition between marsupials, which have a yolk sac placenta, and eutherians, which initially have a yolk sac placenta that is subsequently replaced by a chorioallantoic placenta. The bandicoots (*Parameles* and *Isoodon*, Paramelidae), marsupials of Australia, bear a resemblance to eutherians in their placentation. The amnion replaces the yolk sac contact with the uterine lining. At about day 9 bandicoots develop a yolk sac placenta, and somewhat later a chorioallantoic placenta forms and replaces the function of the yolk sac (Hughes, 1974; Padykula and Taylor, 1977). The fetal-maternal contact in the paramelids is the most intimate of any marsupial. Although gestation in the bandicoots is not appreciably longer than in other marsupials, their young are morphologically more advanced at birth.

In the majority of marsupials the intrauterine duration of development (the length of pregnancy) equals the luteal phase of the estrous cycle. At the end of both the luteal phase and the pregnancy, plasma progesterone declines. Because progesterone is essential for the secretory role of the endometrium, which in turn maintains the embryo, the decline of progesterone may signal parturition.

In the Paramelidae, lactation maintains the corpora lutea as long as young remain in the pouch. Removal of the pouched young results in estrus several days later. In the macropod marsupials the corpora lutea regress at parturition, and lactation (prolactin) prevents activation of the corpus luteum.

7.5 FETAL NOURISHMENT

Following fertilization, which has taken place in the upper oviduct, cells divide. By the time the organism has reached the uterus, the zona pellucida has broken off or eroded, and the hollow ball of cells (the blastocyst) has begun to differentiate. A possible function of the zona is to prevent adhesion to the wall of the oviduct. Unlike embryos that develop from eggs with abundant yolk, the mammalian embryo requires a substantial amount of exogenous nutrition. Before implantation the blastocyst may acquire proteins. Two-cell mouse embryos take up ^{125}I-labeled bovine serum albumin, and the rate of uptake increases with blastocyst development (Pemble and Kaye, 1986).

Generally speaking, whole elements, such as erythrocytes, are not transferred across the embryonic membranes. Other materials are selectively moved between fetal and maternal circulations. Placental transfer is selective, with metabolic capability of modifying materials moving to the fetus. Some materials pass through the placenta in the direction of the lower concentration, and other substances are moved in a selective way against a gradient. There is an imperfect barrier to whole elements: a small amount of "leakage" of fetal erythrocytes to the blood stream of the mother (in sheep) occurred in about 50% of the individuals examined and in about 10% in some human populations, but there is much less movement in the other direction (Cohen et al., 1964; Cohen and Zuelzer, 1965).

Both steroid and protein hormones are transferred in both directions. Not only do gonadal steroids move from the mother to the fetus through the placenta, but androgens (from male fetuses) diffuse across placenta membranes to be taken up by adjacent female siblings (see Chapter 1, Section 1.2.3). Some transfers are facilitated by countercurrent exchange, and others by active transport. Lipids, fat-soluble vitamins, and carbohydrates (such as glucose) diffuse or are transported across capillary membranes to the fetus. In contrast, fetal concentrations of amino acids in the fetus exceed those in the maternal serum, and movement to the fetus is through active transport. From these amino acids the fetus produces proteins for its own needs.

7.6 THE PLACENTA AS AN ENDOCRINE GLAND

The placenta synthesizes a host of hormones that not only promote the healthy development of the fetus but prepare mammary tissue for lactation. The placenta is a powerful endocrine gland, producing hormones that closely resemble those from the hypophysis and gonads. Both steroid and protein hormones are synthesized and released from the placenta, but hormone synthesis varies substantially from one species to another. Steroid hormones are synthesized from cholesterol, which is produced in the placenta. Amino acids provide the materials for protein synthesis within the placenta and for enzyme synthesis and fetal growth. Amino acids and triglycerides are supplied from the maternal circulation.

In some species the placenta synthesizes increasing amounts of progesterone, and eventually this progesterone is sufficient to maintain pregnancy. In such species ovariectomy does not terminate pregnancy. In other species the corpus luteum remains the major source of progesterone, and in these species ovariectomy is followed by abortion. In the golden hamster, the domestic goat, and the domestic rabbit, for example, luteotropic hormones from the placenta maintain progesterone production of the corpus luteum. In pregnant rats LH maintains the secretion of testosterone in both ovaries and placentae, and this testosterone supports the corpus luteum in early and late pregnancy. Estrogen and progesterone from the placenta promote growth of the uterus and breast, and estrogen also suppresses FSH synthesis by the hypophysis.

Apart from the steroid hormones, the placenta synthesizes protein hormones. In the gravid mare, endometrial cups of fetal origin secrete pregnant mare serum gonadotropin (PMSG). Human chorionic gonadotropin (hCG) is produced by the morula and continues throughout pregnancy. hCG and LH both initiate production of estrogen (by aromatization of testosterone and androstenedione) by the placenta. Similar hormones (with LH-like properties) are known from trophoblasts of a few other mammals. Most of them are luteotropins and maintain the corpus luteum. Another protein hormone is chorionic somatomammotropin, which is similar to prolactin and growth hormone.

In the placenta of some mammals, there is a hormone that resembles prolactin, and this has been called placental lactogen (PL). During the course of gestation, PL stimulates development of mammary tissue (see Chapter 10, The Mammary Gland and Lactation, Section 10.2.1.3). PL is secreted by an early fetal component of the placenta. It has been found in the placentae of some primates (*Homo*, *Papio*, and *Macaca*), certain rodents (*Rattus*, *Cavia*, *Chinchilla*, *Mus*, and *Mesocritus*), and some artiodactyls, but it is absent in the placentae of lagomorphs (*Oryctolagus*), carnivores (*Canis*), and some artiodactyls (*Sus*), as determined by effects on cultured mammary tissue (Talamantes, 1975). The occurrence of PL does not necessarily closely follow phylogenetic lines, but may be correlated with the absence of prolactin from other sources.

7.7 THE BRUCE EFFECT

The Bruce effect is the termination of pregnancy, usually before implantation, when a gravid individual aborts after exposure to the odors, especially odors of urine, of an "alien" male—one other than the male with which she had mated. This effect is produced by a reduction in prolactin secretion, presumably via a nervous pathway from the olfactory lobes to the hypothalamus (Bruce, 1966). The Bruce effect was originally described for the house mouse (*Mus musculus*) (Bruce, 1959) and also exists for the deer mouse (*Peromyscus maniculatus*) (Bronson and Marsden, 1964) as well as some voles (*Microtus* spp.). Not only does the presence or odor of an alien male account for a decline in pregnancy in several species of mice, but the presence of the stud, or parental male, in *Microtus montanus* increases the likelihood of preservation of the pregnancy (Berger and Negus, 1982). Pregnancy failure does not follow short-term exposure to alien males in *Microtus ochrogaster* (a pair-bonded species) if the pair bond is preserved (Hofmann et al., 1987). It is possible that the Bruce effect might isolate (reproductively) stable groups of individuals at moderate or low population densities (Rogers and Beauchamp, 1976).

The signals producing the Bruce effect may differ slightly in the European *Microtus agrestis*. When a captive pregnant female was exposed to an alien male within 48 to 72 hours after mating, pregnancy failed. Airborne odors appear to be insufficient in this vole to block pregnancy, but the exact stimulus is unknown (Milligan, 1976).

7.8 PARTURITION

Parturition requires a weakening of the placental attachment, so that fetal tissue may break away from maternal tissue. In many, if not all, mammals progesterone declines prior to birth, while both estrogens and prostaglandins increase. Prostaglandins, especially of the E and F series, increase prior to parturition in the laboratory rat, guinea pig, and hamster and may reduce the threshold of the myometrium to oxytocin (Lanman, 1977). In the miniature pig there are several sequential changes. In the sections of the uteri containing embryos spontaneous electrical activity (induced by the fetus) increases, followed by a decline in progesterone and an increase in estrogen. About nine hours prior to parturition, circulating oxytocin rises (Taverne et al., 1979). With the approach of parturition there is a decline in glucose transfer, oxygen uptake, protein synthesis, enzyme levels, and general placental metabolism. Relaxin, from the corpus luteum, increases with the approach of parturition and assists in the softening of the cervix. Its presence and source vary from one species to another, and it is clearly not essential.

Shortly before the onset of labor contractions in rats, oxytocin receptors increase in the myometrium (Soloff et al., 1979). In women there is a gradual increase in oxytocin receptors throughout gestation. Thus, in the human, even without a great, sudden infusion of oxytocin at the onset of labor, the increasing number of receptors lowers the threshold for oxytocin, which precipitates myometrial contractions. Increased release of prostaglandins enhances the effectiveness of oxytocin, and fetal oxytocin may elicit the prostaglandins that promote labor (Fuchs et al., 1982). Prostaglandins not only stimulate contraction of uterine muscles but assist in the relaxing or softening of the birth canal.

In some mammals activities of the fetus may determine or influence the time of parturition. In the sheep, progesterone synthesis increasingly becomes a function of the placenta. But as the time for birth approaches, there is a rapid increase of fetal corticosteroids, which depresses progesterone secretion and also accounts for an increase in serum estrogen. Two or three days prior to parturition, there is a marked rise in cortisol, stimulated by ACTH from the fetal anterior pituitary (Challis et al., 1977).

7.9 SUMMARY

The synchrony of male courtship and female receptivity is extended to appropriately timed behavior of sperm and egg within the oviduct. In laboratory mammals fertilization follows capacitation of spermatozoa, which occurs in the oviduct under the influence of both estrogen and progesterone.

The placenta ensures the acquisition of nutrients and the elimination of metabolic wastes, while providing an equable environment for the fetus. Placental development is little more than contact between the yolk sac and the endometrium in most marsupials. In the marsupial *Paremeles*, however, the yolk sac is replaced by a chorioallantoic placenta, as in eutherian mammals.

The placenta and the trophoblast are major producers of both steroid and protein hormones. These hormones maintain the corpus luteum and the endometrium and promote mammogenesis.

After mating and internal fertilization, there may be sperm storage. Viable sperm survive in the ovaries of some teleost fish, and many tetrapods store sperm in specialized regions of the female reproductive tract. In birds sperm storage lasts between 10 and 45 days. In some species of bats (Microchiroptera) sperm remain in the cauda epididymides from autumn until the following spring, when mating occurs. After mating in the autumn, most female bats store sperm in their reproductive tract until ovulation the following spring.

In several major taxa of mammals there is a delay in implantation of the blastocyst called embryonic diapause. This diapause extends the duration of gestation and time of parturition, sometimes to almost one year. During embryonic diapause, there is little or no blastocyst growth and development. Blastocyst growth, uterine activity, and corpus luteum metabolism resume together just prior to implantation.

In some mammals embryonic diapause appears to be under photoperiodic or temperature control, and in others it is induced by lactation.

BIBLIOGRAPHY

Amoroso EC (1959). The biology of the placenta. Pp 15–76. In *Gestation* (Ed: Villee CA). Josiah Macy Jr. Foundation, New York.

Austin CR and Short RV (1982). *Embryonic and Fetal Development*, 2nd ed. Cambridge Univ, Cambridge.

Beaconsfield P and Villee C (Eds) (1979). *Placenta—A Neglected Experimental Animal*. Pergamon, Oxford.

Chamberlain GVP and Wilkinson AW (Eds) (1979). *Placental Transfer*. Pitman, Kent.

Faber JJ and Thornberg KI (1983). *Placental Physiology: Structure and Function of Fetomaternal Exchange*. Raven, New York.

Finn CA and Martin L (1974). The control of implantation. *Journal of Reproduction and Fertility*, **39**: 195–206.

Flint APF, Renfree MB, and Weir BJ (Eds) (1981). Embryonic diapause in mammals. Proceedings of a Symposium Held at Thredbo, New South Wales, Australia, February 1980. *Journal of Reproduction and Fertility* (Suppl 29).

Gruenwald P (Ed) (1975). *The Placenta and Its Maternal Supply Line*. Univ Park, Baltimore.

Hamner CE (Ed) (1973). *Sperm Capacitation*. MSS Information, New York.

Johnson AD and Foley CW (1974). *The Oviduct and Its Functions*. Academic, New York.

Loke YW and Whyte A (Eds) (1979). *Biology of Trophoblast*. Elsevier, Amsterdam.

Mead RA (1986). Role of the corpus luteum in controlling implantation in mustelid carnivores. *Annals of the New York Academy of Sciences*, **476**: 25–35.

Mead RA (1988). The physiology and evolution of delayed implantation in carnivores. In *Carnivore Behavior, Ecology and Evolution* (Ed: Gittleman JL). Cornell Univ, Ithaca. In press.

Mossman HW (1987). *Vertebrate Fetal Membranes*. Macmillan, Basingstoke.

Perry JS (1981). The mammalian fetal membranes. *Journal of Reproduction and Fertility*, **62**: 321–335.

Renfree MR (1980). Placental function and embryonic development in marsupials. Pp 269–284. In *Comparative Physiology: Primitive Mammals* (Eds: Schmidt-Nielsen K, Bolis L, and Taylor CR). Cambridge Univ, Cambridge.

Thibault C (1973). Sperm transport and storage in vertebrates. *Journal of Reproduction and Fertility* (Suppl 18).

Tyndale-Biscoe H and Renfree M (1987). *Reproductive Physiology of Marsupials*. Cambridge Univ, Cambridge.

Wassarman PM (1987). The biology and chemistry of fertilization. *Science*, **235**: 553–560.

Wimsatt WA (1975). Some comparative aspects of implantation. *Biology of Reproduction*, **12**: 1–40.

8

LIVE-BEARING ECTOTHERMS

8.1 DIVERSITY WITHIN LIVE-BEARING ORGANISMS

A diverse assemblage of elasmobranchs, teleosts, amphibians, and reptiles have developed means of retaining the embryo until it has achieved a rather advanced stage of development. Many different mechanisms provide for internal embryonic growth, and there are varying degrees of postovulatory maternal contribution to development. These live-bearing species are of special interest to physiologists, ecologists, and evolutionary biologists, because they suggest evolutionary trends to viviparity, and comparison with oviparous species indicates the physiological and anatomical changes required for viviparity.

Any discussion of viviparity includes two central issues: (1) the adaptive advantage obtained by carrying fertilized eggs in utero until young are born and (2) the anatomical and physiological changes required by viviparity. The environments of viviparous species suggest possible adaptive advantages, but the occurrences of oviparity and viviparity in a given region or habitat are not mutually exclusive. The student is led to what appear to be the most plausible conclusions. As pointed out by Shine (1985), the most useful examples for study are those taxa that embrace both oviparous and viviparous forms. This chapter treats the diversity of maternal support and the possible environmental stimuli for these specializations.

The development of viviparity requires modifications, all of which characterize live-bearing vertebrates. Internal fertilization is essential, whether by coition and intromission or by means of a spermatophore. Usually there is a marked reduction in the number of ova released, and there may be either a reduction or a loss of yolk. The egg shell may be lost, and the yolk sac, which is highly vascularized as in oviparous species, frequently transfers nutrients and metabolic wastes between the embryo and the mother. Additionally, a chorioallantoic placenta sometimes develops. In oviparous vertebrates, at least those with no parental care, reproductive hormones may temporarily decline or disappear in circulation after oviposition. In live-bearing vertebrates, however, reproductive hormones serve specific functions in the development of the embryo, and the corpus luteum especially remains functional and secretes progesterone as well as other steroid hormones and peptide hormones.

As a result of these modifications, the embryo not only receives physical protection from potential predators but enjoys a rather equable environment that provides food and the basis for attaining a large size prior to entry into the outer world. For example, the majority, if not all, of live-bearing fish are in some way predatory and would profit from being large at birth.

Live-bearing ectotherms include species in which there is simply an incipient retention of the egg with very little or no additional maternal contribution to that contained in the original yolk. Egg retention is usually accompanied by postovulatory changes in the female. In some teleosts large numbers of larvae are released, having been nourished by only their own yolk. Some oviparous reptiles may retain their eggs, with the result that total incubation time is reduced. Even in these species, however, the embryos must have a supply of oxygen and a means for the removal of metabolic wastes. These processes may not demand great energy from the mother, but there is a significant increase in maternal metabolism in metabolic rate in viviparous lizards and snakes (Guillette, 1982). Internal development is a means of parental care, sometimes including thermoregulation, prevention of desiccation, and protection from predators, and it does require energy.

8.1.1 Separation of Ovoviviparity and Viviparity

Live-bearing ectotherms are sometimes separated into ovoviviparous and viviparous forms, depending upon the maternal support for embryonic growth. The distinction has not always been easily established and has been variously modified to suit different tastes. By definition, ovoviviparous live-bearers retain the ovum until it hatches, with the energy for the growth and development of the young being derived solely from its own yolk. The embryo then emerges from the reproductive tract of the female at any of various stages of development. In contrast, viviparous species pass nourishment to the embryos to augment that contained in their own yolk, and the young are usually born as metamorphosed juveniles.

Conceptually the distinction is simple, but a prenatal nonyolk contribution is not always apparent and cannot be arbitrarily ruled out. Maternal nourishment is provided in different forms and in varying amounts, even within a given species, and some students have defined viviparity not on the existence of maternal contribution but rather on the means by which it is delivered. That is, young nourished within the reproductive tract are deemed to represent viviparity, whereas those embryos housed in specialized dermal or epidermal structures are sometimes considered to represent parental care. Secretions from glands lining the uterus or oviduct, as well as materials received from the skin structures, may all be considered to constitute a maternal contribution, and ovarian contributions, such as additional ova that sustain embryonic growth, are also maternal contributions. Thus viviparity includes such species as marsupial frogs (Hylidae) and pipefish (Syngnathidae).

The existence of an additional contribution can sometimes be evaluated from a comparison of the dry weight of a freshly formed ovum and the dry weight of the young at birth. Approximately 20 to 55% of the dry weight of an egg is used in embryonic maintenance. In oviparous elasmobranchs about 21% of the yolk is used in embryonic metabolism (Amoroso, 1960), and in teleosts the loss of dry weight in metabolism may run as high as 55% (Wourms, 1981). If the dry weight of the young at birth is the same or greater than the dry weight of the egg, it can be assumed that the fetus has obtained some nutrition beyond that contained in its own yolk.

In any event, a possible maternal contribution cannot be easily eliminated. Eggs of trout (*Salmo gairdnerii*), an oviparous teleost, can pick up exogenous nutrients from the surrounding water, and both micro- and macromolecules can be taken in through the chorion (Terner, 1968). Surely the possibility of an ovum acquiring additional nourishment from the uterine lumen or vascularized dermal pockets is very great, even without obvious specializations. This renders the distinction between ovoviviparity and viviparity quite artificial. Viviparity should be considered synonymous with live-bearing. This definition obviates the need to attempt a distinction between ovoviviparity and viviparity and also includes within viviparity such parental contributions as might be made in dermal pouches and the like.

Whether or not amphibian eggs can absorb nutrients through their egg membranes is not known, but fetuses of some of the various marsupial frogs do receive material from the tissues of the mother. Therefore organic materials can clearly pass to the amphibian embryo at some point. Such physiological custody of eggs, when it includes the passage of nutrients from mother to embryo, is included here under viviparity. Parent-young associations that are solely behavioral in expression are considered under parental care.

The issue then becomes the degree to which there is a contribution in addition to the yolk and the means by which that contribution is delivered. Wourms (1981) distinguished between lecithotrophy, which constitutes nutrition from the yolk, and matrotrophy, in which the mother provides nutrients in addition to the yolk. The distinction between lecithotrophy and matrotrophy is clear, although the determination of the degree of matrotrophy is not always apparent.

Lecithotrophy and matrotrophy are not substitutions for ovoviviparity and viviparity, respectively. One can measure or estimate the amount of energy contained in an egg, but it is very difficult to measure, let alone exclude, the degree of matrotrophy.

Wourms (1981) recognized three mechanisms for matrotrophy in fish: oophagy, including adelphophagy (or sibling cannibalism); chorioallantoic placental analogues; and a yolk sac placenta. In elasmobranchs some placental analogues, such as trophonemata, are more efficient than the yolk sac placenta in transferring nutrients (Wourms, 1981). In addition, there is histotrophy, or the consumption of the nutritious lining of the uterus. These modes of nutrition are not mutually exclusive.

In oophagy, ovulation may persist throughout gestation, enabling the embryo to subsist on a continuous supply of yolk from fresh eggs. In adelphagy the embryo may consume its siblings, which, by definition, constitute a maternal contribution, if in a rather indirect mode.

Among the various sorts of viviparous ectotherms, one finds a great array of vascularized expansions of embryonic tissues that lie opposed to maternal capillary systems. Whether or not nutrients can pass from the maternal circulation to the fetal circulation is in most cases unknown, but these junctions unquestionably serve to exchange gases. Embryos of all sorts have an especially high demand for oxygen, greater than that of the parent, and embryos have developed special means for drawing a large amount of oxygen from the maternal (or paternal) circulation. Pond water may be so depleted of dissolved oxygen that it withdraws oxygen from the circulation of larval amphibians and some fish, requiring them to rise to the surface and replenish their oxygen from the air. The procedures for increasing the fetal oxygen supply have been reviewed by Ingermann and Terwilliger (1984).

8.1.2 The Role of the Oviduct and Uterus

The vertebrate oviduct includes two layers of smooth muscle, a rich vascular supply, and frequently extensive glandular tissue. Teleost fish lack the oviduct derived from the embryonic Müllerian duct system, and their gonoduct does not have the supportive features of the oviduct and uterus of other vertebrates. In nonmammalian vertebrates yolk may be exhausted early in embryonic development, after which nutrients enter either as (1) various uterine secretions that are ingested orally by the fetus or (2) vascular contents through a yolk sac and/or chorioallantoic placentae. During pregnancy, the epithelium hypertrophies and the corpora lutea persist. In viviparous teleosts fertilization takes place within the ovary, or sometimes within the follicle, and several mechanisms serve to transmit nutrients to the growing embryos.

8.2 ELASMOBRANCHS

All living elasmobranchs have internal fertilization, and the pelvic fins of the male are modified to introduce sperm into the uterine tract. Pelvic fin modification may include a cartilaginous rod and/or highly vascularized erectile tissue. During mating, the sides overlap so as to form a groove, or they may be rolled in a scroll-like fashion. Details of the pelvic fin structure and function and their roles during coitus are discussed by Gilbert and Heath (1972).

Although oviparity seems to be the primary reproductive mode in elasmobranchs, they lay few eggs at a time, and there is scarcely a reduction in fecundity with viviparity in these vertebrates. Live-bearing elasmobranchs exhibit several different mechanisms for supporting embryonic growth. After the embryo emerges from its temporary shell, it subsists on the energy in the yolk. Yolk digestion is initially performed by blastoderm cells and later becomes extracellular. Eventually yolk platelets are carried to the intestine by ciliated epithelium.

A yolk sac placenta may develop, and it continues to direct nutrients into the embryonic intestine. Interdigitation occurs in some species, whereas in others there is simply an apposition of fetal and maternal tissues. The uterus in viviparous sharks becomes extremely vascularized, and deep furrows develop during pregnancy. Embryos in the Carcharhinidae, the Sphyrnidae, and the Triakidae are nourished by a yolk sac placenta. In *Carcharhinus falciformis* (Carcharhinidae) the yolk sac is apposed to the highly vascularized lining of the uterus, but without interdigitation of maternal and embryonic tissue (Gilbert and Schlernitzauer, 1966). In some species of *Scoliodon* the uterus develops *trophonematous cups*, to which the yolk sac becomes very closely associated. In addition, the surface of the placental cord has expansions (appendiculae) that absorb nutritious uterine secretions (Amoroso, 1960). In *Mustelus laevis* (Triakidae) the very thin egg membrane is bathed by the maternal capillaries. The gestation may be prolonged. The smooth dogfish (*Mustelus canis*) carries embryos for 10 months (Hisaw and Abramowitz, 1939), and gestation in the spiny dogfish (*Squalus acanthias*) lasts for 22 months (Hisaw and Albert, 1947).

After the yolk is exhausted, a nutritious uterine secretion may support embryonic growth, and these substances are absorbed through various placental analogues. This secretion usually contains varying amounts of fat and organic ions, which are taken up by the embryo, so that its dry weight at birth may substantially exceed the dry weight of the egg. These secretions may well occur in all viviparous elasmobranchs, but in the viviparous rays they are the main source of embryonic food. In some rays elongate trophonemata enter the esophagus of the embryo, so that nutrients are placed directly into its digestive tract. In this manner the embryo digests nutrient fluids in its stomach and yolk in its intestine (Amoroso, 1952). The viviparous freshwater stingrays (*Potamotrygon* spp.) give birth to between five and twelve young, which during development are not surrounded by a membrane but are nourished by *uterine milk*.

In the basking sharks (Cetorhinidae) the eggs contain only a small amount of yolk, but the uterus is lined with trophonemata, from which nutritious secretions nourish the embryos (Matthews, 1950). In some species of live-bearing sharks the umbilical cord is covered by foliate projections, which may serve to absorb nutrients.

Oophagy is the mode for embryonic nutrition in several sharks. In the sand shark (*Eugomphodus taurus*), embryos grow without the benefit of uterine secretions. Each uterus contains a single embryo, with its head directed anteriorly toward the ovary. It feeds on ova, which are released continuously throughout an extended gestation, and at birth its stomach is distended with eggs (Springer, 1948). A similar oophagous nutrition is known for the porbeagle (*Lamna cornubica*) (Shann, 1923).

8.3 TELEOST FISH

Although most teleosts are oviparous, among live-bearing taxa there is a broad gradation from little to virtually all embryonic nutrition being matrotrophic. An amazing diversity of mechanisms for delivering nutrition to the embryos exists among teleosts. Fertilization occurs in the follicle in some fish, an almost unique site in vertebrates. In other teleosts the gametes may join within the lumen of the ovary. In follicular gestation the embryos are separated, but in ovarian gestation they may be separated by ovarian septa or all together, and dead individuals may serve as food for survivors. Live-bearing fish dwell in both fresh water and the ocean, and the evolutionary thrust from oviparity to viviparity is not obvious in any group. Because viviparity has developed de novo many times in teleosts, and the modifications are diverse and seem to follow no obvious pattern or trend, each family or group is discussed separately.

8.3.1 Latimeriidae

The coelacanth, *Latimeria chalumnae*, produces as many as 19 rather large eggs (9 cm diameter), with a total mass of 300 g (Balon, 1984). It apparently has developed no special structures for maintaining its embryos. Gestation may last 10 to 13 months, after which an 800-g infant is born (Hureau and Ozouf, 1977). Growth is achieved by consuming eggs and siblings.

8.3.2 Scorpaenidae

The species of the marine teleost genus *Sebastodes*, or rockfish (Scorpaenidae), produce very small young after a gestation of one month. Embryonic development takes place within the cavity of each of the paired ovaries, and young are released as their eggs are about to hatch. In species along the southern California coast, young are born in the winter. There are no obvious specializations for prenatal care of the young. The embryo-filled eggs lie against the vascularized ovarian lining, and the fluid serves as a medium of exchange of oxygen and metabolic waste, and possibly some nutrition. Rockfish are unusual among viviparous fish in their great fecundity: broods may exceed two million in large females. The ovaries have a dual arterial supply in contrast to the single ovarian artery in most teleosts (Moser, 1967).

8.3.3 Cottidae

In the sculpins (Cottidae), many marine species have internal fertilization, and the male has a urogenital papilla as an intromittent organ. Within the family there is a gradation from species with external fertilization (*Cottus* spp.) to species with internal fertilization, that is, from oviparity to viviparity. Fertilized eggs have frequently been found within the reproductive tract of females, but usually cleavage does not begin until spawning (Hubbs, 1966).

The Baikal cod or sculpin (*Comephorus*), which is sometimes placed in a separate family (Comephoridae), is related to sculpins and is viviparous. Internal fertilization is achieved by an anal or urogenital papilla. Young develop within the ovary while the adults dwell in the deep waters of Lake Baikal, but a migration to the surface precedes parturition.

8.3.4 Embiotocidae

Teleost viviparity is best developed in the surfperch (Embiotocidae), a marine family of coastal species that occur from Baja California around the northern Pacific to Japan (Fig. 8.1). The median ovary is actually bifurcate, being joined at the posterior end. It becomes highly vascularized and is divided into compartments. Nutrition is through trophonemata and histotrophy.

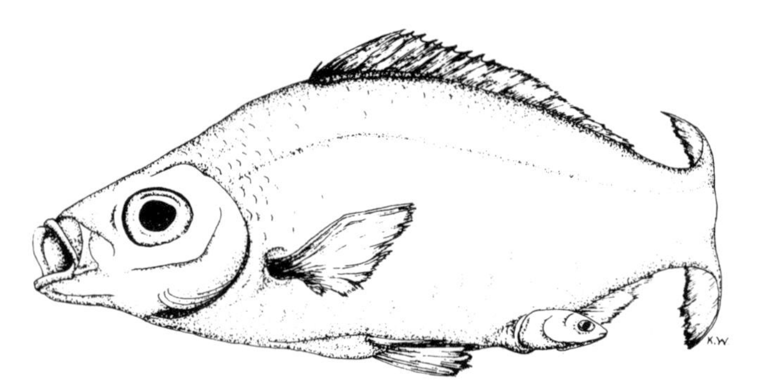

Figure 8.1 Pregnant female of a viviparous teleost. *Phanerodon furcatus* (Embiotocidae), showing the head of a full-term young emerging from the cloaca. (Katherine West.)

In the surfperch the eggs are very unusual among nonmammalian vertebrates in having little or no yolk. Ovarian fluid of *Cymatogaster aggregata* contains peptides that are apparently derived from the ovarian lining, and the absence of these peptides from the circulation suggests that they do not emanate from the liver. Also, the absence of serum progesterone and estrogens in gravid females indicates that these fish may be the only teleosts not producing a vitellogeninlike protein (deVlaming et al., 1983). Surfperch are also unusual among vertebrates in having embryonic spermatogenesis, with testes of the neonate male containing spermatozoa (Hubbs, 1921).

In embiotocid fish, embryonic hemoglobin (at midgestation) has a greater affinity for oxygen than obtains in adult hemoglobin. Differences between fetal and maternal hemoglobins in *Embiotoca lateralis* are both structural and functional. Oxygen affinity is reduced by organic phosphates, and fetal hemoglobin of *E. lateralis* increases its oxygen affinity by reducing adenosine triphosphate and/or guanosine triphosphate (Ingermann and Terwilliger, 1984).

Surfperch have a life cycle that results in females being continuously pregnant. In the well-studied *Cymatogaster aggregata*, mating closely follows parturition in the spring, and sperm lodge externally on the ovarian surface. The sperm of *Cymatogaster aggregata* have a rate of glycolytic metabolism comparable to that of mammals and presumably subsist on ovarian sugars during their storage on the ovarian surface (Gardner, 1978). Folliculogenesis takes place in autumn, and in December sperm are reactivated. The embryos develop in fluid-filled chambers separated by ovigerous sheets. Epithelial cells secrete the nutritious material in which the embryos float. With the approach of parturition, the ovigerous sheets increase in vascularization, and extensions (trophonemata) may enter the branchial cavities of the fetuses (Wiebe, 1986a).

8.3.5 Zoarcidae

The viviparous eel pout (*Zoarces viviparus*) of the eastern North Atlantic produces well-formed young after a gestation of four months. The eggs, which have very little yolk, are fertilized in the ovarian lumen. The ovary becomes lined with highly vascularized villi, which secrete a nutritious fluid, the embryotrophe. The young themselves do not develop any placental analogues, and there is no direct contact between the ovarian lining and the embryos. Nutrients are taken in through the gut, and the embryonic stomach contains erythrocytes and other cellular material, apparently from the ovarian lining. There is absorption through the hind gut, which is expanded and folded; after parturition the hind gut is reduced (Kristoffersson et al., 1973).

8.3.6 Hemirhamphidae

The halfbeaks include both marine and freshwater species. Some brackish water species and some tropical freshwater forms are known to be viviparous, whereas marine species are oviparous. Nutrition is by both lecitrophy and matrotrophy.

8.3.7 Syngnathidae

The pipefish and sea horses are small marine fish in which the male cares for the eggs during incubation. In addition to the established behavioral effects of prolactin in fish, there is an important physiological role in the marsupium, or pouch, of the male sea horse (*Hippocampus*). The marsupium houses the developing eggs, and the lining of this pouch hypertrophies under the stimulation of testicular secretions. A protease secreted from the lining of the marsupium promotes the breakdown of proteins in the marsupium fluid, and these amino acids are presumably taken in by the developing embryos. The activity of the lining declines after partial hypophysectomy, but the activity is stimulated by prolactin in both intact and hypophysectomized males (Boisseau, 1965).

8.3.8 Clinidae

Some clinid blennies have a well-developed anal papilla, and several species are live-bearers. After intrafollicular gestation of a very small egg, there is a substantial increase in fetal weight. Nutrients are secreted by the follicular lining. Food is absorbed via the vascularized unpaired fins, which have conspicuous epidermal ridges that disappear before birth, and the gut (Veith, 1980).

8.3.9 Goodeidae

These little freshwater fish occur in the high plateau region of west-central Mexico and adjacent United States; their unusual viviparity has been discussed in detail by Turner (1940a). Fertilization is by means of a gonopodium formed from the anal fin. The eggs have a small amount of yolk. The embryos develop within the ovarian lumen, which becomes greatly distended during pregnancy (Mendoza, 1940). Nutrients are absorbed by the gut (through gill openings) and by trophotaeniae, which are long projections of the gut into the ovarian lumen (Fig. 8.2). The most interesting specialization of these little fish is the greatly developed pericardial sac, which obtains a superficial vascular supply from the adjacent yolk sac (Fig. 8.2). This extraembryonic tissue is a very effective placental analogue, and it absorbs nutrients secreted from the ovarian lining. The pericardial sac is later withdrawn.

The vascularized trophotaeniae develop as the meager yolk supply is depleted. Ultrastructural study of the trophotaeniae of *Girardinichthys viviparus* reveals two types of epithelial cells. One consists of cuboidal cells with microvilli, many mitochondira, and an agranular tubulolamellar network, suggesting absorptive activity. These absorptive cells show micropinocytic activity, with lipid droplets and cellular debris. The microvilli exhibit alkaline phosphatase activity, characteristically associated with the transfer of low-molecular-weight material across cell membranes. The other type of epithelial cell has microridges and a reticulate arrangement with intercellular spaces, and it apparently functions in gas exchange (Schindler and de Vries, 1986).

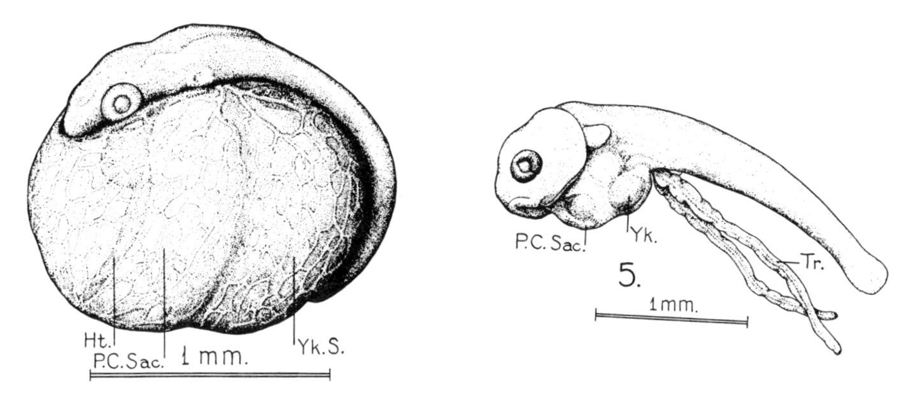

Figure 8.2 *Left*: early stage of an embryo of *Goodea luitpoldii*, showing the relatively small yolk sac (Yk.S), heart (Ht.), and expanded pericardial sac (P.VC.Sac.) with portal network. *Right*: early stage of embryo of *Characodon lateralis*, with yolk sac receding, the maximal development of the pericardial sac, and growing trophotaeniae (Tr.). (From Turner, 1940a.)

In *Ameca splendens* the ovary is divided into two chambers (Lombardi and Wourms, 1985). Five highly vascularized trophotaeniae, derived from the hind gut, lie against the uterine lining and constitute a very effective placental association. The embryo increases 15,000% in dry mass (Lombardi and Wourms, 1985).

8.3.10 Anablepidae

These fish occur in both fresh and brackish waters of Central and South America, and the family now includes the genus *Jenynsia*, which previously had been separated as the Jenynsiidae. Species of *Anableps*, the four-eyed fish, dwell in brackish waters and are best known for their "divided" eyes, which are modified to see in both water and air. In addition, they are bizarre in that the genital orifice in both sexes is placed to one side of center, so that both males and females are either sinistral or dextral. Presumably random mating is impossible.

In *Anableps* the eggs are very small with little yolk, and by the time of parturition there is a 30 to 40 thousand percent increase in body mass of the offspring. There is a large extraembryonic pericardial sac, which functions as a follicular pseudoplacenta (Turner, 1940a). The gut is expanded and vascularized, and the embryo acquires food through both structures.

The eggs of *Jenynsia lineatus* are fertilized with the follicle, but the embryo soon breaks out, and the remainder of its development is in the ovarian lumen. Initially the ovarian epithelium is secretory, but eventually elongate trophonemata grow out from the ovarian lining and enter the opercular opening and even the gut of the fetus (Fig. 8.3). These trophonemata have been called branchial placentae.

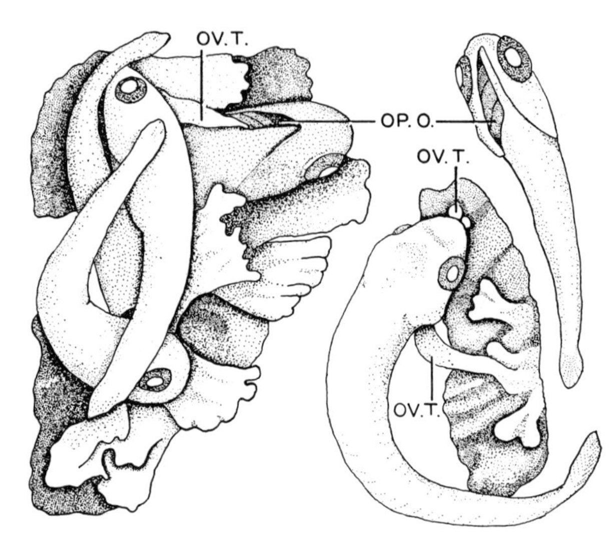

Figure 8.3 Embryos of *Jenynsia lineatus* in late gestation, showing ovarian tissue (OV.T.) developed as trophonemata, which enter the opercular opening (OP.O.). (From Turner, 1940b.)

8.3.11 Poeciliidae

These tropical and subtropical freshwater fish are well known to tropical fish culturists and include the familiar guppies, swordtails, and mollies and their allies. The male has a gonopodium, modified from the anal fin, for the transfer of sperm to the female. Along one side the gonopodium is a groove. Sperm may be placed directly in the genital orifice of the female, or spermatophores may be deposited adjacent to the genital opening (Rosen and Gordon, 1953).

Following mating in the sailfin molly and other poeciliids, sperm are stored in crypts or folds on the surface of the ovary (Tseng, 1972), and they enter the follicle where the ovum is fertilized. In poeciliids the follicular wall and the yolk sac are both highly vascularized, facilitating chemical exchange (Fig. 8.4). As the embryo develops, it breaks out of the follicle, and gestation continues within the ovary for approximately four weeks (Bretschneider and de Wit, 1947).

The amount of yolk varies from one species to another, and in some species reproduction appears to be lecithotrophic, with no measurable amount of maternal contribution. The yolk in the sailfin molly may vary from one year to the next, even in the same population, and in some years reproduction may be entirely lecithotrophic (Trexler, 1985).

The eggs of the guppy (*Poecilia reticulata*) contain substantial yolk, and the embryos are lecithotrophic. Among some 16 species of poeciliids for which dry weights of eggs and young at full term are compared, all appear to receive some maternal nourishment (Scrimshaw, 1945). Some species have small eggs and are highly matrotrophic. In *Heterandria formosa* the dry-weight increase from egg to birth is some 3900% (Scrimshaw, 1945).

Superfetation occurs in some species of poeciliid fish and is apparently unique among live-bearing teleosts. As follicles mature in close succession, fertilization follows the same sequence, so that there may be two or more broods at different levels of development. These broods tend to be small and the times between parturition close together. The ovary of *Poeciliopsis prolifica* may hold four or five broods in different stages of growth, and parturition occurs at intervals varying from two to eight days. The brood size is small (averaging four), and final growth is rapid, so that the mother is not unduly distended at any time. Superfetation allows the female to avoid a sudden major demand on energy just prior to parturition, and the reduction in ova size in such species allows the simultaneous development of two or more broods (Thibault and Schultz, 1978).

8.3.12 Deep-Sea Families

Viviparity is known from benthic members of several families. Much of the information comes from examination of museum specimens, and their physiological modifications for viviparity can only be guessed at. Some of these modifications are discussed by Mead et al. (1964). Other viviparous families, while not truly from deep water, occur beyond tidal waters and are discussed together here.

The Ophidiiformes include the Bythitidae and Aphyonidae, which have both deep-sea and coastal species and which can be either oviparous or viviparous

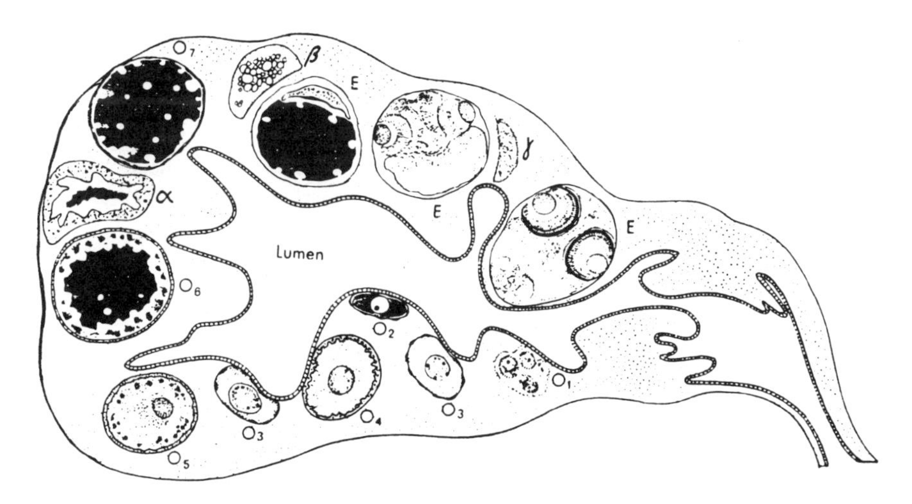

Figure 8.4 Transverse section through the ovary of a guppy (*Poecilia reticulata*). Fertilization takes place within the follicles lining the wall of the ovary. The several stages of development shown do not occur simultaneously in the guppy. O_1 to O_7, oocytes in various degrees of development; E, embryos; and α, β, oocytes in various stages of atresia. (From Lambert, 1970, courtesy of Academic Press.)

(Nybelin, 1957). The Parabrotulidae have one benthic member, *Parabrotula plagiophthalmus*, which lives at 1500 m. The embryos are provided with food by trophotaeniae.

In the Ophidiidae, marine benthic fishes, there are both oviparous and viviparous forms. The males possess a conspicuous intromittent organ, and fertilization occurs within the ovary. The small embryos have trophotaeniae as absorptive surfaces (Turner, 1936). *Microbrotula randalli* is a viviparous species from Samoa and New Hebrides. The embryo bears three long trophotaeniae, which exceed the embryo in length. The epithelium of the trophotaeniae resembles that of the intestine and presumably absorbs ovarian secretions (Cohen and Wourms, 1976).

8.4 AMPHIBIANS

Live-bearing amphibians occur in all orders, and the requirements are essentially the same as those for fish. There must first be some means of internal fertilization. Internal fertilization obtains in all of the caecilians (Gymnophiona or Apoda). An eversible cloaca, inserted into the cloaca of the female, is an intromittent organ. Most salamanders transfer sperm by means of a spermatophore, but external fertilization remains as a primitive trait in the Hynobiidae and the Cryptobranchidae. Although most anurans fertilize their eggs externally, during amplexus, one (*Ascaphus truei*) has a cloacal intromittent organ, and certain terrestrially breeding species achieve internal fertilization without major morphological modifications. The constraint of coelomic space reduces the number of young that can be housed to metamorphosis or beyond. Therefore, in general, viviparity is accompanied by a reduction in fecundity.

8.4.1 Gymnophiona

The pantropical caecilians are both aquatic and terrestrial and are also either oviparous or viviparous. Viviparity in these amphibians has been discussed by Parker (1956) and Wake (1977a, 1980a, 1982). Viviparity is known from three families of caecilians (Caeciliidae, Scolocomorphidae, and Typhlonectidae), but some species of the Caeciliidae are oviparous (Wake, 1977b). From one-half to three-fourths of caecilians are viviparous, and gestation may last from seven to eleven months. During embryonic development, there is a tremendous increase in body mass.

Live-bearing caecilians are mostly dwellers of low elevations, and the departure from egg laying seems not to have been in response to low temperatures (Wake, 1977a). Within the fully aquatic family Typhlonectidae all species are viviparous, but some terrestrial caecilians are viviparous, and the ecological significance of live-bearing in these amphibians is not apparent.

The oviducts of viviparous caecilians are provided with secretory beds that provide nourishment for embryos. The eggs are large (about 4 to 10 mm) with a large amount of yolk. They provide materials for initial embryonic growth, but at birth the young is many times larger than the egg. Thus matrotrophy is important, but there is no placental analogue for exchange of nutrients and metabolic wastes. After the yolk is exhausted, the embryo emerges from the egg membranes and feeds on fat-rich secretions from the secretory beds, and the transitory embryonic teeth may serve to loosen the oviducal epithelium (Wake, 1976). Corpora lutea are active throughout pregnancy (Wake, 1968). Gills develop during the period of matrotrophy and serve for gaseous exchange and possibly the acquisition of nutrients, but they are lost before parturition.

The viviparous caecilian (*Dermophis mexicanus*) of Central America has a gestation lasting one year, and females apparently breed biennially. In this species the ova are small, and the large neonate achieves its growth from extensive maternal contribution (Wake, 1980a).

Amphibian viviparity is reviewed by Wake (1982).

8.4.2 Anura

Among many tropical frogs there are elaborate schemes for parental care. When these methods appear to include parental nutrition, it is most logical to consider these species as viviparous. Other species are discussed in Chapter 10, Section 9.5.4.

8.4.2.1 Bufonidae. The several species of the bufonid genus *Nectophrynoides* illustrate a transition from an aquatic, oviparous species with feeding tadpoles (*N. osgoodi*) to an oviparous species with free-living but nonfeeding terrestrial larvae (*N. malcomi*) to terrestrial, live-bearing species (*N. tornieri*, *N. viviparus*, and *N. occidentalis*), which produce small froglets (Wake, 1980b).

In live-bearing species of *Nectophrynoides* there are from two to ten eggs, which contain little yolk, and the young are nourished by a uterine secretion rich in mucoproteins. The embryonic gut is well developed, and the tail is also long and richly provided with capillaries. The adult females estivate during the dry season, at which time the corpora lutea remain active and secrete progesterone, which inhibits embryonic growth (Xavier and Ozon, 1971). The live-bearing *Nectrophrynoides* are montane in Nigeria. The females become active with the return of the rains, and the corpora lutea regress, progesterone declines, and embryonic growth resumes (Xavier, 1977).

8.4.2.2 Hylidae. Certain hylid frogs have achieved independence from standing water by carrying their eggs either cemented to the dorsum of the female or within a dorsal pouch, or marsupium. The dorsal pouch of the marsupial frogs (*Gastrotheca* spp.) is an adaptation to their arboreal life. Not only is the pouch lining itself well provided with blood vessels, but the dorsal and ventral walls lie in folds that are connected by vascularized threads of tissue (del Pino, 1980). During development, the gills almost completely enfold the embryo and transport gases and fluids between the embryo and parent. When the young are well developed, the female enters water and, with the aid of one or both hind feet, enlarges the entrance of the marsupium, allowing the tadpoles to swim out. In *G. marsupiata* and *G. riobambae* the young emerge as tadpoles, but in *G. ovifera* metamorphosis occurs in the marsupium and emergence occurs on land (del Pino et al., 1975).

The pouch development in *Gastrotheca riobambae* has been discussed by del Pino (1980). Prior to oviposition there is a reduction in the degree of keratinization in the pouch, with an increase in blood supply and the development of mucous glands, which appear to secrete mucopolysaccharides. The walls of the pouch are thick and become closely associated with the developing embryo. The walls remain closed during a pregnancy of some 103 to 120 days (in the laboratory), during which time the corpora lutea remain large (del Pino and Sanchez, 1977). Each egg settles into a vascularized pocket, and early in development the embryos develop expanded gills called *bell gills*, which gradually grow out to envelop the embryo. Both the pockets and the gills are provided with extensive capillary systems.

Gastrotheca riobambae females produce as many as 76 tadpoles (Fitzgerald et al., 1979). Gill circulation ceases at hatching, and the gills are withdrawn into the body 24 hours later. Jones et al. (1973) suggest that in addition to gaseous exchange, embryonic gills might provide nutrients to the embryos. Inasmuch as there is no appreciable change during embryonic development (del Pino et al., 1975), matrotrophy is indicated.

In species of some other hylid genera, *Flectonotus* and *Fritziana*, the pouch is less elaborately developed, does not enclose the eggs, and has weaker vascularization.

8.4.2.3 Myobatrachidae. The Australian *Assa darlingtoni* is unusual in that the male develops lateral pouches that house the eggs during their development. The eggs are large (2.2–2.6 mm). The rather large size (11 mm) of the emerging young (Straughan and Main, 1966) suggests that there is some nutrition from the parent.

8.4.2.4 Leptodactylidae. Perhaps the most bizarre and surprising mode of live-bearing is seen in the Australian gastric brooding frog (*Rheobatrachus silus*). After fertilization, the female swallows the eggs and broods the young in her stomach. The ova are relatively large, from 4.0 to 4.7 mm. After the female swallows her eggs, the stomach lining becomes vascularized and acid secretion ceases. Mucosa cells become atrophied. The developing young appear to secrete a substance (prostaglandin E_2) that inhibits gastric acid secretion (Tyler et al., 1982). Prostaglandin E_2 may emanate from the mucous coating of the eggs and skin of the embryos. Nutrition from the mother is not known, but fragments of squamous epithelium occur from the embryo's stomach to its colon. Gestation is estimated to take from six to seven weeks followed by the emergence of metamorphosed adults (Tyler and Carter, 1981).

Tyler (1983) pointed out that the development of gastric brooding must have occurred in one step, without intermediate or transition stages. Gastric brooding would seem to have evolved by virtue of (1) the female, perhaps through cannibalism, consuming her own eggs and (2) prostaglandin (and its suppression of gastric secretion) protecting the eggs from being digested.

Darwin's frog (*Rhinoderma darwinii*), of Chile, is renowned for its unique method of oral incubation of its eggs during the latter part of their development. The female lays her eggs on the ground, where they are guarded by the male. After

20 days, at which time there is some internal movement, the male picks up the eggs in his mouth, and embryonic development continues within his vocal sacs. Tracers (horseradish peroxidase, ^{3}H valine, and ^{3}H leucine) injected into the paternal circulation were recovered from the larvae, indicating the possibility of paternal nourishment of the embryos; the injection was initially through the skin and subsequently through the gut (Goicoechea et al., 1986).

8.5 REPTILES

As in elasmobranchs, teleosts, and amphibians, there are several degrees of maternal dependence and different mechanisms for delivering nourishment to embryos. Live-bearing occurs only in squamates—the lizards, snakes, and amphisbaenians—and the scattered phylogenetic appearance and diversity of transfer devices indicates the ease with which live-bearing may have developed. The mechanisms for transferring nutrients to the growing fetus are reviewed in detail by Yaron (1985). A detailed review of the phylogenetic occurrence of viviparity and its adaptive significance in reptiles is given by Shine (1985) and Tinkle and Gibbons (1977).

8.5.1 Environmental Significance of Squamate Viviparity

An early body of evidence established a clear association between elevation and viviparity in lizards and snakes in southeastern Australia, Tasmania, the Pyrenees, and the French Alps (Weekes, 1935). The retention of eggs, in itself, requires no specialization of the uteri, but in live-bearing reptiles the uteri have increased vascularization. The subsequent loss of the shell and expansion of vascularization on embryonic membranes and uterine epithelium have recurred many times. Almost one-fifth of the species of lizards are viviparous, and of these approximately two-thirds are skinks (Scincidae). Most of these species have evolved in cold climates (Blackburn, 1982). In the Serpentes, viviparity is seen in about one-fifth of the species (Shine, 1985; Blackburn, 1985).

In addition, some viviparous species adjust gametogenesis so that mating occurs in the spring and ovulation and fertilization in the autumn (see Chapter 13, Patterns of Seasonality, Section 13.6.1). In the viviparous iguanid lizard, *Sceloporus jarrovi*, ovulation and fertilization in late autumn are followed by slow embryonic growth in the winter. In March a chorioallantoic placenta develops, and there is a reduction in mass of the corpus luteum concurrent with a rise in progesterone, presumably of placental origin (Guillette et al., 1981). This schedule allows the gravid female to bask in the winter and the young to be born in the spring, which is presumably a more propitious season than just prior to hibernation. Similarly, in the montane viviparous lizard *Barisia imbricata* of Mexico, maximal testicular mass is seen in the spring, with follicular growth, ovulation, and mating in the autumn (Guillette and Casas-Andreu, 1987).

Retention of the eggs allows the female to regulate their temperature as she adjusts her own body temperature by behavioral means. Additionally, several pythons, which are oviparous, are well known to warm their eggs by becoming thermogenic during incubation. It is quite possible that some lizards also become thermogenic while retaining their eggs. Although no such example has yet been described, it would be a plausible development.

As pointed out by Shine and Bull (1979), viviparity has arisen more frequently in genera in which parental care is known, the transition from brooding eggs to retaining them being a minimal stress on the female.

8.5.1.1 Viviparity in Cold Climates. The cold climate hypothesis has been examined by contrasting the temperatures to which eggs (or embryos) are exposed in viviparous and oviparous Australian lizards and in viviparous and oviparous species of the Nearctic iguanid genus *Sceloporus* (Guillette et al., 1980). The higher temperatures surrounding eggs that are retained in the uteri result in earlier births (Shine, 1983). Thus viviparity would provide an advantage to squamates living in areas in which soil temperatures would be sufficiently low to retard hatching times of oviparous species. This conclusion is compatible with the mid- or late-summer hatching time of oviparous species of *Sceloporus* and the spring parturition of viviparous species.

Viviparity begins with facultative retention of eggs, with little, if any, change in the nature of the shell. This phenomenon is widespread among the squamates, and a large number of lizards and snakes tend to retain eggs for the initial part of their embryonic development (Shine, 1983). Among separate populations of an anole (*Anolis cybotes*) on Hispaniola, egg retention with embryonic development increases from sea level to 1150 m (Huey, 1977). The same trend toward egg retention occurs along latitudinal gradients in some lizards and snakes. *Lacerta vivipara* is live-bearing in northern Europe but oviparous along the Mediterranean countries.

Viviparous squamates are not necessarily at a disadvantage in a warm, humid climate. In southeastern United States, natricine snakes (species of *Seminatrix*, *Liodytes*, *Regina*, and *Virginia*) are all viviparous and today confined to warm, moist habitats. In the very recent past, while the Northeast was covered by Wisconsin ice sheets, the environment of the southeastern states was cool temperate. Some of these viviparous natricines (e.g., *Storeria*) have returned to the north and occupy recently glaciated regions, and others remain in the warm Southeast. The existence of these live-bearing snakes in a warm, moist climate today is not incompatible with the cold climate hypothesis. It is just as plausible that the just-mentioned natricine snakes have never been exposed to a cold climate but simply to a cooler climate than obtains today in the southeastern United States.

Some viviparous species of squamates may have been derived from an oviparous species that had been exposed to a cold climate or to a cold and moist climate (Guillete et al., 1980). There remain squamates for which the cold climate hypothesis would seem not to apply. Two other environmental factors may have played a role in the development of viviparity in squamates, and in some taxa there is at present no apparent ecological significance to viviparity. There are examples of live-bearing reptiles in arid and aquatic environments, but their history is unclear. One can argue that tropical and desert-dwelling live-bearing reptiles became viviparous in a cold climate and subsequently became tropical or desert dwellers. There is, however, no evidence for such a history, and it is important to avoid imagining an adaptive sequence of events simply to conform to a hypothesis.

8.5.1.2 Viviparity in Deserts. Live-bearing is clearly an advantage in arid habitats, regions in which eggs laid in surface soil are at risk of desiccation. Retention of eggs prevents their loss of water. This would be advantageous in an arid environment, and viviparous species would be adapted to desert life (Weekes, 1935). It does not necessarily follow that viviparity has evolved in desert-dwelling live-bearers, but the possibility should not be rejected.

It seems possible that aridity might induce egg retention. The anole, *Anolis aeneus*, retains its eggs during dry periods and proceeds to excavate for nests following rainfall (Stamps, 1976). Two African skinks (*Mabuya homacephala homacephala* and *M. quinquetaeniata margaritifera*) can be oviparous in South Africa (Visser, 1975) but may also be live-bearers (Fitzsimons, 1943). In southwest Tanzania, a region with a wet-dry seasonality, the oviparous *Agama cyanogaster* lays its eggs during the rainy period. The viviparous *Mabuya striata* gives birth mostly in the first half of the dry period (Robertson et al., 1965) but does, in fact, breed throughout the year (Barbault, 1975). The only known viviparous species of *Lygosoma* are confined to the arid regions of Somalia and eastern Kenya, suggesting aridity as a stimulus to their live-bearing mode. These live-bearing skinks represent two stocks, the other members of which are oviparous, indicating that viviparity developed at least twice in these arid regions (Greer, 1977).

Most species of desert squamates remain oviparous. It is not surprising that there are not more viviparous squamates in the world's deserts, for most modern deserts are relatively young (Axelrod, 1979; Sarnthein, 1978; and others).

8.5.1.3 Viviparity in Aquatic Snakes. There is a tendency among some aquatic snakes to be live-bearers. Viviparity is best developed among some of the marine snakes (Hydrophiidae), which are widely distributed in tropical seas. Also, the species of Acrochordidae of the tropical Orient are all live-bearers. Both of these families are well adapted to marine life in many ways and clearly have had a long history in the water. There seems to be no evidence that these reptiles developed viviparity in a cold climate.

8.5.1.4 Viviparity without Environmental Significance. In some live-bearing squamates there is no clear environmental inducement for live-bearing. This is the mode of reproduction of the boas (Boinae), which are mostly tropical snakes of the New World and Madagascar. Inasmuch as Madagascar parted from Africa and Africa separated from South America in the Cretaceous, viviparity in this group must date from at least that period. The live-bearing amphisbaenians represent a pantropical group, most of which are oviparous. At least five species are viviparous and produce well-developed young from eggs rich in yolk. Since the Cretaceous, these two groups of reptiles have occupied the warmest regions of the world and would not seem to have evolved the live-bearing mode as a result of any obvious environmental pressure. Viviparity in reptiles has reached its greatest modification in the Brazilian skink, *Mabuya heathi*. It dwells close to the equator and is discussed in Section 8.5.2.2.

8.5.2 Mechanisms of Nutrition

A number of live-bearing species are largely, and perhaps entirely, lecithotrophic, suggesting how reptilian viviparity might have developed. It has been proposed that the reptilian calcified shell inhibits gas exchange between the embryo and its mother (Packard et al., 1977). Embryos breathe effectively through the shell in oviparous species, and the difficulty would seem to be in providing the uteri with ample oxygen and in enhancing the affinity of the embryonic hemoglobin for oxygen.

8.5.2.1 Squamate Placentae. For many years most students assumed that all live-bearing reptiles were ovoviviparous, but it is now clear that many species obtain some nutrition from the maternal circulation and that the concept of ovoviviparity is artificial and unrealistic. The placentae of a few species are very efficient in the transfer of large molecules. One of the earliest studies implicating the reptilian placentae in the transfer of more than respiratory materials was done by Clark et al. (1955). They pointed out that the egg of the garter snake (*Thamnophis sirtalis*) is noncleidoic, and they indicated that it might transfer amino acids to the fetuses. Even in some oviparous lizards (*Eumeces obsoletus*, *Sceloporus undulatus*, and *Crotophytus collaris*), oviducal vascularity significantly increases during gravidity (Masson and Guillette, 1987).

Among several unrelated taxa of squamates, there is a major infusion of nutrients from the mother to the fetus. This transfer usually involves a highly efficient chorioallantoic placenta and/or a yolk sac placenta, and these structures vary in the roles of the membranes involved. The chorioallantoic placenta of the garter snake (*Thamnophis sirtalis*) lies appressed to the highly vascularized uterine wall, but it lacks interdigitation (Hoffman, 1970). There may or may not be some degree of rugosity of the fetal and maternal surfaces and some moderate interdigitation of fetal and maternal membranes, sometimes with erosion of maternal tissue. Details of these structures are discussed by Yaron (1985).

8.5.2.2 Absorption of Nutrients. In some squamates the transfer of materials may be limited to respiratory gases and inorganic ions. A live-bearing Australian skink, *Sphenomorphus quoyi*, has a gestation of three months, during which time the embryo absorbs inorganic ions (calcium, sodium, and potassium). However, there is no uptake of amino acids, and the embryo at birth has less dry weight than the freshly fertilized egg and about half of the initial lipids. The freshly ovulated egg weighed 286 mg (dry), whereas the full-term embryo weighed 240 mg (dry). When the embryos were removed from the uterus and reared in a 0.6% NaCl or a nutrient medium at 27°C, development sometimes continued to full term (Thompson, 1977). The embryonic hemoglobin during mid-development has a greater affinity for oxygen than the maternal hemoglobin does, but the two become equal by the time of parturition (Fig. 8.5).

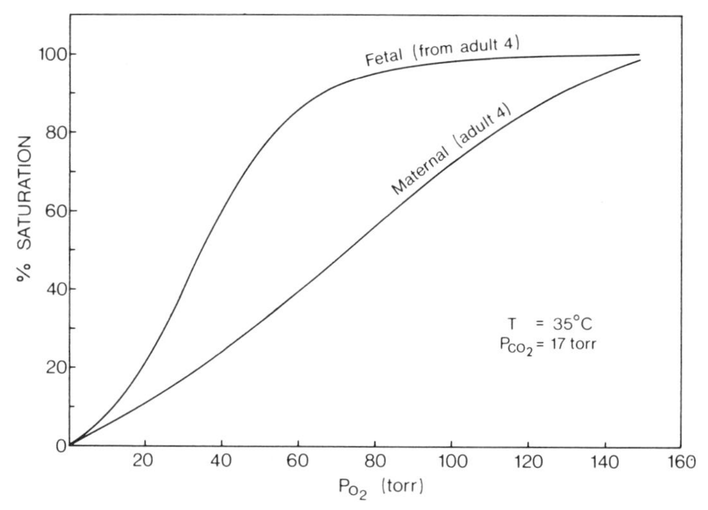

Figure 8.5 Representative oxygen equilibrium curves of whole blood from a gravid *Sphenomorphus quoyi* and one of her embryos. (From Grigg and Harlow, 1981, courtesy of Academic Press.)

Lacerta vivipara is also facultatively viviparous; in northern Europe it is a live-bearer, but in the Mediterranean region it is oviparous. Small amounts of inorganic ions may pass to the embryo through its shell, but fertilized ova have been successfully incubated in physiologic serum (Panigel, 1956).

The placentae of other species have developed a much greater efficiency in delivering nutrients to the embryos. Nutrients and oxygen pass to the embryo through the yolk sac or the chorioallantoic placentae, and metabolic wastes are removed. The placenta of the garter snake (*Thamnophis sirtalis*) allows the passage of urea from the embryo, and the embryo appears to receive approximately 45 mg of protein as amino acids (Clark and Sisken, 1956). Early viviparous embryos are ureotelic and become uricotelic just before birth, but oviparous embryos are uricotelic from the beginning of development (Packard et al., 1977).

The chorioallantoic placenta of the Nearctic water snake, *Nerodia sipedon*, permits passage of both ^{22}Na and ^{131}I. Twenty-four hours after injection of NaI^{131} in gravid water snakes, radioactive iodine appears in fetal thyroids; the introduction of $^{22}NaCl$ into pregnant *N. sipedon* and *N. cyclopion* is followed by the presence of ^{22}Na and ^{131}I in fetal livers and entire fetal bodies (Conaway and Fleming, 1960).

Matrotrophic nutrition is indicated in the natricine snake *Regina grahami*: the eggs are without a shell, and the oviducts become hypertrophied during pregnancy. The dry weight of embryos increases 30% during pregnancy (Hall, 1969). The transfer of ^{14}C-labeled leucine from the maternal circulation of the chameleon (*Chamaeleo pumilus*) suggests that amino acids are carried from the maternal to the fetal circulation (Veith, 1974).

The night lizard (*Xantusia vigilis*) is a desert-dwelling species of the Xantusiidae, a family occurring from southern California south to Panama. All known species are live-bearers. The yucca night lizard (*Xantusia vigilis*) has a brood of only one or two young. After parturition, the mother pulls the fetal membranes from her cloaca and eats them (Cowles, 1944). The wet weight of the neonate is twice that of the freshly ovulated egg (Miller, 1951). In *X. vigilis*, absorption of materials across the placenta is rapid: labeled leucine moved from the maternal circulation to the yolk, amniotic fluid, and embryo in 15 minutes, having first been concentrated in the maternal liver (Yaron, 1977). The absorption of amino acids is quite possible.

The best-developed dependency on matrotrophy is shown by the tropical Brazilian skink, *Mabuya heathi*. From extremely small eggs, after a gestation of nearly one year, rather large young are born, with a dry mass increase of 38,400% (Vitt and Blackburn, 1983). The dry mass of the egg is approximately 0.4 mg, and the dry mass of the near-full-term fetus is 154 mg (Blackburn et al., 1984). This skink is also unusual in that the very young females (three to four months of age) mate while they still have most of their growth before them, and they continue to grow rapidly prior to the rapid growth of their embryos. The yolk sac placenta is replaced by a chorioallantoic placenta in the early stages of development, after which most of the fetal growth is seen. *Mabuya heathi* is similar to the teleost family Embiotocidae in its eutherianlike adaptations for viviparity.

In most species of viviparous squamates in which the corpora lutea have been examined, it appears that corpora lutea are active and serum progesterone is present (see Chapter 3, Section 3.2). The function of progesterone in viviparous squamates remains equivocal. Steroidogenesis in the placenta of squamates is, as yet, unknown (Yaron, 1985). Clearly the squamate placenta can pass inorganic ions and amino acids to the fetus, and waste products can be removed from the embryonic circulation.

8.6 SUMMARY

Viviparity in the elasmobranchs must be extremely old. Its irregular appearance within families and its frequency throughout the class indicate that it has developed independently many times. In the mechanics of providing nutrients, there is considerable convergence. Unlike the pattern seen in most vertebrates, in sharks and rays fecundity is not reduced with the acquisition of viviparity. As pointed out by Wourms (1981), live-bearing rays (Rajiformes) are mostly tropical and subtropical, but the environmental or evolutionary significance of this fact is not apparent. Viviparity in sharks is most characteristic of the larger species. Viviparity could be an advantage for a predator, because greater size at birth is surely an asset: its natural enemies would be fewer and its predatory prowess would be enhanced.

The development of viviparity in teleost fish, although seen in only 13 families, has occurred in both freshwater and marine species of diverse habitats. Internal fertilization is essential to the retention of ova for additional matrotrophic nourishment. Following internal fertilization, there tends to be a reduction in both egg size and egg number and a development of mechanisms to provide the embryos with food and exchange of gases.

The degree of matrotrophy varies from slight, as in the guppy (*Poecilia reticulata*) and rockfish (Scorpaenidae), to virtually complete, as in the surfperch (Embiotocidae). The mechanisms for matrotrophy vary according to the structures produced by the embryo, such as the pericardial sac and trophotaeniae in many cyprinodontiform fish to the trophonemata in many viviparous teleosts. There appears to be no single advantage to viviparity in teleosts, for they commonly occur with oviparous forms.

Viviparity is apparently very old among some caecilians: the matrotrophy is very well developed, and fetal dentition is modified for stimulation of epithelial cells from the lining of the ovary. Some salamanders that are viviparous apparently consume eggs and perhaps siblings, with little or no postovulatory matrotrophy. Several species of anurans have evolved unusual modifications for protecting their young, including placing the eggs in vascularized dermal pockets within the stomach and within vocal sacs. There seems to be no clearly apparent environmental significance or evolutionary advantage to live-bearing in amphibians, beyond that inherent in extended parental care.

Viviparity in reptiles is confined to squamates (snakes, lizards, and amphisbaenians). Nutrition is both lecithotrophic and matrotrophic, and well-developed yolk sac placentae and chorioallantoic placentae provide the young with both food and gas exchange. In viviparous squamates the corpus luteum remains active and apparently secretes progesterone, but the role of progesterone is not known in reptiles.

The geographic and ecologic distribution of squamates indicates that a greater proportion of viviparous species lives in high latitudes and high elevations, indicating that low temperatures have provided an incentive for viviparity. There is also evidence that viviparity has developed under the stimulus of an aquatic life and a desert habitat.

BIBLIOGRAPHY

Amoroso EC (1960). Viviparity in fishes. Zoological Society of London Symposium No. 1:153–181.

Guillette LJ Jr (1987). The evolution of viviparity in fishes, amphibians and reptiles: An endocrine approach. Pp. 523–562. In *Hormones and Reproduction in Fishes, Amphibians and Reptiles* (Eds: Norris DO and Jones RE). Plenum, New York.

Guillette LJ Jr, Jones RE, Fitzgerald KT, and Smith HM (1980). Evolution of viviparity in the lizard genus *Sceloporus*. *Herpetologica*, **36**:201–215.

Hogarth PJ (1976). *Viviparity*. Edward Arnold, London.

Matthews LH (1955). The evolution of viviparity in vertebrates. Pp. 129–144. In *The Comparative Physiology of Reproduction and the Effects of Sex Hormones in Vertebrates* (Eds: Jones IC and Eckstein P). Cambridge Univ, Cambridge.

Rosen DE and Gordon M (1953). Functional anatomy and evolution of male genitalia in poeciliid fishes. *Zoologica*, **38**:1–47.

Scrimshaw NS (1945). Embryonic development in poeciliid fishes. *Biological Bulletin*, **88**:233–246.

Shine R (1983). Reptilian viviparity in cold climates: Testing the assumptions of an evolutionary hypothesis. *Oecologia*, **57**:397–405.

Shine R (1984). Physiological and ecological questions on the evolution of reptilian viviparity. Pp. 147–154. In *Respiration and Metabolism of Embryonic Vertebrates* (Ed: Seymour RS). Junk, Dordrecht.

Shine R (1985). The evolution of viviparity in reptiles: An ecological analysis. Pp. 605–694. In *Biology of the Reptilia*, vol 15, *Development B* (Eds: Gans C and Billett F). Wiley, New York.

Shine R and Bull JJ (1979). The evolution of live-bearing in lizards and snakes. *American Naturalist*, **113**:905–923.

Turner CL (1947). Viviparity in teleost fishes. *Scientific Monthly*, **65**:508–518.

Wake MH (1968). Evolutionary morphology of the caecilian urogenital system. I. The gonads and the fat bodies. *Journal of Morphology*, **126**:291–332.

Wake MH (1982). Diversity within a framework of constraints. Amphibian reproductive modes. Pp. 87–106. In *Environmental Adaptation and Evolution* (Eds: Mossakowski D and Roth G). Gustav Fischer, Stuttgart.

Wake MH (1985). Oviduct structure and function in nonmammalian vertebrates. *Fortschritte der Zoologie*, **30**:427–435.

Weekes HC (1935). A review of placentation among reptiles with particular regard to the function and evolution of the placenta. *Proceedings of the Zoological Society of London*, 1935:625–645.

Wourms JP (1977). Reproduction and development in chondrichthyan fishes. *American Zoologist*, **17**:379–410.

Wourms JB (1981). Viviparity: The maternal-fetal relationship in fishes. *American Zoologist*, **21**:473–515.

Yaron Z (1985). Reptilian placentation and gestation: Structure, function, and endocrine control. Pp. 527–603. In *Biology of the Reptilia*, vol 15, Development B. (Eds: Gans C and Billett F). Wiley, New York.

9

PARENTAL CARE

9.1 GOALS OF CARE

Parental care is most logically defined as the postovipositional or postparturitional expense that enhances survival of the young. In its broadest sense parental care is risk or energy expended for survival and development of the next generation after oviposition of the eggs or birth of the young. It may also include selection of a site for oviposition or housing for the newborn young. This definition is close to that of Shine and Bull (1979). Parental care is shown in many ways and varies with the most critical requirements of the eggs or young. The needs of vertebrate eggs or infants vary greatly from class to class, but there may be broad similarities within a class: amphibians laying eggs on land face desiccation, whereas birds' eggs are threatened by chilling.

This diversity has spawned conflicting conceptual notions concerning the underlying causes and benefits of different sorts of parental behavior. Parental care provides increased safety and augments the likelihood of continued development for the most vulnerable stages. In birds and mammals, which normally provide food for their offspring, nutrition may be a major component leading to growth and maturation.

This chapter is an overview of parental care. It considers some of the major benefits as well as the costs of parental care, but it is not presented as a catalogue of the diversity of modes in which care can be found.

9.1.1 Costs and Gains

The benefits derived from parental care must justify the costs. A benefit is implicit with an increase in care and a decrease in fecundity. It is easy to imagine a variety of possible costs of parental care but difficult to measure them. The reduction in nutrition during incubation of a skink almost certainly accounts for its biennial reproduction. Other possible costs of care assume increased exposure to predation, but although this assumption is reasonable, it might be difficult to test and more difficult to measure. A cod spawning millions of eggs, which subsequently drift passively at the surface, exerts no effort in their welfare or preservation. The single young of an elephant represents the totality of reproductive investment up to the point of parturition, and it is obviously imperative that young elephants experience minimal mortality. No one would question that the elephant expends more energy in care than the cod does.

Clearly there is a relationship between fecundity and a need for parental care. As pointed out in this chapter and in Chapter 14, fecundity responds to factors other than parental care. Parental care is sometimes a response to oxygen requirements, danger of desiccation, surface-mass ratio of eggs, and so forth, which, together with parental care, affect survival. The entire matter is complex and should not be summarized in simplistic rules and expressions.

Many years ago Professor Carl Hubbs, the eminent ichthyologist, described the behavior of the gray whale during parturition. A guest in the audience asked if the gray whales went into shallow waters to avoid attacks of killer whales. Professor Hubbs replied, "It's difficult to get the whale's point of view on that." Herein lies a great difficulty faced by the behaviorist. Much of the phraseology in behavior tends to be anthropocentric, which tends to guide interpretations. Certain expressions, especially those with sexual application, are inappropriately applied to nonhuman behavior (Gowaty, 1982). Although it may be impossible to eliminate expressions used in human behavior, one should strive to avoid words or expressions that carry emotional connotations and one should make every effort to view the world from an animal's perspective.

Various hypotheses purport to predict patterns of parental care. Most concepts interpret behavior in a rather anthropocentric fashion. Different predictions may result from the assumption that parental care serves the same function in all vertebrates. If a parent remains with the eggs and/or young, it is in a position to defend them against predators, but such protection may not constitute the main function of care for that particular species. Williams (1966) suggested that small, short-lived species may adopt risk-prone procedures, such as parental care. In contrast, Tinkle (1969) predicted that parental care would be found to characterize larger, long-lived species. These hypotheses and others have been treated by many researchers (*e.g.*, Shine & Bull, 1979; Trivers, 1972; Williams, 1969b).

It has been suggested that the cost of parental care comes from future investment in reproduction and/or growth. This was proposed by Williams (1966a, 1966b), and it is reviewed by Sargent and Gross (1986) as it applies to teleost fish. The concept is that if total energy resources are allocated to (1) reproduction and (2) growth and maintenance, an increase in one should cause a compensatory reduction in the other. That is, energy devoted to reproduction removes energy for growth and maintenance. This generality would seem to be self-evident. This concept is predicated on an assumed independence of available resources, so that, as resources increase, there is a corresponding increase in present reproduction and/or future reproduction as well as growth. Although this thesis has seldom been documented, it is embraced in many discussions. Among some reptiles the trade-off between reproduction and survival may be more significant than that between reproduction and growth (Shine, 1980).

There are difficulties in applying this concept. As fish grow and age, their egg number tends to increase and their rate of growth declines. In the American plaice (*Hippoglossoides platessoides*), for example, production of eggs takes an increasing proportion of energy, while there is less and less used for somatic growth (MacKinnon, 1972). As some species of teleosts become older and senescent, however, fecundity declines, either relatively or absolutely, and it no longer bears a direct relationship to body mass. With increasing age, moreover, much ovarian growth represents connective tissue, and consequently increased ovarian size does not necessarily indicate a comparable increase in capacity to form ova (Woodhead, 1979).

Under extreme variations of availability of food in nature, however, shortages prevent reproduction, with all energy being devoted to maintenance and some perhaps to growth. Environmental conditions of one year may or may not relate to biological events in subsequent years. Among iteroparous teleosts, which reproduce several times, over several years, environmental resources vary from year to year, and the expenditure of energy for reproduction and growth in one year should not necessarily dictate performance for the following year, when resources may be rather different. This varying fecundity has been reported for many kinds of reptiles. In species with determinate growth, such as birds and mammals, the energy devoted to parental care would have no appreciable effect on subsequent growth. In birds reproduction begins after the cessation of growth, and energy expended on care of the young would not affect growth of the parent. This tends to be somewhat less true in mammals. Many small rodents become reproductively active before growth ceases, but in other rodents, such as many or most shrews (Soricidae), reproduction rarely occurs before the attainment of maximal body mass. In some small mammals, such as the marsupial mouse (*Antechinus*) of Australia, reproduction occurs only after growth is complete, and it is closely followed by death. Similarly, in the desert-dwelling cyprinodont "instant fish," oviposition is followed shortly by death. With the vesper bats (Vespertilionidae), which may live 30 or more years in the wild, there is annual reproduction but no subsequent growth after the initial reproductive year.

9.1.2 Measure of Parental Investment

The degree of parental care exercised by one sex or the other is sometimes said to be predicated upon the relative amount of parental investment made by each sex. Determining parental investment is very difficult. If one assumes investment to represent the caloric content of the eggs or, in mammals, the energy required to synthesize milk, the female makes the greater parental investment.

A major constituent of most nonmammalian eggs is the lipid component, which is usually comparable to the lipid content of the species' food. That is, although eggs are energy-rich, the cost of formation is not consistent with their energy content: the energy content of an egg is transferred from the food of the female. There are valid reasons for crediting the male with substantial parental investment. In the ejaculate alone large numbers of sperm are released, in addition to fluid products, and all are necessary to effect fertilization. In some kinds of mammals, moreover, successive copulations are needed to achieve ovulation, and closely spaced matings result in temporary exhaustion of semen (Dewsbury, 1982).

The energy required in the synthesis of the very substantial amount of DNA that is contained in millions of sperm constitutes parental investment. This synthesis requires an increase in oxygen consumption and is frequently accompanied by a seasonal loss of body fat. It appears unrealistic to assume that males make only a trivial parental investment, and hypotheses predicated on this assumption are suspect.

9.2 COOPERATIVE BREEDING

The cost and the effectiveness of parental care can be altered by the presence of *helpers*. There are examples of care exercised by related adults, by offspring of the previous year or brood, and sometimes by unrelated individuals. Among deviations from a clearly monogamous mating pattern are many examples that are sometimes called cooperative. These arrangements may or may not involve some degree of polygamy, but they include indications of mutual care of eggs and/or young belonging to another individual.

Grimes (1976) reviewed observations of cooperative behavior in 52 species from 30 families of African birds. Some examples will illustrate the diversity of these relationships. The ostrich (*Struthio camelus*) commonly nests in groups of one male and two or three females; the females lay in a communal nest and share the responsibilities of incubation. Among several species of rails, immature young may assist in the care of the subsequent brood. Males of the pied kingfisher (*Ceryle rudis*) consistently outnumber females, and up to four or five "unmated" males have been observed to assist in the feeding of young. The superb starling (*Spreo superbus*) nests in cooperative groups, but each female lays her eggs in a separate nest. Nonbreeding individuals share in feeding the nestlings, and even recently fledged immatures share in this activity. In these situations it is difficult to distinguish simple communal nesting from the more inexplicable cooperative arrangements. In some species care of the young is shared by adults that are not the parents. In others immature young assist in the feeding of the next generation.

Descriptions of some of these cooperative societies have an underlying theme of apparent altruism, which, conceptually, should enhance the fitness of the family. The helper may be considered as increasing its own fitness by promoting survival of its siblings or half-siblings, who contain copies of its own genes. This concept of *inclusive fitness* may be ample evolutionary justification for a sib to become a helper. Helpers occur in reproductive groups in a number of species of birds and in some mammals. These examples all involve related individuals assisting members of the group in rearing young and sometimes caring for the breeding adults.

In two species of North American jays of the genus *Aphelocoma*, related assemblages defend a territory and assist in parental care. The scrub jay (*A. coerulescens*) in Florida is monogamous, but the young of one year may remain with their parents and assist in caring for the young of the following year (Woolfenden, 1981). In southwestern United States the Mexican jay (*A. ultramarina*) forms groups of from one to three pairs of adults together with young of the previous two years. These young are nonreproductive helpers and assist in providing food for breeding females and nestlings. The group in the Mexican jay remains a cooperative unit that shares and defends a common territory (Brown and Brown, 1981).

The Tasmanian native hen (*Tribonyx mortierii*), a rail, breeds in pairs and trios. The latter are formed by sibling males that form a bond with one female. This polyandrous assemblage cares for the young and reportedly achieves greater survival of the young than pairs do (Ridpath, 1972).

The groove-billed ani (*Crotophaga sulcirostris*), a cuckoo (Cuculidae) of Middle America, nests communally, with females laying in a common nest in which they share incubation (Vehrencamp, 1978). Females that are last to lay in the communal nest tend to eject previously laid eggs. The nocturnal incubator male is usually the mate of the female that has laid the last eggs. In pairs in which the nocturnal incubator male is the mate of the first laying female, that female sometimes either lays additional eggs or tries to abandon the nest (Vehrencamp et al., 1986).

The acorn woodpecker (*Melanerpes formicivorous*), of western North America, lives in small groups that breed cooperatively. Several adult males join with one or two females, and young of the previous year assist in care of the hatchlings (MacRoberts and MacRoberts, 1976). The breeding females are full or half-siblings. There is some destruction of eggs by females, either by siblings of the parent or the parent herself (Mumme et al., 1983). In this species, survivorship is 70% in nests of single females and 79% in nests of cooperative groups (Mumme et al., 1983). Considering the greater number of individuals in a cooperative group, there would appear to be no *per capita* increase in survivorship in group nesting.

In some waterfowl, broods from several families assemble into *creches*, which are attended by several adult females or by males and females together. This is cooperative care, for adults are protecting young, most of which are not their own, and the adults themselves are replaced by other adults. In the common eider (*Somateria mollissima*), creches may consist of up to 500 young ducks, which are cared for by a succession of adults, each one of which remains only four or five days. Gulls (*Larus* spp.) prey upon these ducklings, and although larger creches are more frequently attacked than smaller ones, survival of young is greater in the larger aggregations (Munro and Bedard, 1977). A constant influx of new broods compensates for a high mortality of ducklings, so that the size of the creche and the ratio of adults to young do not change greatly (Gorman and Milne, 1972).

Young of the shelduck (*Tadorna tadorna*) assemble in their first week of life in the estuaries of Scotland. These creches include many broods and are guarded by adults of both sexes. Recorded survival of ducklings to the fledgling stage in creches is approximately 22%, in contrast to 34% when broods are reared solitarily by their parents (Williams, 1974). In this species, creche formation does not seem to enhance offspring survivorship. Perhaps the present condition of reduced survival may be the result of high density, although it may, in fact, have evolved as an adaptation to low density (Patterson et al., 1982). In the Firth of Forth in Scotland, productivity (number of young fledged) of the shelduck was greater when reared solitarily (0.72–1.22%) than when reared in creches (0.04–0.32%). Herring gulls preyed upon ducklings in creches during fights among the adult shelducks (Pienkowski and Evans, 1982).

Several species of bee eaters (Meropidae) breed cooperatively. In addition to the known parents, other adults have been observed to enter active nest holes and to share in the incubation of the eggs and feeding of the young. In the rainbow bee eater (*Merops ornatus*) of Australia, some nests have been found to be attended by three adults (of unknown sex), all with brood patches (Filewood et al., 1978). The white-fronted bee eater (*Merops bullockoides*) in Kenya is a colonial species that breeds cooperatively. The colonies, which may include up to 450 individuals, consist of "clans" of from two to eleven birds, which include one or several monogamous pairs. These clans form discrete units that share reproductive responsibilities as well as defend common foraging territories (Hegner et al., 1982).

Helpers also have been reported to assist in the care of young in some mammals, mostly carnivores and especially canids. Young of the black-backed jackal (*Canis mesomelas*) sometimes remain with their parents until and beyond the birth of the litter of the next year. These young helpers assist in providing food for the new pups, and litters with helpers are larger than those tended only by their parents. It has been suggested that helpers increase survival of the young litter (Moehlman, 1979) and, alternatively, that the greater supply of food functions as a proximate factor that increases the litter size (Montgomerie, 1981). Both suggestions are plausible and not necessarily mutually exclusive.

The brown hyena (*Hyaena brunnea*) hunts in groups of related individuals, and adults share food indiscriminately with the young (Owens and Owens, 1984). The spotted hyena (*Crocuta crocuta*) also lives in groups, of some 5 to 20 related individuals. In the southern Kalahari they hunt together, but perhaps because of the large size of the prey, the spotted hyena does not carry food back to the small young (Mills, 1985).

Juvenile Mongolian gerbils (*Meriones unguiculatus*) may remain with their parents after the arrival of another litter. They are tolerated by the father, but the parents inhibit sexual activity of the juveniles. It is questionable that they are helpers, for their presence retards the growth of the young (Ostermeyer and Elwood, 1984).

The clearest example of a social mammal is the naked mole rat (*Heterocephalus glaber*) of the hot, arid regions of Kenya, Somalia, and Ethiopia. This rodent is both fossorial and colonial. Members of the colony perform separate tasks, and reproduction is confined to a single female, which is also the largest female. Smaller mole rats perform such tasks as nest building and foraging for underground plant parts and are called either *frequent workers* or *infrequent workers*, according to their degree of activity. *Nonworkers* are large and do little work. Male nonworkers mate with the breeding female and assist in care of the young (Jarvis, 1981). In nature food is limited and food sources are scattered in patches in the desert. Extensive burrow systems connect the food sources with the center of the system, and the tunnels are of small diameter and adequate for the passage of the worker groups (Jarvis, 1978).

In discussions of cooperative breeding, the designation of such nonbreeding young adults as helpers may have conditioned some behaviorists to consider their role as altruistic. Had these individuals originally been called slaves or dupes, the question of altruism may never have occurred. The substitution of the emotionally neutral *alloparent* for *helper* lessens the need to search for an altruistic explanation (Emlen and Vehrencamp, 1983).

As pointed out by Emlen and Vehrencamp (1983), independent breeding is frequently restricted by environmental features, which may limit dispersal of the young and allow only an experienced adult to mate, construct a nest, and rear young. Both breeding parents and alloparents are exposed to identical environmental conditions of food and weather, but possible psychological restraints imposed by the breeding parents could delay reproduction by the alloparents. Such suppressive effects are known for some mammals. It would seem very possible that independent nesting by alloparents, which have attained a subfunctional reproductive state, is somehow restrained by the older adults. If this is correct, the only choice alloparents have is to assume a tangential role and participate in posthatching care of young.

Common to many of the populations in which alloparents occur are stability or predictability of the habitats and scarcity of appropriate or suitable habitats into which offspring can disperse. This situation clearly indicates an imbalance between fecundity on the one hand and mortality on the other.

There have been several in-depth discussions on the adaptive and evolutionary significance of cooperative breeding (Emlen and Vehrencamp, 1983; Koenig, 1981; Moehlman, 1979; and Woolfenden and Fitzpatrick, 1984). Generally, in species with alloparents, there is a per-pair increase of survival of the young but no per capita increase of the entire group.

It has been suggested that the lifetime fitness of an alloparent is enhanced through the experience of assisting an older, more mature adult in breeding (Koenig and Pitelka, 1981). Although breeding success in long-lived birds increases over time, it is not apparent that a nonproductive year spent as a helper would be any more beneficial than an initial year spent as a paired breeder, and the latter experience would seem to be more productive.

The advantage in cooperative breeding remains somewhat equivocal.

9.3 SOCIAL PARASITISM

Certain vertebrates have evolved procedures by which they transfer their young to be reared by adults of other species. In a few species, young are raised in nursery communities or in colonies so that parental chores are shared by adults of the same species. Thus the transition from sharing responsibility to shedding the need to provide care is gradual. Most vertebrates do not readily adopt a parasitic way of reproduction, and it occurs only among oviparous species, especially birds and fish.

9.3.1 Fish

Inasmuch as many fish construct a simple gravel nest on a stream or lake bottom, it would appear to be easy for other species to use this nest for their own eggs. This behavior has long been known among minnows (Raney, 1947). Indeed, it is surprising that more fish have not developed parasitic spawning habits.

The exposed nest and territoriality of North American bass and sunfish (Centrarchidae) make these fish especially vulnerable to social parasitism. One of the well-established examples is the redfin shiner (*Notropis umbratilis*), which lays its eggs in the nest of the green sunfish (*Lepomis cyanellus*). The reproductive cycles of the two species occur together, and the shiner is drawn to the nest of the sunfish through the odor of its milt (Fig. 9.1); moreover, the sunfish does not attack the shiner (Hunter and Hasler, 1965). Functionally, this behavior may not be parasitism, because there seems to be no harm to the sunfish. The shiner probably profits from the fanning of the nest by the attendant sunfish.

A very similar relationship exists between the rosefin shiner (*Notropis ardens*) and the longear sunfish (*Lepomis megalotis*). The odor from the milt of the sunfish attracts the shiners, which then spawn over the nest of the sunfish. As much as 85% of the growing larvae in shared nests may be shiners, and males of both species may guard such nests simultaneously (Steele and Pearson, 1981).

The smallmouth bass (*Micropterus dolomieui*), another centrarchid, is sometimes host to the longnose gar (*Lepisosteus osseus*). The bass is active during the day and quiet at night, but the gar, spawning at night, seems not to be driven off by the guarding bass; after the gar spawns, its eggs are cared for (fanned and defended) by the bass. This is only one of a number of examples of social parasitism mentioned by Goff (1984).

A more clear-cut example of teleost social parasitism is that of the mochokid catfish (*Synodontis multipunctatus*), of Lake Tanganyika. This catfish manages to place its eggs, or have them placed, within the mouth of a mouthbreeding cichlid. At least six species of cichlid fish are known to brood the eggs and young of the catfish. The catfish eggs hatch before those of the host, and the catfish fry feed on the yolk sac and whole bodies of the host fry (Sato, 1986).

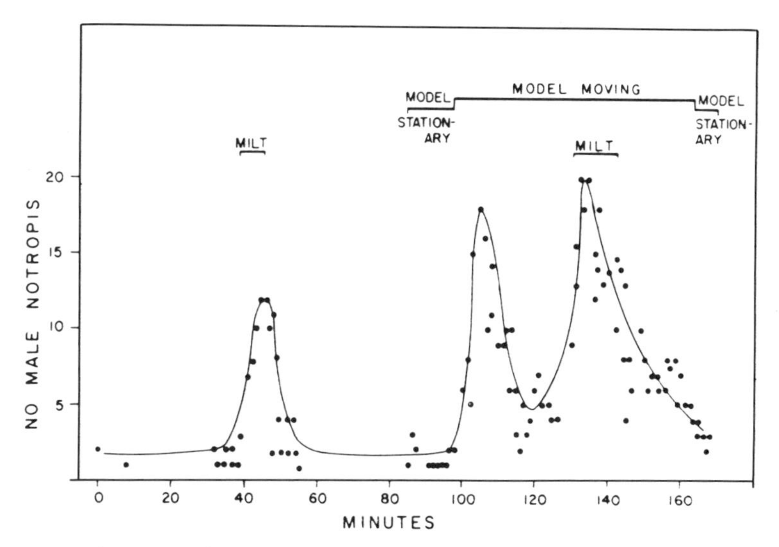

Figure 9.1 Clustering of male redfin minnows (*Notropis umbratilis*) over an unoccupied nest of a sunfish with the presence of milt and a model. (From Hunter and Hasler, 1965, courtesy of the American Society of Ichthyologists and Herpetologists.)

Some Central American cichlid fish are known to adopt and defend young of other species. In Lake Jiloa, Nicaragua, adoptions have been reported for at least three species of *Cichlosoma* and for *Neetroplus nematopus*. It is suggested that the inclusion of adopted fry into the brood by the defending parents may reduce predation on their own fry (McKaye and McKaye, 1976). *Cichlosoma nicaraguense* may defend the fry of *C. dovii*, a large predator, but it is not clear what benefit this apparent altruism brings to the adopting parent species (McKaye, 1977).

9.3.2 Reptiles

Reptiles seem rather free from interspecific social parasitism. Most reptiles, moreover, do not guard their nests or provide parental care; so opportunities for social parasitism are few. Among lizards, amphisbaenians, and chelonians the eggs are buried, and such concealed nests are not used by other individuals or species. Typically snakes may deposit their eggs rather close to the surface, such as beneath a rock or log. Unusually large clutches of eggs, guarded by two snakes, would appear to be the products of the two individuals, with no implication of anything but a fortuitous relationship (Shine, 1980).

9.3.3 Birds

Many species of birds free themselves from the costs of parental care by depositing their eggs in nests of other species. The host species incubates the eggs and feeds the young of the parasitic species. It is not surprising that a bird may sometimes lay an egg in the nest of another bird, because the presence of a nest, either with or without eggs, may trigger oviposition of an ovulated egg. Indeed, it seems logical that some of these accidents have evolved into fairly well-established patterns of social parasitism, for once having rid herself of her egg, the female parent is thereafter free of all responsibilities of caring for that egg and the chick that is destined to emerge from it. These duties are assigned to the host parent(s).

As with other types of parasitism, some rather refined adaptations have evolved that ensure the success of the alien infant. Certain physical and physiological features of the host become adopted by the parasite. The parasite must undergo folliculogenesis and ovulation more or less concurrently with the host. Not surprisingly, both parasite and host may respond to similar photoperiodic regimens (Payne, 1973), and obviously the adaptation is on the part of the parasite to become synchronized with the reproductive cycle of the host. Incubation may require less time for the parasite's eggs than for the host's eggs, and posthatching development of the alien infant is usually more rapid than that of the host's young.

The host, for its part, may recognize the parasite and harass the adults and eject their eggs. In some examples, the host may, at the expense of its own initial investment, abandon its nest and first clutch. Although the parasite may become adapted to several aspects of the biology of the host, the hosts responses are rather rudimentary, and thus the relationship has only initial elements of coevolution.

Some of these birds have developed specificity to hosts. With the ploceid finches, a number of species are colonial nesters, with breeding pairs in close physical proximity. In the African subfamily Viduinae all species are social parasites. Host specificity among the viduine finches is well developed, and certain species typically parasitize one species of estrildine finch. Young males of viduine species mimic the song of the host species, and female viduines favor such males, suggesting that parasitic young become imprinted to the song of the host (Payne, 1973).

In the cuckoo (*Cuculus canorus*) there is a general mimicry of the size and color of the host's eggs. If the resemblance is sufficiently close, the host is presumably less likely to reject the alien egg, and the trait for similarity of eggs will spread with the eventual development of a race or gens of the cuckoo that favors a given host species. Other species of cuckoo are known to have developed gentes. The concept has been reviewed by Southern (1954).

Avian social parasitism may have begun by accidental deposition of eggs in nests built by other individuals. Probably, in many species, territoriality would tend to retard, but not prevent, such occasional intraspecific parasitism, but there would be more opportunities for interspecific parasitism. In either case survival of young would perhaps result in the transmission of genes coding for the same tendency as an adult. The taxonomic diversity of social parasitism among birds indicates that it has developed independently several times.

Avian social parasitism is known to occur in several passerine families (Ploceidae, Icteridae, Indicatoridae, and, most notably, Cuculidae) and ducks (Anatidae). With the possible exception of the Indicatoridae, the species of these families are largely nonparasitic. There is a wide range of host species, and they generally are of a comparable size and have a similar range of food items as the parasitic species do.

9.4 STIMULI TO PARENTAL CARE

The totality of parental care consists of a series of sequential events. For a bird, parental care usually starts with nest building, which is followed by egg laying, brooding, feeding of young, and finally some postfledgling care. Other classes of vertebrates have different care activities throughout the reproductive season. These activities may vary dramatically with the needs of the developing young.

A complex web of relationships among endocrine secretions, exogenous stimuli, and inhibitions controls both reproductive activity and general motor activity. Reproductive behavior is initiated by exogenous signals, as discussed in Chapters 11 and 12, and exogenous stimuli induce changes in hormonal levels. Hormonal signals from one individual elicit behavioral responses from another individual. Hormonal signals usually originate with some exogenous stimulus (or the elimination of exogenous restraint, as in mammals). Behavioral responses, produced by offspring or mate, may also constitute stimuli. Gathering of nesting material by female canaries is estrogen dependent but is also stimulated by the song of male canaries (Fig. 9.2). Although reproductive hormones are remarkably similar from one class of vertebrate to another, their roles in parental care can be rather different. Prolactin, which has so many functions, may figure prominently in supporting parental care in some vertebrates and play no apparent role in others.

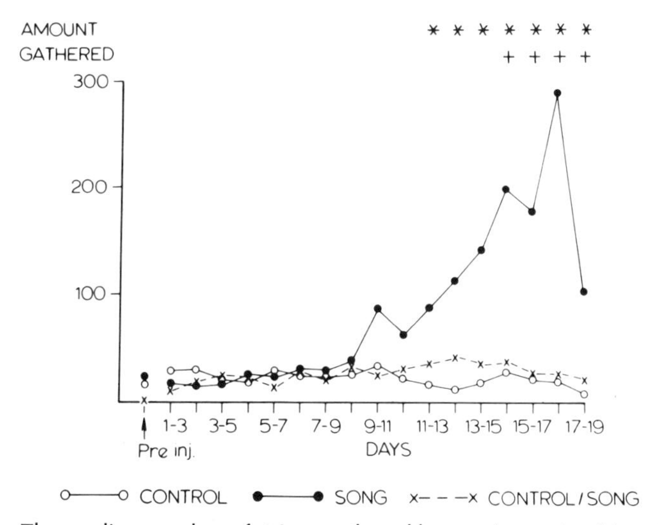

Figure 9.2 The median number of strings gathered by ovariectomized female canaries on a 12L:12D schedule and exposed to male canary song (SONG), budgerigar song (CONTROL/SONG), or to no song (CONTROL). Gathering was measured 3 days before the first estrogen injection and for 19 days thereafter. Data are running 3-day totals. (*) Difference between control and song groups are significant, $p < 0.05$. (+) Difference between control and control/song groups are significant, $p < 0.05$. (From Hinde and Steel, 1976, courtesy of Academic Press.)

During the various phases of parental care, changes in hormonal production occur. In identifying the hormonal stimulus underlying parental care, one should remember that different specific phases may have different stimuli. The aggressiveness female parent mice show toward males is independent of prolactin levels. When prolactin production is suppressed by ergot, prolactin levels and lactation decline, but aggressiveness remains unchanged (Mann et al., 1980). In addition, timing of specific events is frequently determined by nervious stimuli, usually initiated by the young.

In the reproductive cycles of fish, some species are known to exhibit parental care following injections of ovine prolactin. Among several species, injections of prolactin are followed by the stylized fanning behavior, which serves to increase the flow of water over the eggs (Muller, 1981). In addition, prolactin induces mucus secretion, and such mucus provides food for small young in a natural reproductive cycle (Blum and Fielder, 1965).

Stimuli have not received much attention in reptilian parental care. It is interesting to note that five-lined skinks (*Eumeces fasciatus*) brood their eggs without the stimulus of ovarian steroids, for ovariectomy does not decrease their attentiveness to their clutch (Stewart and Duvall, 1982).

Virtually all species of birds incubate their eggs. Broodiness, or the tendency to remain on the nest, is accompanied by a decline in gonadotropins and ovarian steroids and a sharp rise in prolactin (Sharp, 1981). In many species of birds prolactin rises during incubation and then drops abruptly upon hatching. Most altricial birds have a peak of prolactin during incubation and a gradual decline during the care period (Fig. 9.3), but in most precocial species there tends to be a rapid decline immediately after hatching (Figs. 9.4 and 9.5). In altricial species brooding continues until the young acquire an adequate degree of thermoregulation, but in precocial birds brooding may diminish sharply at hatching.

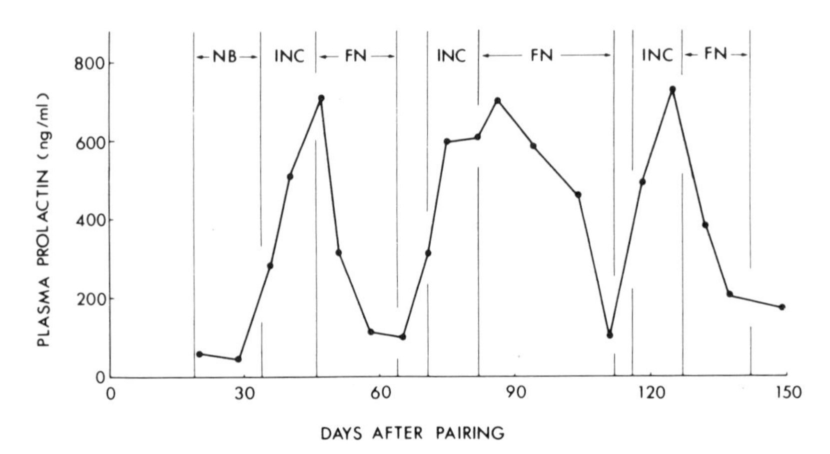

Figure 9.3 Concentrations of plasma prolactin in a female canary through three successive breeding cycles. NB, duration of nest building; INC, incubation; FN, feeding nestlings. (From Goldsmith, 1982.)

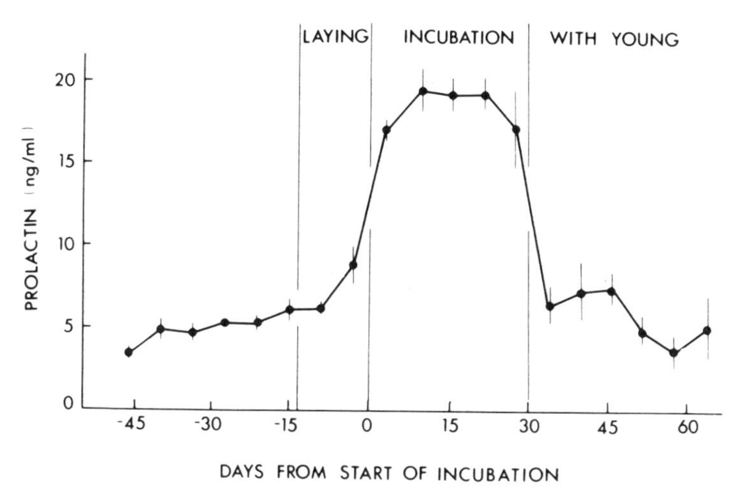

Figure 9.4 Concentrations of plasma prolactin in female mallard ducks. Circles are means; vertical lines are SEM. (From Hall and Goldsmith, 1983.)

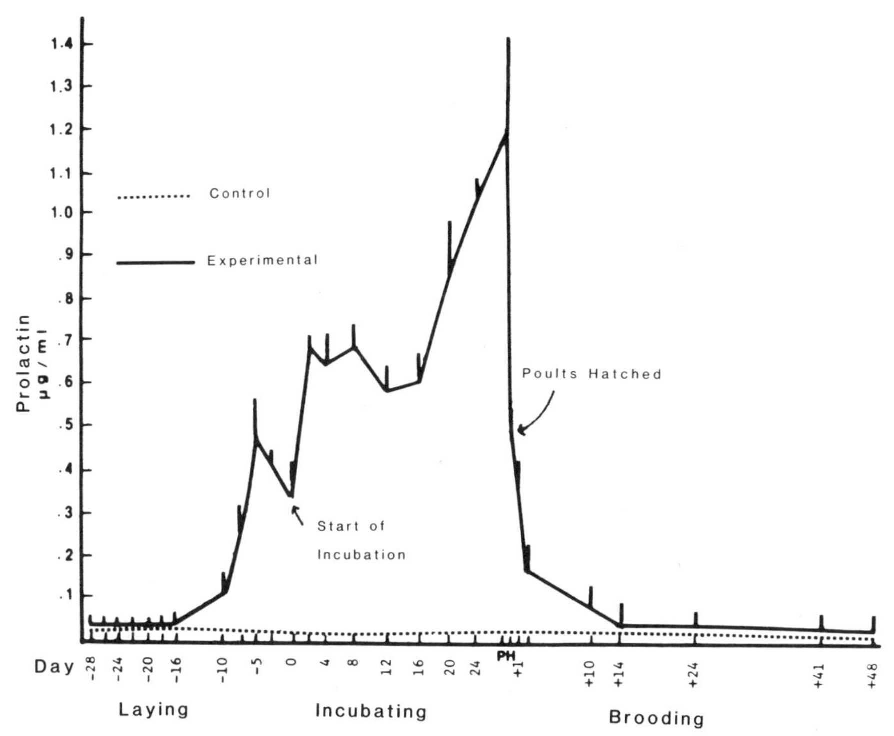

Figure 9.5 Mean plasma prolactin concentrations during laying and incubating and after hatching in the domestic turkey hen. The dotted line shows levels in controls (nonbreeding hens). Vertical lines represent SEM when it exceeds the width of the solid line; $n = 4$. (From Wentworth et al., 1983, courtesy of *Biology of Reproduction*.)

It is not clear that prolactin induces the urge to incubate, for the contact of eggs against the brood patch promotes prolactin release. Some experiments indicate that exogenous prolactin may not induce broodiness in all species. The subject has been reviewed by Goldsmith (1983), and the question remains unanswered.

Prolactin has long been known to induce crop gland development and pigeon milk secretion in pigeons and doves of both sexes (Silver et al., 1974). The presence of eggs against the brood patch stimulates prolactin secretion, and stimulation of the crop by the young also causes the release of prolactin. Thus, not only does the behavior of the young provide the stimulus for their own nutrition, but the prolactin may also be the mechanism for suppression of ovulatory cycles in these birds (Friedmann, 1977). In pigeons and doves prolactin remains high until the young cease to take pigeon milk (Goldsmith et al., 1981).

The brood patch in birds forms in one or both sexes, shortly before incubation. Prior to incubation, the individual develops a brood patch, a temporarily defeathered area characterized by extensive vascularization, warmth, and tactile sensitivity. These areas develop in apteria, which are regions of the skin where there are no contour feathers, but sometimes apteria develop down. The brood patch develops under the stimulation of progesterone (Lehrman and Brody, 1961). It is normally a development in females but also occurs in males of some species. In hypophysectomized fringillid finches, incubation patch development follows injections of estrogen and prolactin together. During incubation, down is lost in the ventral apterium, and dermal temperature is increased by both enlargement and proliferation of the internal pelvic artery. In the Adele penguin (*Pygoscelis adeliae*), for example, brood patches develop in both sexes soon after their arrival at the rookery, and down returns soon after the eggs hatch (Ainley, 1975).

Following the fixed sequence of courtship, nest building, mating, and egg laying, both sexes of the ringdove (*Streptopelia risoria*) incubate. Injections of progesterone increase the brooding drive in female ringdoves, and this response is greater when in the presence of a male (Bruder and Lehrman, 1967). Castrated males do not incubate. Testosterone or progesterone alone administered to castrated males does not produce the usual brooding drive. Brooding behavior can be produced, however, by the administration of progesterone *after* the male has been conditioned by a treatment of testosterone (Stern and Lehrman, 1969).

At parturition in mammals there is a marked reduction in gonadal steroids and an increase in prolactin. If postpartum estrus occurs, there are temporary high serum levels of estrogen and progesterone, but prolactin continues to remain high. Nursing neonates exert a number of stimuli that may affect the mother. There are various types of contact, as well as differences in temperature, odor, and sound. Some of these stimuli may exert a direct effect, and others may elicit maternal behavior through hormonal release.

Maternal care in the laboratory rat is initially (prepartum) under hormonal control, and this care includes not only such internal matters as mammogenesis but also behavioral activities such as nest building. After parturition, however, the mother-infant bond is supported and strengthened by stimuli from the young (Rosenblatt and Siegel, 1980). At the time of parturition there is a rapid change in the importance of these stimuli (Fig. 9.6).

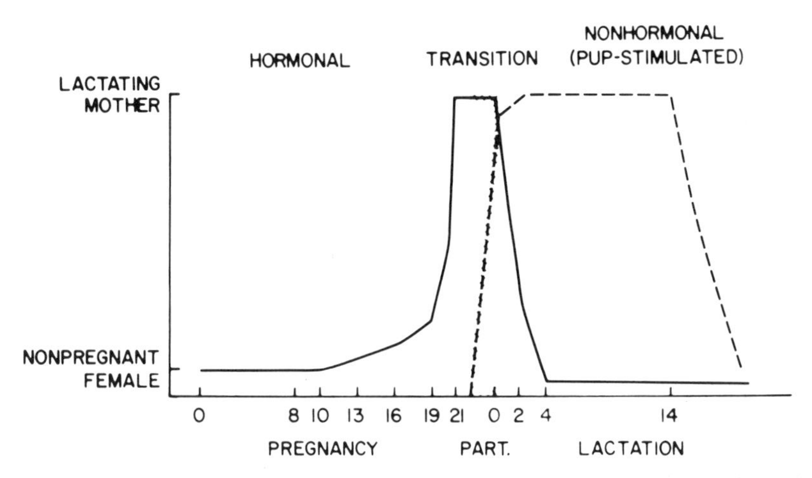

Figure 9.6 Stimulation of maternal behavior of the female laboratory rat. The ordinate indicates levels of maternal activity represented by the lactating female (high) and the nonpregnant female (low). The abscissa shows the three phases of the maternal cycle. (From Rosenblatt and Siegel, 1980, courtesy of Spectrum Publications.)

Nest building is one of the earliest manifestations of maternal care in many mammals. The gravid female gathers fine-shredded material within which the neonate can lie, and some rabbits will pull out fine fur from their abdomen to line a nest. This behavior is apparently supported by progesterone, which in nature may occur with relatively low levels of estrogen from the postpartum ovary. Also, inasmuch as progesterone synthesis begins to increase prior to parturition, it is difficult to separate the roles of the two steroids. Estrogen by itself does not stimulate nest building (Lisk, 1971).

In ovariectomized virgin female laboratory mice primed with estrogen, intracerebroventricular injections of oxytocin and prostaglandin E_2 can induce shredding of paper and other possible nesting materials (Pedersen et al., 1982). In the golden hamster, which has a relatively brief history of domestication, nest building is reinforced by prolactin (Wise and Pryor, 1977).

Maternal care may develop either gradually or suddenly during pregnancy, and it differs among several families of rodents studied in the laboratory. When day-old pups are presented to pregnant females, the response of the females depends upon the stage of pregnancy. Initially they may eat the infants, but as parturition approaches, there is an increased tendency to pick up the pups and build nests (Rosenblatt and Siegel, 1983). In the golden hamster, maternal behavior develops about one day prior to parturition (Siegel and Greenwald, 1975). Apparently the gravid female is modified by circulating hormones by the time term is reached, after which there is a response to the stimuli provided by the infants.

In mammals at the time of parturition, there is a sharp drop in serum levels of progesterone. Many studies suggest that progesterone inhibits parental care in rodents, and its decline at parturition facilitates parental care (Rosenblatt and Siegel, 1980). In some species of placental mammals and perhaps all marsupials, parental care appears after the mother is exposed to small young. Three-day-old rats, for example, are more effective in eliciting care in females than are young 12 days old or older (Stern and MacKinnon, 1978).

Organ removal and replacement studies of a number of rodents and lagomorphs show little widespread consistent responses in maternal care. There is frequently a prepartum behavioral response to estrogen or estrogen with progesterone in several rodents and the ewe and to prolactin in rabbits (Rosenblatt and Siegel, 1981). In rodents prolactin seems to play no stimulatory role in parental care, although nursing pups will elicit a sporadic high level of serum prolactin, which tends to suppress subsequent ovulatory cycles. Vocal, olfactory, and tactile stimuli all promote parental care in laboratory mice (Rosenblatt and Siegel, 1980). Postpartum production of heat is enhanced by adrenocortical hormone (Leon et al., 1978).

Serum prolactin is elevated in male marmosets (*Callithrix jacchus*) while they are associated with their young (Dixson and George, 1982). It is not apparent if the prolactin induces care of the young or if contact with the young raises prolactin levels.

Following birth of young, laboratory rats display both maternal care, toward their young, and maternal aggression, toward adults of either sex. Maternal aggression appears by the fifth day postpartum and is probably initiated by estrogen, progesterone, and possibly also prolactin. After its development, it is subsequently preserved by contact with the young (Rosenblatt and Siegel, 1981). In mice, lactation is needed for the first two days for maternal aggression to develop (Svare and Gandelman, 1976). In these rodents aggression apparently results from both hormonal and behavioral stimuli, and association with suckling young promotes the condition. There is a marked increase in maternal aggression in lactating individuals, and this aggression may result from high hormonal levels at this time. This complex of hormones includes high levels of prolactin.

9.5 AQUATIC ENVIRONMENTS—ASSURING OXYGEN

There is a relationship between the characteristics of the environment and the feasibility of parental care. The ocean surface is well oxygenated and rich in plankton. This is not an environment in which a fish can exercise the parental care of its eggs and larvae, yet it would be astounding indeed if some fish had not become adapted to produce eggs that float and hatch in the highly productive surface layers of the open sea. When large numbers of eggs float at the surface, they lie at the air-water interface, where dissolved oxygen is maximal. Oil droplets within the egg contribute to buoyancy, and parental care, behaviorally, is lacking. In departing from pelagic spawning, fish need to make provisions for oxygen for embryonic growth. If the eggs are submerged, oxygen will be less available than at the surface, and there must be some provision to ensure a continuous supply of oxygen.

Because of the mass of water, environmental changes are much slower than in aerial habitats. Daily and seasonal variations in the physical features (oxygen, mineral content, and heat) are rather slight in a given area but conspicuously large from one region to another. The maximal amount of dissolved oxygen in water is approximately one-thirtieth the oxygen content of air at sea level, but carbon dioxide is approximately 28 times as soluble in water as oxygen is. In shallow, sunlit waters photosynthesis releases oxygen in the daytime, and there may be substantial daily cycles of dissolved oxygen; but where light penetration is weak, oxygen is sparse. Dissolved oxygen is important in spawning, for developing eggs and larvae have high oxygen demands. Because of the slow changes in dissolved oxygen content in any one aquatic site, fish tend to move to and spawn in sites with maximal amounts of oxygen. Such spawning migrations, however brief, do constitute part of parental care.

The demand for oxygen by eggs is greatest when they are first laid. Gradually tolerance for lower oxygen concentrations increases as growing individuals develop opercular movements. Some benthic-spawning fish have demersal eggs that are carried in the buccal cavity of one of the parents, and in this manner the eggs are bathed in a continuous current of water. In a detailed classification of the reproductive guilds of fish, the requirement of oxygen has been emphasized (Balon, 1975; Potts, 1984).

Among teleost fish the male, more often than the female, is likely to assume care. Some reproductive patterns seem to favor male care. There is an approximate dichotomy in the division of care, predicated on the nature of fertilization. Paternal care is said to characterize species with external fertilization, and maternal care predominates species with internal fertilization (Dawkin and Carlisle, 1976), but exceptions may be more common than generally presumed. Paternal care is proportionally as common in teleosts with internal fertilization and prompt oviposition as among those species with external fertilization. Paternal care in teleosts is provided by the male in those species in which oviposition is within the territory of the male (Gross and Shine, 1981). In some polygynous teleosts the male guards eggs that may have been contributed by several females, and the male is the only parent of all the zygotes. With internal fertilization, internal development usually follows, so that there may be a substantial lapse between mating and birth. In internally fertilizing vertebrates, moreover, there may be some element of doubt as to the paternal parent.

Having been depleted of a substantial score of lipids and proteins during yolking, the female would seem to need to replenish her energy. Conceptually, fitness is increased if the female, after spawning, can accumulate energy for growth and subsequent reproduction.

9.5.1 Marine Fish

Those coastal fish with pelagic eggs spawn near or at the surface, usually near the end of a daylight period, and the floating eggs become scattered during the darkness (Keenleyside, 1979). Species that spawn in deep water, such as eels (Anguillidae) and cod (Gadidae), have pelagic eggs that drift at the oxygen-rich surface. Most large species of teleosts are pelagic spawners, consequently parental care is usually noted among smaller species (see also Chapter 13, Section 13.3.4).

Many shallow-water marine species build nests where the supply of oxygen is greatest, which may also increase the exposure to potential predators. These nests may be covered, but any concealing covering requires that the eggs be fanned to prevent them from being silted (Potts, 1984).

While widespread geographic occurrence characterizes many coral reef fish, larger species disperse in the egg stage, but the smaller fish (less than 100 mm standard length) disperse in the early, immature stage. Among many smaller species of coral reef fish, the eggs are demersal and defense is the norm. Parental care ceases at hatching, and the young then enter a planktonic life (Barlow, 1981).

Many species spawn along the shore, in shallow waters, and usually lay demersal eggs. For example, the buffalo sculpin (*Enophrys bison*), a coastal cottid of western North America, normally lives at depths of 20 m or more but spawns near the shore, in about 1.5 m of water. Males guard the eggs, and the fry are pelagic (DeMartini, 1978). The Pacific herring (*Clupea pallasi*) spawns demersal eggs over beds of algae. These eggs adhere to the algal thalli, which move with the shoreline and tidal currents, enhancing aeration of the developing embryos.

9.5.2 Freshwater Fish

A strategy of minimal care is adopted by species such as salmon, trout, and many minnows, which build an undefended nest in riffles with a strong water flow. Other options include attaching the spawn to anchored vegetation, which is frequently in motion: perch (*Perca flavescens*) spawn in beds of attached vascular plants. In such examples the parents manage to place their eggs so that they are constantly bathed by movement of water, and there is no additional parental care. Although they are not fanned, these eggs are nevertheless situated so as to obtain fresh water continuously.

Many freshwater fish release demersal eggs, and frequently these eggs are dropped over a rocky substrate or in a nest of pebbles. Such fish are called lithophils. These demersal eggs may be exposed to view, in which case they are almost always fanned by one or both parents. Oxygen provision is quite apparent in fanning, for this activity not only prevents debris from accumulating on the nest but also provides a constant replacement of fresh water over the eggs. The largemouth bass (*Micropterus salmoides*), a New World freshwater species, moves to warm, shallow water for spawning, and the male parent guards the nest and fans the eggs constantly. The larvae are very sensitive to oxygen deprivation until about six days of age. On day 7 the beginning of opercle movement provides for irrigation of the gills and allows the larvae to cope with low levels of dissolved oxygen (Spoor, 1977, 1984).

A very common strategy is for fish to take up the eggs in the mouth, which then becomes a brooding chamber. Even small young may seek refuge in the mouth of a parent, prior to developing full independence from the brooding parent.

Anabantids build nests of bubbles, and the eggs and young are provided with oil droplets, which provide buoyancy. Gouramis and Siamese fighting fish similarly build bubble nests. These species are members of a major group (Anabantoidei) that lives in ponds and small lakes of southeast Asia, in waters very low in dissolved oxygen. These little fish are endowed with specialized linings of the gill chamber, so that by gulping air they can enhance their oxygen intake. The intake of air also provides for the air bubbles that constitute the nest of the fish. The male picks up a fertilized egg in his mouth and spits it into the nest. He remains below the nest, occasionally replenishing the bubbles. In this manner the eggs and larvae are kept in the top stratum of water, where oxygen is greatest.

Among the Cichlidae are both substrate spawners and mouthbrooders. Some substrate spawners pick up the small larvae just as they emerge from their egg membranes and thereafter brood them in the buccal cavity. These fish are called larvophilic brooders, and this behavior may represent a step toward ovophilic brooders. Both parents of the Neotropical cichlid, *Aequidens paraguayensis*, oversee the development of eggs and fry, the provision of oxygen apparently being the major goal. After laying eggs on a loose, submerged leaf, both parents remain with the leaf and fan the eggs; they may even move the leaf and eggs if the eggs are disturbed. Prior to the hatching of the eggs, both parents transfer the eggs to their mouths, where they brood the young for several weeks (Timms and Keenleyside, 1975).

Mouthbrooding in fish has arisen independently several times and is well known in ariid catfish, brotulids, and some cichlids, as well as in the anabantids. It has been suggested that in the mouthbrooding fighting fish (*Betta* spp.), mouthbrooding evolved from bubble-nest-breeding species. In *Betta splendens*, a bubble-nest-breeding species, the female releases her eggs into the cup-shaped anal fin of the male, where they encounter spermatozoa. The female then picks them up and spits them into the mouth of the male. The male then rises and spits the eggs one by one into the nest (Fig. 9.7). In the absence of a bubble nest, as in moving waters, the male might simply retain the eggs in his mouth. Species of *Betta* in slow-moving streams are mouthbrooders, and the spectrum of habits of this genus suggests such an evolution of mouthbrooding (Oppenheimer, 1970).

The spraying characid (*Copeina arnoldi*), a small fish in the quiet waters of rivers in northern South America, has succeeded in escaping oxygen-deficient waters by depositing its eggs in air on the underside of leaves of emergent vegetation. In Guyana, in waters where dissolved oxygen may be as low as 4.0 ppm, this fish places it spawn on creekside vegetation, mostly from 60 to 90 mm above the surface of the water. The male splashes water on the egg masses about once a minute during the brief developmental period (about three days) (Fig. 9.8). Atmospheric humidity is critical, for spawning is confined to the two annual rainy periods, and splashing ceases temporarily during times of actual rainfall (Krekorian, 1976). This seemingly unique aerial egg deposition for teleosts provides the eggs and growing embryo with abundant oxygen, and the equally unique care ensures adequate humidity.

The Asiatic anabantid fish *Badis badis* spawns in a darkened crevice of a cave. After extrusion of the eggs, the male picks them up and places them about the roof of his cave, fanning them and replacing them as they fall to the bottom. He guards them until they reach the postlarval free-swimming stage (Buckley, 1966). In *Amblyopsis spelaea*, a cave-dwelling amblyopsid fish, the eggs and larvae are kept

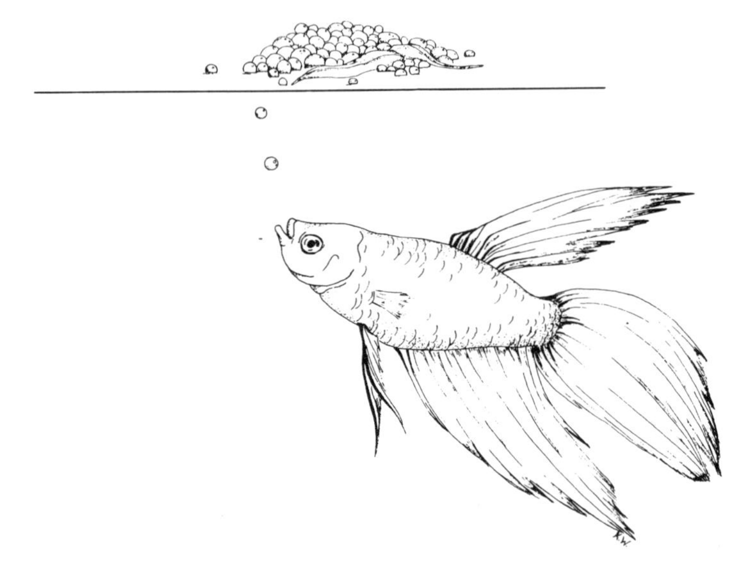

Figure 9.7 A male fighting fish (*Betta splendens*) remains below the floating nest of bubbles, which contains the eggs and small young. If an egg drops from the nest, he promptly spits it back into the nest. (Katherine West.)

under the gill covers, adjacent to the gills, where they are exposed to a continuous movement of water (Poulson, 1963).

Although provision for oxygen must be a paramount requirement for developing embryos, danger of predation is quite real. Both parents of the Neotropical cichlid *Cichlasoma maculicauda* guard their nest. This species breeds in colonies, but individual nests are spaced so that one pair can defend the eggs and fry from attacks of neighboring adults. Guarded nests do not suffer from predation, but when defending parents are removed, the neighboring adults then attack and eat the small fry (Perrone, 1978). On the other hand, both sexes of the Indian murrel (*Ophiocephalus punctatus*) care for eggs and young. The male is the primary guardian, but the female remains very close except when feeding. Both parents feed during the helpless period of the young but never attack immatures of their own species (Qayyum and Qasim, 1962).

9.5.3 Amphibians

Most aquatic-breeding species practice little or no care. Parental care in amphibians seems to develop with the oviposition of submerged eggs and, more conspicuously, with the deposition of eggs on land. On land there is contact between the brooding female and her eggs, virtually precluding the opportunity for another female to intrude and eat the eggs.

The participation of male amphibians in the care of eggs appears to depend upon the nature of mating. In most salamanders fertilization is internal, and oviposition may not occur for days or even months; in such species the female is the guarding parent. In some species the male is known to remain with the eggs, and in these species fertilization is external. In the hellbender (*Cryptobranchus alleganiensis*) the eggs are fertilized externally and deposited in a saucerlike depression in a stream. The male assumes a position below the eggs, so that his movements assist in stirring the water about them. In contrast, fertilization in the mudpuppy (*Necturus maculosus*) is internal and follows an autumnal courtship and mating. Oviposition occurs the following spring. The female attaches her spawn to the underside of a stone and remains with them (Bishop, 1941). In anurans fertilization is nearly always external, and participation by sexes is a specific trait.

As in fish, developing amphibian eggs have a high oxygen requirement. Many anurans spawn egg masses that lie at the surface of a pond and receive no care. Some species of salamanders (*Ambystoma spp.*) attach their eggs to vegetation beneath the surface but do not guard their eggs. Algae grow within the egg envelopes, releasing oxygen for the developing egg. These algae also remove ammonia, which is released by the embryo; eggs that have algae growing within their capsules have an accelerated rate of development (Goff and Stein, 1978). Stream-spawning forms of *Ambystoma texanum* lay larger eggs than do pond-spawning individuals (Petranka et al., 1987). An increase in egg size is probably facilitated by enhanced aeration of moving water.

9.5.4 Transitions to Terrestriality

Within many families of amphibians there is a transition from aquatic to terrestrial oviposition. This transition is accompanied by several important changes. There is

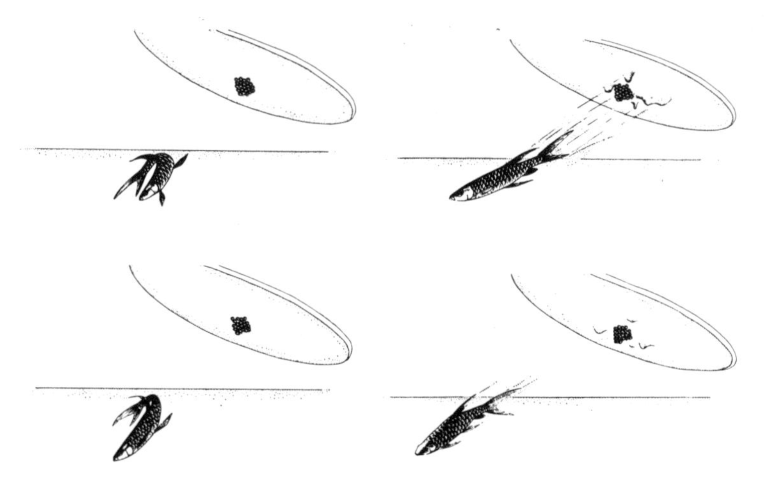

Figure 9.8 Spraying of eggs by the spraying characid (*Copeina arnoldi*). The spawning fish attaches the eggs to the underside of overhanging vegetation. The entire process takes less than 0.1 second. (From Krekorian and Dunham, 1972, courtesy of the *Zeitschrift für Tierpsychologie*.)

a reduction in number of eggs and an increase in size of eggs. The increase in size amounts to greater nutrient content, which allows direct development of the embryo into a miniature adult. Also, with the increase in size is the reduction of surface-mass ratio. For eggs laid in water, a reduction of the relative amount of surface might reduce the availability of oxygen to the extent that the embryo would suffocate. It is surely no accident that eggs laid in water are small and invariably develop into larval forms and that eggs laid on land frequently develop into fully formed, but small, adults. It is clearly an advantage if the terrestrial amphibian can eliminate the larval stage. There remain many intermediate species, which lay on land but produce larvae. These intermediate species exhibit some of the most interesting transitions to a totally terrestrial life.

The transition from aquatic spawning to terrestrial oviposition has been made among caecilians, anurans, and salamanders. Intermediate stages are exhibited by frogs that build foam nests that float on the surface of a pond and by frogs and salamanders that suspend their eggs over water, so that the larvae drop into the water. Another strategy involves embracing eggs within folds or pits in the skin.

Foam nests are made by some anurans in tropical ponds. While in amplexus and as eggs are extruded and fertilized, the female kicks up a froth with her hind legs. The eggs become held in the foam and remain elevated above the surface of the pond. The foam nests of the leptodactylid frog *Physalaemus pustulosus* not only retard desiccation but account for a temperature increase of 8.2°C above the ambient (Dobkin and Gettinger, 1985).

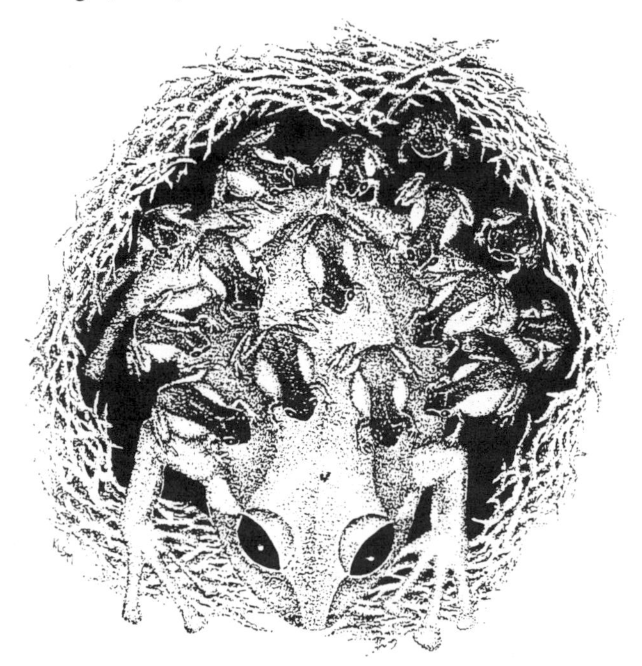

Figure 9.9 Adult *Cophixalis parkeri* (Microhylidae), together with froglets. Light areas indicate yolk still present. (From Simon, 1983, courtesy of Springer-Verlag.)

Almost without exception terrestrial eggs are guarded by a parent. Guarding the eggs provides at least four potential benefits: (1) the proximity of the parent promotes high humidity about the eggs, (2) the skin of the parent may produce a fungicidal substance, (3) the parent sometimes eats eggs infected by fungi, and (4) the presence of an adult is likely to deter predators. As an example, during the 85- to 100-day development of *Cophixalus parkeri*, a microhylid frog of New Guinea, if a parent is removed, there is a very substantial increased loss of eggs to cannibalism, predation by arthropods, fungal infection, and other factors (Simon, 1983). After hatching, the newborn froglets remain with, or even on, the parent for up to 30 or 40 days (Fig. 9.9).

Brooding of terrestrial salamander eggs appears to reduce water loss. Desiccation of eggs of the dusky salamander (*Desmognathus ochrophaeus*) is prevented because contact with the brooding female reduces their evaporative surface. Experimentally, the presence of a piece of polyethylene tubing of the size and shape of a brooding adult retarded water loss (Forester, 1984).

9.6 CHANGES WITH THE CLEIDOIC EGG

The cleidoic egg of reptiles and birds has a hard shell and contains adequate yolk to nourish the embryo past the larval stage. Metabolic wastes are converted to urea, which concentrates nitrogen and is nontoxic to the growing embryo. Beneath the hard shell is a membrane, which retards the passage of water. Depending on the species, however, varying degrees of both water and gases pass through the egg shell and membrane (Carey, 1983). The relatively large size of the eggs of birds and reptiles greatly reduces the surface-mass ratio. Deposited in an aerial environment, the egg's relative reduction in surface does not restrict its access to oxygen. Because of their relatively large size, the eggs of reptiles and birds are relatively few in number, in contrast to the usually abundant eggs of fish and aquatic amphibians. Thus the large cleidoic egg allows embryonic development to proceed beyond the gill cleft stage, but the large investment in energy requires some procedures to ensure survival.

9.6.1 Reptilian Care

The nature of reptilian care may be limited to protection from disturbance by other vertebrates, which is inferred from the female's remaining with her eggs. Maternal protection appears to be more common in venomous and very large species (Shine, 1980), supporting the concept of protection. Thermogenesis among the pythons (Old World Boidae) promotes rapid and proper development of the embryo, in addition to protection (Van Mierop and Barnard, 1978).

In reptiles virtually all parental care is performed by females. In view of sperm storage and the duration between ovulation and oviposition, the male is usually not near the female when eggs are laid. Care by males seems to occur only in the Crocodilia.

In oviposition reptilian eggs are almost always concealed, and in some species they are guarded. If they are buried, as in crocodilians, care may be the presence of one or both adults, with no contact until hatching. Nevertheless, many reptiles do care for their eggs, and there is an impressive variety of custodial modes (Shine & Bull, 1979).

Among the orders of reptiles, the constant attention shown by crocodilians is the most complex. The alligator (*Alligator mississipiensis*) not only remains at the nest but removes covering material as the time for hatching approaches. The female picks up each egg when hatching is imminent and manipulates it between her tongue and palate, apparently assisting the escape of the young from the shell (Kushlan and Simon, 1981). She may carry the newly hatched young to water and remain with them for several weeks.

Nests of crocodilians provide thermal stability by minimizing fluctuations of daily temperatures (Fig. 9.10). High diurnal temperatures are transferred slowly to the core of the nest of *Crocodilus porosus*, so that maximal nest temperatures are usually nocturnal, and endogenous heat may account for a continuous increase in core temperature (Magnusson, 1979). Nests of *Paleosuchus trigonatus* in Amazonas, Brazil, are most frequently placed over or next to termite mounds, which maintain eggs from 28.4 to 32.1°C (Magnusson et al., 1985).

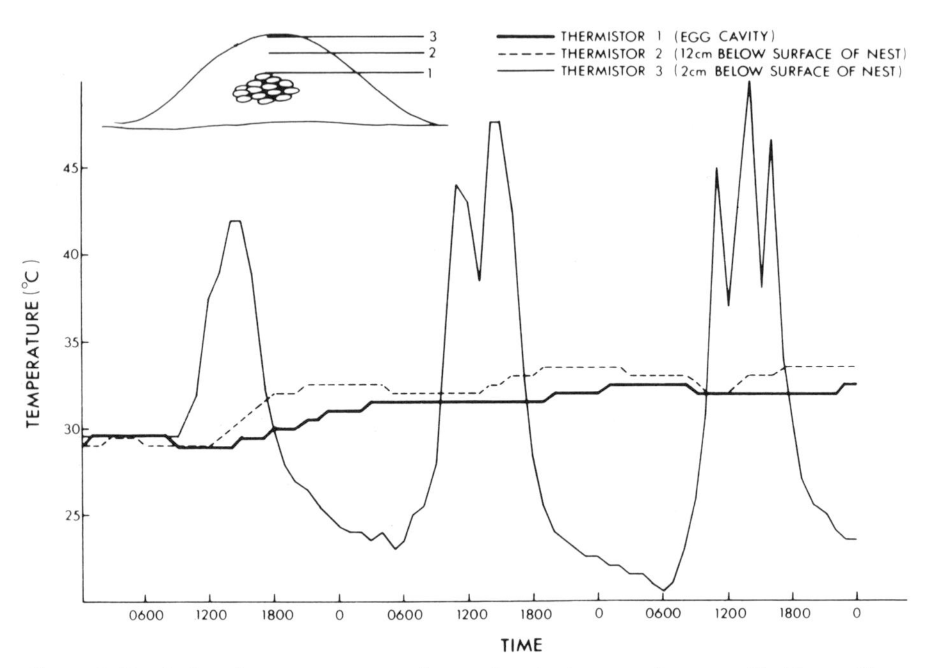

Figure 9.10 Soil and nest temperature fluctuations in the nest of a crocodile (*Crocodilus porosus*), indicating stability within the egg cavity. (From Magnusson, 1979, courtesy of the *Journal of Herpetology*.)

Among squamates a number of well-known species of snakes remain with the eggs during incubation. Fewer lizards show any concern for their eggs after oviposition, and care may be absent among the amphisbaenians.

Among lizards the best known care is provided by skinks (Scincidae), in which the female remains with her eggs, turns them, and may even moisten them with her urine. The skink *Eumeces obsoletus* assists the young in emerging from their eggs and remains with them for up to 10 days or more after hatching (Evans, 1959), but young of *E. fasciatus* are reported to disperse shortly after hatching (Vitt and Cooper, 1986). *E. okadae* of Japan broods its eggs for some 37 days (Hasegawa, 1984). The two-legged *Neoseps reynoldsi* remains with its clutch of two eggs until hatching (Telfoid, 1959). Parental care is not universal among the skinks.

9.7 CHANGES WITH ENDOTHERMY

In endotherms the nature of initial parental care depends upon the condition of the young at birth. In altricial species, which are nude, blind, and capable of little locomotion, the parent must provide for protection against chilling. In precocial species the body is clothed and the danger of chilling is minimized. Poor thermoregulatory ability is due not only to a lack of an insulative cover but also to the small size of most altricial species. Generally, the danger of hypothermia is greater in altricial birds than in altricial mammals, for the latter are usually provided with brown adipose tissue (or "brown fat"), which endows the infant with considerable thermogenic ability. In addition, nests of mammals are frequently rather well insulated from sudden changes in temperature. In either group, the altricial-precocial dichotomy is associated with the relative need of a nest, and a nest is provided for virtually all altricial young, prior to hatching or birth. The nest of precocial birds may serve as little more than a depression to retain the eggs during incubation, and precocial mammals are frequently born in simple and temporary structures that provide for some concealment but little protection. To protect her fragile altricial offspring, the mother usually makes a nest, which serves not only to ameliorate environmental changes in temperature and moisture but also to conceal the young. The pregnant rabbit (*Oryctolagus* and *Sylvilagus*) provides her infants with a fur-lined nest cavity to protect her altricial young from hypothermia. In contrast, hares (*Lepus*), which give birth to precocial young, place them in an unlined depression on the surface of the ground, and young hares are exposed to changes in ambient temperature and are often in full view.

9.7.1 Avian Incubation

Incubation of eggs by birds is primarily thermoregulatory, but there are ancillary gains such as control of humidity and protection from potential predation. A brood, or incubation, patch develops in the brooding sex or sexes and enhances the transfer of heat. The temperature of the eggs can be regulated by the duration of the incubation period as well as the degree of contact between the brood patch and the eggs.

There is a fairly narrow temperature range within the nest preserved by incubating birds, which is surprising in view of the extremes of ambient temperature under which incubation may occur. A broad taxonomic spectrum of birds maintains a core nest temperature of 34°C ± 2.38 (SD), but the temperatures of individual eggs within a single nest vary by as much as 4.6°C, depending upon their position in the nest (Huggins, 1941). Thus eggs appear to develop through a rather restricted range of temperatures.

There is a trend from precocial to altricial birds, which includes a reduction in body mass, a reduction in relative egg mass, a reduction in the relative amounts of solids to liquids in egg contents, and an increase in parental care (Carey, 1983). Precocial young rapidly develop an adequate thermogenic ability, so that brooding is minimal. At hatching the precocial young of the red junglefowl (*Gallus gallus*) are brooded almost 100% of the time, but by day 6 less than 25% of the time is spent being warmed by the parent (Sherry, 1981).

Sharing of parental duties is very common among birds. This is partly due to the need for the eggs to be kept warm and the poor thermoregulation in hatchlings, consequently requiring warmth from the parent. The requirement of constant warmth cannot always be met by one parent alone if the parent is to be allowed an opportunity to feed. Only in such birds as penguins, which accumulate an appreciable supply of fat, can there be prolonged care by one parent; but even then, care is shared. With biparental care, one parent can brood while the other forages. It must also be remembered that most birds build nests in bushes or trees or, less commonly, on the ground, in contrast to many mammals that nest under the ground. This leaves eggs and nestlings exposed to potential predators, in contrast to the relatively well-concealed and protected nest of a mouse, mole, or shrew. Thus the sharing of parental care, in addition to the thermoregulatory advantage it brings, provides for protection against enemies.

Precocial birds endow their eggs with substantially more yolk than altricial birds of similar size do (Ar and Yom-Tov, 1978), and the eggs of precocial birds have about twice the caloric content as the eggs of altricial birds of a similar size (Romanoff and Romanoff, 1949). Moreover, for eggs of comparable sizes the incubation times are similar (Rahn and Ar, 1974).

In addition to incubation or the maintenance of a certain temperature within the egg, birds sometimes face the problem of overheating and desiccation. Care may be directed to lowering the nest temperature and raising the humidity. The African skimmer (*Rynchops flavirostris*) wets its abdominal plumage before brooding its eggs. The incubating bird calls until its mate dips into the water and dampens its feathers while in flight, at which point it will assume incubation and relieve the incubating bird (Roberts, 1976). Some plovers wet themselves from time to time during incubation. The white-crowned plover (*Vanellus albiceps*) may cover unattended eggs with wet vegetation (Begg and MacLean, 1976). Both activities are seen in hot weather and serve to lower temperature as well as raise humidity. Belly soaking would also tend to cool the incubating bird. A number of plovers and ploverlike birds cover their eggs when they leave them unattended. Frequently the brooding bird scatters sand over the eggs, and some nests are sometimes half buried even when brooded. This habit seems to be most common in the low latitudes (MacLean, 1972).

A unique form of care is seen in the mound builders, bush fowl, and brush turkeys (Megapodiidae), which occur from the Philippine Islands south to Celebes and from the Moluccas to New Guinea and Australia. These birds place their eggs in a cavity in the ground, and incubation is promoted by heat from decaying vegetation, geothermal heating, or solar radiation. The covering also provides insulation against sudden daily changes in temperature. The male attends the nest and modifies the protective cover so as to preserve a nearly constant temperature for the eggs. The eggs are very well provided with yolk, and the precocial young are quite independent and receive no care from the parents (Diamond, 1983).

9.8 PARENTAL CARE IN MAMMALS

Among mammals parental care comprises many separate but associated activities (Fig. 9.11). Lactation is universal, but nest building, maternal aggression, licking, and thermoregulatory aspects are distinct and are not necessarily found in the mother-neonate relationship in all species of mammals. These separate activities may be overlapping, and some are sequential, and behaviorists have devoted a great deal of research effort to reveal the underlying stimuli for these various kinds of maternal

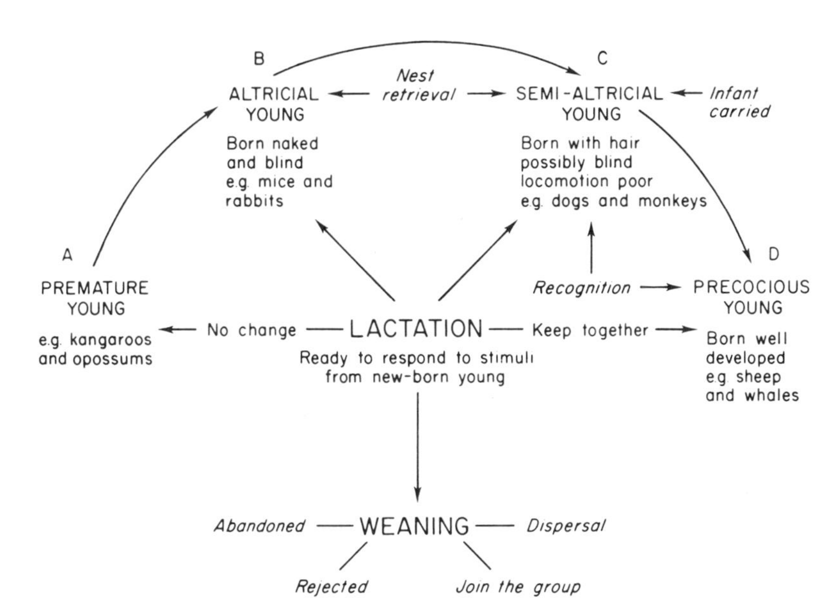

Figure 9.11 Types of maternal behavior in mammals. Lactation is the unifying activity, and divergence represents conditions of young at birth. Patterns A to C include the young being housed in a pouch or nest. In C and D there is a strong infant–mother bond. (From Walser, 1977, courtesy of the Zoological Society of London.)

care. Mammalian parental care may be inseparable from lactation, which induces prolactin secretion, which in turn may promote nonlactational events of maternal care.

Infant mammals, if altricial, are almost always born in a lined nest, so that difficulties of thermoregulation are minimized. Multiple births characterize altricial mammals, so that huddling of infants also tends to reduce heat loss. Newborn bats may cling to the mother, even to the time at which they begin to fly themselves. Young of the yellow-winged bat (*Lavia frons*) are covered by the wings of the mother at relatively low temperatures (20–25°C) but are exposed when the air is warmer (Vaughan and Vaughan, 1987).

9.8.1 Mother-Offspring Bond

The degree of mother-offspring attachment varies with the condition of the young at birth and the duration of lactation. Among murid and cricetid mice there is generally not a firm attachment between infant and mother. Presumably because the young remain secluded in a nest for most of the suckling period, the infant-mother bond would serve no purpose in nature (Gubernick, 1981). What attachment may exist is normally ended before the weaning of the next litter, for in many mice there is a strong tendency for early dispersal.

Mother-offspring bonds are especially critical in precocial mammals, particularly in those species that travel in herds with an assemblage of mothers and young moving together. In these species the bond is established, frequently through odor, in the first few hours after birth. Among artiodactyls the mother indulges in thorough licking of the newborn infant, which must be done in the immediate postpartum period if it is to be maximally effective. The odor of the young then forms the basis for the female's later recognition of her own young and rejection of other infants. In pinnipeds recognition between mother and infant is mutual and may be based mostly on vocalizations (Trillich, 1981).

Most of the research done on artiodactyls has been carried out with domestic sheep. Because of the similarity of habits between sheep and many wild species, there may be similarities in maternal behavior and its underlying causes. In the ewe there is a brief period of sensitivity, during which she may establish a bond with the newborn lamb. If the ewe is deprived of association with the neonate for as long as 24 hours, she is likely to reject it (Poindron and LeNeindre, 1980). The capacity to establish the bond seems to depend upon high serum levels of estradiol. As in other precocial mammals, the mother extensively licks the infant, and later recognition is predicated on odor. Maternal care of the lamb and recognition of an individual lamb are two distinct and separate phenomena.

Among smaller mammals the hystricomorph rodents must be the most precocial. Domestic guinea pigs, for example, will eat green vegetation before their fur is dry of amniotic fluid. Pinniped young are born in a precocial condition, with well-developed thermoregulation. Some young sea lions may be left alone for several days, without milk, while the mother returns to sea.

9.8.2 Paternal Care

The role of males in mammalian care is variable. Maternal care is frequently adequate for the needs of mammalian infants. Except for the oviparous monotremes, which care for their eggs, the embryo is housed within the mother. In all mammals the female alone provides milk, which is not only a means of nutrition but also a means of conveying warmth to the infant. The role of the male, when present, is to provide food, warmth, and protection against potential danger.

Male participation in parental care is frequent among primates. Males of the chacma baboon (*Papio ursinus*) are often seen carrying infants during encounters

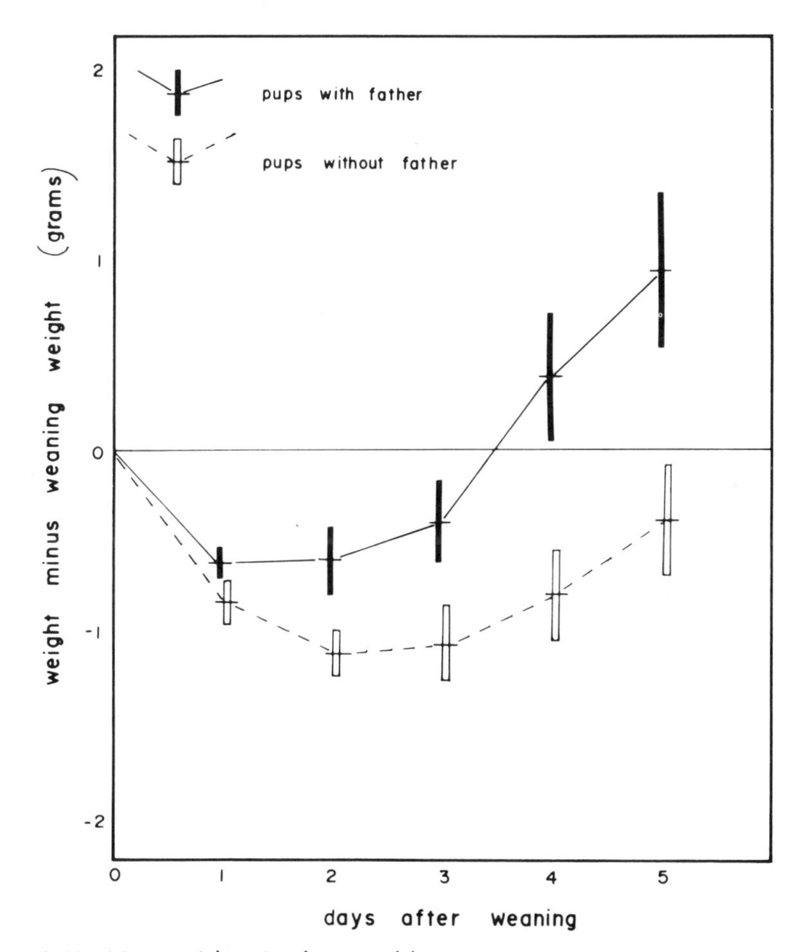

Figure 9.12 Mean weight gain of young of the parasitic mouse (*Peromyscus californicus*) in litters raised with and without the male ($p < 0.01$). Some pups with no male present died less than five days after weaning, and their last live weights were used in calculating the above means. (From Dudley, 1974, courtesy of Academic Press.)

with other males, and the underlying cause is subject to speculation. The presence of an infant, especially if it is nursing, tends to delay a resumption of fertility of its mother. Presumably the carrying male is the parent of the infant, and by protecting his own young, he is also minimizing the likelihood of his mate's being taken by another male (Busse and Hamilton, 1980).

Various degrees of association between male and female range from apparent monogamy to an opposite extreme in which adults are mostly solitary. In many taxa the two sexes associate only at the time of mating. At the monogamous end of the spectrum, in which a male and female remain associated during the breeding season, each parent may contribute to the security, welfare, and development of the young. Among some gerbils (e.g., *Rhombomys* spp. and *Meriones* spp.) there are social groups in which males stay with the females, and these bisexual reproductive groups are quite amicable in captivity (Elwood, 1983). An advantage to the male is his presence to impregnate his mate immediately at the time of postpartum estrus.

Inasmuch as the male cannot provide milk, what contribution can he make to offspring survival? The male's presence with the young can provide a material advantage if he brings food to the nest and if he remains in the nest for extended periods, especially when the female is absent, to provide heat and protection from potential predators. In nature *Peromyscus californicus* forms persistent pair bonds. In captivity survival and growth of young are enhanced by the presence of the male when the female is removed periodically from the young (Fig. 9.12). Presumably the advantage is derived from heat provided by the male parent while the female is absent. Additionally, the presence of the male would allow the female to forage for longer periods.

An advantage of paternal care has been demonstrated in house mice (*Mus musculus*) acclimated to low temperatures for 10 generations. Fifteen pairs were born and reared at 3°C for 10 generations, and fifteen control pairs were from stock kept at 23°C but placed at the lower temperature when mated. Young from these matings provided four groups: (1) control males and control females, (2) control males and cold-acclimated females, (3) acclimated males and control females, and (4) acclimated males and acclimated females. Mated pairs in which the male was from the cold-acclimated stock reared a higher proportion of their young than did pairs in which the male was the control stock (Barnett and Dickson, 1985).

9.9 SUMMARY

Parental care serves functions specific to the needs of early vulnerable stages. Vulnerability varies from one major taxon to another. In most groups care requires the attendance of one or both parents. Physical presence of an adult is tantamount to protection from predators, although this role may not be the apparent primary function of care for that group. *Guarding* is usually mentioned in the literature on care, but care frequently provides for particular requirements not otherwise available.

Although aggression toward intruders is typically part of parental care, this aspect is quite difficult to quantify.

Teleost fish and amphibians require an abundance of oxygen for their eggs and early larval stages. At the surface of the water, where movement of air and water tends to increase the oxygen content of the latter, the oxygen is usually sufficient for pelagic eggs. Care in these vertebrates commences when eggs are placed beneath the surface. Some fish and amphibians place their submerged (or demersal) eggs in moving waters, which bring oxygen and remove metabolic products. Submerged eggs in quiet waters are attended by one or two adults, who create a movement of water over the nest. In warm, stagnant waters, such as prevail in tropical ponds, oxygen is assured by incubating the eggs in a bubble nest or froth nest, or by oral brooding.

The transition to terrestriality requires a major change in egg size and care needed. Aerial incubation provides an abundance of oxygen, and the increased size, which reduces the surface-mass ratio of the egg, does not result in oxygen deprivation. A larger egg provides for additional growth, and in terrestrial amphibians the emerging individual is a miniature adult in form, having passed the larval form within the egg. In many amphibians that oviposit on land, however, the young are still larvae and either are taken to water by the parent or remain in a humid environment until transformation. In air the egg risks desiccation and fungal growth. The attendant parent usually coils about the eggs and preserves high humidity. There is some evidence that materials from the adult's skin may reduce fungal growth.

The cleidoic egg, of reptiles and birds, retards water loss, and the shell seems to exclude entrance of fungal spores. Among reptiles there is still a need for some humidity, and eggs are placed where there is soil or atmospheric moisture in addition to warmth and insulation from rapid daily changes in temperature. Reptiles, except for some boids, do not incubate their eggs.

Documented examples of parental care in reptiles are best known for the crocodilians and the skinks. Crocodiles care for the nest covering so as to preserve a near-constant temperature for the eggs. They also provide assistance to the young in their escape from the egg and movement to water. Skinks remain in contact with their eggs and seem to provide some moisture for them. Among the Old World boas the female becomes thermogenic during the incubation period and transfers heat to her eggs.

Infant birds of many species are hatched with an abundance of down, some thermogenic ability, and strong legs; such precocial young are guided and, usually for a few days, brooded by the parent(s). Altricial young, which are nude, weak, and blind at hatching, are thermolabile and helpless. They are brooded and fed while they are nestlings, and only gradually do they become independent of parental care. Not only do birds incubate their eggs and brood the young, but they commonly feed them. Thus birds present a much more complex pattern of care than do reptiles.

Newly born mammals are similarly divided into precocial forms, which are fully furred and alert at birth, and altricial forms, which are born nude, blind, and helpless. All neonate mammals take milk at birth, but the transition to solid food and the development of thermogenesis occurs very early in precocial species. There is usually a strong bond between infant and mother in precocial species; and the relationship may be prolonged in some herbivores. Altricial infants are kept in a nest, which insulates them from temperature changes and conceals them from predators. Milk not only provides their only initial nutrition, but transfers heat from mother to infant.

Cooperative breeding has been observed in a number of species of birds and mammals. Young of one year remain with their parents, and assist in providing food for the young of the following year. Such behavior of the assisting individuals has been called "altruistic." The evolutionary significance of cooperative breeding has been extensively discussed and remains enigmatic.

Social parasitism, in which adults of one species care for the eggs and sometimes the young of another species, is known for some kinds of fish and several families of birds. Several fish are known to spawn over the nest of a fish that fans and protects its eggs. This behavior apparently does no harm to the young of the host fish. One fish, a mochokid catfish, manages to place its eggs in the mouth of a mouthbrooding cichlid. The eggs of the catfish hatch before those of the cichlid, and the catfish fry feed on the cichlid eggs and larvae. Several families of birds contain parasitic species. The host adults incubate the eggs of the parasite together with their own eggs, and growth of the alien young is usually at the expense of the host young.

Parental care is controlled either by hormonal stimuli, cues from mates or neonates, or a combination of the two. Investigations on stimuli have involved mostly domesticated vertebrates, and the applicability of the conclusions is not known.

Nest building by birds may be stimulated by estrogen and singing of a male. In mammals nest building is, logically, promoted by hormones of late pregnancy—oxytocin, progesterone, and prostaglandin E_2. After parturition these hormones may have no role in parental care, and stimulation by neonates becomes important.

Prolaction stimulates various types of parental care in many kinds of vertebrates, from fanning eggs in some fish to care of young in birds. Prolactin usually appears in increased levels after oviposition or parturition. The effects of prolactin may be enhanced by priming with estrogen or, in males (as in doves and pigeons), testosterone. Prolactin release may be stimulated by the presence of eggs or young, but the actual role of prolactin is not always apparent.

BIBLIOGRAPHY

Balthazart J, Prove E, and Gilles R (Eds) (1983). *Hormones and Behaviour in Higher Vertebrates*. Springer-Verlag, Berlin, Heidelberg.

Bermant G and Davidson JM (1974). *Biological Bases of Sexual Behavior*. Harper & Row, New York.

Elwood RW (Ed) (1983). *Parental Behaviour of Rodents*. Wiley, London, New York.

Emlen ST and Vehrencamp SL (1983). Cooperative breeding strategies among birds. Pp 93–133. In *Perspectives in Ornithology* (Eds: Brush AH and Clark GA Jr). Cambridge Univ, Cambridge.

Krebs JR and Davies NB (Eds) (1984). *Behavioural Ecology. An Evolutionary Approach*, 2nd Ed. Blackwell, Oxford.

McDiarmid RW (1978). Evolution of parental care in frogs. Pp 127–147. In *The Development of Behavior: Comparative and Evolutionary Aspects* (Eds: Burghardt GM and Bekoff M). Garland, New York.

Perrone M and Zaret TM (1979). Parental care patterns in fishes. *American Naturalist*, **113**: 351–361.

Potts GW (1984). Parental behaviour in temperate marine teleosts with special reference to the development of nest structures. Pp 223–224. In *Fish Reproduction. Strategies and Tactics* (Eds: Potts GW and Wootton RJ). Academic, London.

Ridley M (1978). Paternal care. *Animal Behaviour*, **26**: 904–932.

Shine R (1988). Parental care in reptiles. In *Biology of the Reptilia* (Eds: Gans C and Huey RB), vol. 16, Ecology B. Academic, New York.

Silver R (Ed) (1977). *Parental Behavior in Birds*. Dowden, Hutchinson & Ross, Stroudsburg, Pennsylvania.

Smotherman WP and Bell RW (Eds) (1980). *Maternal Influences on Early Behavior*. Spectrum, New York.

Walser ES (1977). Maternal behaviour in mammals. Pp 313–331. In *Comparative Aspects of Lactation* (Ed: Peaker M). *Symposium of the Zoological Society of London*, No. 41. Academic, London.

Weir BJ (1974). Reproductive characteristics of hystricomorph rodents. Pp 265–301. In *The Biology of Hystricomorph Rodents* (Eds: Rowlands IW and Weir BJ). *Symposium of the Zoological Society of London*, No. 34. Academic, London.

Woolfenden GE and Fitzpatrick JW (1984). The Florida scrub jay. Demography of a cooperative-breeding bird. *Monographs in Population Biology*, 20. Princeton Univ, Princeton, New Jersey.

10

THE MAMMARY GLAND AND LACTATION

10.1 INTRODUCTION

Mammals develop mammary glands, and female mammals universally feed their newborn young on mammary secretions. The mammary structures and the nursing of young are unique to mammals and have profound effects on mammalian social structure. Although the tissue occurs in both sexes, it is genuinely functional only in females.

Successful nourishment of the suckling mammal is dependent upon a series of signals and responses between mother and young. The suckling young of some rodents produce vocal, tactile, and olfactory stimuli, all of which cue the mother for lactational events. There are maternal activities in addition to providing nourishment. The mother may prepare a nest if the young are altricial. In some species (some bats) there is preparturition segregation and migration. In many species of mammals mother-offspring recognition is based upon odor (Doty, 1976; Stoddart, 1980). This is apparently established by licking, and anosmic mice and sheep experience difficulty in nursing their young. Normally lactation very early becomes a mutually exclusive relationship. In wild populations of the peccary (*Tayassu tajacu*), however, young may suckle from more than a single female, and several struggling juveniles are not repulsed by a lactating female (Byers and Bekoff, 1981). Likewise, lions form loose social groups, and cubs of a similar age may suckle indiscriminately from two or more females (Bertram, 1875). In large colonies of some bats when milk-laden females return from feeding, the waiting infants hastily attach to the nearest nipple, regardless of parentage.

10.2 GROSS MAMMARY STRUCTURE

The mammary gland and lactation distinguish the Class Mammalia from other vertebrates. There are no fossil records of such soft parts of mammary tissue, and lactation in early mammals must be inferred from the occurrence of skeletal canals for nerves and blood vessels, which might indicate movable lips in early mammals. Such features suggest lactation. It is logical that concurrent with the development of mammary tissue and the production of milk, the rather precocial young of reptiles became increasingly altricial. Because altricial young are now supported for long periods through lactation, one is led to conclude that substantial mammary development must have come before the protomammalian infant became very altricial.

Mammary tissue consists of alveoli arranged in clusters called lobules. These alveoli open into a confluent system of ducts (Fig. 10.1). In most mammals the ducts lead directly to a nipple, but in ruminants they open into a cistern, a cavity that constitutes a storage area leading to false nipples, or teats (Fig. 10.2). In marsupials and placental mammals an eversion of the adjacent skin results in a nipple. Nipples enlarge during pregnancy in some taxa but are permanent protuberances in others. The nipple or teat serves not only to conduct milk to the mouth of the infant but also to prevent loss of milk when the neonate is not suckling.

Mammogenesis in marsupials has been reviewed by Tyndale-Biscoe and Renfree (1987). The marsupial mammary gland begins to develop at birth as epidermal thickening in the region of the future pouch. Hair follicles develop adjacent to mammary tissue analogues, so that sebaceous glands surround the median mammary

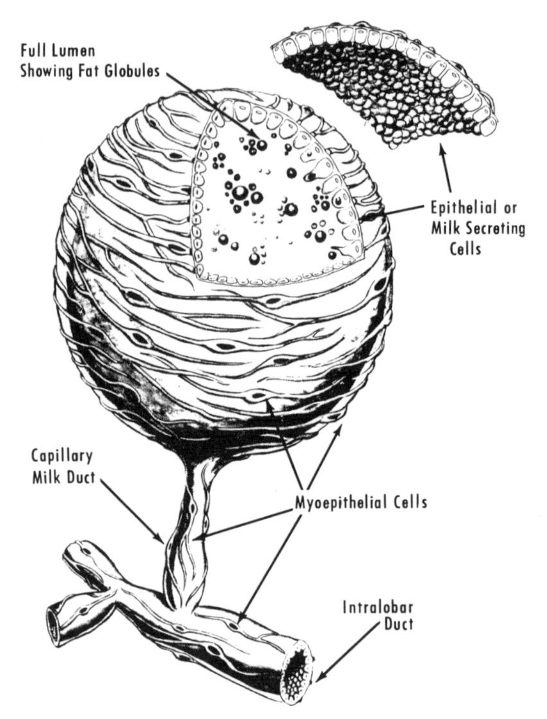

Figure 10.1 The alveolus, connected to a duct, with a section showing milk-secreting tissue and the lumen with milk and fat globules. Myoepithelial contractile cells surround the outside of the alveolus. (From Turner, 1973, courtesy of Babson Brothers.)

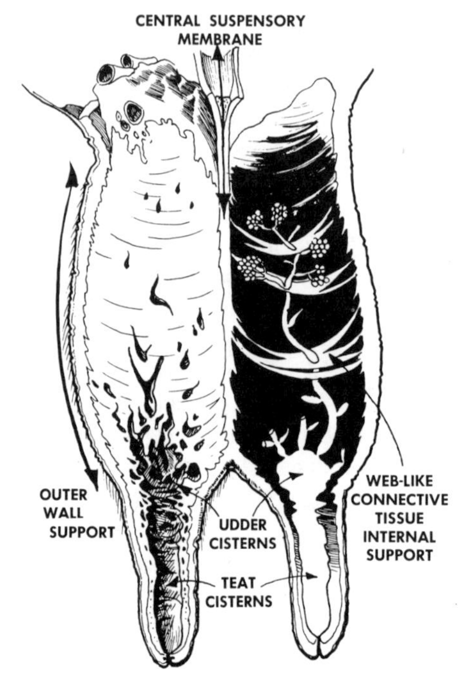

Figure 10.2 Cross section of the rear half of an udder of a cow, showing the center of the udder on the left and a single duct with its alveoli on the right. The ducts lead to a cistern, where the milk accumulated between periods of suckling or milking. (From Turner, 1973, courtesy of Babson Brothers.)

gland. Each hair follicle becomes a duct, and the number of ducts for one nipple varies among species. The nipple everts at sexual maturity. In monotremes mammary tissue consists of two flattened lobes, which release milk onto a central areola on the skin. Mammary development in monotremes does not include nipples, but each areola has a median elevation from which the infant does suck. In the echidnas these areolae lie in the marsupium, or pouch, but the duckbill lacks a pouch.

In cetaceans there is a pair of nipples or mammary cones recessed in two submedian grooves or gutterings, situated in front of the anus. Several pairs of mammary anlagen have been reported from some cetacean embryos. The mammary tissue itself is elongate and is scarcely apparent externally in lactating individuals. Thus there is little interference with the passage of the animal through water. Projecting tissue and nipples or cones would not only offer resistance but would be subject to abrasion (Arvy, 1974) (see also Section 10.5).

10.2.1 Hormones Promoting Mammary Development

Prepubertal growth of mammary tissue is isometric with that of body mass, but with the onset of ovarian steroidogenesis, mammogenesis is rapid. Normally this rapid growth is concurrent with the first pregnancy in wild mammals, but it occurs with menarche in the human.

Mammary ducts require growth hormone, estrogen, and adrenal steroids for full development in the laboratory rat, and prolactin and progesterone seem not to be essential. On the other hand, lobulo-alveolar development is stimulated by prolactin, growth hormone, progesterone, adrenal steroids, and estrogen together, with the greatest role apparently played by prolactin, growth hormone, and placental lactogen (Fig. 10.3). These conclusions follow from replacement therapy after ovariectomy, hypophysectomy, and adrenalectomy (Cowie et al., 1980). Placental lactogen plays a major role.

Prolactin increases casein mRNA activity, and the effect of prolactin during pregnancy is countered by progesterone and enhanced by adrenal corticosteroids (Devinoy et al., 1978). It has been suggested, however, that adrenal secretions (perhaps epinephrine), produced under stress, may inhibit lactogenesis (Cross, 1961). As steroid hormones are secreted, a major proportion becomes chemically bound to protein molecules, and in this biologically inactive state they are transported to the target organ. The hormone becomes dissociated at the target site and is then available (Westphal et al., 1977; Liang-Tang and Soloff, 1972).

10.2.1.1 Early Development. By destroying fetal gonads with localized irradiation, one can observe the effect of gonadectomy on the mammary anlagen. Such fetal gonadectomy does not change the course of mammary bud development in the female laboratory mouse, indicating that early prenatal changes occur independently of the ovaries. In the castrated male fetus, however, mammary bud development parallels that of the ovariectomized female fetus, indicating that fetal testes normally inhibit mammary growth in the male (Raynaud, 1971). With the postnatal growth of mammary tissue, ducts arise with alveoli and lead to the surface. In marsupials

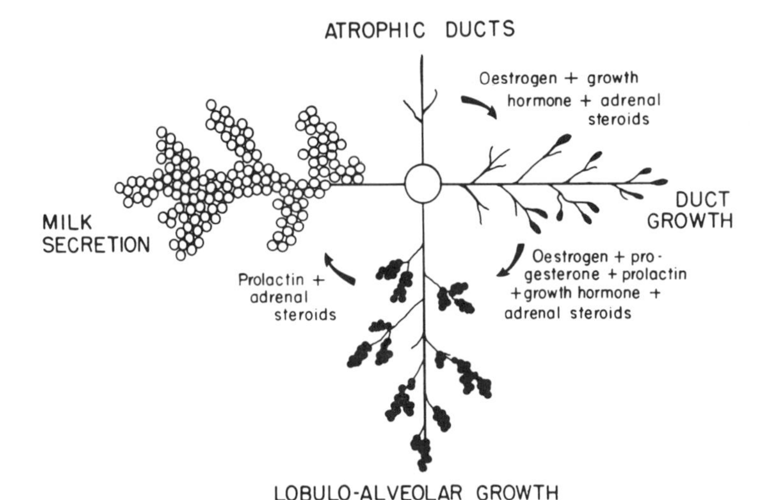

Figure 10.3 The separate effects of some of the hormones that influence development of mammary tissue and lactation in the laboratory rat. (From Cowie, 1972, courtesy of Cambridge University Press.

alveolar growth follows in the event of pregnancy, and by the time of parturition, fluid has accumulated in the alveoli. Early duct growth is isometric. The functional state of mammary tissue of some species at birth results from combined effects of maternal hormones during pregancy. In a number of mammals, including humans, infant mammary tissue secretes what is referred to as *witch's milk*.

10.2.1.2 Pubertal Development. With the approach of sexual maturity, mammary growth greatly accelerates. In the nonpregnant individual the state of development of the gland fluctuates slightly, with a detectable increase during the luteal phase of the ovulatory cycle. Some mammogenesis is seen in the postovulatory phase of the ovulatory cycle in marsupials, and development is sufficient in a sterile cycle so that a female can provide milk for a foster infant. Estrogen, progesterone, and prolactin, acting together synergistically, all promote mammogenesis, and growth hormone is also stimulatory. Much of the growth of breast tissue in the human female represents an increase in adipose and connective tissues. As with many hormonal effects, experimentally administered hormones do not have results in proportion to the dosage, which adds to the difficulty of separating the roles of the several hormones involved. Low dosages of estrogen, for example, promote alveolar growth in rats, whereas high dosages are inhibitory (Fulkerson, 1979).

10.2.1.3 Development during Pregnancy. Mammary tissue resumes growth and differentiation after the release of ovarian and placental hormones. In a continuous-cycling species the elaborately branched system of mammary ducts remains prepared by estrogen and growth hormone. The alveoli enlarge during the active period of the corpus luteum and regress with the decline in progesterone. Maximal development occurs in the presence of both steroids (Jacobsohn, 1961). In mice, estrogens promote mammogenesis not only by a direct effect but also by causing the establishment of prolactin receptors in mammary tissue (Nagasawa and Yanai, 1971).

Patterns of steroid levels during pregnancy differ greatly among mammalian taxa, illustrating the danger of broad application of patterns from conventional laboratory animals. For example, there is a general similarity of steroid hormones in the chimpanzee and the human female during pregnancy, which contrasts with the pattern in the laboratory rabbit (Fig. 10.4). There are also marked individual and hourly variations in levels of some of these hormones, and occasional samples may have little meaning. Levels of prolactin in a pregnant woman exhibit a continuous rise during gestation, but they have an episodic pattern, with peaks during sleep, regardless of the time of day. During the course of pregnancy in laboratory mice, there is great development of mammary tissue without any increase, or even with a decrease, in levels of pituitary prolactin (Nagasawa and Yanai, 1971).

Inasmuch as mammary tissue develops to its maximal extent during pregnancy in eutherians, it is natural to suspect the placenta of additional stimulation. Both progesterone and chorionic gonadotropin are candidates. Another hormone, placental lactogen, is important in mammary development in some mammals and promotes alveolar growth (Holcomb et al., 1976). Placental lactogen is secreted early in the life of the human embryonic trophoblast (the outer layer of the blastocyst) and reaches the maternal circulation following implantation. In both the domestic rabbit and the laboratory rat, insulin together with ovarian steroids promotes the development of the alveoli. Insulin is stimulatory to mammary development in laboratory mouse tissue in vitro and works synergistically with prolactin and cortisol (Turkington et al., 1967).

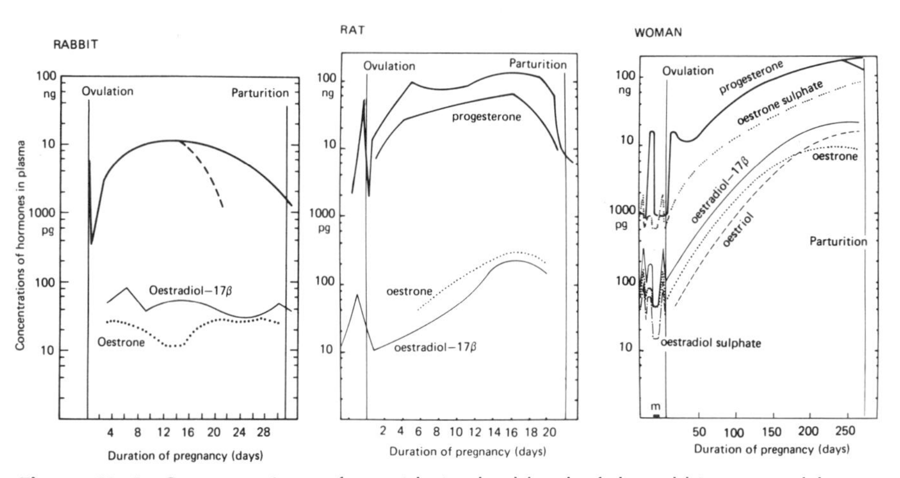

Figure 10.4 Concentrations of steroids in the blood of the rabbit, rat, and human female during pregnancy. m, time of menstruation. (From Cowie et al., 1980, courtesy of Springer-Verlag.)

In marsupials, in which gestation is very brief, there is no extensive mammary activity. Following parturition in marsupials, stimulation from the attached neonate promotes additional mammogenesis and lactogenesis, and any vacant nipples and their underlying glands regress. Postparturition lactogenesis is stimulated by prolactin, without the action of other hormones required in eutherian lactogenesis (Nicholas and Tyndale-Biscoe, 1985).

In placental mammals there are substantial postovulatory and preparturition changes in mammary cells and tissues. In the domestic rabbit, which is an induced ovulator, estrus is continuous and accompanied by mammary duct development. Lobules of alveoli form early in pregnancy, and in the latter half of gestation there is a synthesis of lactose, fat, and casein (Mellenberger and Bauman, 1974). Mammary cell number (as indicated by DNA content) in the domestic rabbit increases markedly before parturition, and there is a conspicuous fall in RNA following birth of the young (Cowie et al., 1980). In the golden hamster (*Mesocricetus auratus*) there is a similar preparturition rise in DNA, but in contrast to the rabbit, a sudden decline follows birth of the young. This difference probably reflects the rather brief period of lactation in the hamster.

In addition to these physiological changes during pregnancy, there are also behavioral activities that promote mammary development. Female rodents of some species spend time licking their nipples and surrounding areas as parturition approaches (Fig. 10.5) (Roth and Rosenblatt, 1968). Mechanical stimulation by sable-hair brushes of breasts of pregnant mice proved to be an adequate substitute for self-licking in females prevented from reaching their own breasts (Herrenkohl and Campbell, 1976).

10.2.1.4 Steroid Synthesis by Mammary Glands. In at least three domestic mammals estradiol is synthesized and released by the mammary gland, but the extent of this activity among mammals in general is unknown. During late pregnancy in the goat there is a marked rise in production of estradiol, as indicated by a rise

Figure 10.5 During pregnancy a house mouse (*Mus musculus*) grooms her nipples, a procedure necessary for the proper postpartum function of mammary tissue. (Katherine West.)

in levels of mammary venous plasma in contrast to levels of mammary arterial plasma. This estradiol from mammary tissue accounts for an estimated 90% of the late prepartum increase of this steroid. This increase is apparently unnecessary for lactogenesis and for prolactin production (Maule Walker and Peaker, 1978), but may be needed for normal ovarian function. Following mastectomy in the goat, estrus is prolonged and recurs more frequently than in intact goats (Peaker and Maule Walker, 1980). The mammary gland seems to have a role during pregnancy as well, for mastectomy followed by pregnancy results in shorter gestation and prolonged parturition with weaker uterine contractions than in sham-operated controls (Maule Walker and Peaker, 1981). The mammary gland of the goat is also capable of synthesizing progesterone and may, itself, aid in suppression of prepartum lactogenesis (Slotin et al., 1970).

10.2.2 Mammary Glands in Males

Mammary development in males has been experimentally induced in several species of laboratory rodents. Mammary tissue develops from a synergistic combination of estrogen and progesterone, and the result is most clearly seen in castrates (Folley, 1956). Some mammary tissue develops as a normal event with an increase in testosterone, but nipple development remains slight in males of almost all species and is absent in some.

Inhibition of mammogenesis in males occurs in fetal development. Considering the elaborate hormonal sequence involved in prepartum mammogenesis, it is clear that there are profound physiological obstacles to lactation in males. In most groups of mammals, moreover, the male has no role in caring for the young. Theoretically, a male increases his chances for mating if he is not tied to a routine of parental care. This topic and its ramifications are discussed at length by Maynard Smith (1977) and Daly (1979). Males of many species, however, do participate in other aspects of parental care (Elwood, 1983).

10.3 PARTURITION, LACTOGENESIS, AND THE MAINTENANCE OF LACTATION

There are important changes in serum hormone levels associated with parturition and the initiation of lactation. When the placental attachment is broken and the placenta and fetus are discharged, there is an abrupt decline in progesterone and placental lactogen concurrent with an increase in prolactin, adrenal corticoids, and estrogen. Progesterone is a strong inhibitor of the effects of prolactin during pregnancy, and after parturition the overall hormonal balance favors the synthesis and release of milk. Prolactin release is effected by the stimulation of the nipple by the nursing neonate.

This hormonal background to the secretion of milk is based on laboratory and domestic mammals, but variations exist. Despite the general importance of prolactin in promoting milk secretion, it may not be essential for the maintenance of lactation in all mammals. There is no discrete correlation between circulating prolactin and milk yield in domestic cattle, although prolactin is essential in the laboratory rat. In domestic ruminants growth hormone may replace prolactin as the stimulus for milk secretion, but in the domestic rabbit, prolactin alone simulates and maintains milk flow (Cowie and Tindal, 1971). Administration of corticosterone from day 7 to day 19 of lactation in the laboratory rat is followed by increased synthesis of milk from days 14 to 20 (Hahn and Turner, 1966); cortisol and prolactin act synergistically to promote milk secretion (Korsrud and Baldwin, 1972).

10.3.1 The Role of Prolactin

Tactile stimulation of the nipple results in the release of both prolactin and oxytocin, and in ruminants growth hormone may be released during suckling. Release of prolactin and oxytocin is probably promoted not only by stimulation of the nipple itself but also by the infant butting its head against mammary tissue while suckling. Plasma levels of prolactin rise shortly after young rats start to suck, and experimentally an increase in serum prolactin occurs upon exposure to odors or sounds of young even in the absense of tactile stimuli (Mena and Grosvenor, 1971).

Prolactin circulates at low levels during pregnancy in the cow, and shortly before parturition there is a sharp rise in both prolactin and growth hormone (Fig. 10.6). The significance of prolactin to the initiation of copious milk secretion has been questioned. Although lactation does occur just after a conspicuous rise in prolactin, it has been suggested that the prepartum rise is symptomatic of a sequence of hormonal changes at this time and is not functional (Karg and Schams, 1974). Some support for this proposal is provided by the marked seasonal fluctuations of

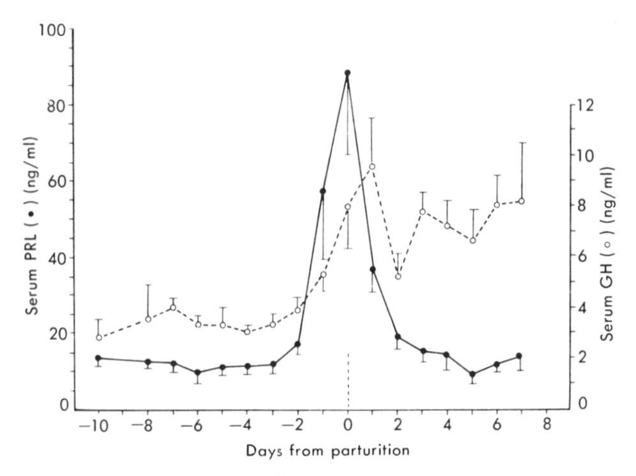

Figure 10.6 Serum prolactin (PRL) and growth hormone (GH) concentrations in dairy cows in late pregnancy and early lactation. The cows calved in September. Means ± SE. (From Johke, 1978, courtesy of Japan Scientific Societies Press.)

prolactin in both sexes of several mammals. For some mammals, lactation normally increases at a time of increasing day length. This may account for high serum levels of prolactin in cows and goats in summer and low levels in winter, independent of other factors (Johke, 1978).

There are clearly two explanations for the initiation of lactation: (1) the secretory tissue, following parturition, is relieved of hormonal restraints present during pregnancy and is *allowed* to function; (2) the secretory tissue, exposed to an assemblage of hormones that are absent or at low levels during pregnancy, is *stimulated* to function. These two explanations are not mutually exclusive. Also, the events leading to lactation vary among different groups of mammals. The complicated aspects of this question are reviewed by Kuhn (1977).

Evidence exists for the presence of a prepartum milk secretion inhibitor in some ruminants. Prepartum milking in goats and other domestic ruminants results in lactogenesis. This lactogenic response to prepartum milking in goats can be prevented if the secretion from one teat is reinjected into the contralateral teat canal, indicating the presence of an inhibitor in the secretions of prepartum milking. It has been suggested that prostaglandin $F_{2\alpha}$ is the local inhibitor of prepartum lactogenesis in goats (Maule Walker, 1984).

10.3.2 Milk Secretion and Milk Removal

The secretion or synthesis of milk (lactogenesis) within mammary tissue is distinct from milk ejection or letdown, which is the release of milk previously secreted from the alveoli. Milk secretion proceeds, presumably at a fairly constant rate, under hormonal stimuli, whereas milk ejection is a response to exogenous factors, the most powerful of which is suckling. Contractile myoepithelial cells form a branching network over the surface of the alveoli and serve to expel milk into the ducts (see Fig. 10.1). The nursing neonate provides tactile stimuli to nerves in the nipple. Not only does suckling provide the stimulus to eject milk but also the absence of this stimulus further delays milk secretion when there are no hungry mouths to accept milk. With the onset of suckling there is an increase in circulating oxytocin, for the stimulation of the nipple is transmitted neurally to the hypothalamus, triggering a release of oxytocin. It is oxytocin, in turn, that causes the contraction of the myoepithelial cells about the alveoli, expelling milk into the ducts; oxytocin stimulates milk ejection in monotremes, marsupials, and eutherians.

Although monotremes lack a discrete nipple, the mechanism is essentially the same. The young echidna (*Tachyglosses aculeatus*) butts the areolar area, and a sucking sound is clearly audible (Griffiths, 1978). The neonate monotreme does not lick or lap up milk from the areola or surrounding fur, as is sometimes supposed. Injected oxytocin is followed by the expulsion of milk in lactating monotremes.

The neurohumoral mechanism explains the expulsion of milk from the alveoli into ducts and eventually into the nipple or teat. Additionally, a conditioned reflex, stimulated by the sight or sound of an infant, may result in milk flow. In the absence of actual suckling by the infant, there may be some discharge from the nipple. A common impression is that the infant's sucking withdraws milk by establishing a differential in pressure between the mouth and the inside of the breast. Removing milk from the nipple seems to involve a combination of negative pressure and the action of the tongue, which strips the milk as the hand of a milker forces milk down the teat. Suckling in the kid establishes a negative pressure in the mouth of from 150 to 200 mm of mercury (Blaxter, 1961; Cowie et al., 1980). Thus as the infant strips milk from the nipple, it also stimulates the neurohumoral loop, which results in expulsion of milk from the alveoli. Nursing stimulates the secretion of oxytocin, prolactin, and corticosteroids—hormones that are necessary for further secretion of milk.

10.4 COMPOSITION OF MILK AND DURATION OF NURSING

Quality or composition of milk and the term of lactation are both related to growth rate, and all three are related to major features of the annual cycle of mammals. Among different species of mammals, milks have very different proportions of lactose, protein, and fat; moreover, milk changes in its composition from its initial secretion to weaning.

10.4.1 Colostrum

The first postpartum mammary flow has a distinct chemical composition and is called colostrum. Generally colostrum contains higher concentrations of sodium, calcium, and proteins; but these concentrations vary considerably among different taxa. Colostrum is low in lactose but contains a higher concentration of fat-soluble vitamins than milk does.

Colostrum is frequently the vehicle for introducing protective immunoglobulins into the body of the neonate. The diversity of means of immunity transfer, however, illustrates the variety of pathways by which animals may accomplish the same objective. The transfer of most immunoglobulins in the rabbit, guinea pig, and at least some primates is through the placenta, but in many ungulates virtually all immunoglobulins enter the infant in colostrum. In some mammals (e.g., dogs, cats, and some rodents) immunoglobulins enter both through the placenta and in colostrum. Colostrum of domestic cattle contains agglutins for *Escherichia coli, Brucella abortus*, and *Trichomonas foetus* (Lovell and Rees, 1961). In the pig, transfer of unaltered immunoglobulins is assured by the presence of a trypsin inhibitor in colostrum (Jensen, 1977). One might assume that some such provision occurs in those mammals that use colostrum to pass immunities to their offspring.

In many mammals iron is transferred across the placenta, but in at least some marsupials and monotremes the initial milk is rich in iron.

10.4.2 Trends in Milk Composition and Lactation

The production of milk soon follows the flow of colostrum, and dependency on milk varies greatly with its composition and the life history of the species. There is a very complex relationship between body mass, growth rate, length of gestation, condition of young at birth, food, migratory patterns, and environmental factors and the quality of milk and duration of lactation. Data on milk composition are difficult to evaluate, because procedures for analysis vary, and the stage in the lactational cycle may not be stated.

In spite of the bewildering variation of milk compositions, growth rates, and durations of lactation, one can make several generalizations. Changes in milk lipids and proteins tend to occur together and are opposite to trends in sugar. There appears to be no general correlation between protein content of milk and growth rate of neonates (Oftedal, 1984). The energy (fat) content and protein content tend to be correlated with body mass of the female parent (Martin, 1984). Mammals, such as phocid seals, that carry substantial stores of body fat and eat infrequently during lactation produce milks of high fat content. Also, the water content of milk is reduced in those species synthesizing milk with a high lipid content.

There is substantial interspecific variation in the amount of dry matter of milk. Generally those species that suckle their young at frequent intervals synthesize milk with a relatively low content of dry matter. Mammals that suckle infrequently, at intervals of 24 hours or longer, have a high content of dry matter in their milk (Oftedal, 1984). Milks of two desert artiodactyls, the ibex (*Capra ibex nubiana*) and the gazelle (*Gazella dorcas*), are high in both fat and protein and low in water. The percentage of total solids is 23.3 $\pm$ 2.1 and 24.1 $\pm$ 0.6 for the ibex and the gazelle, respectively. Even during lactation, the gazelle drinks infrequently but conserves water by consuming the urine of its young (Maltz and Shkolnik, 1984).

Dogs, cats, and bears characteristically have milk with a high protein content, varying from 7 to 12%. For most known milks of terrestrial carnivores, fat content values lie between 6 and 11%, with polar bear milk, with 30% fat, being a clear exception; lactose fluctuates between 2 and 4% (Ewer, 1973).

The milks of several species of bats have been discussed by Jenness and Studier (1976). Some species synthesize milk with a rather high fat content. Samples from *Glossophaga soricina* had a mean of 5.2% fat, but in *Vampyrodes caraccioloi* the mean was 29.0%. In *Artibeus jamaicensis* fat constituted 18.6% and in *A. cinereus* 23.0%. Although most bats are relatively small mammals, the duration of lactation is much longer than for most rodents of similar size. Inasmuch as Microchiroptera are rather altricial and moreover cannot obtain their own food until capable of sustained flight, a long period of lactation is to be expected. The high fat content of milk, however, is surprising in mammals with prolonged lactation.

There is a tendency for smaller species with short periods of lactation to have a high energy and protein content in their milk, but the milk of cetaceans and pinnipeds, which tend to be rather large, has a high energy content. Also, milk content varies greatly among related species: domestic ruminants (goats and cows) have about a 4% fat content in their milk, but the caribou (*Rangifer tarandus*) or reindeer has about a 20% fat content (Fig. 10.7). In eastern Fennoscandia the highly precocial young are born in April, and the rich milk supports a neonatal growth of 300 to 400 g/day (Soppela and Nieminen, 1985). The pattern exhibited by the reindeer is repeated by many wild species and constitutes adaptations to environmental

Figure 10.7 The caribou or reindeer (*Rangifer tarandus*) has a milk much richer in fat than that of domestic ungulates, perhaps a provision for rapid growth needed for a late-summer migration. (Katherine West.)

pressures. When rapid growth is essential, milk has a high energy content and the duration of nursing is abbreviated. The calf of the caribou in northwestern Canada is born in May or June and begins to migrate in August, while it is still nursing, and rapid growth is essential to survival on migration. Lactation continues into September (Harper, 1955).

10.4.2.1 Temporal Changes in the Composition of Milk. Generally, the fat content of milk is minimal when lactation begins, and as fat and protein contents rise, the proportion of lactose tends to decline. The milk of the black-tailed deer (*Odocoileus hemionus*) is characterized by a decline in lactose and increases in fat and protein after the middle of lactation (Mueller and Sadleir, 1977). Milk of domestic rabbits has been well studied and clearly exhibits the same sort of change (Fig. 10.8). Milk of the elephant seal over a 30-day period becomes richer in fat and contains decreasing amounts of water, while the protein proportion remains unchanged (Fig. 10.9). As is the case in milk of other pinnipeds, the milk of the elephant seal has no lactose (Riedman and Ortiz, 1979). Essentially the same trends are seen in marsupials, but the changes are of a greater magnitude (Fig. 10.10), probably reflecting the very long duration of lactation of these mammals.

In the nursing laboratory rat, however, the initial milk flow is high (24%) in fat. The newborn pup has glycogen stores to last only half a day, and the fat content of the milk meets the infant's demand for energy for the initial hours of life (Luckey et al., 1954).

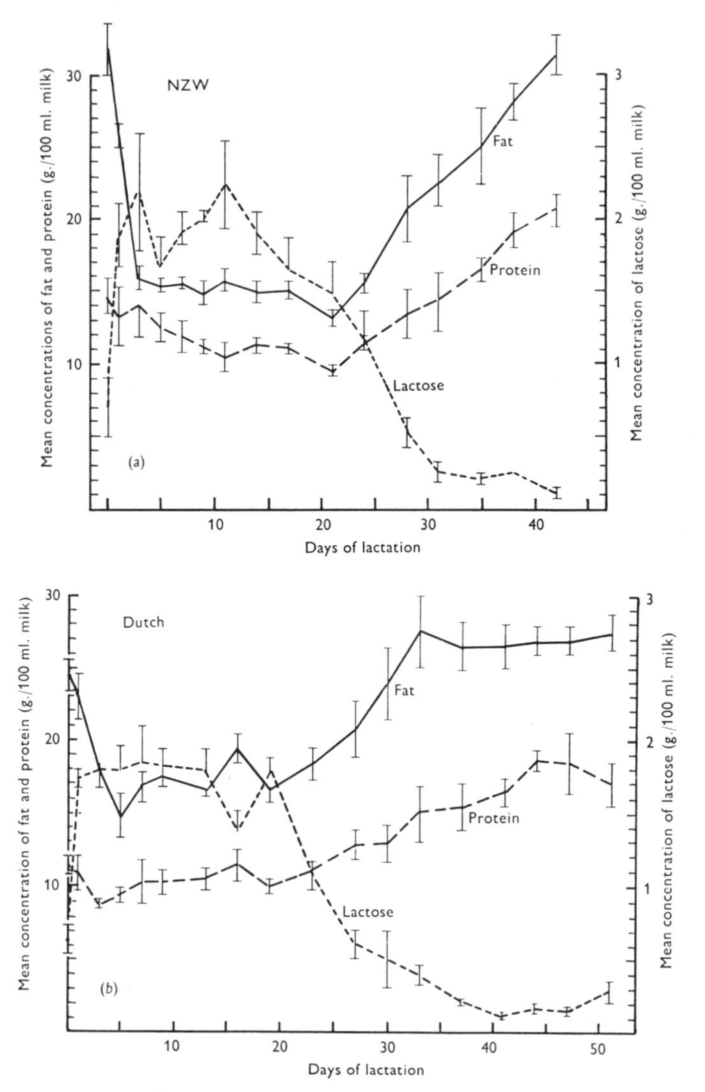

Figure 10.8 Secular changes in chemical composition of milk of New Zealand white (NZW) and Dutch rabbits throughout lactation. Vertical bars indicate SE. (From Cowie, 1969, courtesy of the *Journal of Endocrinology*.)

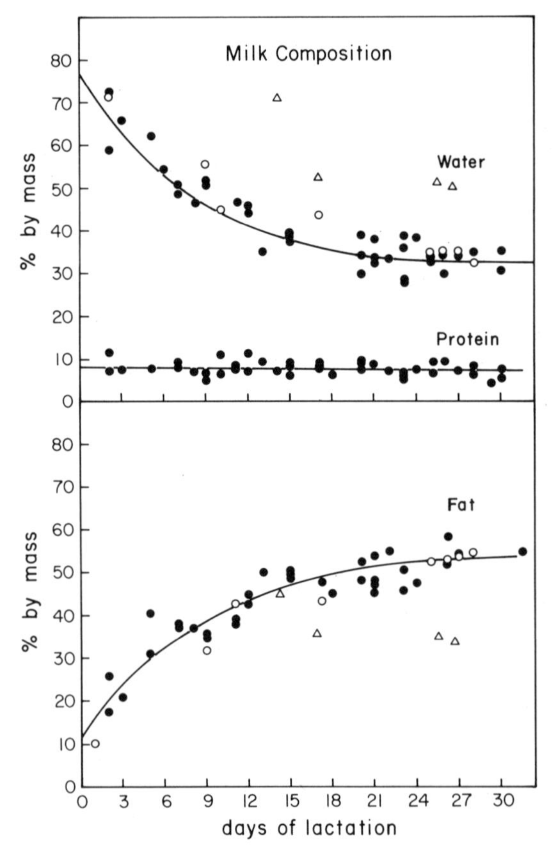

Figure 10.9 Secular changes in chemical composition of milk of the northern elephant seal (*Mirounga angustirostris*). Closed circles, samples from 18 cows; open circles, repeated samples from one cow; and triangles, repeated samples from one cow. (From Riedman and Ortiz, 1979, courtesy of the University of Chicago Press.)

10.4.3 Milk of Marine Mammals

There is much literature on the milk of pinnipeds and cetaceans, with emphasis on the high proportion of fat in the milk of these mammals. In discussing the milk content of cetaceans and pinnipeds, one must remember that both contain families that are not closely related. The pinnipeds contain three families (Phocidae, Otariidae, and Odebenidae), which became similar in form after they became marine. Similarly, the whales consist of two large suborders, the whalebone whales (Mysticeti) and the toothed whales (Odontoceti), which are not closely related. There is no phylogenetic basis, therefore, in considering these marine mammals as two discrete groups. It is widely known that whales and pinnipeds produce milk with a very high fat content, frequently as much as 50%. Also, in the milk of these mammals lactose is very low or absent.

The high fat content of milk of marine mammals is most characteristic of those migratory species that spend the summers in high latitudes and in which lactation is rather brief. The arctic and antarctic waters, exposed to greatly extended periods of daylight in the summer, are extremely productive. High fat content characterizes the planktonic organisms that constitute the base of the food chain, which probably enables high-latitude marine vertebrates to concentrate fat in their tissues. Those species at lower latitudes are not very migratory, and the duration of lactation is longer than that for most species at high latitudes. The species at the lower latitudes tend to have less fat in their milk.

10.4.3.1 Milk of Whales. The two major groups of whales differ in their term of lactation and also in the energy content of their milk. In the whalebone whales nursing lasts from 5 to 10 months, whereas in toothed whales it tends to be much longer. Two of the large toothed whales, the sperm whale (*Physeter catodon*) and the white whale (*Delphinapterus leucas*), may suckle their young for up to two years or longer. In the blue whale (*Balaenoptera musculus*), a baleen whale, fat may constitute up to 50% of the total milk content (Arvy, 1974; White, 1953). The blue whale, like most of the Mysticeti, is highly migratory. On the other hand, most of the toothed whales, which have rather prolonged periods of lactation, are generally found at lower latitudes and have considerably less fat in their milk.

10.4.3.2 Milk of Pinnipeds. At birth pinnipeds are more advanced than most mammals are. There is nevertheless a great variation in the duration of lactation,

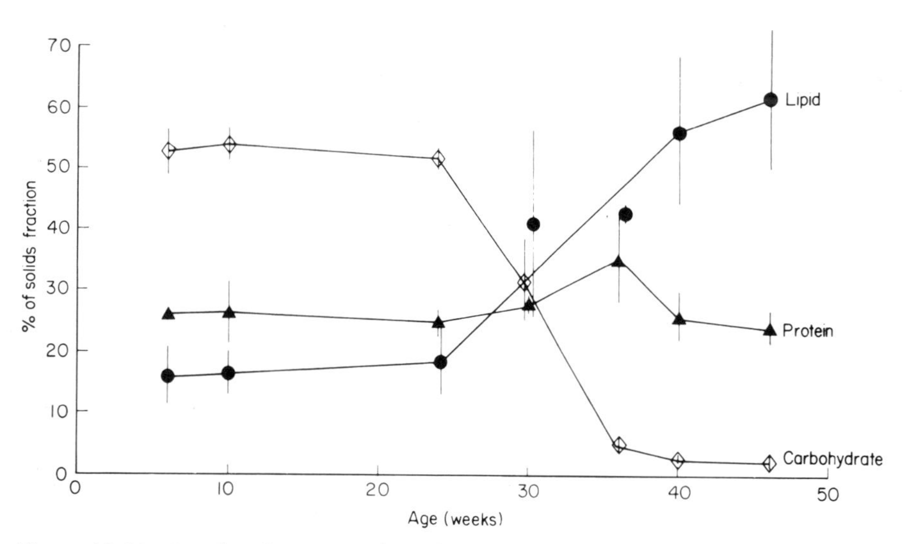

Figure 10.10 Secular changes in the relative concentrations of lipids, proteins, and carbohydrates in the solids fraction of the milk of the tammar wallaby (*Macropus eugenii*). (From Green, 1984, courtesy of the Zoological Society of London and Academic Press.)

which is correlated with the composition of their milk, their migration, and their geographic distribution.

Phocidae. Among the phocids lactation is rather brief, usually lasting between two and three weeks. Lactation is especially brief in those species that haul up on ice floes at the time of parturition (Bonner, 1984). This habitat is precarious and cannot be relied upon to persist for long periods. The harp seal (*Phoca groenlandica*) and the hooded seal (*Cystophora cristata*) both have milk with more than 50% fat content before weaning, and both nurse for less than two weeks (Sivertsen, 1941). The harp seal has a body mass of 10.8 kg at birth, which increases to 34.4 kg at weaning, most of the increase being in fat (Stewart and Lavigne, 1980). The harp seal is arctic and the hooded seal subarctic. Similarly, the grey seal (*Halichoerus grypus*) has milk with more than 52% fat, and the young take milk for only 17 days (Amoroso et al., 1951). The grey seal is subarctic in North America and occurs south to the British Isles in the eastern Atlantic area.

In contrast, those phocids that give birth to their young in lairs in snow banks have much more extended periods of lactation (Bonner, 1984). The ringed seal (*Phoca hispida*) may nurse for 42 days (McLaren, 1956).

Among midlatitude phocid seals, all of which haul up on land prior to parturition, there is a suggestion of a longer term of lactation and a greater variation in fat content of milk.

Lactation in the elephant seal (*Mirounga angustirostris*) lasts four weeks, during which time the pup takes milk two to four times daily. The samples of milk were taken at Año Nuevo Island (approximately 34° north); the maximal fat content was 54% in the fourth week of lactation (Fig. 10.9). As in other phocids, the lactating elephant seal mother does not feed. The milk is synthesized from the mother's stores. Relieved of the need to forage, she is able to conserve energy needed for her own maintenance (Costa et al., 1986).

The southern elephant seal (*Mirounga leonina*) nurses for some 23 days. The fat content of its milk varies from 10 to 13% at the time of birth to 46% in the second week and declines to 16% at weaning (Bryden, 1968). The milk of the Weddell seal (*Leptonychotes weddelli*) of the Antarctic varies in its fat content, from 26% (at the onset of lactation) to 40% 18 days later (Kooyman and Drabek, 1968; Stull et al., 1967).

The mother phocid seal is with the young almost constantly until weaning. The rich milk in this family is an important vehicle for transferring energy to the infant, for this fat must sustain the infant from the time of weaning until it captures its own food.

Otariidae. The milk of eared seals or sea lions may have a very high fat content, and pinnipeds have a shorter period of lactation in the high latitudes. The northern fur seal (*Callorhinus ursinus*) at 57° north latitude, for example, has milk with a fat content of from 43 to 60% (Ashworth et al., 1966). In contrast, the allied California sea lion (*Zalophus californicus*), at 35° north latitude, has milk with 15% fat (Schroeder and Wedgeforth, 1935), and the Cape fur seal (*Arctocephalus pusillus*), at 35° south latitude, has milk with 19% fat (Rand, 1955). In contrast, milk of the southern fur seal (*Arctocephalus tropicalis*) from South Georgia (55°

south latitude) has milk with 40 to 50% fat (Peaker and Goode, 1978). In this family lactation may last from four months to a year, or sometimes more.

Odobenidae. In contrast to the high-latitude seals and sea lions, the walrus (*Odobenus rosmarus*) is not migratory, and the young may nurse for up to two years. As one might expect, the milk of the walrus is not especially rich; its fat content varies from 13 to 32% (Fay, 1982).

10.4.4 Differential Mammary Development in Macropod Marsupials

Among the wallabies and kangaroos there are frequently two lactating young of greatly dissimilar ages: one is a newborn young, which forms a temporary attachment to a nipple, and the other is an older joey, which continues to nurse from a different nipple but no longer occupies the pouch. The different nutritional requirements of the young of two ages are met by milk of differing qualities. The gland serving the nipple used by the older young is richer in fat and protein and contains less lactose and iron than the milk taken by the younger individual (Sharman, 1970; Griffiths et al., 1972). The qualitative changes that occur in the milk over the duration of lactation (from parturition to weaning) parallel those known to occur in eutherians, but are greater in degree of change (Fig. 10.9). Macropod marsupials normally have four nipples, and two are never used and are in an inactive state. Suckling induces proclactin receptors only in those glands serving the nipples in use (Stewart, 1984).

The mammary tissue under the nipple used by the younger joey has a lower threshold of response to oxytocin. Thus in the *Macropus agilis*, stimulation by suckling of a newborn results in milk ejection only to the nipple to which it is attached. However, when the older joey suckles, the stimulation elicits milk ejection from both glands (Lincoln and Renfree, 1981).

10.5 FREQUENCY OF NURSING

Considerable variation exists in the frequency and duration of nursing sessions. The frequency of suckling varies from the near-constant intake of a young marsupial to the sporadic nursing of a young northern fur seal (*Callorhinus ursinus*), which may go without milk (or any other nourishment) for as long as 10 days. These variations indicate adjustments between the nutritional demands of both mother and infant, and these adjustments include the quality and amounts of milk.

The young of many kinds of mice become rather firmly attached to the nipple, and a mother suddenly disturbed may leave her nest with her brood clinging to her. The domestic rabbit nurses only once a day, and whether the infants are with the mother once a day or constantly, their growth rate is the same (Zarrow et al., 1965). Neonate pikas (*Ochotona princeps*), which together with rabbits constitute the order Lagomorpha, nurse about once every two hours for about 10 minutes (Whitworth, 1984). Less frequent is the feeding of suckling tree shrews (Tupaidae). Infant tree shrews (*Tupaia berlangeri*) take milk once in every 48-hour period. The first meal follows birth immediately, after which suckling occurs at dawn, every other day; between these periods the young are left alone. The infants have a well-developed thermoregulatory ability at birth, and the mother occupies a separate nest (Martin, 1968). A 48-hour interval between suckling is also characteristic of *Tupaia minor* and *T. tana* (D'Sousa and Martin, 1974).

A neonate marsupial attaches and adheres to a nipple for a long period. Depending upon the developmental rate of a given species, the attachment may be for many weeks, and nipples not in use cease to deliver milk. Later the young release the firm attachment to the nipple and move out from the pouch, but they continue initially to subsist solely on milk. Weaning is gradual in most marsupials.

There are rather brief periods of milk expulsion in whales and their allies. The mammary cone or nipple of cetaceans becomes erect during suckling, and milk is extruded through a single orifice (in contrast to up to 20 openings in some mammals). The mammary tissue is surrounded internally by the striated muscle of the ventral abdominal wall and externally by the mammary compressor muscle (Fig. 10.11), apparently a segment of the panniculus carnosus (Howell, 1930). This muscle may account for the very rapid discharge of milk frequently reported for cetaceans, but the role of oxytocin is probably similar to that of other species. Because of the ventral position of the mammary cones, either the suckling young or the mother

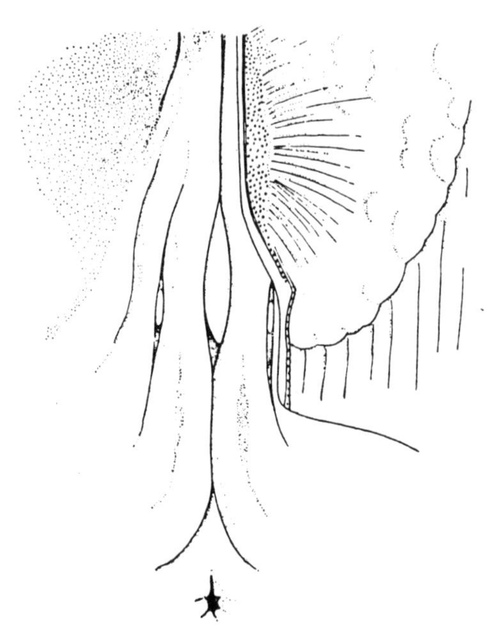

Figure 10.11 Dissected genital area of a 2.1-m embryo of a fin-backed whale (*Balaenoptera physalus*), showing the exposed mammary tissue to the right of the midline and the fan-shaped compressor muscle that covers it. The vulva and clitoris lie in the bottom center. (From Arvy, 1974, courtesy of Prof. Dr. G. Pilleri.)

Figure 10.12 Young dolphin nursing. This is a rapid process, the expulsion of milk being assisted by contraction of muscles about the mammary tissue. (Katherine West.)

must submerge its blowhole during the nursing process (Fig. 10.12). Moreover, the young cetacean does not possess the movable lips that contribute to milk ejection in other mammals. Rapid expulsion of milk in whales is a logical development, and slight pressure about the cone may cause the muscular contraction that expels a stream of milk (Arvy, 1974).

10.6 REGULATION OF MILK QUANTITY

There is a tendency in some mammals for mammary development to be commensurate with the litter size, and this relationship is most apparent in those species with litters of varying size. That is to say, regardless of the number of nipples characteristic of a given species, the underlying mammary development increases with the number of embryos. When embryos are removed surgically in the laboratory mouse, mammary size is reduced, with the result that the secretory tissue is still positively correlated with the number of young at birth (Nagasawa and Yanai, 1971). This number is presumably determined by placental lactogen, the serum levels of which are directly proportional to litter size (Soares and Talamantes, 1983).

Such adaptive mammary development ensures adequate nutrition for the neonate but avoids waste. Among litters of the white-footed mouse (*Peromyscus leucopus*) ranging in size from two to eight, daily individual growth rate from birth to weaning at day 21 did not vary. Among five species of *Peromyscus* that have contrasting mean litter sizes, no significant relationship was found between energetic cost of individual young and litter size, but there was a positive correlation between maternal energy intake and litter size (Glazier, 1985). The desert wood rat (*Neotoma lepida*) has litters that vary in size from one to five. Among young born in captivity, the duration of lactation varied from 16.0 ± 0.37 days in litters of one young to 41.8 ± 0.25 days in litters of five, but the mean weight (32.0–35.5 g) of individuals did not vary significantly (Cameron, 1973). Finally, there is no difference in growth of individual young in pigs of large and small litters (Blaxter, 1961). This tends to confirm that the number of embryos determines the output of placental lactogen and mammary development during gestation and that the female is prepared to provide adequate quantities of milk for the number of young that she has produced, assuming that she herself can obtain sufficient nourishment for lactogenesis.

In litters in which the number of young was altered by the removal or addition of neonates, larger "adjusted" litters had slower growth rates (Fleming and Rauscher, 1978). Manipulation of litters in *Peromyscus polionotus* resulted in retarded growth of young in artificially enlarged litters and greater size at weaning in artificially reduced litters (Kaufman and Kaufman, 1987). At the time of parturition and subsequently, the female makes no lactational adjustment for the number of young.

10.7 THE ECONOMY OF LACTATION

Lactation greatly increases the nutritional needs of the mother. Embryonic growth in eutherians is relatively slow compared with the rapid growth and development during the period in which it takes milk. This requires a substantial increase in the nutrition of the nursing mother.

Lactating red deer (*Cervus elaphus*) augmented their food intake more than twofold over maintenance needs (Arman et al., 1974). When food is drastically reduced in lactating captive white-tailed deer (*Odocoileus virginianus*), milk is reduced but the composition is unaltered (Youatt et al., 1965). In captive pine voles (*Microtus pinetorum*) food consumption (measured as metabolizable energy intake) is directly related to litter size: an average lactating female showed an increase of 47.5% in metabolizable energy needs over maintenance (Lochmiller et al., 1982). Food consumption may not always compensate for the use of energy in lactogenesis. In the pika (*Ochotona princeps*) there is a depletion of body fat during the entire period of lactation, indicating that food intake is inadequate to provide for the synthesis of milk (Millar, 1973).

The increase in energy devoted to lactogenesis is adjusted so as to provide nourishment to the neonates without impinging on the maintenance needs of the mother. Moreover, the lactational demands, which increase with the number of young to be fed (or litter size), are met by an adjustment to mammogenesis during pregnancy. An increase in litter size does not usually constitute a commensurate demand on the body condition (or somatic costs) of the mother.

In the Great Basin of North America, feral horses on better range sites suckle their young for a longer overall period, and at weaning the foals are larger than the foals on inferior ranges (Berger, 1986). In the red deer (*Cervus elaphus*) there is a relationship among nutrition of hinds, milk yields, and serum prolactin levels. In two groups of lactating hinds (of eight and nine hinds each), those on a pasture of permanent grass produced significantly more milk than those on a hill with low-density and low-quality dwarf shrubs and grasses (Fig. 10.13). Serum prolactin levels were significantly higher in the hinds on hill pasture than in those on permanent grass, and the higher prolactin levels were positively correlated with more frequent attempts at suckling by calves on the lower-quality habitat. It was suggested that

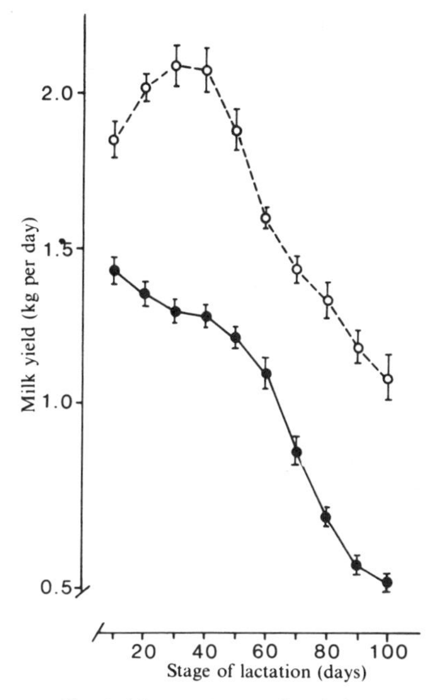

Figure 10.13 The mean milk yields (± SEM) of red deer (*Cervus elaphus*) on hillside pasture of inferior-quality forage (solid circles) and permanent pasture (open circles). (From Loudon et al., 1983, *Nature*, vol. 302, p. 145, © 1983, Macmillan Magazines Limited.)

frequent attempts at suckling, which seemed to result in higher prolactin levels, might also delay the onset of the subsequent reproductive cycle (Loudon et al., 1983).

As the suckling young develop, milk synthesis and flow gradually decline, although the rate varies greatly among different species. Changes in lactation seem to vary partly with the development of the young. When young laboratory mice were replaced weekly with neonates, lactation was prolonged for more than eight weeks (Nagasawa and Yanai, 1976). The photocycle also has an influence on lactation. In domestic cattle, milk production is 7 to 10% greater under a 16L:8D schedule than under a natural photocycle (Fig. 10.14), and this increase is unaffected by temperatures as low as 5°C (Tucker, 1981).

In most mammals there is a gradual rise in milk production, followed by a peak and an inevitable decline as the young take solid food. In New Zealand white rabbits and Dutch rabbits, there is a maximal peak just prior to the midpoint in duration of lactation (Fig. 10.15). The red deer produces peak yields of between 1400 to 2000 g of milk daily for 190 to 280 days, the total daily yield declining gradually (Fig. 10.16) as solids (fat and protein) increase (Fig. 10.17).

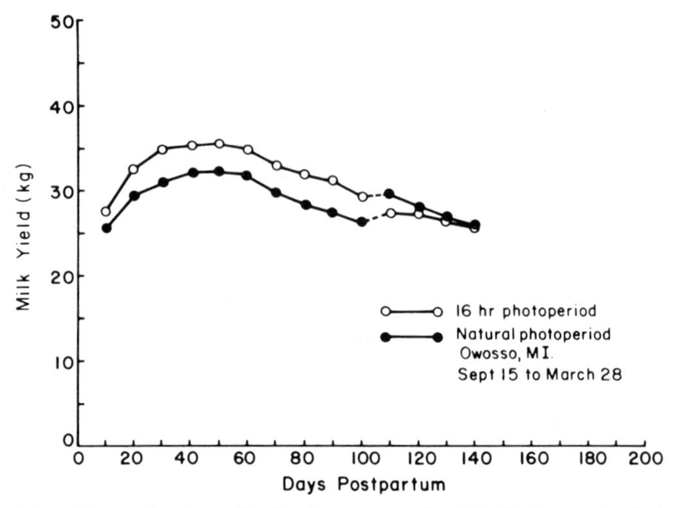

Figure 10.14 Milk production of holstein cows under 16L:8D (open circles) and natural photocycle (closed circles) between 15 September 1976 and 28 March 1977 in Owosso, Michigan. At day 100, the light schedules for the two groups were switched. (From Peters et al., 1978, *Science*, 199: 911–912, © 1978, courtesy of the American Association for the Advancement of Science.)

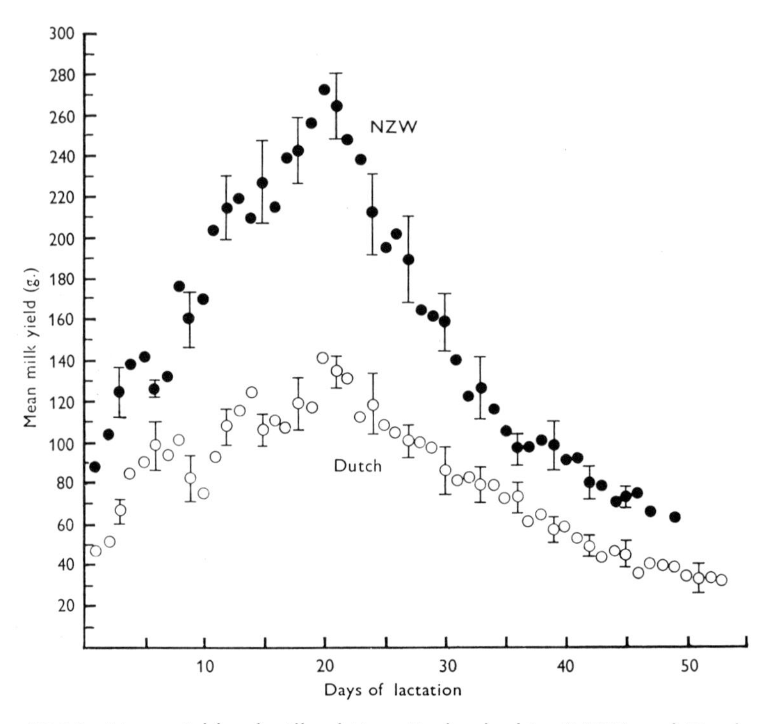

Figure 10.15 Mean yields of milk of New Zealand white (NZW) and Dutch rabbits throughout the duration of lactation. Every third mean has a vertical bar, indicating its SE. (From Cowie, 1969, courtesy of the *Journal of Endocrinology*.)

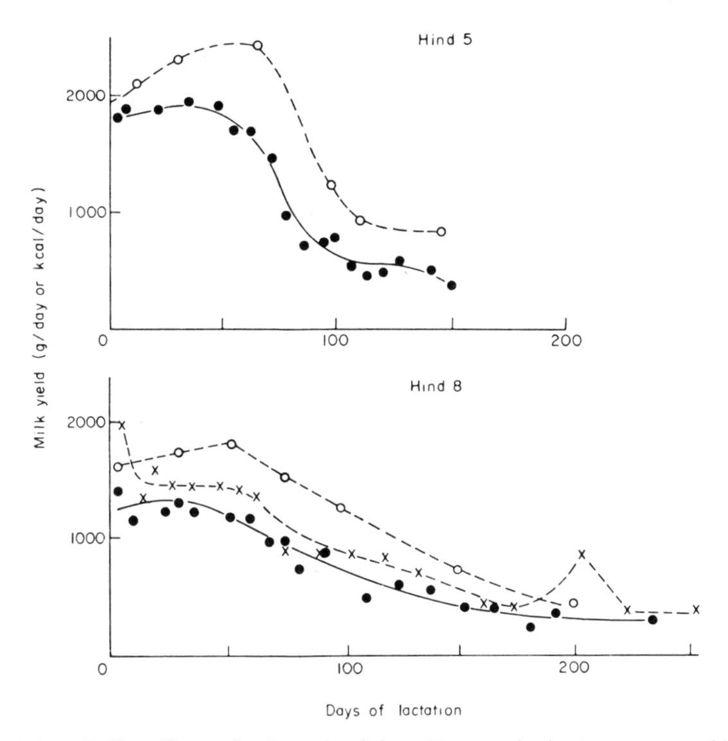

Figure 10.16 Daily milk production of red deer (*Cervus elaphus*) as measured by yield in mass (g/day), solid circles; energy (kcal/day), open circles; X, approximate yields (g/day) from milking. (From Arman et al., 1974, courtesy of the *Journal of Reproduction and Fertility*.)

In the harp seal (*Phoca groenlandica*) the lactating female loses about 3.17 ± 0.52 kg/day (or 250,000 kcal in the nine days from birth to weaning), and the mean gain in mass of the pup is 2.78 ± 0.19 kg/day (or 194,000 kcal from birth to weaning) (Stewart and Lavigne, 1984).

The initial costs of lactation are incurred during gestation: prepartum mammogenesis uses energy consumed during the last trimester of pregnancy. In those mammals that do not eat between parturition and weaning (such as phocid seals), the cost of lactation is totally accounted for prior to birth of the young. In contrast, in marsupials, in which gestation is very brief, there is no prepartum mammogenesis or lactogenesis, and even virgin females produce milk when provided with the stimulus from a newborn infant.

10.8 CONCURRENT LACTATION AND PREGNANCY

In those mammals that experience a postpartum estrus and mating, there is often pregnancy during lactation. In the early stages of pregnancy, implantation may be

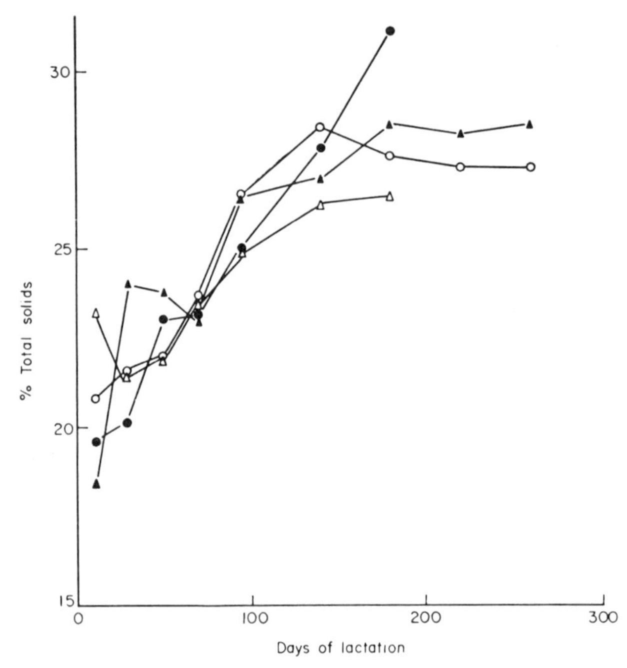

Figure 10.17 Change in solid content of the milk of the red deer (*Cervus elaphus*) during lactation of four different hinds. (From Arman et al., 1974, courtesy of the *Journal of Reproduction and Fertility*.)

delayed; in the later stages, lactation may be curtailed. This has been explored in laboratory mice by comparing lactation in concurrently pregnant and lactating mice in primiparous females in which nursing young on day 14 were replaced with 5-day-old young. There was a resulting delay of eight to nine days before implantation in these lactating females. Immediately prior to the birth of the second litter, lactation decreased markedly. The young of the first litter could obtain virtually no milk just

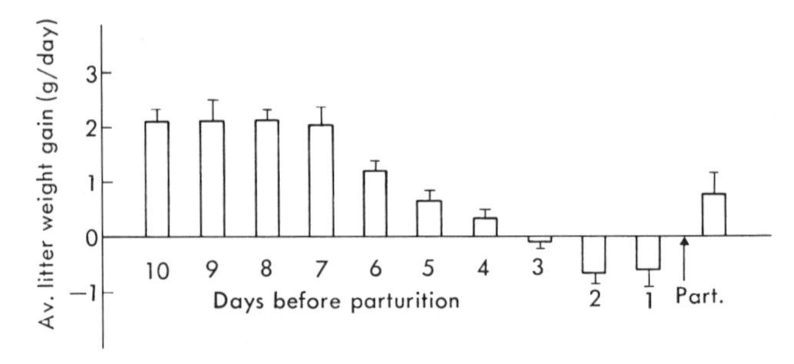

Figure 10.18 The mean daily gain in litter weight of concurrently lactating pregnant mice during a 10-day period before the next birth; mean ± SE, n = 12. (From Mizuno and Sensui, 1978, courtesy of the Japan Scientific Societies Press.)

before parturition (Fig. 10.18), but a control group of young from nongravid females (with prolonged lactation and also with substitute young) continued to gain weight, although at a slowed rate (Mizuno and Sensui, 1978).

10.9 SUMMARY

Mammary tissue develops early in embryonic life, and the sexual differentiation of mammary glands is apparent at birth. The secretory tissue forms the walls of the alveoli that empty into ducts. In most mammals the ducts converge and direct milk through the nipple, which conveys it to the infant. In some ungulates the ducts open into a cistern, where the milk is briefly stored until the infant removes it by suckling at the teat.

Mammary tissue undergoes rapid development, first with sexual maturity and second with pregnancy. Alveolar development is stimulated by prolactin, growth hormone, progesterone, adrenal steroids, estrogen, and insulin. Growth of ducts is induced by growth hormone, estrogen, and adrenal steroids; ducts connect the alveoli with the nipple. During pregnancy, growth hormone, prolactin, and placental lactogen are especially important.

Copious mammary secretion starts after parturition and with the changes in serum levels of hormones that occur at the time of birth. With the discharge of the young and the placentae, progesterone levels decline. Experimentally, high levels of progesterone prevent lactogenesis and also inhibit the stimulatory effects of prolactin. Nursing and the tactile stimulation of the nipple cause a rise in serum prolactin, oxytocin, and possibly growth hormone, all of which promote milk secretion; but the exact roles of the pituitary hormones vary among different taxa. Stimulation of the nipple initiates a sensory signal to the hypothalamus, which in turn causes a release of oxytocin. Oxytocin induces contraction in the myoepithelial cells about the alveoli, forcing milk into the ducts.

The initial mammary secretion is called colostrum. It is rich in fat-soluble vitamins, sodium, calcium, proteins, and immunoglobulins, but it is low in lactose. The composition of milk (which soon follows colostrum) varies with the species of mammal and with the stage of lactation. Generally, milk is high in fat content in those mammals with relatively brief periods of lactation. Lactose levels, which are very low in colostrum, are high in the early milk and gradually decline as the infant grows. Lactose levels are very low or absent in the milk of many marine mammals. As the young develop, proportions of fat and protein increase.

The quantity of milk increases after parturition and later declines with the approach of weaning. Milk output is known to increase with day length in cattle, which might result from the known increase of prolactin with increase of day length.

With concurrent lactation and pregnancy, a phenomenon that occurs in many mammals, milk output declines and may cease as the second parturition nears. If a postpartum estrus does not result in pregnancy, lactation may prevent ovulation until after weaning.

BIBLIOGRAPHY

Cowie AT, Forsyth IA, and Hart IC (1980). *Hormonal Control of Lactation*. Springer-Verlag, Berlin, Heidelberg, New York.

Drewett RF (1983). Sucking, milk synthesis, and milk ejection in the Norway rat. Pp 181–203. In *Parental Behaviour of Rodents* (Ed: Elwood RW). Wiley, Chichester.

Forsyth IA (1982). Growth and differentiation of mammary glands. Pp 47–85. In *Oxford Reviews of Reproductive Biology*, vol 4 (Ed: Finn CA). Oxford Univ, Oxford.

Larson BL and Smith VR (Eds) (1974–1978). *Lactation: A Comprehensive Treatise*. 3 volumes. Academic, New York.

Mepham TB (Ed) (1983). *Biochemistry of Lactation*. Elsevier, Amsterdam.

Neville MC and Neifert MR (Eds) (1983). *Lactation: Physiology, Nutrition and Breast-Feeding*. Plenum, New York.

Peaker M, Vernon RG, and Knight CH (Eds) (1984). *Physiological Strategies in Lactation. Symposium of the Zoological Society of London*, No. 51. Academic, London.

Sulman FG (1970). *Hypothalamic Control of Lactation*. Springer-Verlag, New York, Heidelberg.

Tyndale-Biscoe H and Renfree M (1987). Lactation. Pp 343–371. In *Reproductive Physiology of Marsupials* (Eds: Tyndale-Biscoe H and Renfree M). Cambridge Univ, Cambridge.

Yokoyama A, Mizuno H, and Nagasawa H (Eds) (1978). *Physiology of Mammary Glands*. Japan Scientific Societies Press, Tokyo.

PART III

ENVIRONMENTAL RESPONSES

An individual is an integrated network of hormonal and neural responses to internal and external stimuli. It is obvious from the previous chapters that the reproducing vertebrate does not operate in a physiological vacuum. This section develops the roles of exogenous factors. The harmony between reproductive cycles and the environment results from a concerted but flexible set of responses. The integrated mechanism consists of the animal, which reproduces itself, and the environment, which provides the fuel and the stimulus for reproduction.

Essential to a comprehension of life history theory are the underlying physiological processes (Stearns, 1976). The following five chapters are primarily concerned with these mechanisms.

Most vertebrates appear to be endowed with sufficient flexibility in their responses to survive and reproduce in variable environments. That is to say, they are not genetically constrained to live solely under one specific set of conditions, but rather are programmed to exist within a certain range of temperatures, precipitation, food supply, and photocycle. Since the onset of the Pleistocene some three million years ago, modern species of vertebrates have survived climatic changes by virtue of their phenotypic plasticity, and life histories of modern vertebrates constantly remind us of their broad adaptive nature. This does not minimize the role of the genotype in natural selection, but the capacity to respond within a range of a given parameter, such as clutch size or seasonality, provides for survival over short-term environmental variations, without precluding genetic modifications over the long term.

Exogenous factors are conventionally divided into proximate and ultimate categories. These terms were originally proposed by Baker (1938) and since then have been used, with various definitions, by many authors. A proximate factor, or cue, provides environmental information that is (1) not generally of immediate survival value but (2) indicates the direction of environmental change. An ultimate factor is usually more directly significant to the survival of the individual and is likely to precipitate a prompt behavioral response.

Changes in day length, or in the regular pulsation of the annual photocycle, induces different kinds of physiological changes that prepare an animal for future events. In most vertebrates light, or the absence of light, can cause changes in the release of gonadotropins, eventually producing the cyclical changes in the gonads, which in turn bring about behavioral and morphological changes. In ectotherms changes in temperature may induce gonadal growth and may constitute a proximate factor. Increased temperature may also cause the emergence of amphibians in the spring, and rising temperatures may induce movement of frogs to a spawning pond. In these cases temperature is an ultimate factor. The proper allocation of environmental factors or cues assists in understanding the totality of reproductive behavior, but the reader should be aware that not all students employ the same criteria for distinguishing proximate and ultimate factors.

Fecundity is sometimes discussed in terms of *r*- and *K*-selection. This concept states that contrasting influences of two sorts of environment—stable (predictable) or variable (unpredictable)—select for rapid population growth (*r*-selection) or for slow or no growth (*K*-selection) and that certain reproductive parameters are associated with these two sorts of environment. Stated most simply, *r*-selection promotes high fecundity with an early age of sexual maturity and a relatively short generation time. *K*-selection favors low fecundity with late sexual maturity and a greater interval between generations. For a detailed treatment the student is referred to the discussion by Stearns (1976).

The concept of *r*- and *K*-selection has been widely discussed in both the technical literature and in texts, but in the opinion of many very competent students (e.g., Hairston et al., 1970; Colwell, 1974; and Shapiro, 1986), it has rather limited usefulness and applicability.

Population levels reflect the presumed balance among reproductive potential, mortality, and the effects of exogenous factors on each. The exogenous factors change over the short-term span, so that the density of a species varies from season to season and from year to year. The extremely high incidence of extinctions over the long term testifies to the sometimes persistent greater mortality over natality. Past extinctions indicate that a species that thrives today may disappear tomorrow. The underlying causes are rarely known, but one assumes that the extinct species could not adapt to relatively sudden changes in exogenous factors, and mortality exceeded natality. Within a human life span one can observe numerous environmental changes, and since the close of the Pleistocene, climatic changes have been substantial. Their effect on vertebrate faunas is sometimes of great magnitude and is never ending.

The inescapable conclusion is that animals vary profoundly in the degree to which they are adapted to their environments. Some species thrive and abound in several habitats over a broad geographic range, and others are always rare, confined to a relatively small area, and seem only to have a toehold on survival. The following five chapters review mechanisms by which environments modify vertebrate reproduction, and the kinds of responses of which vertebrates are capable.

11

PHOTOPERIODIC RESPONSES OF ENDOTHERMS

11.1 PHOTOPERIODIC RESPONSES OF BIRDS

Knowledge of the influence of day length on avian physiology is not new. Artificial extension of daylight was used to promote laying in domestic fowl long before there was electricity. The exact mechanism of this response, however, has not been understood until rather recently, and research continues to provide explanations for such events as seasonality, photorefractoriness, molt, and migration.

11.2 EXTRARETINAL PHOTORECEPTION IN BIRDS

Extraretinal reception of light in birds appears to lie in the hypothalamus. The expected gonadal response of birds to long days is seen in bilaterally enucleated individuals, but only if the top of the head receives full light. When light is obscured or blocked from reaching the crown of the bird, the photoinduced response is lacking. Even in intact (sighted) sparrows (*Passer domesticus*) the eyes do not transmit the visual signals that result in gonadal development (McMillan et al., 1975), and the same phenomenon is seen in the white-crowned sparrow (*Zonotrichia leucophrys gambelii*) and the Japanese quail (*Coturnix coturnix*).

When deprived of their vision, domestic ducks respond to long days by gonadal recrudescence and to short days by gonadal regression (Benoit, 1970). This response does not necessarily preclude a role for the paired eyes and retinal stimulation, but it does indicate the presence of other photoreceptors. After the realization that the paired eyes are not the essential photoreceptors that affect gonadal activity in birds, there followed a search for other sensory tissues or structures. Light directed to the pineal gland of blinded ducks stimulates testicular growth (Hisano et al., 1972), but light directed to the avian pineal may well strike adjacent structures. The mechanism for this response is not apparent. Luminous beads implanted in the brains of sexually mature Japanese quail (*Coturnix coturnix*) had an effect when the birds were reared and continued under a nonstimulatory (8L:16D) light schedule, suggesting that the hypothalamus is photosensitive. Within three weeks after the beads were implanted, the testes had achieved a greatly enlarged state in contrast to the minute testes of controls without implanted beads (Homma et al., 1980). A more precise determination of the photosensitive area was made by the use of fiber optics. By directing light-conducting fibers to basal hypothalamic areas, it is possible to illuminate specific sites. In the male white-crowned sparrow (*Zonotrichia leucophrys gambelii*), photosensitive areas, as revealed by fiber optics, lie at or near the ventral hypothalamus; these photosensitive areas may also produce gonadotropin-releasing hormone (GnRH) (Yokoyama et al., 1978). There appears to be a rhodopsinlike visual pigment in the tissues of this region (Foster et al., 1985).

11.3 THE EFFECT OF LIGHT ON THE AVIAN REPRODUCTIVE SYSTEM

Studies on the responses of birds to varying light schedules have been conducted mostly on small passerine species and some domestic birds, such as the fowl, the Pekin duck, and the Japanese quail. Most research concerns males, in which the results are more readily observed.

In male birds the first effect of long days is the rise of serum luteinizing hormone (LH) and follicle-stimulating hormone (FSH). Secretion of gonadotropins in the Japanese quail is proportional to day length and is not a simple "on-off" response. The development and activity of the Leydig cells are promoted by LH, and an increase in testicular size follows the increase in serum levels of FSH and testosterone.

This, in turn, is paralleled by growth and spermatogenic activity of Sertoli cells. If the bird is soon returned to a short-day schedule, LH levels remain high for from 4 to 10 days (Follett and Robinson, 1980), but the cause of the "carryover" effect is unknown. In the Japanese quail the increase in testosterone occurs about a week after the onset of long day lengths, suggesting that LH receptors develop in about one week (Maung and Follett, 1978). After day length has reached a critical minimal stimulatory period, the rate of gonadal recrudescence has a log-linear relationship to the duration of light in passerine birds (Farner, 1964) and in the Japanese quail (Follett and Maung, 1978).

There is no question that long days are stimulatory to gonadal growth in males of many passerine birds. In females light has a stimulatory role on early follicular development (Farner, 1975). In females, long days are essential for development of follicles up to the onset of yolk deposition, and this is the state at which female white-crowned sparrows arrive on their breeding grounds. Thereafter supplementary information, such as singing males, courtship, and proper trophic conditions, is needed to bring the ovary into a functional state.

11.3.1 The Photosensitive Phase

There is a period of sensitivity, the photosensitive phase (PSP), during which long days can induce the release of gonadotropins, and light is stimulatory only during this period. The photoperiodic control of the endocrine system results from the coincidence between a photoinducible phase of an entrained circadian oscillation (which determines the PSP) and the natural photocycle (Fig. 11.1). Thus stimulation

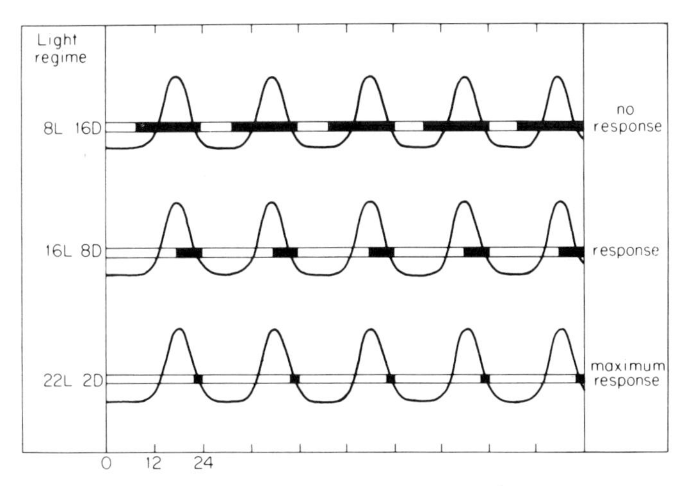

Figure 11.1 Suggested diagrammatic explanation of the external coincident model for birds. The daily photoperiodic regimen entrains the circadian rhythm of sensitivity (the curved lines on the graph). The extent of the response to light depends upon the degree to which the light period (clear bar) intrudes into the photosensitive phase (PSP), or the part of the curve above the horizontal. (From Farner, 1975.)

of gonadotropin release is predicated on the duration of the coincidence of light with the PSP (Farner, 1980b).

11.3.2 Photoperiodic Responses of Birds at Low Latitudes

Miller (1958) and Lofts and Murton (1968) have suggested that the role of photoperiod as a cue is insignificant in low latitudes and that daylight itself does not set the nesting schedule. The change in day length is, indeed, rather slight near the equator, but this does not preclude an important role for the photocycle in tropical avian reproduction. The tropical weaver finch (*Ploceus philippinus*) can discriminate between small differences in day length and has the potential to detect minor photoperiodic changes in nature (Chandola et al., 1976). Although tropical day lengths are presumably adequate to bring the avian reproductive system to a "subfunctional" state, at least for species native to the tropics, actual nesting may depend upon other exogenous factors.

Several tropical birds exhibit increased testicular growth when exposed to long days. This has been shown in the rufus-collared sparrow (*Zonotrichia capensis*) by Miller (1959), in the zebra finch (*Poephila guttata*) by Marshall and Serventy (1958), and in several other species (Marshall, 1970). The zebra finch, in fact, rarely has conspicuously regressed gonads (Farner and Serventy, 1960; Serventy, 1971).

There is no reason to assume that low-latitude vertebrates have lost the ability to respond to changes in day length or that tropical species have been immobile in the low latitudes for endless millions of years. Pleistocene faunal movements have been abundantly documented for the temperate regions, and in the low latitudes of both Africa and South America there were concurrent faunal movements in both latitude and elevation (Hamilton, 1976; van Zinderen Bakker, 1976; Muller, 1973). In evolutionary time these events occurred only yesterday, and it would be very surprising if sensitivity to day length had been lost in this short time unless it had been deleterious.

It will become clear, however, that a basic difference between the tropical environment, in which the annual photocycle maintains the near-tonic subfunctional state, and the temperate latitudes, in which day length promotes vernal recrudescence, is the influence of many nonphotoperiodic factors. Although it would be unrealistic to underestimate the role of the highly contrasting photocycle in the mid- and high latitudes, one should not be so blinded by the effect of day length that the non-photoperiodic factors are neglected. In the tropics a broad spectrum of nonphotoperiodic factors is scattered throughout the year, but in temperate latitudes these factors appear close to the maximal influence of increasing day length in a vernal crescendo. These factors are discussed in Chapters 13 (Section 13.8) and 14 (Section 14.5) Because of the ease with which day length can be studied in the laboratory, our understanding of proximate factors is almost totally confined to the annual photocycle (Farner and Wingfield, 1978).

Passerine birds have a multifaceted response to day length. Not only are there several gonadal, migratory, and feeding responses to day length, but there is a complex of interrelated feedback mechanisms that modify the effects of the exogenous factors (Fig. 11.2). This arrangement provides for a considerable latitude of physiologic

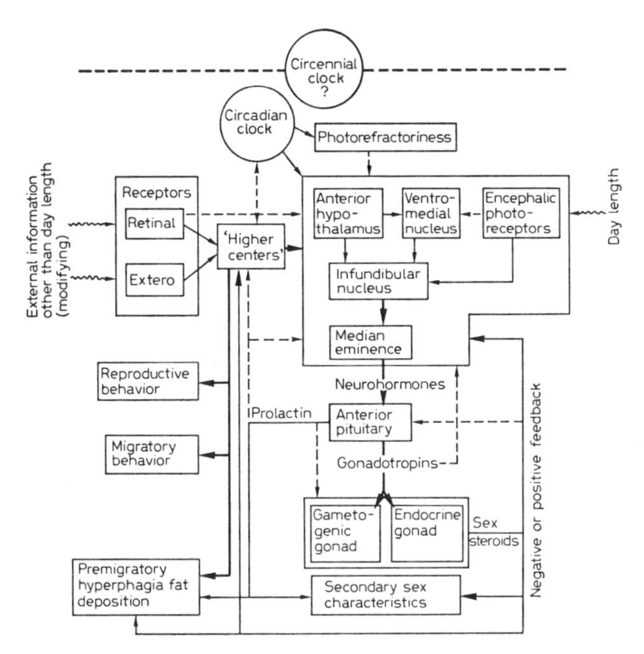

Figure 11.2 The transduction of exogenous influences into endocrine messages controlling reproduction and associated activities. (From Wingfield and Farner, 1980, *Progress in Reproductive Biology* 5, courtesy of S. Karger AG, Basel.)

and behavioral pathways, so that the individual is constantly adapted to a broad spectrum of environmental conditions.

11.3.3 Weakly Photosensitive Birds

Among both migratory and sedentary birds are some whose gonadal cycles seem to be indpendent of photoperiod. The short-tailed shearwater (*Puffinus tenuirostris*) is unusual in having a transequatorial migration and breeding in the Southern Hemisphere: it nests in Tasmania and annually migrates to the Aleutians. When kept under unnatural photoperiodic schedules for up to six months, the birds showed no change in gonadal development (Marshall and Serventy, 1958). Similarly, the spotted munia or mannikin (*Lonchura punctulata*), a nonmigratory ploceid finch of the tropical Orient, does not respond to photoperiodic stimuli but appears to have a tendency to follow an endogenously controlled reproductive pattern (Chandola and Thapliyal, 1978). For more examples see Chapter 13, Section 13.8.3.

11.4 REFRACTORINESS

Many vertebrates have a refractory period, a time during which the reproductive system does not respond to exogenous cues. This period is best known for birds but apparently occurs in all classes. Refractoriness prevents the continuation of reproduction into a season that would be unfavorable for the survival and growth of the next generation, as well as the survival of the adults.

Although the avian reproductive system is very sensitive to the increase of day length in the spring, it becomes unresponsive to the same day lengths later, usually in the late spring or summer. This refractory condition constitutes the postreproductive failure to be stimulated by factors that were stimulatory prior to breeding. The refractory period varies with respect to the time at which it ends reproduction and the duration of its effect.

Refractoriness is best known in birds. There is an increased sensitivity of the hypothalamus (and perhaps also the anterior pituitary) to the negative feedback of gonadal steroids; such increased sensitivity accounts for the decline in gonadotropins in late summer. In addition to a decline in gonadotropins, refractoriness can also bring about a decline in gonadal steroids. Thus the drop in plasma steroids does not automatically result in a rise of gonadotropins. Photorefractory red grouse (*Lagopus lagopus scoticus*) had mean plasma LH between 0.28 and 0.34 ng/ml under a 16L:8D schedule. After castration (and the elimination of gonadal steroids), circulating LH increased markedly in a 20L:4D schedule and increased only slightly in a 6L:18D schedule (Fig. 11.3). Photosensitive and photorefractory male white-crowned sparrows (*Zonotrichia leucophrys gambelii*) treated with synthetic GnRH

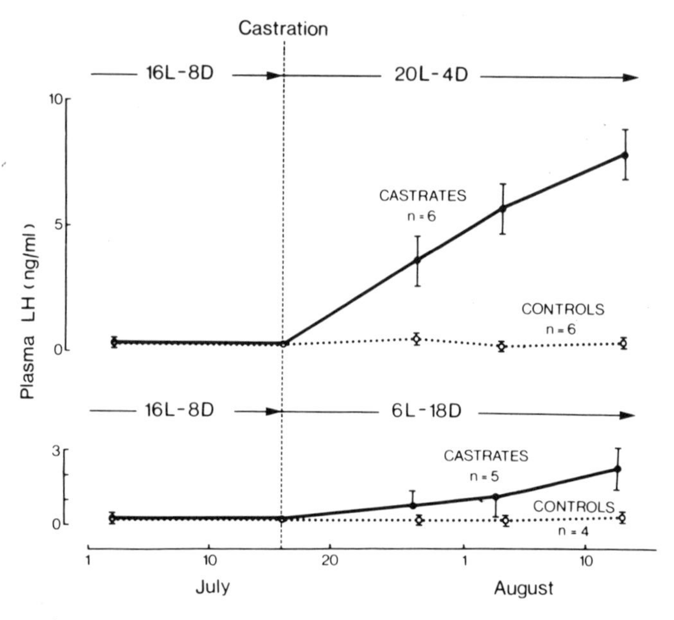

Figure 11.3 Changes in plasma LH in intact (open circles and dotted lines) and castrated (solid circles and continuous lines) red grouse, following exposure to increased (*upper*) or decreased (*lower*) day lengths. The birds were kept under natural day lengths until July 1, and half in each group were castrated on July 15. The experimental photoperiods began on July 16. Vertical lines indicate SE. (From Sharp and Moss, 1977, *General and Comparative Endocrinology*, vol. 32, courtesy of Academic Press.)

responded positively and equally, suggesting that refractoriness occurs at a level above the anterior pituitary (Wingfield et al., 1979).

There is also a negative feedback from other steroids, perhaps from the adrenal cortex (Wilson and Follett, 1975). This feedback is suggested by castration of photosensitive and photorefractory willow grouse (*Lagopus lagopus lagopus*). Castration of photosensitive individuals was followed by an immediate rise in LH, but in the photorefractory birds the increase in serum LH did not occur until some five weeks later, at which time they may have no longer been refractory. The continuation of refractoriness in castrated grouse suggests that steroids other than testosterone continued to depress the activity of the hypothalamus and perhaps also the anterior pituitary (Stokkan and Sharp, 1980).

An alternative explanation for the development of refractoriness lies in a proposed seasonal time shift in the PSP. When exposed to light early in the natural dark period in October, house finches (*Carpodacus mexicanus*) did not experience gonadal recrudescence, but exposure at the same time in January was followed by gonadal growth (Hamner, 1968). It is likely that different physiological conditions account for refractoriness among the many families of birds in which it is known.

The development of photorefractoriness in the Japanese quail depends upon serum thyroxin during exposure to long days. Thyroidectomized males, moreover, respond only slowly to a reduction in day length (Follett and Nicholls, 1985). In the starling (*Sturnus vulgaris*), thyroidectomy prevents photorefractoriness and subsequent molt, but short days nevertheless induce testicular regression (Dawson et al., 1986).

A conspicuous characteristic of the refractory period is the eventual increasing sensitivity to stimulatory day lengths. This is seen most clearly when refractory individuals are kept under a schedule of short days. Hermit thrushes (*Hylocichla guttata*), when maintained on a 9L:15D schedule for six weeks, became responsive to long days. Gonadal growth followed exposure to 15L:9D days, whereas another group remained photorefractive from October to February when kept on a 15L:9D schedule (Annan, 1963). Similarly, the slate-colored junco (*Junco hyemalis*), when subjected to short days (9L:15D) for five to six weeks in July and August, became sensitive to long days. By breaking refractoriness with exposure to short days, it became possible to obtain five annual periods of gonadal growth and fat deposition and two molts (Wolfson, 1960).

11.5 THE ROLE OF THE PINEAL IN BIRDS

Because of the importance of the pineal in mediating light in mammals, it is reasonable to explore its role in birds. There is little evidence, however, to suggest a major role for the pineal in fecundity of birds. As previously indicated, the hypothalamaus is the structure of primary importance in the response of birds to the photocycle.

The avian pineal gland is nevertheless photosensitive (Binkley et al., 1978; Deguchi, 1981) and is an important source of melatonin. Melatonin is synthesized from serotonin through the action of *N*-acetyltransferase. An intermediate product is *N*-acetylserotonin, which, in the presence of hydroxylindole-O-methyltransferase (HIOMT), is changed to melatonin. HIOMT is localized in the pineal gland of both birds and mammals (Axelrod, 1974). The light-dark rhythmicity of *N*-acetyltransferase known for the laboratory rat is also seen in the avian pineal (Binkley et al., 1974). The depressing effect of light is seen in the diurnal elimination of *N*-acetyltransferase, which appears in a circadian pattern in constant darkness (Fig. 11.4). There is also a circadian rhythm of melatonin in the pineal of the Japanese quail, with an increase in the dark phase (Cockrem and Follett, 1985). In the rat this rhythm originates in the suprachiasmatic nuclei and is transmitted via the superior cervical ganglia, but in the chicken the rhythm arises in the pineal (Deguchi, 1979). Unlike the mammalian pineal, which follows retinal signals, the pineal of the domestic fowl is cued by an extraretinal photoreceptor (Binkley et al., 1975). Similarly, the domestic duck, which also has a light-sensitive pineal gland, remains responsive to light when enucleated and pinealectomized (Hisano et al., 1972).

Although the avian pineal is sensitive to light and exhibits nocturnal peaks of *N*-acetyltransferase, it is not known to affect gonadal activity in birds maintained

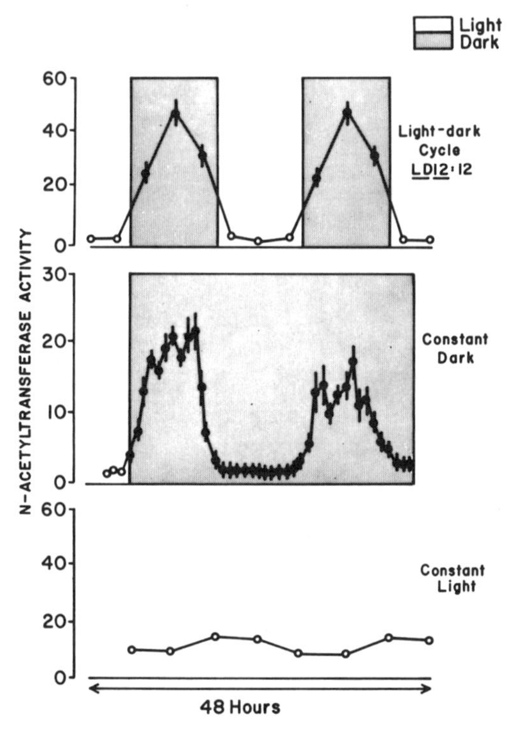

Figure 11.4 The in vitro endogenous rhythmicity of *N*-acetyltransferase activity in the pineal of the chicken. Constant light suppresses this rhythmicity, indicating photosensitivity of the pineal of the fowl. (From Binkley, 1976, *Federation Proceedings*, vol. 35, fig. 2, p. 2349.)

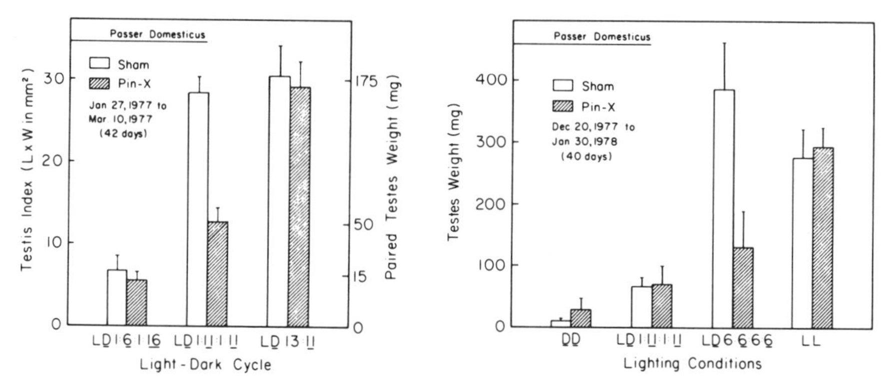

Figure 11.5 The decline in testicular size of the sparrow (*Passer domesticus*) after pinealectomy (Pin-X) is striking, but it varies with the time at which pinealectomy occurs. *Left.* Mean testis index ± SEM. *Right.* Mean paired testes weights. (From Takahashi et al., 1978, courtesy of Elsevier Biomedical Press B.V.)

under natural photocycles (for further comments on this enzyme, see Section 11.6). In the sparrow (*Passer domesticus*) kept under certain "skeletal" photoperiods (1L:11D:1L:11D) and 6L:6D:6L:6D) following pinealectomy, there is a marked reduction of testicular growth (Fig. 11.5). This effect is not consistent and may depend upon a seasonal change in sensitivity (Takahashi et al., 1978). The pineal in the house sparrow preserves circadian rhythms of motor activity (Zimmerman and Menaker, 1975), a function seen in some reptiles. In many nonmammalian vertebrates the pineal is important in generating endogenous circadian rhythms and may function in controlling the cycle of photosensitivity of the hypothalamus.

11.6 PHOTOPERIODIC RESPONSES IN MAMMALS

In many major taxa of each of the classes of nonmammalian vertebrates, the pineal gland is photosensitive, but in mammals the pineal receives light signals from the retinas of the paired eyes. In all classes, therefore, the pineal gland is potentially a structure that can transmit photic information to the endocrine system. The pineal may be extremely important to the seasonal adjustment of the annual photocycle and may also have roles in other seasonal cycles of nonreproductive activity. A major product of the pineal is melatonin, a hormone that depresses gonadal activity in mammals.

11.6.1 Historical Aspects

Early in the twentieth century there were suggestions that the pineal gland was somehow related to a suppressed gonadal development. In the first 30 years of the

1900s several researchers proposed that pathological changes in pineal size and activity were inversely correlated with sexual development and precocity in humans and that pineal products were antigonadal (Kappers, 1979).

Until the early 1960s the pineal "body," as it had come to be known, had no clearly established functions, and it was certainly not accepted as an endocrine gland. From the early 1970s, research on the mammalian pineal gland has demonstrated that it is the source of substances that are carried to other regions of the body, with a subsequent change in metabolic rates of specific tissues. The pineal gland of mammals is indirectly cued to light, and the daily photocycle is a timegiver or *Zeitgeber* that entrains rhythmic production of some pineal enzymes and hormones. This has been shown by extensive experimental work on the laboratory rat, the hamster, and to a lesser degree on the ferret and some wild mice.

The concept of an antigonadal role for the products of the pineal gland and its relationship to the mammalian pituitary had long been expressed (e.g., Simmonet et al., 1954) if not always accepted; but the effect of light on the pineal and the relationship of pineal secretions, especially melatonin, to reproduction remained obscure for many years. The demonstration that subcutaneous injections of melatonin depressed ovarian metabolism in the rat (Wurtman et al., 1963) suggested a regulatory role of reproduction for the mammalian pineal. By the late 1960s the role of the pineal as an endocrine gland was suggested by the substantial amount of work conducted on the laboratory rat (Wurtman et al., 1968). Also, it was found that the preovulatory surge of LH is depressed by melatonin in the laboratory rat, even though the reproductive system of that rodent is not especially sensitive to day length (Wurtman et al., 1963; Reiter, 1968). It later became apparent that although the gonads of the laboratory rat are only slightly responsive to photoperiodic changes, some other kinds of rodents clearly reflect changes in day length by fluctuations in pineal activity. The enzyme HIOMT leads to the formation of melatonin from serotonin, and this process is known to occur in both the retina and pineal gland of fish as well as in the pineal of fish, amphibians, reptiles, and birds in response to changes in light. Although the mammalian pineal is not photosensitive, both HIOMT and melatonin occur in the mammalian pineal and respond to light signals from the retina (Quay, 1965b). In a wild population of the Uinta ground squirrel (*Spermophilus armatus*), HIOMT increased after reproduction to its highest levels just before hibernation. Testicular size was maximal at emergence the following spring, when HIOMT levels were lowest (Ellis and Balph, 1976).

Gradually the relationship of the occurrence of the pineal enzymes and their products to the daily light cycle was elucidated in the laboratory rat. During the daytime, serotonin levels are high; but its conversion to melatonin, in the presence of the enzymes *N*-acetyltransferase and HIOMT, is a nocturnal reaction inhibited by light. The pineal of the laboratory rat has long been known to increase HIOMT formation and melatonin secretion in the dark (Wurtman and Axelrod, 1964). Also, early studies on the white-footed mouse (*Pero myscus leucopus*) suggested a reduction in pineal activity as a consequence of exposure to long day lengths (Quay, 1956a).

11.7 POSITION AND FUNCTION OF THE PINEAL GLAND

The pineal gland (also known as the epiphysis cerebri) is an outgrowth of the posterior section of the dorsal surface of the diencephalon. As pointed out previously, the pineal is photosensitive in nonmammalian vertebrates (Deguchi, 1981; Oksche, 1965; Quay, 1970). Blood enters the pineal from branches of the posterior choroid arteries and leaves via venules to adjacent menigeal tissue. Studies of pineal ultrastructure reveal endocrine secretory tissue with sympathetic innervation. The development and relationships of the pineal gland are described by Oksche (1965) and Quay (1974).

11.7.1 The Effect of Light on Pineal Activity

In mammals the pineal receives light information via a complex neural route from the retinas of the paired eyes. Impulses travel from the retina to the suprachiasmatic nuclei (SCN) of the hypothalamus. A second part of the pathway involves neurons going to the medial basal hypothalamus (Moore, 1978). The message then travels down the upper thoracic spinal nerves and, via the superior cervical ganglion, returns to the skull and to the pineal (Fig. 11.6). During darkness norepinephrine is released by the sympathetic neurons within the pineal, where it stimulates the conversion of serotonin to melatonin, at least in the laboratory rat (Zatz, 1981). In nonmammalian vertebrates the pineal is itself photosensitive and releases melatonin in darkness (Fig. 11.7).

Although the pineal is clearly involved in reproductive responses to day length in some mammals, there are many unanswered questions regarding the pathway

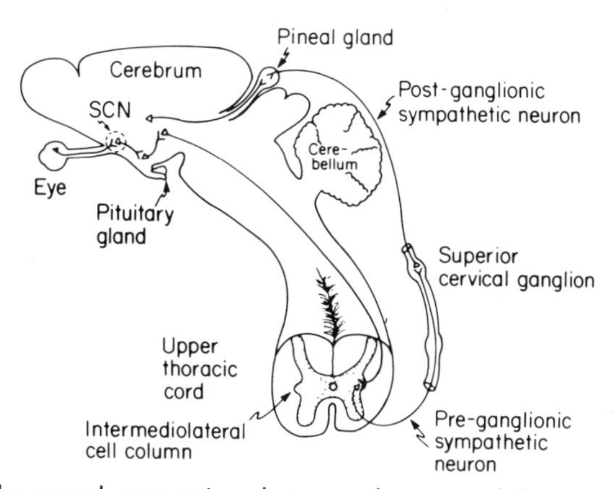

Figure 11.6 The neural connections between the eye and the pineal as they appear to exist in mammals. SCN, suprachiasmatic nuclei. (From Reiter, 1983, courtesy of Plenum Publishing Corporation.)

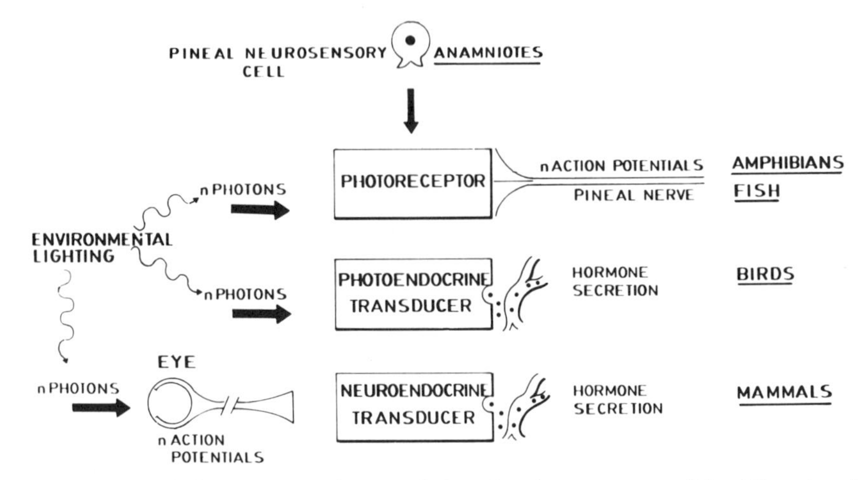

Figure 11.7 Phylogenetic evolution of the pineal response to light. The pineal is photosensitive in fish, amphibians, reptiles, and birds, and it transduces light signals into neural impulses. The pineal receives light signals from the retina in mammals, and, indirectly, light suppresses the secretory activity of the pineal. (From Cardinali, 1981, *Endocrine Reviews*, 2: 327–346, © Williams & Wilkins, 1981.)

through which melatonin becomes effective. Experiments indicate that different species of mammals are not equally responsive, and the basis for some observed differences are gradually being revealed by surgical elimination or destruction of the various critical structures.

Experimental work on the mammalian pineal clearly established it as the organ that, controlled by signals from the eye, reflects day length for the reproductive system. Although research has been confined to only a few species of mammals, it is the only structure in mammals known to have such a role. Studies of ectotherms suggest that the pineal, which is photosensitive in those vertebrates, also affects their reproductive activity. The pineal produces indoleamine and polypeptide hormones, and they probably have a variety of effects that are not well understood. The indoleamines in the pineal are derived from the amino acid tryptophan. In mammals at least, melatonin is perhaps the most important pineal secretion affecting reproductive activity.

Pineal activity is monitored by observing changes in levels of the enzymes that lead to the production of melatonin. There is a moderate peak of HIOMT and a very conspicuous peak of *N*-acetyltransferase during dark periods. The enzyme *N*-acetyltransferase and melatonin both respond to the cessation of light: their production commences soon after the onset of darkness and ceases promptly with the return of light. The circadian rhythm of *N*-acetyltransferase in the pineal has long been known to have a nocturnal peak as much as 30 to 50 times the diurnal low (Klein and Weller, 1970; Axelrod, 1974). Therefore, pineal activity is usually indicated by changes in levels of *N*-acetyltransferase.

11.7.2 Responses to Light

The testicular cycle of the golden, or Syrian, hamster (*Mesocricetus auratus*) is much like that of many wild mice. Testicular growth follows periods of long days, in the spring, and regression follows a decrease in day length, in late summer (Fig. 11.8). In a regimen of changing light, the golden hamster requires at least 12½ hours of light daily to preserve testicular function (Elliott, 1976). Concomitant with the short-day induction of testicular regression in hamsters, there is a gradual decline in copulatory effort (Moran and Zucker, 1978). Apparently the melatonin increase under short days (< 12.5 hours) causes a lower threshold (increased sensitivity) of the hypothalamaus to gonadal steroids, thus effecting a decline in secretions of LH, FSH, and prolactin from the anterior pituitary (Blask et al., 1979; Reiter, 1980b).

Like the gonads of many wild cricetid mice, the testes of captive hamsters remain small for the duration of most of a natural (or simulated) winter. In the Djungarian hamster (*Phodopus sungorus*), day length affects gonadotropin secretion, independent of gonadal feedback (Simpson et al., 1982). In that species the critical day length is about 13 hours, and gonadal recrudescence is not induced by shorter day lengths (Hoffmann, 1981). In the vole *Microtus agrestis* the critical day length is approximately 13.5 hours, and after six weeks testicular growth is greater in males kept at 15 and 16 hours than at 14 hours (Grocock, 1981). In the white-footed mouse (*Peromyscus*

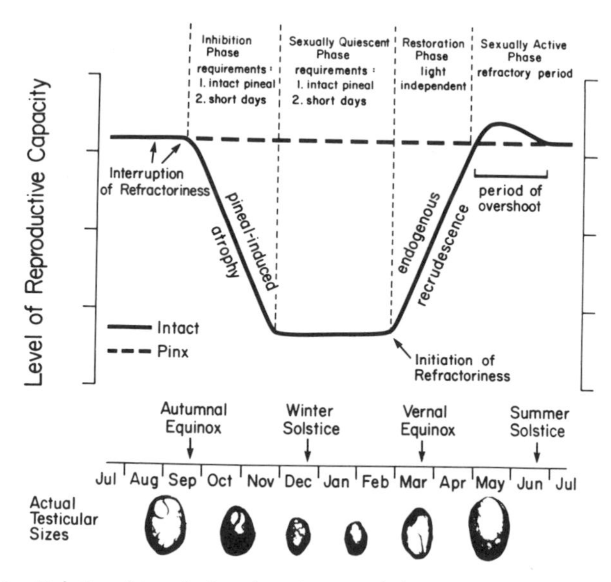

Figure 11.8 Relationship of pineal activity and the annual reproductive cycle of the golden hamster, based on experimental studies of the hamster. Scotorefractoriness occurs at the onset of testicular recrudescence and terminates near the end of the summer. Testes at the bottom are from intact animals collected at the times indicated. (From Reiter, 1980b, *Endocrine Reviews*, 1: 109–131, © Williams & Wilkins, 1980.)

leucopus), which has an extensive latitudinal distribution, the critical day length increases with latitude (Lynch et al., 1981a).

It is not always clear how the annual photocycle is used as information in control of the annual reproductive cycle of some species of wild mammals. Many mammals are more or less nocturnal and spend daylight hours in burrows or hollow trees, in which they are well shielded from daylight. Nevertheless, in a strictly nocturnal species, such as the southern flying squirrel (*Glaucomys volans*), there is a clear testicular response to changes in day length in the laboratory (Muul, 1969). In such species of mammals, the critical aspect of the photocycle is not the period of day length but the period of darkness.

The active pineal is antigonadal in female hamsters as it is in males, but although there is both uterine regression and a reduction in ovarian activity, ovarian interstitial tissue increases and the ovaries are actually heavier than in sexually active females (Reiter and Johnson, 1974). Also, melatonin has some effects in the human female. Just prior to the estimated day of ovulation, there is a drop in circulating melatonin, based on daily measurements taken at 8 A.M. (Arendt, 1979). Also, exposure to 24-hour light stimulation at the midpoint of the menstrual cycle reduced variance in the duration of the cycle and increased the occurrence of 29-day rhythms (Dewan and Rock, 1969).

In the ferret (*Mustela putorius*), which is a strongly seasonal species in nature, pinealectomy alters the timing of estrus and reduces the animal's responsiveness to light. Pinealectomized female ferrets fail to respond to artificially early exposure to long days or short days, although estrus occurs at the normal time for those under a natural photocycle. The effect of pinealectomy on the ferret, however, varies with the time at which the operation is performed: pinealectomy early in one estrous cycle delays estrus the following year. The pineal is essential for the female's reproductive cycle to become entrained to a natural photocycle (Herbert, 1981).

A photoperiodic effect has been suggested in female monkeys. In Rhesus monkeys sexual activity generally increases during the short-day part of the annual photocycle, but social influences are also very important, for mating efforts do not necessarily parallel hormonal cycles (Michael and Zumpe, 1981). Although most conceptions in macaques (*Macaca* spp.) occur under regimens of decreasing day lengths (October–December), experimentally they seem not to be cued by short days (Wehrenberg and Dyrenfurth, 1983). The ring-tailed lemur (*Lemur catta*) breeds in the winter of the Southern Hemisphere in Madagascar. A light schedule of 14L:10D inhibits ovulatory cycles, and a 5.5-hour reduction in the light period was followed by a return to ovarian cyclicity (van Horn, 1975).

11.7.3 Circadian Sensitivity to Melatonin

In addition to the rhythm of melatonin production, there is a circadian pulsation to the sensitivity to melatonin, and this sensitivity occurs in the dark period or in the hours immediately prior to expected darkness (Fig. 11.9). For example, both male and female hamsters experience gonadal growth under long days (e.g., 14L:10D), and injections of melatonin are effectively antigonadal only if given in the latter

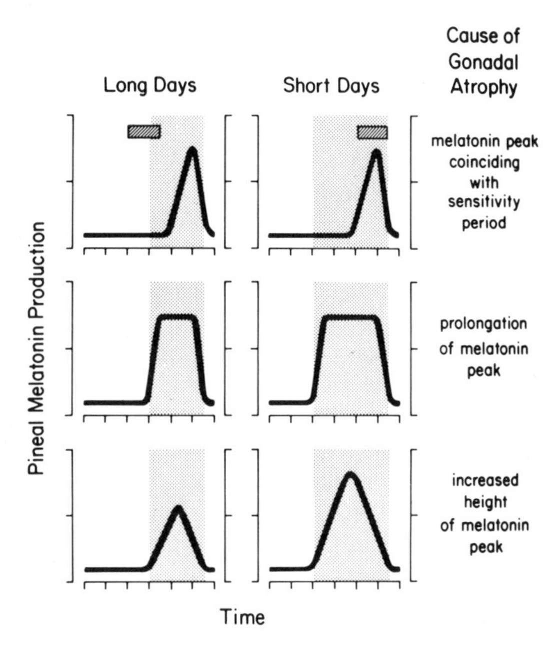

Figure 11.9 Conceptual mechanism of the effect of melatonin on the reproductive system in mammals, as synchronized with the period of sensitivity (shaded areas). (Compare with Fig. 11.1.) The effect may vary with the prolongation of the peak of melatonin production or the duration of sensitivity. (From Reiter, 1983, courtesy of Plenum Publishing Corporation)

part of the light cycle. The photosensitive phase of the golden hamster begins about 30 minutes prior to the individual's daily activity period and extends to approximately 12 hours. The circadian period of sensivitity to melatonin in mammals seems to be comparable with the PSP of birds, and it provides for a restricted period of responsiveness corresponding to the times during which melatonin is secreted. These afternoon injections of melatonin do not depress gonadal development in pinealectomized animals (Tamarkin et al., 1976). This suggests the possible secretion of additional melatonin at a later time in the daily photocycle. There is also the possibility that the pineal produces other antigonadal hormones.

Because the rhythm of greater sensitivity is circadian, any light schedule other than one totaling 24 hours, or in 24-hour increments, will introduce part of a stimulatory period in the unresponsive phase of the animal. That is to say, a light-dark schedule repeated every 36 hours might expose the individual to light in its responsive phase (or PSP) every other day. Thus, ahemeral regimens (other than a nycthemeron, or 24-hour, period) of 12 or 36 hours, for example, produce results inconsistent with a 24-hour schedule with the same light-dark ratio (Grocock and Clarke, 1974).

To explore the effects of melatonin on the fecundity of the golden hamster, exogenous melatonin has been administered both by injections and by subcutaneous implants. Results vary, and melatonin does not always cause the gonadal regression seen in hamsters kept under a regimen of short days. Hamsters under a 14L:10D schedule are unresponsive to melatonin when it is given by injection early in the light period, but injections are antigonadal when administered late in the light period (Tamarkin et al., 1976). Pinealectomy or superior cervical ganglionectomy results in continued fecundity in male hamsters (Fig. 11.10), and when males and females that have undergone pinealectomy or superior cervical ganglionectomy are housed together, fertile matings follow (Reiter et al., 1976; Reiter, 1980b). Thus pinealectomized hamsters resemble those kept under a long-day regimen, and destruction of the superior cervical ganglia produces the same effect.

Similarly, subcutaneous implants of melatonin in beeswax in hamsters kept under short day lengths are counterantigonadal (Hoffmann, 1974; Reiter et al., 1974). Apparently chronic administration of as little as 3.6 μg/day of melatonin suppresses the antigonadal effect of pineal melatonin and is tantamount to pinealectomy. Chronic exposure of receptor organs may maintain refractoriness to the suppressive effect of pineal melatonin (Reiter, 1980a). In other words, subcutaneous implants of melatonin have a counterantigonadal (instead of progonadal) effect and therefore resemble pinealectomy.

Thus, although short days and melatonin administered late in the light period both depress gonadal activity, this effect disappears when the animal is continuously exposed to melatonin by subcutaneous implants. It seems plausible that above a certain exposure to melatonin, its receptors become insensitive to the melatonin's effect, accounting for the failure to respond early in the daily light period after a nighttime of exposure to melatonin (Reiter, 1980a, 1981a). Hamsters kept under

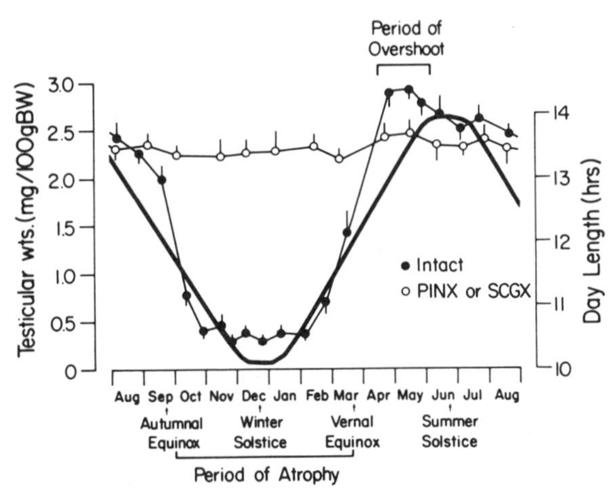

Figure 11.10 Testicular weights in intact and pinealectomized (PINX) hamsters under natural day lengths. Either pinealectomy or superior cervical ganglionectomy (SCGX) prevents gonadal atrophy induced by short days. (From Reiter, 1980b, *Endocrine Reviews*, 1: 109–131, © Williams & Wilkins, 1980.)

stimulatory schedules (14L:10D) and given 25 μg at 1700 hours underwent gonadal involution when saline had been administered at 1100 hours, but not when 1 mg of melatonin had been injected at 1100 hours (Chen et al., 1980). These results support the concept that exposure to melatonin renders receptors temporarily insensitive and also that injections of melatonin early in the light period prolong this insensitivity. The nature or location of these receptors is still undetermined. In the white-footed mouse (*Peromyscus leucopus*), receptors appear to be in the suprachiasmatic nuclei (Glass and Lynch, 1981).

11.8 THE SUPRACHIASMATIC NUCLEI

The role of the hypothalamus in mammalian rhythmicity was suggested by the loss of discrete circadian rhythms in laboratory rats following lesions in the ventral median area of the hypothalamus (Richter, 1967). More recent studies of circadian control have centered on the specific region of the suprachiasmatic nuclei (SCN). The SCN of the rat lie in front of the anterior hypothalamus above the rear margin of the optic chiasma. The SCN receive a major afferent nerve tract from the retina through the optic chiasma. Other afferent nerve tracts are received from the midbrain raphe nuclei and from the ventral lateral geniculate nuclei. Thus there is a variety of nervous input to the SCN, and within the SCN there is a large complement of synaptic types. Efferent fibers from the SCN are not well known (Rusak and Zucker, 1979).

The SCN seem to preserve circadian rhythms in some mammals, for its destruction is generally followed by arrhythmicity, including the loss of pineal rhythms (Rusak and Zucker, 1979). There is a clear correlation between circadian rhythmicity of motor activity (such as wheel running) and the regularity of estrous cycles in the golden hamster, and they appear to be regulated by the SCN (Fitzgerald and Zucker, 1976).

The SCN of the domestic rabbit contain neurosecretory cells that show increased activity following ovariectomy and also after copulation (Clattenburg et al., 1972, 1975). These neurosecretory cells lie adjacent to capillaries and in some rodents may be the source of GnRH (Rusak and Zucker, 1979).

11.9 AUTUMNAL BREEDING MAMMALS

In contrast to many species of mammals of the temperate regions, several kinds of ruminants mate in the autumn, under a regimen of decreasing day lengths. Autumnal mating occurs in several kinds of deer (*Odocoileus hemionus, O. virginianus*, and *Cervus elaphus*) and domestic sheep. Perhaps because of their rather large size and the precocial condition of the young at birth, fall mating allows the development of young by the following spring. The roe deer (*Capreolus capreolus*) constitutes a conspicuous exception. It mates in the summer, but because of delayed implanation, the young are nevertheless dropped the following spring (see Chapter 4, Section

7.2.3). In the roe deer, testicular growth and high levels of testosterone are concurrent with maximal day lengths (Sempère, 1978).

Experimental work with domestic sheep has been designed to clarify the pathway of testicular recrudescence under short days. When Soay rams were moved abruptly from four months of long days (16L:8D) to short days (8L:16D), there was a resurgence of gonadotropins in less than two weeks and an increase in testicular size in from two to four weeks. LH and FSH peaked daily in the dark part of the 24-hour period, which may reflect a circadian rhythm of the hypothalamic-pituitary-gonadal axis (Lincoln and Peet, 1977). In another group of Soay rams kept under alternating periods of long (16L:8D) and short (8L:16D) days, gonadotropin release and gonadal activity were conspicuous under the short-day schedule. At the ninth week of short days there was a marked rise in both LH and FSH with a subsequent increase in both testicular growth and testosterone output; after week 9 gonadotropin levels declined, but testicular activity continued to increase (Fig. 11.11). When the rams were returned to long days, gonadotropins quickly dropped to a minimum level, but testicular activity declined only slowly (Lincoln et al., 1977). In the red

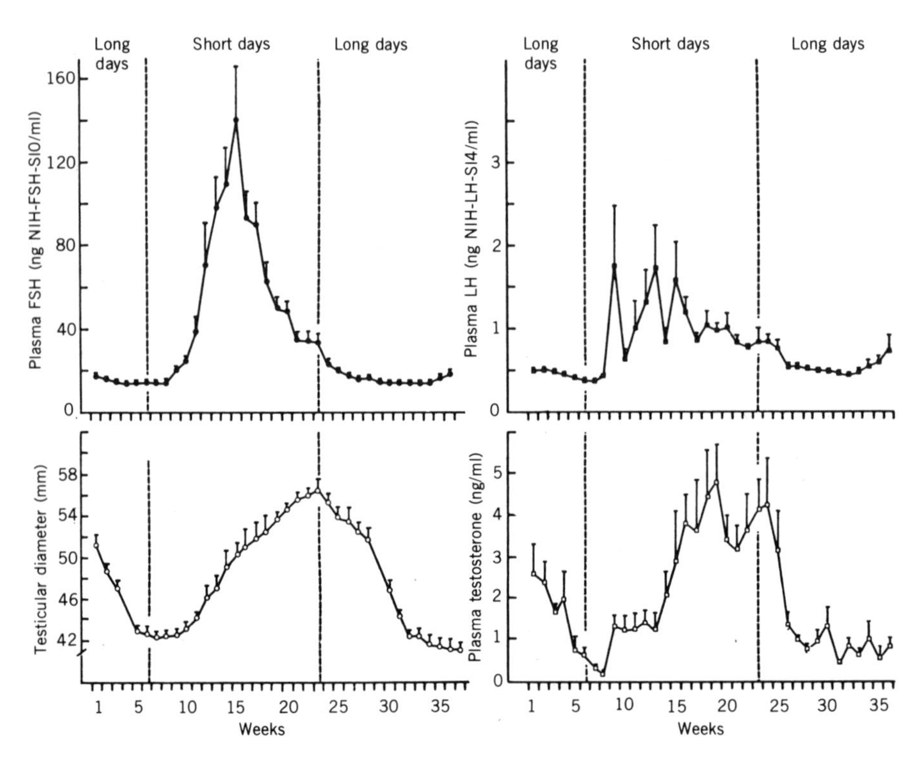

Figure 11.11 Changes in levels of plasma FSH (solid circles), LH (solid squares), and testosterone (open squares) in six adult Soay rams under long days (16L:8D) and short days (8L:16D) ± SE. Testicular diameter (open circles) follows the appearance of gonadotropins. (From Lincoln et al., 1977, courtesy of the *Journal of Endocrinology*.)

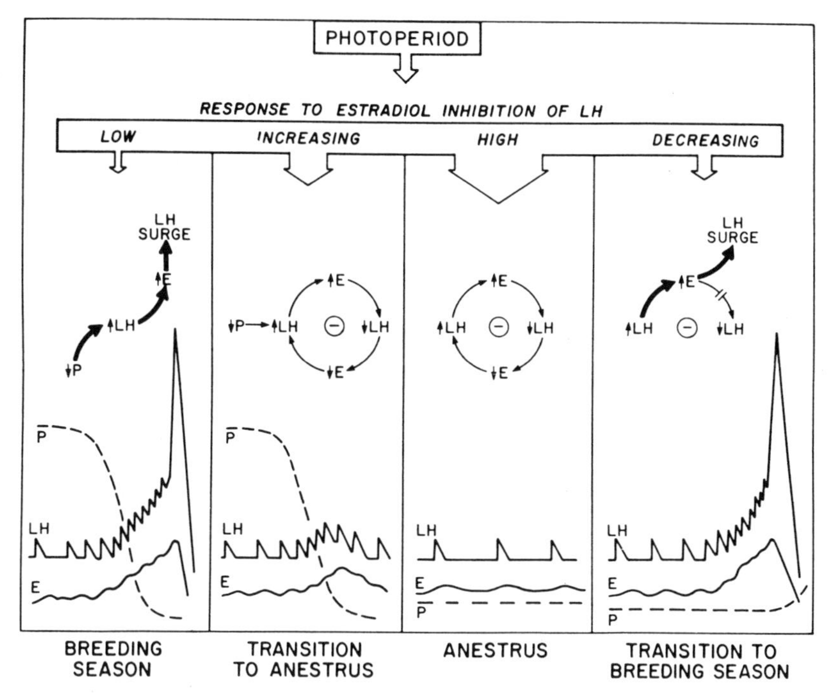

Figure 11.12 A proposed hypothesis for the effect of short day lengths on gonadotropin (luteinizing hormone or LH) secretion, estrogen (E), and progesterone (P) secretion and resumption of reproduction in the ewe. (From Goodman and Karsch, 1980, modified from Legan et al., 1977.)

deer (*Cervus elaphus*) serum LH levels rise to a peak in August, and testosterone levels are maximal from September to November, the period of mating (Lincoln and Kay, 1979).

In sheep, a decrease in response of the pituitary and the hypothalamus to the negative feedback of estradiol is associated with decreasing day length (Fig. 11.12). Once this response decrease results in a sustained rise of LH for two to three days in autumn, ovulation will ensue (Legan et al., 1977; Legan and Karsch, 1978; Legan and Winans, 1981). Conversely, in the nonbreeding season of the domestic sheep, estradiol suppresses the release of LH (Legan and Karsch, 1979) through an increased response of the hypothalamus and anterior pituitary to the negative feedback of estradiol (Fig. 11.13).

To examine the role of pineal melatonin in the photoperiodic control of seasonal breeding, pinealectomized and ovariectomized ewes were provided with physiological levels of estradiol from Silastic capsule implants, and serum LH was monitored. Pinealectomy eliminates melatonin secretion, and therefore melatonin was infused intravenously. A simulated long-day (or short-night) secretion of melatonin was followed by an LH decline typical of long days and similar to the decline seen in pineal-intact ovariectomized ewes. The pattern of melatonin secretion is an aid in

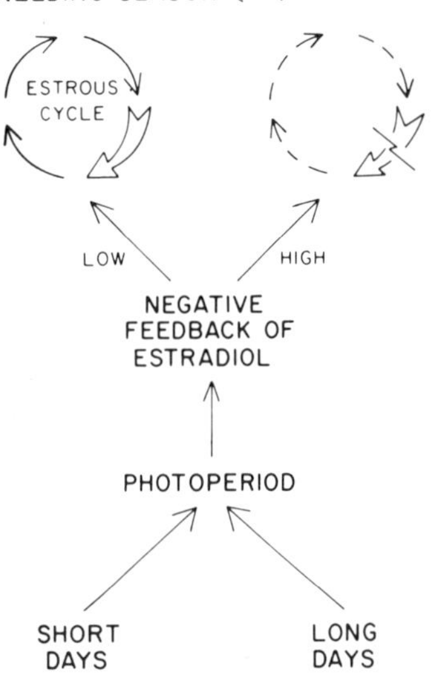

Figure 11.13 Schematic proposal for seasonal changes in neuroendocrine regulation of the estrous cycle of the ewe. (From Legan and Karsch, 1979.)

measuring day length and transduces the photoperiodic signal into a change in the response of LH to estradiol negative feedback (Bittman and Karsch, 1984). It has also been suggested that gonadal activity depresses melatonin secretion, for in ovariectomized ewes melatonin is three times that in intact ewes (Arendt et al., 1983).

Many vespertilionid bats (Vespertilionidae) have a modified autumnal mating, with spermatogenesis in the spring and summer and copulation in the fall, by which time the testes have regressed and spermatozoa are stored in the cauda epididymides. The male pallid bat (*Antrozous pallidus*) maintained from April or May until August on either short (10L:14D) or long (14L:10D) schedules revealed a sensitivity to day length. By August, males on short days had regressed testes, and all had epididymal spermatozoa. Males on long days showed some spermatogenesis, and only 55% had epididymal spermatozoa. By September all males had regressed testes and epididymal sperm (Beasley and Zucker, 1984). The effect of day length may be a result of melatonin secretions in this bat. When males were provided in July with melatonin subcutaneous Silastic capsules or with empty capsules and thereafter maintained on short days (10L) or long days (14L), the long-day, melatonin-treated individuals had smaller testes and were more likely to have epididymal spermatozoa than those long-day bats that had received empty capsules. Melatonin seems to mimic the short-day effect on testicular activity (Beasley et al., 1984).

11.10 REFRACTORINESS

In mammals refractoriness is usually a failure to respond to the suppressive influence of darkness, which is called scotorefractoriness. This results in a spontaneous recrudescence of the reproductive system in individuals kept under a schedule of long nights (or short days).

When kept under short days (less than 12.5 hours of light), hamsters undergo testicular regression. Eventually, after about 30 weeks, there is spontaneous recrudescence, indicating that the reproductive system had become insensitive to the suppressive influence of short days. In the late winter and early spring the animals have already become scotorefractory (Fig. 11.14). Sensitivity to short days returns, however, after activity is restored to the reproductive system. Exposure to long days (14L:10D) for 20 weeks resulted in renewed sensitivity to the suppressive influence of short days in the golden hamster (Bittman, 1978). Spontaneous testicular recrudescence has been found to occur also in the white-footed mouse (*Peromyscus leucopus*) when kept for 10 to 20 weeks under short days (10L:14D) (Johnston and Zucker, 1980). The hamster (Reiter, 1972) and the grasshopper mouse (*Onychomys leucogaster*) (Zucker et al., 1980) showed similar responses.

After three months of exposure to long (16L:8D) or short (8L:16D) days, two groups of pine voles (*Microtus pinetorum*) did not differ in weights of body, testes,

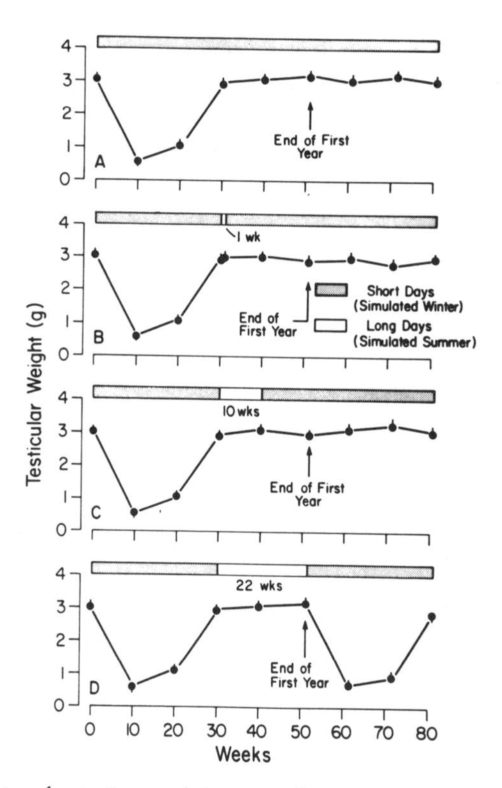

Figure 11.14 Scotorefractoriness of the reproductive system of hamsters is uninterrupted by brief exposures to long days, but 22 weeks of summer (long) days cause a return to sensitivity to short days (<12.5 hours). (From Reiter, 1980b, *Endocrine Reviews*, 1: 109–131, © Williams & Wilkins, 1980.)

or seminal vesicles or in testosterone levels (Lepri and Noden, 1984). Voles (Arvicolidae) tend to be rather weakly responsive to seasonal changes in daily photocycle and are clearly capable of becoming scotorefractory.

The mechanism by which mammals become scotorefractory is not yet established. There may be an increasing threshold of sensitivity of the neuroendocrine axis to melatonin. When given melatonin injections (25 g) daily just before the end of the light period of a 14L:10D schedule, hamsters experienced a sudden decline in testicular size. However, in males that had either been (1) pinealectomized or (2) exposed to 30 weeks of a 10L:14D schedule and had spontaneously recrudesced testes, refractoriness to the antigonadal effects of melatonin was exhibited in both groups. The intact individuals had regressed testes after seven weeks of melatonin treatment, but their testes recrudesced spontaneously after an additional fourteen weeks of melatonin treatment (Bittman, 1978a). This suggests that spontaneous testicular recrudescence in the hamster results from an eventual failure of melatonin to suppress gonadal activity.

Although most research on the relationship of light on gonadal recrudescence has emphasized specific photoperiods, in mammals as in birds, there is an apparent sensitivity to *changes* in duration of light. Mating in the Australian dasyurid *Antechinus stuartii* is correlated with the rate of change rather than with the length of the light period, and the response to rate of change is significantly different in two populations from different latitudes (Fig. 11.15).

A return to fecundity in the absence of long days suggests a mechanism for the appearance of an endogenous circannual rhythm. Fecundity returns not only after

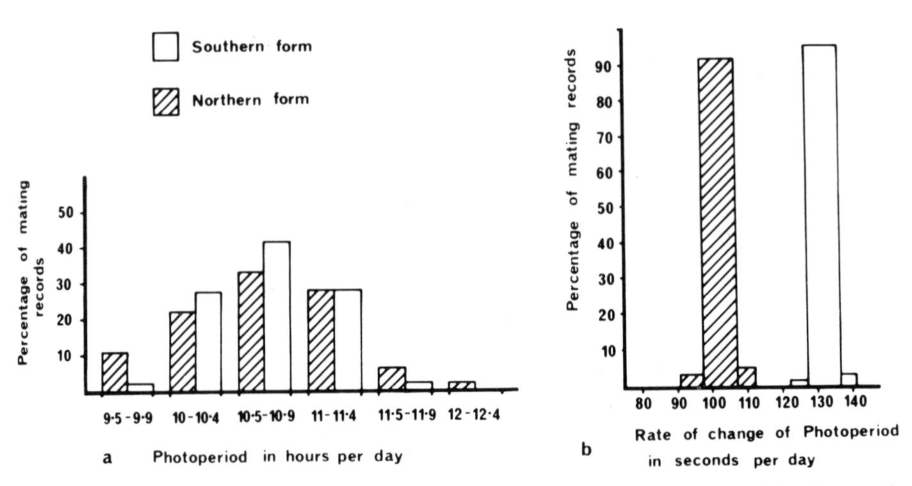

Figure 11.15 The association of mating in northern and southern forms of the dasyurid marsupial *Antechinus stuartii*, shown as (a) hours of daylight (n = 162) and (*b*) rate of change of photoperiod. Intermediate localities (at 114–120 seconds/day) are not shown, and 154 localities are included. (From McAllan and Dickman, 1986, courtesy of Springer-Verlag.)

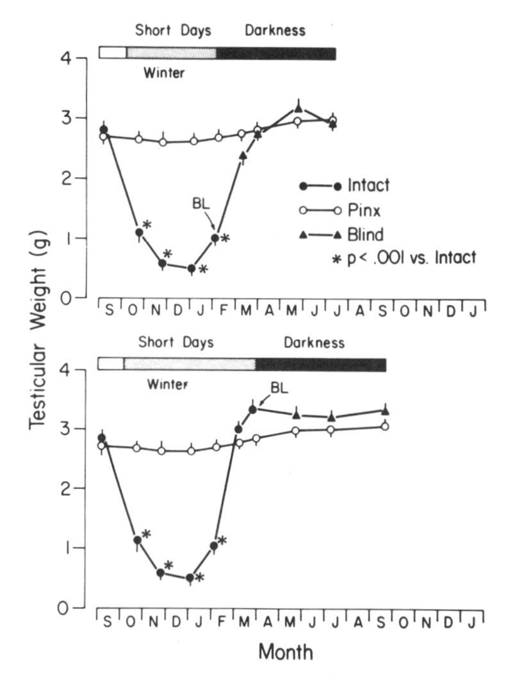

Figure 11.16 The effect of blinding (BL) on gonadal recrudescence in hamsters in the springtime. Blinding does not cause testicular atrophy at this season because the animals have become scotorefractory. (From Reiter, 1980b, *Endocrine Reviews,* 1: 106–131, © Williams and Wilkins, 1980.)

a prolonged exposure to short days (or long nights) but also, in hamsters, when blinded shortly after the winter solstice (Fig. 11.16). It appears that during the winter, hamsters develop a gradual increase in scotorefractoriness and a concomitant increase in sensitivity to long days. Although photorefractoriness (to long days) and scotorefractoriness (to short days) might be viewed as opposite sides of the same coin, the spontaneous return of testicular activity under short days (< 12.5 hours) indicates an increase in scotorefractoriness in the absence of long days. Brief interruptions of short (winter) days by long (summer) days (> 12.5 hours) do not reverse the scotorefractory condition, but a sensitivity to short days returns after exposure of between 10 and 22 weeks of summer days.

As previously pointed out, sheep breed as the days become shorter—at the end of summer and early autumn. Sheep appear to become alternately refractory to both long and short days. Soay rams experience testicular regression under long days but become photorefractory after about 16 weeks, and thus testicular recrudescence begins somewhat prior to the onset of short days. Similarly, after some 16 weeks of exposure to short days, the system becomes scotorefractory and involution commences (Lincoln, 1978). The role of melatonin release in Soay rams is equivocal. It is not clear whether the effect of melatonin depends upon the duration of the dark period or the time of the melatonin peak relative to the light-dark cycle. After

25 weeks of exposure to short days (8L:16D) or to long days (16L:8D), the peaks of melatonin in Soay rams became somewhat dissociated from the dark period, perhaps reflecting slow alterations in the timekeeping mechanism. This slow adjustment might underlie scotorefractoriness in mammals in both short-day and long-day breeders (Almeida and Lincoln, 1984).

11.11 WEAKLY PHOTOSENSITIVE MAMMALS

Some mammalian reproductive seasons depart from oscillations of the natural photocycle. This does not necessarily mean that such species are not sensitive to the effects of light observed in other mammals, but simply that they may be cued by several exogenous factors. Voles and lemmings (Arvicolidae) are notoriously variable in their fecundity, both in time and magnitude. Voles (*Microtus* spp.) do respond to day length, and gonadal systems do recrudesce under long days. When subjected to several light schedules in the laboratory, the field vole (*Microtus arvalis*) displayed an increase in testis mass, seminal vesicles, and epididymal sperm as well as increased ovulation rates with increased day length (Lecyk, 1962; Thibault et al., 1966), but it appears that light alone did not control reproductive seasonality in the field (Martinet, 1970).

The photoperiodic response of *Microtus agrestis* in the British Isles resembles that of *M. arvalis*. When reared from birth under long days (16L:8D), *M. agrestis* becomes sexually mature at 40 days; when kept under short days (8L:16D), fertility was attained at six months of age (Grocock, 1979). Although testicular regression occurs in voles kept under short day lengths, regressed testes of captive voles do not reach the extremely small size of those of wild mice (Grocock and Clarke, 1974; Clarke, 1977). Factors other than day length promote seasonal changes in gonadal size and activity.

Lemmings live in the Arctic and are subnivean for much of the year. Although they live where photoperiodic changes are extreme, lemmings are virtually unresponsive to changes in day length. The collared lemming (*Dicrostonyx groenlandicus*), when reared under contrasting light regimens (6L:18D and 20L:4D) from weaning to day 70, exhibited no difference in gonadal development in either sex, and individuals under both sets of light schedules became sexually mature (Hasler et al., 1976).

The cotton rat (*Sigmodon hispidus*), a crecitid rodent of the New World, experiences little effect from various photoperiodic regimens. When kept on short days (19L:14D) from birth to day 50, there was slower gonadal growth than in animals on a long-day schedule (14L:10D). Gametogenesis, however, was not statistically different in the two groups (Johnston and Zucker, 1979). Quite likely, reproductive seasonality in the cotton rat in nature is controlled by factors other than the photocycle. Similarly, seasonal reproduction is not modified by day length in the feral house mouse (*Mus musculus*), and the house mouse is nonseasonal in its reproduction in nature. Wild house mice respond negatively to strong light, whereas laboratory stocks (CF-1) are unaffected by light intensity. Captive-born young of wild stocks breed in constant darkness (with only a very dim red light), but when exposed to a 14L:10D schedule including light of 1000 lux, there is a marked reduction in litter size. There is no apparent mechanism for this response, and domestic (CF-1) mice are unaffected (Bronson, 1979).

Although reproduction is associated with seasonal or sporadic rainfall in many Australian marsupials, the pouched "mouse" (*Sminthopsis crassicaudata*) experiences gonadal recrudescence under a long-day schedule (15L:9D) (Godfrey, 1969). The hopping mouse (*Notomys alexis*), a murid rodent of Australian deserts, is only slightly affected by photoperiod: short days prolong estrus, but do not influence ovarian or uterine weights (Breed, 1975). The degree of photosensitivity is unknown in most desert rodents, but as we shall see in Chapter 13, Section 13.9.2, desert-dwelling rodents at both low and midlatitudes appear to breed in response to factors other than day length.

It has sometimes been suggested or assumed that domestication somehow dilutes the sensitivity of the gonadal system to day length. As we have seen, however, some wild rodents are not cued by the photocycle, and some domestic animals (e.g., sheep) are quite dependent upon day length for gonadal stimulation.

The mare experiences a winter anestrous phase (Osborne, 1966; Hughes et al., 1975), presumably reflecting the biology of wild or feral individuals. Exposure to a 16L:8D cycle in midwinter is followed by estrus almost two months prior to that of controls, with a simultaneous increase of LH and estradiol (Oxender et al., 1977). Data on the hamster suggest recent changes in its photosensitivity, perhaps a result of its history of domestication (Reiter, 1980a).

Testicular activity of the laboratory rat is known to be minimally responsive to variations in photoperiod. The pineal gland in rats does, however, have a depressing effect on the gonads when rats are kept under short days, but there is no effect in intact animals. When male rats are deprived of their olfactory bulbs, however, the typical short-day response of testicular regression and decreased testosterone levels occurs; but this effect does not occur when bulbectomy and pinealectomy are performed concurrently (Nelson and Zucker, 1981). That is, olfaction in this rodent may limit the antigonadal activity of the pineal. This state of affairs raises more questions than it answers. Possibly there are olfactory signals that are of social significance and more powerful than day length in influencing gonadal activity.

Weakly photosensitive species are also discussed in Chapter 12, Section 12.6.

11.12 SUMMARY

The great emphasis placed on the photocycle in regulating seasonal reproduction is based on a large number of experiments conducted on a broad variety of light regimens with many species, and the experiments are almost always designed to maintain uniformity of other potential cues. In nature, however, other factors are potential cues, and their roles are rarely as carefully explored as the photocycle has been.

In some passerine birds laboratory studies indicate that photoperiod alone may be sufficient to bring the male gonads to a completely functional state (Farner and

Wingfield, 1978). These species (e.g., *Zonotrichia leucophrys*, *Fringilla chloris*, and *F. coelebs*) are generalized feeders that can quite readily shift their foraging efforts among a variety of foods, and they will not suffer if one of their favorite foods fails to appear. Possibly many small passerine species fall into this category. Some midlatitude species that are specialized feeders, however, depend upon cues in addition to day length. That is, some midlatitude endotherms appear to be brought by photoperiod to a near-tonic or subfunctional reproductive state but require additional exogenous stimulation before reproduction will occur. In this sense, they resemble some tropical species that are cued by nonphotoperiodic factors. The action of the photocycle does not differ with latitude: its effect on the hypothalamic-pituitary-gonadal axis is the same.

There is a striking dichotomy between birds and mammals not only in the mechanisms of the reception of photic signals but also in their influence over the hypothalamic-pituitary-gonadal axis.

In birds long days stimulate the light-sensitive hypothalamus directly, with light penetrating the feathers, skull, and brain to the hypothalamus. Destruction of the paired eyes does not affect the avian reproductive system. Although the avian pineal gland is light sensitive, it seems not to have the primary role in avian reproduction. Following a prolonged exposure (several months) to long, stimulatory days, a bird will become photorefractory and the gonads will regress. Under a prolonged exposure to short days, the gonads recover their sensitivity to light and will recrudesce when exposed to long days.

In mammals light signals are received through the paired eyes and transmitted via a neural pathway to the pineal gland. Light depresses the activity of the pineal, which becomes increasingly active under a regimen of short days. A main secretion of the pineal is melatonin, which depresses gonadal activity. When pineal metabolism is depressed, as under a schedule of long days, the release of melatonin declines, and the hypothalamic-pituitary-gonadal axis resumes activity and restores fecundity. Destruction of the paired eyes in mammals is followed by a marked decline in gonadal activity and, temporarily, a loss of fecundity.

In nature, because the production of melatonin occurs during dark periods, the antigonadal effect of melatonin increases as the daily duration of night increases. With the approach of winter, the seasonal regression of mammalian gonads ensues. It is conventional to think of photoperiod in terms of the amount of light needed in a 24-hour period to stimulate the release of gonadotropin and eventually gonadal development. In birds this convention is correct. In mammals, however, in which gonadotropin release and gonadal activity appear to occur spontaneously unless suppressed, the photoperiodic effect on reproduction is determined by the duration of darkness.

Prolonged exposure of a mammal to a regimen of short days is followed by scotorefractoriness, a loss of sensitivity to short days. The return of fecundity can occur either in blinded mammals or in individuals maintained in continued darkness.

Some mammals, such as artiodactyls (deer and sheep and their relatives), are frequently short-day breeders, both in nature and in domestication. This behavior seems to result from a gradual loss of sensitivity of the negative feedback during the long days of summer, with the consequent increase of gonadotropins.

Other species, such as voles and lemmings (Arvicolidae) and also some well-studied diurnal squirrels, either fail to respond to photic signals or have a weak response, and their reproductive seasonality is controlled by other exogenous factors or perhaps solely by endogenous cues.

It is quite logical that birds should be stimulated by an increase in day length. Birds evolved from a group of reptiles that was primarily heliothermic and sun-loving. Birds today are primarily diurnal, and it is not surprising that they have retained the capacity to be stimulated by the length of a light period. Mammals, on the other hand, are blessed with a reproductive system that becomes active spontaneously, except when depressed by prolonged periods of darkness. Most modern mammals are essentially nocturnal, and the earliest forms were also probably nocturnal (Crompton et al., 1978). Thus, being active in the time of darkness, mammals, not surprisingly, are influenced by the duration of the period of darkness and become scotorefractory and capable of reproducing in extensive periods of darkness. It is perhaps also significant that those mammals that appear to reproduce independently of photoperiodic cues are some species of diurnal squirrels.

In both birds and mammals the photocycle provides proximate cues so that the reproductive system is in a subfunctional or functional state when it received exogenous ultimate cues. Because day length and other exogenous cues at mid- and high latitudes occur close together, there is no simple means in nature of separating their effects. Experimental exploration of the photocycle almost always includes food of high nutritional content with unknown content of secondary plant compounds, and even in experimental situations the precise role of day length cannot always be distinguished from some other exogenous factors. Because the seasonal changes in the quality and quantity of food in nature do affect gonadal development, the roles of nonphotic cues, discussed in the next chapter, should always be kept in mind.

The difference in the photoperiodic response between birds and mammals has important ecological significance. Because the reproductive cycle of most birds is stimulated by the progressively longer day lengths in the spring and also because birds become photorefractory in the summer, their reproductive calendar is rather consistent and predictable from one year to the next. At least some species of mammals are reproductively suppressed by long nights and become reproductively active in seasons of short nights (the spring). Mammals, however, become scotorefractory to long nights and may then either resume gonadal activity spontaneously or respond to a decrease in the duration of darkness or to nutritional cues. For these reasons mammalian reproductive seasonality (especially for some rodent and lagomorph species) is less predictable and more variable from year to year than is the rather regular seasonality observed for most species of birds.

BIBLIOGRAPHY

Assenmacher I and Farner DS (Eds) (1978). *Environmental Endocrinology*. Springer-Verlag, Berlin.

Axelrod J, Fraschini F, and Velo GP (Eds) (1983). *The Pineal Gland and Its Endocrine Role*. NATO Advanced Study Institute, Series A, Plenum, New York.

Barrington FJW (Ed) (1975). Trends in comparative endocrinology. *American Zoologist* (Suppl 1), **15**.

Bergmann M (1987). Photoperiod and Testicular Function in *Phodopus sungorus*. Springer-Verlag, Berlin, Heidelberg.

Cardinali DP (1981). Melatonin. A mammalian pineal hormone. *Endocrine Reviews* 2:327–346.

Farner DS (1964). The photoperiodic control of reproductive cycles in birds. *American Scientist* 5:137–156.

Follett BK and Follett DE (Eds) (1981). *Biological clocks and seasonal reproductive cycles*. Proceedings of the 1980 Colston Symposium. Wright, Bristol.

Gilmore D and Cook B (1981). *Environmental Factors in Mammal Reproduction*. Macmillan Publ Ltd, London.

Kappers JA and Pevet P (Eds) (1979). *The Pineal Gland of Vertebrates Including Man. Progress in Brain Research 52*. Elsevier/North Holland Biomedical Press, Amsterdam.

Lincoln GA (1981). Seasonal aspects of tersticular function. Pp 255–302 In *The Testis* (Eds: Burger H and de Kretser D). Raven Press, New York.

Lofts B and Murton RK (1968). Photoperiodic and physiological adaptations regulating avian breeding cycles and their ecological significance. *Journal of Zoology*, London 155:327–394.

Reiter RJ (1978). The pineal gland and reproduction. *Progress in Reproductive Biology 4*. S Karger, Basel.

Reiter RJ (1981). *The Pineal Gland, II. Reproductive Effects*. CRC, Boca Raton, Florida.

Reiter RJ and Follett BK (1980). Seasonal reproduction in higher vertebrates. *Progress in Reproductive Biology 5*. S Karger, Basel.

Short RV (1985). Photoperiodism, melatonin and the pineal. *Ciba Found Symposium* 117. Pitman, London.

Tanabe Y, Tanaka K, and Ookawa T (Eds) (1980). Biological Rhythms in Birds. Neural and Endocrine Aspects. Japan Scientific Societies Press, Tokyo; Springer-Verlag, Berlin.

Zucker I, Johnston PG, and Frost D (1980). Comparative, physiological and biochronometric analysis of rodent seasonal reproductive cycles. *Progress in Reproductive Biology*, **5**:102–133. S Karger, Basel.

12

OTHER ENVIRONMENTAL FACTORS

12.1 INTRODUCTION

A seemingly infinite variety of environmental variables are associated with mating and the emergence of the next generation. Some of these variables are internal stimuli, which determine gonadal activity, and others are environmental factors or cues to which the reproductive cycle is adapted. This chapter deals with the roles of these factors or cues. The apparent adaptive significance of reproductive seasonality is reviewed in the next chapter.

12.2 NUTRITIONAL EFFECTS ON REPRODUCTION

Food has nutritional effects concerned with growth and energy metabolism and also specific effects on the reproductive system. Levels of quality as well as quantity affect reproductive ability in both sexes. In malnutrition or starvation the retarded output of pituitary hormones, including gonadotropins, causes inanition (exhaustion through starvation) in both sexes (Leathem, 1966). Sadleir (1969) emphasized the role of nutrition in reproductive seasonality of wild mammals, both in producing adequate growth in young individuals and in achieving proper support for gametogenesis in adults. Malnutrition delays sexual maturity and fecundity by depressing the activity of the hypothalamus (Srebnik et al., 1978). The study of nutritional effects on reproduction has generally emphasized the well-recognized needs for growth and maintenance and their relationship to the functions of the various parts of the hypophyseal-pituitary-gonadal axis (Scott, 1973; Wootton, 1973a, 1973b; and others).

There are many examples of proper nutrition promoting fecundity and of inadequate nutrition preventing reproduction. In most taxa food intake increases with need, but the sequential relationship of nourishment and reproduction, however, varies widely among groups of vertebrates. In midlatitude reptiles, for example, nutrition in summer provides much of the energy for egg formation the following spring, and food intake may actually decline during the reproductive period. In most mammals both pregnancy and lactation place a high demand on the female, but the compensatory increase in nutrition must be synchronized not only with demand but also with hibernation, migration, and food availability. Some bears produce young during periods of fasting (during periods of lethargy in the winter), and a number of pinniped species fast for several weeks while lactating. The actual details of nutritional requirements for reproduction are known for only a few species (Widdowson, 1981), but the variation is substantial.

The nutritional quality of food has a paramount effect on reproductive performance. In the laboratory rat, inadequate protein content arrests both steroidogenesis and gametogenesis. The protein component, however, may not be absolutely essential in all kinds of mammals. In the domestic rabbit, for example, gonadotropins are synthesized even on a protein-free diet, and such protein as is needed is apparently derived from nitrogenous products from the rabbit's own tissue (Friedman and Friedman, 1940). Vitamin E is essential for full development of reproductive structures in the laboratory rat. A deficiency causes sterility, and such a sterility is permanent (Raychaudhuri and Desai, 1971). Vitamin E is not known to function as a cue for reproduction in nature. Experimentally, the collared lemming (*Dicrostonyx groenlandicus*) exhibits reduced fertility when maintained on a vitamin D–depleted diet. During the winter, when there is no sunlight and when lemmings are subnivean, vitamin D must come from dietary sources (Hickie et al., 1982).

Nutrition is a direct reflection of the individual's environment. The protein content of foods of the red-backed vole (*Clethrionomys rufocanus bedfordi*) appears in their serum protein and is related to the productivity of their habitat. The fecundity of this vole, moreover, varies with the protein content of its natural foods: protein content in vole foods is markedly lower on mature forest land than on plantations of young seedlings. In forests, where the protein of vole foods falls below maintenance levels from October to April, the voles cease to breed in late spring. In plantations, where protein in vole foods is optimal for six months of the year, populations are higher and reproduction extends into summer (Maeda, 1973).

Specific differences in sensitivity to malnutrition are seen among species of comparable size and diet. A 30% reduction of an ad libitum diet was followed by reduced testicular activity in deer mice (*Peromyscus maniculatus*), amounting to azoospermia in some individuals, but with no measurable effect in house mice (*Mus musculus*) (Blank and Desjardins, 1984). Hamsters exhibit an effect similar to that seen in deer mice: a reduction of food resulted in a reduction in testis size, independent of an intact pineal gland (Eskes, 1983).

The domestic horse, a seasonally reproductive and photosensitive species, responds also to changes in nutrition, but this may be masked by changes in ambient temperature, reflecting fluctuations in energy demands. An increase in energy content in the diet of mares in winter was followed by an earlier return to estrus than in controls (Belonje and van Niekerk, 1975). On the other hand, mares kept in winter at 10 to 15°C ovulated earlier than a control group kept at −15 to 9°C (Oxender et al., 1977). One must conclude that horses are sensitive to nutritional levels and that nutritional needs are modified by ambient temperature, as in some wild birds.

In a broad spectrum of both freshwater and marine teleost fish, there is a general positive correlation of quality and quantity of food with the number of ova (Wootton, 1979). The correlation is apparent in captive fish maintained on controlled diets as well as those in the wild. Food supply is usually density dependent, and in wild, free-ranging populations food abundance may be deceptive, for greater densities of fish mean less food for each individual. Wild populations of fish regularly face periods of plenty alternating with periods of poverty. The Atlantic herring (*Clupea harengus*), like many temperate ectotherms, has a marked annual cycle of fat accumulation during a nonreproductive period and fat depletion at spawning. Herring captured after spawning and kept without food for 78 days not only survived but also suffered no irreversible damage to reproductive capacity, although there was no gametogenesis as in well-fed fish (Wilkins, 1967). In *Gillichthys mirabilis,* a goby of brackish water, starvation results in prompt gonadal regression but does not prevent seasonal gonadal recrudescence if there has already been an accumulation of energy (De Vlaming, 1971). The three-spined stickleback (*Gasterosteus aculeatus*), when kept on diets that varied only in amount and frequency, grew more rapidly with increased nutrition, and the weight of mature fish was correlated with increased frequency of spawning as well as number and mass of eggs produced (Fig. 12.1). The convict (*Cichlasoma nigrofasciatum*), a Central American cichlid fish, becomes increasingly fecund with more frequent feeding: spawning is positively correlated with ratios of 1 g of food given daily, three times a week or once a week (Townsend and Wootton, 1984). In the African catfish (*Clarias lazera*) there is a sexually dissimilar gonadal response to food and temperature. Males are more sensitive to

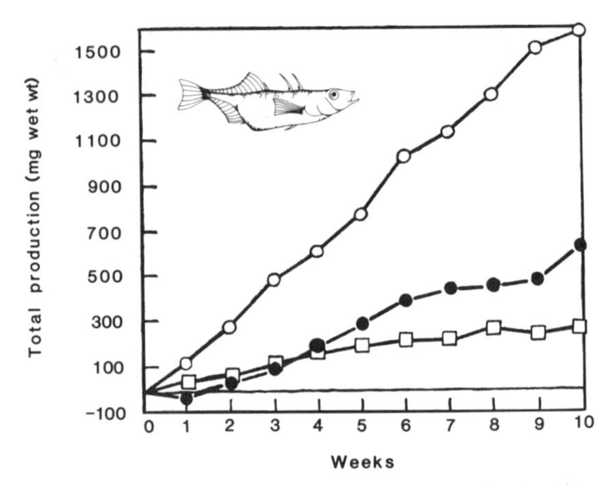

Figure 12.1 Cumulative egg production per tank (a total of eight fish per tank) of three-spined sticklebacks (*Gasterosteus aculeatus*), kept at three nutritional levels. (From Wootton, 1973, courtesy of the *Journal of Fish Biology*.)

temperature than to food, but females are more responsive to varying levels of nutrition (Richter et al., 1982). Captive haddock (*Melanogrammus aeglefinus*) produced a mean of 16.6 (10–25) batches of eggs over a mean period of 33.2 (19–59) days while maintained under controlled feeding of diets that contained from less than 5 kcal/day to more than 13 kcal/day. Egg production and food intake were positively correlated (Hislop et al., 1978).

Egg number does not always relate directly to nutritional levels. When kept on diets of various protein levels (15%, 31%, and 47%), the guppy (*Poecilia reticulata*) showed greater somatic development in response to greater protein intake. Although gonadal weights were positively correlated with body weight, an increase in ovarian size was not accompanied by an increase in the number of ova in females fed higher protein diets (Dahlgren, 1980).

12.2.1 Nutritional Variations in Nature

There are many examples of wild vertebrates failing to reproduce in years of general food scarcity. A few examples will illustrate adjustments in fecundity following natural variations in food supply. Many of these species live in high latitudes, where there may be few different kinds of food and the decline or failure of one food source or prey species may be catastrophic for the predator. Some raptors lay large clutches in years of prey abundance, and some passerine birds raise large broods when their favorite food abounds (see Chapter 14, Section 14.5.3).

The lapwing (*Vanellus vanellus*) arrives on its nesting ground in southern Sweden at the same time every year, but the length of the prelaying period is negatively correlated with the abundance of lumbricid worms, the main fare of this plover (Högstedt, 1974).

12.3 SECONDARY PLANT COMPOUNDS

The effect of food is complex and may reflect not only its general nutritional value but also, for herbivores, specific influences of secondary plant compounds on the reproductive system. Many plants contain materials that structurally and functionally resemble gonadal steroids (Fig. 12.2). Some of these compounds are chemically unlike estrogens of animal origin, and many are not steroids, but they do stimulate or depress the reproductive system. Some have been compared with gonadotropins, but it is not known how their effects are produced. Although some of these secondary plant compounds are known not to function as estrogens, the term *phytoestrogen* is well established in the literature and has long been used as an inclusive name for sexual stimulants, as well as suppressants, that occur in plants (Labov, 1977).

Studies in the late 1930s revealed that extracts from fresh alfalfa leaves stimulate ovulation in the domestic rabbit, and it was noted that commercially prepared alfalfa feed is much less effective (Friedman and Friedman, 1939).

Early Chinese books included many botanical drugs that were administered to treat reproductive disorders (Needham and Lu, 1968). Although much of this information lies in the realm of folklore, some of these substances are known to be toxins, such as alkaloids, and sexual stimulants, such as isoflavones (Kong et al., 1976). The most famous Oriental herb is ginseng (*Panax ginseng*), which appears to be a general tonic, without discrete effects on the reproductive system.

Several of the precursors of gonadal steroids of vertebrates occur regularly in plants. Cholesterol is known to be widely distributed among plant taxa. In plants,

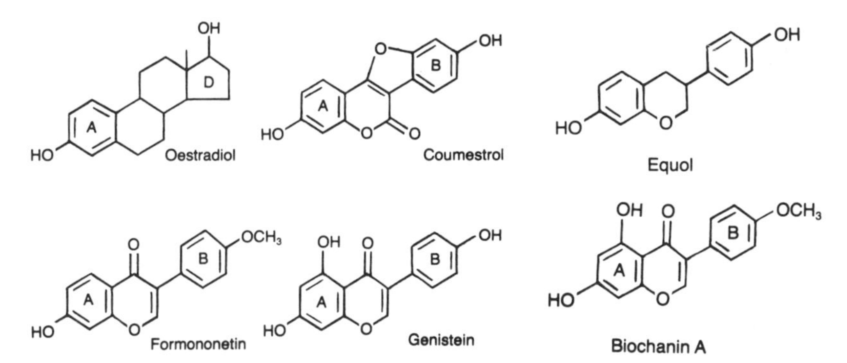

Figure 12.2 Structure of ovarian estradiol as compared with several phytoestrogens and equol, products of the digestion of formononetin. (From Shutt, 1976, courtesy of *Endeavour*, Pergamon Press, Ltd.)

as in animals, progesterone is derived from cholesterol, with pregnenolone as an intermediate product (Bennett and Heftmann, 1965). Moreover, sitosterol, a common sterol in many plants, can also be converted to progesterone (Bennett et al., 1969). Progesterone, a gonadal steroid as well as a precursor of other gonadal steroids in vertebrates, is found in plants (Heftmann, 1975; Grunwald, 1980) and presumably gives rise to estrogens and androgens that have been reported to be in plants. Thus there exists in some plants the chemical sequence for deriving the steroid hormones that are produced by vertebrate gonads. It will be seen, however, that not all sexual stimulants in plants are steroids.

The existence of estrogenlike and androgenlike substances in plants has been known since about 1926, and their chemical similarity to animal steroid hormones was demonstrated in the early 1930s. A most surprising similarity was shown, experimentally, when animal hormones sprayed on a normally bisexual species, the caryophyllaceous herb *Silene=Melandrium dioecum*, determined the sex of the flowers that subsequently developed (Löve and Löve, 1945).

12.3.1 Phytoestrogens and Mammalian Reproductive Systems

There has been a sporadic history of plant-caused reproductive disorders in humans and domestic animals, but most effects are rather mild and therefore go unnoticed. For example, tulip bulbs were sometimes eaten in the Netherlands during food shortages in World Warr II, and this tissue is rich in phytoestrogens that cause menstrual irregularities (Bickoff, 1963). Also, certain strains of subterranean clover (*Trifolium subterraneum*) contain substances that alter reproductive performance in sheep.

A wide variety of succulent and rapidly growing plant materials stimulate or depress various aspects of ovulation, pregnancy, and lactation when taken orally. Many plant taxa have in their leaves, stems, roots, flowers, and fruits substances that stimulate the gonads, endometrium, and mammary tissue. Some of these secondary compounds in legumes and grasses are known to be especially powerful.

Phytoestrogens from some plants may cause irregularity in implantation, distochia (or distokia, difficulty in parturition), and failure in sperm transport in sheep. They are deleterious because sheep are almost constantly feeding on these plants and because the effects are cumulative and irreversible. These substances occur in leaves as glycosides and, when crushed, are released as free isoflavones (formononetin, genistein, and biochanin A). They are not always potent in laboratory mice but may have a strong stimulatory effect in sheep (Beck, 1964). Coumestrol, a phytoestrogen in some clovers and alfalfa, is about 30 times as powerful as genistein when taken orally (Millington et al., 1964). Coumestans are of greater estrogenic potency than isoflavones are and appear to depress levels of ovarian estrogens in sheep (Cox, 1978), suggesting a negative feedback on the hypothalamus and/or the anterior pituitary. In most examples, the active compounds have a rather prompt effect and in this way differ from changes resulting simply from improved body condition and increased weight. They cause hypertrophy of the endometrium, presumably due to chronic stimulation caused by continuous ingestion of clover. After grazing on "estrogenic" clover for three years, ewes experienced a significantly higher ovulation rate than was observed in a control group, although there was no difference in body mass (Adams et al., 1984). Some of these materials in clovers induce lactation in nonpregnant ewes and even in wethers (castrated rams).

A difference in the estrogenic activity of some phytoestrogens results from the degrees of digestion. In the ruminant stomach of the sheep, both genistein and biochanin A are changed to estrogenically inactive phenols, but formononetin is degraded to equol, which is mildly estrogenic (Shutt, 1976). Relatively weak phytoestrogens such as equol become stimulatory when concentrated in the plasma as a result of frequent ingestion. Thus, despite their rather low affinities at estrogen binding sites (Fig. 12.3), their estrogenic effects are strong because of their continuous ingestion and high serum levels, and they may inhibit binding of estradiol to uterine binding sites in vitro (Shutt and Cox, 1972). Phytoestrogens are rather similar in structure to estrogens of vertebrate origin and seem to function in a similar manner when ingested. Both genistein and coumestrol, for example, compete with estradiol at estrogen binding sites on the rabbit uterus (Farnsworth et al., 1975).

In Australia the inclusion of lupin grain in the food of sheep significantly increases their fecundity. This effect is independent of the caloric content or nitrogen content of the feed. Lupin grain added to supplement normal feed of Merino ewes increases both ovulation rate and percentage of ewes twinning (Lightfoot and Marshall, 1976; Reeve et al., 1976). Grain of uniwhite lupin (*Lupinus angustifolius*) increases the ovulation rate by 8 to 25 ovulations per 100 ewes, without an accompanying increase

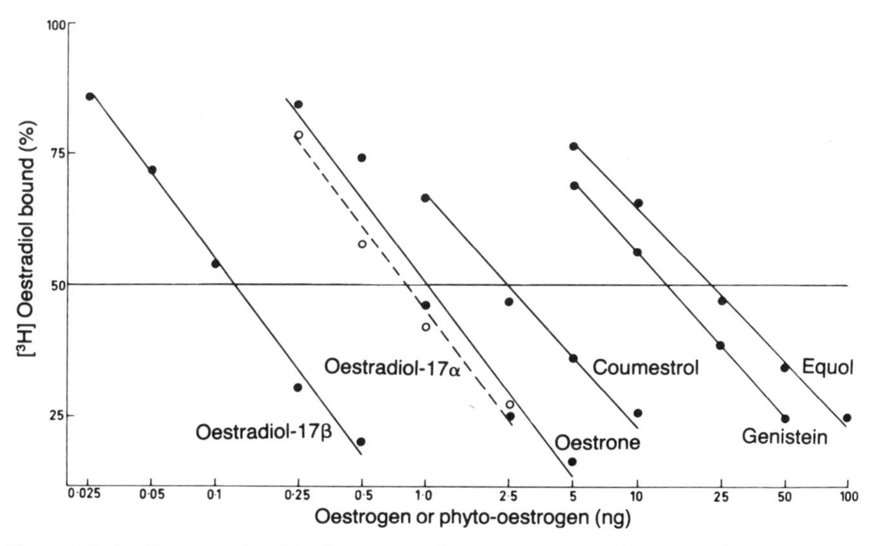

Figure 12.3 Comparative binding curves for some mammalian steroid estrogens and some phytoestrogens. Test materials plus (^{3}H)-labeled estradiol were incubated with estrogen receptor protein in sheep uterine cytosol. (From Shutt and Cox, 1972, courtesy of the *Journal of Endocrinology*.)

in body mass of the ewes. If 50% more nitrogen (in the form of urea) and 25% more digestible energy is in the feed, there is no increase in ovulation (Knight et al., 1975). The effect of lupin on ovulation and overall fecundity seems to be directly on the reproductive system and is not an indirect effect of nutrition.

Isoflavones have been found to affect reproduction in wild populations of the California quail (*Lophortyx californicus*). In 1972 in San Luis Obispo County, California, when there was light rainfall and a meager growth of spring forbs, which contain high levels of genistein and formononetin, the reproduction level was low: there were 25 offspring per 100 adults. Isoflavones were nearly absent in the lush spring growth a year later, and there were 325 offspring per 100 adults (Leopold et al., 1976). Isoflavones reduced fecundity in quail as in sheep, and the pathway in quail may be the same.

12.3.2 Occurrence of Phytoestrogens

The phytotoestrogen content of plants varies with both the season and the growth stage. In very early growth of alfalfa there is a slight increase in levels of coumestrol, which soon declines to undetectable amounts until a very marked buildup from the dough stage of the seed through to the development of the mature seed (Bickoff et al., 1960).

There appear to be local and yearly variations in the presence of steroid hormones in plants, for efforts to locate them are not uniformly successful. Estrone, for example, has been reported to occur in high concentrations in seeds of the pomegranate (*Punica granatum* var. *nana*) by some researchers (Bennett et al., 1966). The occurrence of isoflavones in plants is known to increase with a deficiency of phosphorus (Rossiter and Beck, 1966). Fungal infections also account for marked increases in the coumestan content of white clovers or lucernes (*Medicago* spp.). Methylated coumestans and coumestrols occur in both leaves and stems, and levels are higher in those plants growing on sands than on gravelly soils. In *Medicago polymorpha* var. *denticulata* infected with a rust (*Uromyces striatus* var. *medicaginis*), the coumestan content of leaves of equal age increases over that found in uninfected leaves. Those methylated coumestans are probably not demethylated in the sheep's gut as isoflavones are known to be (Francis and Millington, 1971).

Several native African grasses have been found to contain reproductively stimulating compounds. Laboratory mice experienced a marked increase in uterine weight with only a trivial increase in body mass when fed extracts of veld hay that included *Eragrostis curvula* (Altona and Tilley, 1963). Three common veld grasses (*Hyparrhenia filipendula*, *Setaria ciliolata*, and *Cynodon dactylon*) grew rapidly and developed a high level of estrogenic material, and their estrogenic activity was positively correlated with their growth rates. The active materials were found to be genistein and diadzein; biochanin A, formononetin, and coumestrol were absent (Millar, 1967). Among several British pasture grasses, such as ryegrass (*Lolium perenne*) and cocksfoot (*Dactylis glomerata*), estrogenic activity is highest in the spring, during early growth of the plant. In red clover (*Trifolium pratense*), however, which grows in the summer, there are repeated high concentrations of estrogens (Legg et al., 1950). In beans (*Phaseolus vulgaris*) phytoestrogens develop at the time of flower bud formation and are at maximal levels during bud growth and pod formation, at which time the greatest concentrations are in the leaves (Kopcewicz, 1971).

Some of the stimulating plant compounds are not estrogenic in their activity. There is in some grasses and germinating grass seeds a material, 6-methoxybenzoxazolinone (6-MBOA), that is a powerful stimulant to reproduction in voles (Sanders et al., 1981; Berger et al., 1977, 1981). This material is highest in the growing tissue of corn seedlings and other grasses (Klun and Robinson, 1969). This and similar compounds in corn protect the growing plant from attacks of the corn borer (*Ostrinia nubilalis*) (Klun and Robinson, 1969; Russell et al., 1975). The presence of this reproductive stimulus in the early stages of grasses increases the likelihood of these compounds being proximate cues, for they would announce the abundance of food for young voles a few weeks later. When 6-MBOA was injected subcutaneously in laboratory rats or administered in Silastic capsules, there was a significant increase in ovary weight due to an increase in number of enlarged follicles. This increase occurred within 24 hours following subcutaneous treatment and three days after treatment with Silastic capsules. There was also a 100% increase in uterine weight in animals 30 days of age. Following 6-MBOA treatment, the number of corpora lutea increased from 13.4 ± 1.3 in controls to 19.2 ± 0.9 in those females treated with 6-MBOA (Butterstein et al., 1985).

12.3.3 Effects of Secondary Plant Compounds in the Field

One of the earliest reports of environmental variation in litter size in wild mammals was the observation that the Levant vole (*Microtus guentheri*) of the Near East has a mean litter of 8.5 on irrigated land but only 5.2 on nonirrigated fields (Bodenheimer, 1949). This effect was attributed to "gonadotropins" in growing leaves of legumes (Bodenheimer and Sulman, 1946), and similar effects are now known to occur in other voles.

Populations of a California vole (*Microtus californicus*) on perennial grasslands have smaller litters than those on annual grasses. When the two populations are maintained under the same conditions in the laboratory, however, there is no significant difference in mean litter size, indicating that differences observed in the field do not have a genetic basis. When the same voles are kept in outdoor enclosures of either annual or perennial plant cover, the mean litter sizes observed in the field are replicated: on perennial grasses the mean litter size was 3.45 in contrast to a mean of 5.23 for voles on annual grasses (Krohne, 1980). In undisturbed wild populations of *Microtus montanus*, yearly variations in reproductive seasonality are correlated with the appearance of green shoots and freshly growing vegetation. In addition, winter breeding, although the exception, occurs with the unusual presence of green sprouts under snow cover (Negus et al., 1977).

12.3.4 Experimental Studies on Fecundity in Rodents

The experimental observations on *Microtus californicus* clearly indicate that at least some variation in mean litter size in voles is not determined genetically. Laboratory studies on *Microtus montanus* suggest pathways through which 6-MBOA may alter fecundity. This vole lives in the Great Basin of North America, a semiarid region characterized by year-to-year variation in rainfall and in the growth of forbs and grasses on which voles feed.

A diet including extracts from sprouting wheat produced a 41% increase in delta cells of the anterior pituitary of *Microtus montanus* (Hinkley, 1966). Delta cells are a site of gonadotropin production, and their increase may enhance the pathway by which certain foods stimulate reproduction. Moreover, when sprouted wheat was given to a nonbreeding wild (field) population of this vole, all of 11 females captured two weeks later were gravid, but there were no pregnancies in 11 females in a control area 200 yards distant (Negus and Berger, 1977). In captive voles of this species, sprouted wheat stimulated immediate onset of estrus in immature females, and a diet of spinach extract fed to 4-week-old females was followed by an increase in uterine weight and number of developing follicles (Negus and Pinter, 1966).

An extract of growing wheat (20 cm) fed to *Microtus montanus* stimulated the reproductive tract of both sexes, and the active material was identified as 6-MBOA (Sanders et al., 1981). Also, 6-MBOA was effective in the field when applied to rolled oats. When 6-MBOA was given to free-ranging populations of voles in the field in November and December or in January, males showed testicular hypertrophy and females became pregnant; a nearby control group given untreated oats showed no stimulation (Berger et al., 1981).

When intact and ovariectomized voles were fed rabbit chow supplemented with extracts from fresh lettuce, there followed a highly significant increase in uterine weights in intact and sham-operated voles but not in ovariectomized females, suggesting that the active substance in lettuce is not estrogenic (Berger and Negus, 1974). Lettuce is also stimulatory to testes in pocket mice (*Perognathus formosus*), without a significant increase in body mass (Kenagy and Bartholomew, 1981). In captive *Microtus californicus* fresh green spinach enhances the stimulatory effect of long days (17L:10D) and overcomes the suppressive effect of short days (10L:14D) (Nelson et al., 1983).

The effects of phytoestrogens and such materials as 6-MBOA are most apparent in deserts where rainfall is irregular and on agricultural land with a mosaic of irrigated and unirrigated cropland. Stimulatory qualities, however, are found in leaves, fruits, flowers, and underground parts of many sorts of plants and have long been known to occur in such diverse materials as palm nut oil and willow catkins. Phytoestrogens are in tissues of such common foods as spinach, apples, parsley, garlic, and cherries (Labov, 1977).

12.3.5 Negative Cues That May Terminate Reproduction

In many vertebrates the reproductive season is assumed to cease as a consequence of a photorefractory period, but there is evidence that some environmental factors have powerful suppressive effects on the reproductive system. Extracts from several species of *Lithospermum* (a genus of Boraginaceae) are antigonadotropic to equine chorionic gonadotropin. The effect occurs when extracts are administered subcutaneously to laboratory rats and is greatest when injected in moderate amounts. The antigonadotropic material is most concentrated in stems in autumn (Graham and Noble, 1955). Injections of water-soluble extracts of *L. ruderale* roots, consisting of polyphenolic acid salts, inhibit both gonadotropin and thyrotropin secretion in day-old cockerels. The mechanism of this inhibition is not known (Breneman et al., 1976).

Toward the end of the growing season, some grasses accumulate compounds that retard or suppress activity of the reproductive tract. Inhibitors such as phenols and cinnamic acids (4-vinylguaiacol, ferulic acid, and *p*-coumaric acid) increase as grasses mature, develop fruiting heads, and start to brown (El-Basouni and Towers, 1964; Kuwatsuka and Shindo, 1973). The dried and brown culms of salt grass (*Distichlis stricta*), an important food of the vole *Microtus montanus,* build up high levels of 4-vinylguaiacol, which experimentally suppresses both follicular and uterine growth but without any deleterious or apparently toxic effects (Berger et al., 1977).

Powerful ergot alkaloids, such as ergocryptine, produced by the fungus *Claviceps purpurea,* cause a marked suppression of prolactin secretion, causing hypoglactia (reduced milk flow). This fungus grows on cereal grains, especially rye, and also on wild grasses. It is sporadically encountered in cultivated fields. Ergocryptine is specific in its suppression of prolactin secretion and is about 100 times as potent as dopamine (Malcolm, 1981). As little as 1% barley ergot can inhibit lactations in sows (Nordskog and Clark, 1945). Bromocryptine, a synthetic ergot alkaloid, is sometimes used as a suppressant of prolactin, but the side effects are unpredictable, for this material resembles the hallucinogenic drug known as LSD. Ergot alkaloids, especially ergocornine, bromocryptine, and ergocryptine, inhibit implantation and lactation in the laboratory rat, and both effects are caused by a suppression of prolactin secretion. Ergocryptine is about twice as potent in this effects as bromocryptine is.

Although the ergot alkaloids are dramatic and extremely powerful, other fungi also have toxic effects. Toxic fungi (*Epicoccum nigrum*, *Cladosporium* sp., and *Fusarium tricinctum*) that occur on the tall fescus (*Festuca arundinacea*) are known to be deleterious to cattle on rangelands and to mice experimentally. These fungal infections are common in some pastures in the Mississippi Valley, and infected grass is not visually different from uninfected plants (Yates, 1971).

The evidence for nonphotic negative (inhibitory) factors is strongest in rodents, but probably most or all herbivores are affected. Negative factors or cues were not considered by Baker (1938), and they appear to be ignored by modern ecologists.

The response to these secondary plant compounds, regardless of whether they are estrogenic or not in their mode of action, suggests that they might be important regulators of the breeding season in some vertebrates. The stimulatory materials appear in some grasses during early growth, and in legumes they seem to be produced over a greater part of the growing season. In herbivores the stimulation for reproduction occurs as succulent forage becomes available for both adults and young. In contrast, suppressive materials develop at the end of the summer period of plant growth, when there are few tender plants to serve as food for young voles and other small mammals. It is also of interest that the wild mammals known to respond to the effects of secondary plant compounds are also those that are only weakly responsive to the photocycle.

12.4 TEMPERATURE

Ambient temperatures are usually reflected in the body temperatures of ectotherms. Temperature not only affects the teleost hypothalamus but may directly influence testicular steroidogenesis (Kime, 1982). In most fish, ambient and body temperatures are the same or nearly so. Amphibians, through evaporative cooling, are likely to be cooler than the ambient when in air. Some reptiles have the capacity, mostly through behavioral thermoregulation, to raise their core temperatures well above the ambient.

Despite some degree of control over their body temperatures, metabolism in these three vertebrate classes is by no means independent of ambient temperatures. These animals are generally exposed to rather regular annual changes in environmental temperatures, and it is not surprising that their annual reproductive cycles are frequently adjusted to seasonal temperature cycles. They are also usually exposed to the annual photocycle, and in some species both the pineal gland and the parietal eye are known to be photosensitive. The role of day length can be important in a nocturnal species as well as in a diurnal animal, for the lengths of both the light and the dark periods can affect gonadal activity. The annual photocycle of ectothermic vertebrates acts together with thermal influences, in addition to the role of nutrition discussed earlier in this chapter.

12.4.1 Interactions between Daylight and Temperature

In water, light intensity is reduced and red light penetrates least. By secreting themselves at will, a fish may restrict its exposure to light within the light periods of the natural photocycle (Scott, 1979; Wiebe, 1968a). Nevertheless, many examples of fish and other ectotherms are sensitive to changes in day length. One obstacle to the study of light and temperature as reproductive factors appears to be that responses vary in the captive bred forms that are frequently employed in experiments (Lam, 1983). Photoperiodic responses of ectotherms are very commonly related to ambient temperatures, or responses to temperature are related to photoperiod, and it is helpful to consider the two combined as photothermal factors.

12.4.1.1 Fish. Numerous lines of evidence indicate that the pineal gland plays an important role in the annual cycles of fish (Kavaliers, 1980). Pineal influence in fish is, however, strongly affected by water temperature (de Vlaming and Olcese, 1981). The pineal and the retina, together with temperature, may mediate photoperiodic responses and thus suppress or limit gonadal activity during seasons of short days. In cyprinodont fish, however, temperature is an important regulator of gonadal activity, and both day length and temperature may be paramount in salmonid and gastereosteid fish (de Vlaming, 1974). In vernal-spawning species gonadal growth is most rapid in cool water and with a schedule of increasing day lengths, and gonadal recrudescence may be greatly retarded if day length is shortened or temperature raised (Peter and Hontela, 1978; Peter and Crim, 1979).

When pinealectomy is performed at the onset of the breeding migration in the lamprey (*Lampetra fluviatilis*), there is a delay in gonadal development and appearance of secondary sexual characteristics in both sexes (Joss, 1973). As in some endotherms, levels of hydroxylindole-Q-methyltransferase (HIOMT) in larvae of the lamprey (*Geotria australis*) have a peak in the dark phase of the photocycle some four or five times greater than in light (Joss, 1977). Nevertheless, in lampreys (Petromyzontidae) gonadal growth and reproduction seem closely cued by water temperature, and light does not appear to affect gametogenesis or spawning (Larsen and Rothwell, 1972).

Observations on elasmobranchs indicate effects of both light and temperature on their gonadal growth. The pineal of the dogfish (*Scyliorhinus caniculus*) contains receptor cells with the morphological features of retinal cones, and the organ appears to be a photoreceptor (Rüdeberg, 1968, 1969). The dogfish pineal, moreover, appears to have an antigonadal role, which can be greatly reduced by pineal extract (possibly melatonin) injection, resembling the complex role of the pineal in endotherms. The apparent role of the pineal in the dogfish, however, is secondary to that of temperature (Dobson and Dodd, 1977). In nature there is a broad spectrum of reproductive seasonality in elasmobranchs. Some breed continuously throughout the year, and others have a rather discrete period of reproductive behavior (Wourms, 1977).

The anterior brain region of at least some teleosts is sensitive to light, so that pinealectomy and enucleation together do not destroy light-induced responses. Thus it is difficult to isolate the role of the pineal gland in photoperiodic phenomena (Matty, 1978). The rainbow trout (*Salmo gairdneri*), for example, can detect light even when deprived of its eyes and/or the pineal gland (Hafeez and Quay, 1970), and plasma melatonin increases significantly in the dark phase of a 12L:12D cycle, independent of temperature (Gern et al., 1978). The rainbow trout is a spring-spawning species and experiences autumnal gonadal recrudescence following exposure to gradually *shortening* periods of light at both 8°C and 16°C, but no change in gonadal growth occurs when this fish is maintained at *constant* contrasting light schedules (8L:16D and 16L:8D) at these temperatures (Breton and Billard, 1977; Billard & Breton, 1978). The rainbow trout (*Salmo gairdneri*) has cycles of ovulation and testosterone peaks approximately every six months when kept on a 18L:6D schedule (Scott and Sumpter, 1983). Melatonin occurs in the pineal of the salmon (*Oncorhynchus tshawytscha*), and levels in sea-dwelling immature fish were found to be about six times as high as in adults that had begun a breeding migration in fresh water (Fenwick, 1970a). After they enter fresh water on their spawning migration, salmon are exposed to much brighter light than they experience in the sea, regardless of the photocycle. The spawning of the shelly (*Coregonus lavaretus*), a salmonid of the English Lake district, is temperature induced, and its eggs require cold water. The eggs are demersal and develop at surface temperatures of from 4 to 6°C, but there is no survival at 10°C (Bagenal, 1970).

Many salmonid fish have a breeding migration, and it is not surprising that both migration and gonadal growth are affected by the same exogenous factors. Among several species that spawn in the autumn, gonadal activity is promoted by both short days and low temperatures (Billard et al., 1978). The brook trout (*Salvelinus fontinalis*) is a well-known autumnal spawner, and the gonads seem to be stimulated by short day lengths. Gonadal recrudescence in nature is apparently initiated under long days and accelerated by shortened day length. This has been confirmed many times (Bye, 1984). Experimental evidence in salmonids indicates that increasing day lengths trigger spermatogenesis and that decreasing day lengths promote sperm maturation (Scott et al., 1983).

Among minnows (Cyprinidae) there is a general pattern of spawning during warm seasons, and experimental evidence indicates that minnows seem to be stimulated by increases in both daylight and temperature. One of the earliest demonstrations of photosensitivity in minnows showed a clear positive response of follicular growth independent of temperature (Fig. 12.4).

A Eurasian minnow, *Phoxinus phoxinus*, has a temperature-modified response to day length. When these minnows are maintained at low temperatures and long days, there is no gonadal growth, but when kept at higher temperatures, the same fish experience a significant increase in gonadosomatic index in response to long days (Fig. 12.5). In nature there are year-to-year variations in the time of vernal recrudescence, such variations being correlated with annual variations in vernal water temperatures (Scott, 1979). In the golden shiner (*Notemigonus crysoleucas*), a small minnow of eastern North America, long days and warm temperatures stimulate gonadal development. Under these same conditions, gonads show arrested or reduced development after pinealectomy, suggesting a stimulatory role for the pineal gland. Also, under a schedule of long days and low temperatures, pinealectomy was followed by retarded ovarian development. Under short days and warm temperatures, however, pinealectomy seemed to account for an *increase* in gonadal development, suggesting that pineal activity during times of short days suppresses gonadal growth (de Vlaming and Vodicnik, 1977). The pathway of pineal influence in fish is unknown, but the pattern seen in the golden shiner resembles that seen in hamsters—the pineal is active in darkness, and long days result in "physiological pinealectomy" (Reiter, 1980a).

Among Japanese species of bitterlings (Cyprinidae), *Rhodeus ocellatus* spawns from spring through summer, *Acheilognathus tabira* spawns in the spring, and *Pseudoperilampus typus* spawns in the autumn. Experimentally, gonadal recrudescence is induced by rising temperatures, regardless of photoperiod, in *Rhodeus* and *Acheilognathus*, and regression follows a shortened day length without a temperature change. In contrast, in the autumn-spawning *Pseudoperilampus*, gonadal maturation is in response to a decrease in day length, and subsequently regression is induced by a lowering of temperature (Hanyu et al., 1982).

The goldfish (*Carassius auratus*) is sensitive to day length and temperature. Goldfish ovulate spontaneously when maintained at 20°C on a 16L:8D cycle (Stacey, 1984). Volumetric and morphologic changes in photoreceptor cells of the pineal gland of the goldfish when exposed to a long-term 16L:8D schedule (over six

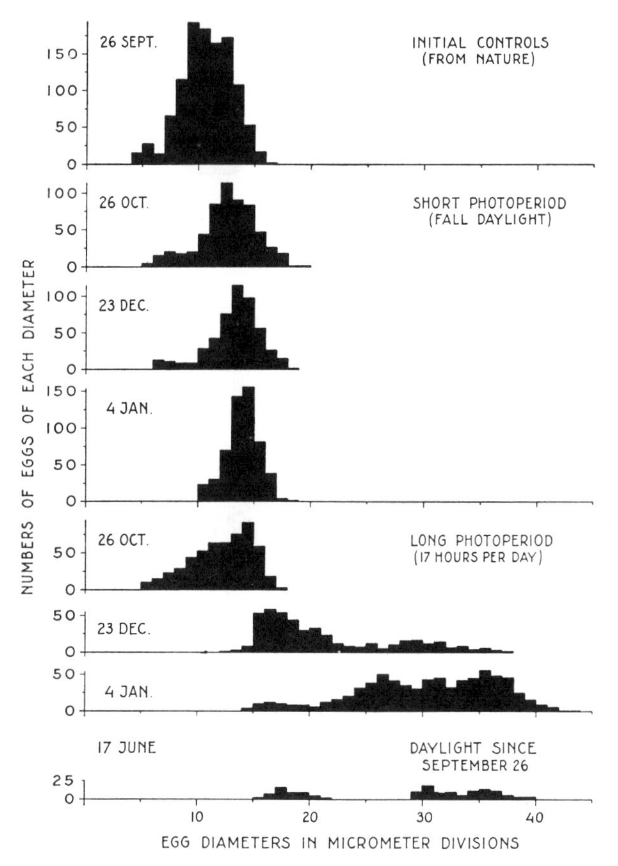

Figure 12.4 Follicular growth of *Notropis bifrenatus* under an autumnal photocycle (9L:15D) and a long day schedule (17L:7D). The diameters of the largest 50 eggs in one ovary of each fish were measured. One micrometer division = 21 μm. This is one of the earliest clear demonstrations of the effect of the photocycle on an ectotherm. (From Harrington, 1957.)

months) indicate that secretory cell organelles (rough endoplasmic reticulum and Golgi bodies) are affected by light (McNulty, 1982). When exposed to 20°C and long days in the spring, goldfish experience gonadal enlargement, but when pinealectomized at this season and maintained under short days, goldfish have smaller gonads than sham-operated controls do (Fenwick, 1970b; de Vlaming and Vodicnik, 1978). Maintaining goldfish in total darkness is, however, more suppressive to gonadal development than pinealectomy is (Vodicnik et al., 1979).

As in salmonids, gametogenesis may be initiated under one light-temperature regimen, and gamete maturation may be initiated six months later under a contrasting light-temperature environment (Bye, 1984).

The medaka (*Oryzias latipes*), a cyprinodont fish of Japan, responds to long days (16L:8D or 14L:10D) with increased gonadal development, and blinded in-

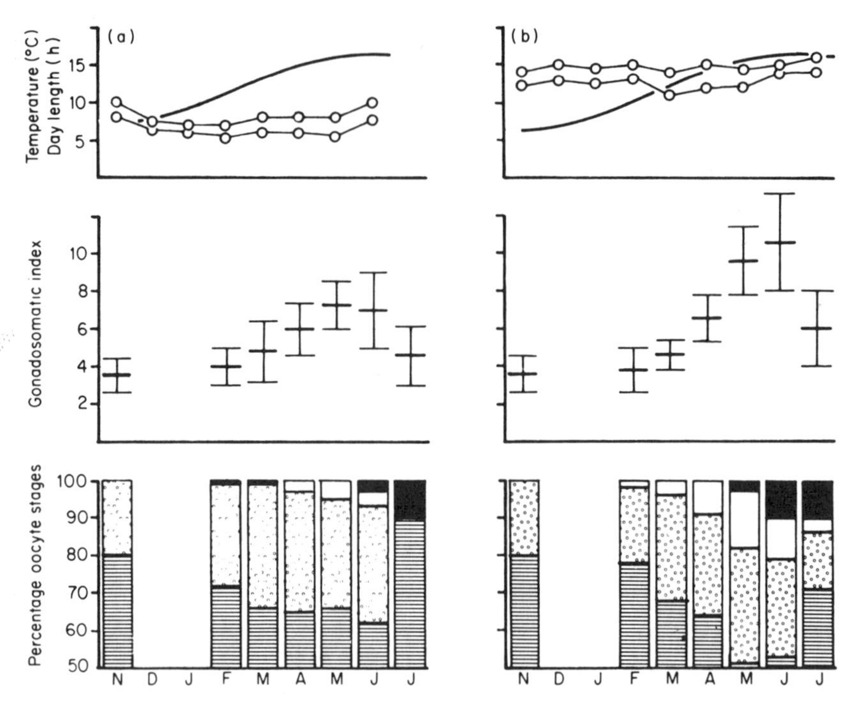

Figure 12.5 The effect of day length on gonadal recrudescence of *Phoxinus phoxinus* at different temperatures. *Top*. Connected circles indicate the monthly minimum and maximum temperatures, and solid lines show the day length. *Middle*. The gonadosomatic index ± 1 SD. *Bottom*. Stages of ova development: previtellogenic (cross-lined), early (stippled), final (white), and atretic (black). Photoperiod is gradually increasing in both (*a*) and (*b*), but in (*b*) the gonadosomatic index changes rapidly in response to day length when the temperatures are warmer. (From Scott, 1979, courtesy of The Zoological Society of London.)

dividuals show a similar but reduced effect, including a distinct reduction in number of eggs laid. The pineal in this species is photosensitive. When kept under long days, blinded and pinealectomized individuals have reduced ovarian growth, suggesting a nonsuppressive or permissive role for the pineal gland under long days. Under a natural short-day photocycle (November to December), however, blinded and pinealectomized medaka had greater gonadal development than controls did, which indicates a reduction of the antigonadal effect of both eyes and the pineal under short days (Urasaki, 1976, Urasaki et al., 1981). Melatonin inhibits ovarian growth when injected into medaka kept on a 14L:10D schedule at 26°C, but melatonin had no effect at 17°C (Urasaki, 1977). Similarly, injections of melatonin (3 g/day) for 15 days two hours after the onset of daylight retards gonadal growth in the killifish (*Fundulus similis*) in contrast to controls (de Vlaming et al., 1974).

A possible error in the technique of pinealectomy in fish may account for the occasionally reported "progonadal" effect. A part of the pineal gland may adhere to the parietal bone as the latter is raised to expose the pineal for removal, and with replacement of the parietal bone, this segment of the pineal remains. Thus, what appears to be a pinealectomy may leave some of the secretory tissue remaining (de Vlaming and Olcese, 1981).

The mummichog (*Fundulus heteroclitus*), an estuarine cyprinodont of the North American Atlantic coast, is well known as a tidal spawner and lays eggs with a semilunar frequency in synchrony with spring tides. It is also stimulated by a long-day (L15:D9) schedule and fails to develop ova when exposed to a short (L9:D15) photocycle (Taylor, 1986).

Several poeciliid fish, introduced in northern Australia, have seasonal patterns of breeding. These patterns are controlled by day length and, to a lesser extent, temperature. *Gambusia affinis*, *Xiphophorus helleri*, and *X. maculatus* show an increase in the rate of pregnancy corresponding to increased day length, but reproductive activity declines with day length in midsummer while temperatures are still favorable (Milton and Arthington, 1983). Fecundity of poeciliid fish in nature tends to be maximal in summer. Two species, *Poeciliopsis gracilis* and *Poecilia sphenops*, in El Salvador (in nature) are exposed to natural extremes in day length from "short" days of 11.3 hours to "long" days of 12.9 hours. When these species are kept in these two photoschedules and at similar temperatures, the longer day length increases vitellogenesis and number of embryos, independent of the size of the female, and there was also an increase in testicular size (Burns, 1985).

The testes of the marine live-bearing teleost *Cymatogaster aggregata* are affected by both temperature and light, and the gonadal response varies with the time of year. In early winter increased day length, from 8L:16D to 16L:8D, concurrent with a change in water temperature from 10 to 20°C was followed by an increase in testicular growth and spermatogenesis, the change being due to temperature increase independent of day length. In late winter and spring the stimulatory effect of higher temperature was enhanced by increased day length, and shorter day length resulted in gonadal regression (Wiebe, 1968a).

The role of photoperiod in tropical ectotherms is difficult to evaluate, for if a long day (12 hours) is stimulatory, tropical species are in a continuously stimulatory photocycle. It is plausible that the tropical photocycle brings an individual ectotherm to a subfunctional reproductive state (as seem to occur in birds and mammals), and it is then prepared to respond to one or more ultimate factors.

The pineal organ of the angel fish (*Pterophyllum scalare*), a Neotropical cichlid fish, is photosensitive, and its electrical activity varies with exposure to light (Morita and Bergmann, 1971). The four-eyed fish (*Anableps dowi*) breeds throughout the year in a tidal estuary in El Salvador. There is a correlation, however, between day length and number of embryos and the gonadosomatic index of the male, suggesting a photoperiodic response in this fish (Burns and Flores, 1981). Several species of *Sarotherodon* (*Tilapia*) have distinct periods of spawning in tropical regions of relatively constant schedules of light and temperature. In these fish gonadal activity can be experimentally induced by warm temperatures, and in nature spawning occurs at the beginning of the seasonal rains (Sundararaj, 1981).

In tropical species that inhabit waters of constant flow and composition, spawning may occur more or less continuously throughout the year. Very commonly, tropical waters are altered by periods of drought and rains, and resulting conditions control reproduction in many low-latitude teleosts. These seasonal changes not only affect the physical and chemical aspects of the waters but also may time seasonality of planktonic bloom. Rainfall is frequently followed by changes in pond levels of stream flow, which is associated with spawning of a number of African and Asiatic teleosts (Lam, 1983). In the Murray-Darling River system of New South Wales, several species of native fish in earth-lined ponds were induced to spawn by replenishing water that had been lost to evaporation. Apparently there were cues associated with water entering the dry soil (Lake, 1967).

The Neotropical electric fish *Apteronotus albifrons* and *Eigenmannia virescens* live in freshwater watercourses where there are only slight seasonal changes in light and temperature. In these species spawning can be experimentally induced by (1) a decrease in conductivity of water (2) an increase in water level, and (3) an imitation of rain (Kirschbaum, 1975, 1984).

12.4.1.2 Amphibians. Inasmuch as amphibian behavior is responsive to both light and temperature and inasmuch as temperatures tend to fluctuate with the daily and annual photocycles, separate roles of temperature and light are not always apparent. Experimentally, few investigations on the role of the photocycle are designed with "cold" light, so that a cycle of artificial light usually produces a cycle of temperature as well. In some salamanders (Salamandridae) the hypothalamus appears to be the thermosensitive center controlling gonadotropin and prolactin release (Mazzi, 1970). There is evidence, moreover, that daily activity may have a thermoregulatory significance, and the pineal may be important in entraining daily cycles of activity (Ralph et al., 1979). The association of peaks of melatonin with periods of rest in some amphibians does not preclude a direct role of this hormone in depressing gonadal activity of amphibians.

In addition to the pineal, anurans possess a frontal organ, also known as the *stirnorgan*, which is a homologue of the parietal eye. The frontal organ is absent in salamanders and caecilians. In anurans both the frontal organ (or parietal eye) and the pineal gland are photosensitive and respond both to visual signals (Korf et al., 1981) and to ultraviolet wavelengths (Adler, 1976). As in fish (and in a great many tetrapod vertebrates) the pineal gland secretes melatonin. A well-known effect of this hormone in amphibians is the contraction of melanocytes in the skin, producing a light hue. This response accounts for lightening of skin color of some frogs in the dark.

Melatonin also retards gonadal development in some amphibians. Injections of melatonin (4 μg) given to tree frogs (*Hyla cinerea*) daily for 28 days, 10 hours after the onset of daylight, resulted in significantly smaller gonads than in controls (de Vlaming et al., 1974). In the leopard frog (*Rana pipiens*), HIOMT levels are greater in the retina than in the frontal organ (Eichler and Moore, 1975), suggesting a regulatory role for the paired eyes. In the larval tiger salamander (*Ambystoma tigrinum*) the pineal responds to darkness in a manner observed in some teleost fish, and there is a nocturnal peak of circulating melatonin in animals kept on a

12L:12D schedule. The melatonin peak disappears after pinealectomy, but melatonin still remains in the light period, possibly from the retina (Gern and Norris, 1979).

Gametogenesis in amphibians and reptiles depends on gonadotropin release, as in other vertebrates, and the hypothalamus in amphibians is responsive to both light and temperature. During early winter (November and December), gonadotropin in *Rana temporaria* increases with rises in temperature, but there is no change in spermatogenesis at that season. From January to March, however, primary spermatogonia become increasingly sensitive to gonadotropin support, and in the springtime the spermatogenic cycle progresses with increases in temperature (van Oordt, 1956). In the bullfrog (*Rana catesbeiana*), spermiation in response to gonadotropin hormone (GtH) is depressed at low temperatures (15–20°C), and optimal spermiation occurs at about 25°C, with sensitivity to temperature being unaffected by day length (Easley et al., 1979). Similarly, under natural conditions day length seems not to be critical in timing of the ovarian activity of the toad (*Bufo bufo*) in Denmark. Temperature is a major influence on oocytes: warm summer temperatures seem to be essential to their growth, but low temperatures that obtain during hibernation are required for their final maturation (Jørgensen et al., 1978).

Day length plays a role in gametogenesis in at least one salamander. In the red-backed salamander (*Plethodon cinereus*), spermatogenesis proceeds rapidly at 20°C and is retarded at 10°C. At 20°C the process is accelerated in increasing day length (through 8L:16D, 12L:12D, and 16L:8D), but temperature is clearly more important than light (Werner, 1969). As pointed out previously, the hypothalamus of salamanders is temperature sensitive and monitors the release of both gonadotropins and prolactin. In the Salamandridae, prolactin initiates the movement to water, and its release is triggered by a sudden change in temperature (Mazzi, 1970).

12.4.1.3 Reptiles. In many species of lizards and in the tuatara (*Sphenodon punctatus*), there is a light-sensitive parietal eye immediately beneath a translucent scale on the top of the head. The parietal eye is lined with a cup-shaped retina, and a nerve leading from the retina courses along the pineal gland. The parietal eye is found in all reptilian orders except the crocodilians. The parietal eye occurs irregularly in lizards but is absent in snakes. In *Lacerta viridis*, a lizard of the Old World, the pineal is both photosensitive and secretory (Kappers, 1967). Lizards of several families can be entrained to light schedules when blinded, and testicular recrudescence occurs as in sighted individuals; the extraretinal receptor is not known (Underwood and Menaker, 1976). The parietal eye and/or the pineal might both serve as this receptor.

Reptiles have a complex of gonadal responses resulting from manipulation of temperature and day length, and the roles of each are not always clearly separable. The parietal eye of the horned lizard (*Phrynosoma douglassi*) affects both thermal and photic responses, and because body temperature is altered by the lizard's photic responses, the two roles are difficult to distinguish (Phillips and Harlow, 1981). The entrainment of a daily rhythm of motor activity, including basking and rest, results in important seasonal trends in changes of body temperature. In at least one lizard the effects of light and heat are sequential: temperature is the initial factor involved in the gonadal activity of the anole (*Anolis carolinensis*), but later photoperiod becomes dominant (Licht, 1971a, 1971b).

Under natural conditions, however, a pond turtle, Blanding's turtle (*Emydoidea blandingi*), nests annually at some time within a 17-day period from late May to early June, and oviposition is independent of mean May conditions but correlated with mean April temperatures (Congdon et al., 1983). Apparently April temperatures constitute a proximate cue in this species, but there remains a possible role of light in the field.

There is a rhythmic secretory activity of the pineal in some squamates and chelonians, with HIOMT and serotonin increasing in the dark phase of the cycle (Quay, 1965; Joss, 1978). It has been suggested that serotonin in the sauropsidian pineal is converted to melatonin as in the mammalian pineal (Kappers, 1967). Little experimental work has been done on the pineal gland of most reptiles, but the majority of observations suggest an antigonadal role for the reptilian pineal. In the Asian agamid lizard, *Calotes versicolor*, there is an inverse cycle of size and activity of the pineal on the one hand and of gonads on the other hand, and pineal hypertrophy in nature occurs in seasons of short day lengths (Misra and Thapliyal, 1980). In this species pinealectomy accelerated recrudescence of the testes and also delayed their autumnal regression (Haldar and Thapliyal, 1977). Melatonin appears to depress gonadal development in the zebra-tailed lizard (*Callisaurus draconoides*) when maintained on a natural vernal light schedule (Packard and Packard, 1977). On the other hand, the role of subcutaneous melatonin may be to alter daily rhythms (and thereby body temperature) and affect gonadal growth only indirectly (Underwood, 1979). As in amphibians, however, reproduction in the reptilia reflects appropriate combinations of day length and temperature, and most investigators indicate that temperature has the greater role (Aleksiuk and Gregory, 1974; Crews, 1975; Licht, 1971a, 1971b, 1972; Lofts, 1977, 1978).

Temperature seems to control the secretion of reptilian gonadotropins and may also differentially alter sensitivity of target organs (Licht, 1972). Experimentally, gonadotropin and testosterone reach higher levels at 20°C than at 3°C in *Anolis carolinensis* (Pearson et al., 1976).

12.4.2 Endotherms

Although the body temperatures of most endotherms are usually sufficiently constant that they do not fluctuate with the ambient temperature, changes in the latter can alter the rate of depletion of energy stores. Exogenous heat does not, of itself, alter gonadal activity, but by altering the amount of energy, environmental heat may alter fecundity.

In northwestern Europe, nesting of the great tit (*Parus major*) is closely correlated with temperature as well as with leafing of birch (*Betula* spp.) and the fall of the caterpillars of the winter moth (*Operophthera brumata*), both of which are related to temperature (Slagsvold, 1976). Thus temperature indirectly provides for a greater food supply for these tits, which allows females to accumulate adequate fat for egg formation early in the year. There is an additional effect of heat that strikes the nest boxes in which the tits sleep. Those boxes exposed to late-afternoon sun have increased nocturnal temperatures, and females roosting in "warm" boxes during their prelaying period lay earlier than do those roosting in "cool" nest boxes. Presumably the lower nighttime temperatures place greater demands on the stored energy of the sleeping bird and retard accumulation of fat for egg formation (Dhondt and Eyckerman, 1979).

There is apparently a relationship between ambient temperature and dates of oviposition in the rockhopper penguin (*Eudyptes chrysocome*). This species breeds annually, from late September until early December, depending on the region, and laying dates are closely correlated with mean annual seawater temperatures (Fig. 12.6). It is not known if the marine temperatures relate to the abundance of food of the penguin or to thermoregulation of the penguins themselves (Warham, 1975).

Mammals generally do not seem to be extremely sensitive to changes in ambient temperatures. Perhaps this apparent immunity from the effects of exogenous heat

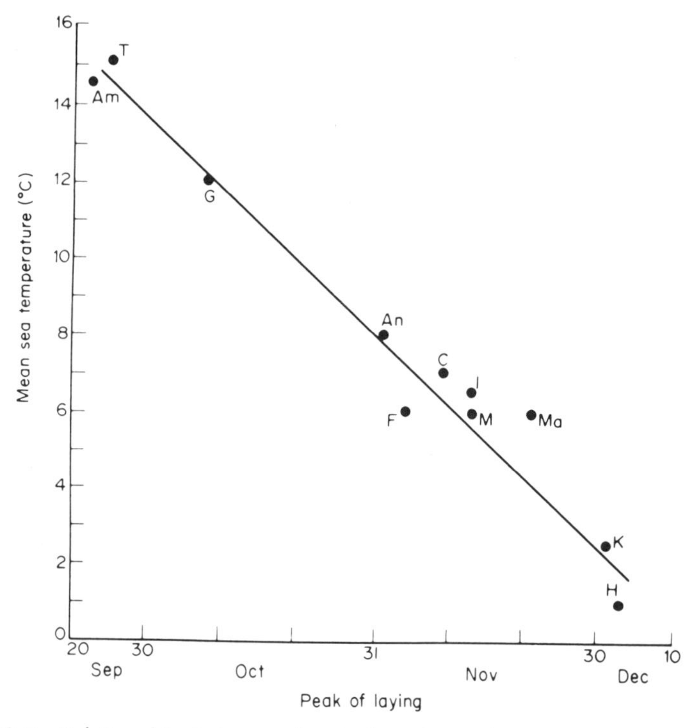

Figure 12.6 Relationship between dates of peak egg laying and mean annual sea temperatures in the rockhopper penguin (*Eudyptes chrysocome* on several Antarctic islands). AM, Amsterdam I.; An, Antipodes I.; C, Campbell I.; F, Falkland I.; G, Gough I.; H, Heard I.; I, Ildefonso I.; K, Kerguelen I.; M, Macquairie I.; Ma, Marion I.; T, Tristan da Cunha. [From Warham, 1975, courtesy of *The Biology of Penguins* (Ed: B Stonehouse).]

is due to the nocturnal activity patterns of most species of mammals. This topic has not been a major realm of study. Deer mice (*Peromyscus maniculatus*) seem to be unaffected, reproductively, by differences in temperature. Among populations of deer mice from contrasting latitudes, ambient temperature has no role in reproduction (Bronson and Pryor, 1983). In the European vole (*Microtus arvalis*), however, fertility is lowered by high temperatures (Daketse and Martinet, 1977).

12.5 HUMIDITY

Environmental humidity can have distinct effects as either rainfall per se or the accumulation of water on or in the ground. There is also evidence that atmospheric humidity can regulate gonadal development and behavioral activity.

12.5.1 Rainfall

Many anecdotal accounts of reproduction of both birds and mammals in the tropics indicate that rain itself may be a proximate cue for reproduction. The exact nature of the stimulus or the mechanism of its action is not known. Even at the sound of thunder and before there could be any effect of rain (such as plant growth), some Australian passerine birds have been known to build nests and lay eggs (Immelmann, 1970). The zebra finch (*Poephila guttata castanotis*), which occurs widely in arid and semiarid regions of Australia, is not affected by variations in the photoperiodic cycle, but it nests after, and sometimes before, rains. Experimentally, after the birds have been deprived of free water, the addition of drinking water is followed by testicular growth. Testicular growth is also induced by the increase of atmospheric humidity or the addition of fresh green grass, which is an important food of these birds during the nesting season. Water in one form or another seems to induce testicular growth in this finch (Priedkalns et al., 1984). The European rabbit (*Oryctolagus cuniculus*), feral in Australia, seems to be preadapted to irregularity in rainfall: when rain broke an 18-month drought, there were suddenly pregnancies in rabbits within two weeks after the rain and before the resurgence of plant growth (Poole, 1960).

Some ectotherms are known to be both sensitive to standing water and dependent on soil moisture or standing water. The spadefoot toads (*Scaphiopus* spp.), which live in arid and semiarid areas of North America, spawn in temporary ponds and must oviposit promptly with the accumulation of standing water. During the nonreproductive seasons adults lie buried deep in the ground. They are stimulated to emerge by low-frequency vibrations (probably below 100 Hz) in the soil (which can be produced by an electric motor, simulating the sound of falling rain) and also by the sound of rain on a roofed enclosure, but not by soil moisture (Dimmit and Ruibal, 1980). The African viviparous toad (*Nectrophrynoides occidentalis*) estivates in dry ground in nature, and in the laboratory saturated atmospheric humidity induces spermatogenesis (Gavaud, 1977). Although the triggering mechanism by which rainfall initiates a gonadal response may well vary among different classes, it is clear that rainfall constitutes a proximate cue.

Rainfall and the consequent soil moisture can also be an ultimate cue. The annual cyprinodont fish of South America and East Africa die as their ponds dry up with the approach of annual droughts, and the species survive as partly developed eggs in the dry soil. The eggs are embryonated, with a half-developed embryo, but they enter a diapause and cease growth until triggered by the return of the rains and temporary ponds. This is one of the very few cases of diapause in vertebrates. When rainfall restores these ponds, the eggs hatch, and the adults quickly reach adulthood and spawn (Bay, 1966; Simpson, 1979). Not all eggs resume growth at first. While one segment of the population matures and spawns, others develop some time later, thus protecting the species against the danger of having the entire population maturing in a pond that is too short-lived to provide water for oviposition. This topic is also discussed in Chapter 13, Section 13.4.1.1.

12.6 ENDOGENOUS REPRODUCTIVE CYCLES

The persistence of a reproductive cycle in the absence of exogenous cues has been taken as evidence for an endogenous reproductive cycle. It has been suggested that some vertebrates have a naturally recurring circannual cycle in which the photocycle is unnecessary. This circannual cycle does not preclude the annual photocycle as a *Zeitgeber* but simply means that the gonads will undergo more or less annual recrudescence without photic stimulation. Although an endogenous circannual rhythm of reproduction and gonadal regression is apparently uncommon, it has been reported to occur in a number of diverse taxa.

Although the Indian catfish (*Heteropneustes fossilis*) is sensitive to day length, there is a strong endogenous circannual element to its ovarian cycle. Captives in continuous light or darkness at 26°C have an annual cycle of ovarian growth comparable to those in nature (Sundararaj, 1978; Sundararaj et al., 1978). This fish is not insensitive to light, for if kept under a schedule of 6L:18D with the dark phase interrupted with one hour of light, there is a significantly greater ovarian enlargement than occurs in a control group of 7L:17D (Vasal and Sundararaj, 1975).

The extent of endogenous circannual rhythmicity among reptiles is unknown. In the desert iguana (*Dipsosaurus dorsalis*) there is a circannual reproductive cycle in individuals maintained under constant stimulatory conditions of light and temperature (Licht, 1972). *Cnemidophorus uniparens*, a parthenogenetic teiid lizard, also has a circannual breeding cycle (Cuellar, 1981). In the field, however, any existing rhythms are cued by exogenous *Zeitgebers*.

The existence of endogenous circannual rhythms in birds has been questioned, but experimentation suggests that some species have reproductive cycles that recur at nearly annual intervals. An African weaver finch (*Quelea quelea*), for example, preserved its reproductive cycle for two years when kept on a 12L:12D schedule (Lofts, 1964). There are also apparently endogenous rhythms of migration of Old World warblers (Sylviidae). Circannual rhythms of testicular size and molt in a warbler (*Sylvia borin*) and a starling (*Sturnis vulgaris*) persisted up to four years when kept under light schedules of 11L:11D, 12L:12D, and 13L:13D (Gwinner, 1977, 1981).

It has been suggested that birds maintained on a 12L:12D cycle for a prolonged period may be able to record the passage of days (i.e., they can, in effect, count the days) and thus keep track of the seasons (Farner and Wingfield, 1978). This ability is clearly not present in all birds, for the white-crowned sparrow (*Zonotrichia leucophrys gambelii*) kept under constant stimulatory photocycles does not spontaneously recover from photorefractoriness (Sansum and King, 1976).

Endogenous circannual reproductive cycles obtain for several species of ground squirrels. The reproductive system of the antelope ground squirrel (*Ammospermophilus leucurus*) is unresponsive to short-day displacement of the photocycle. Annual changes are unaffected by experimental exposure to extremes of daily cycles (16L:8D and 8L:16D) at all times of the year (Kenagy and Bartholomew, 1979). Animals held in constant laboratory conditions (both in 12L:12D and in continuous light) for about two years showed spontaneous cycles of gonadal function with an approximately annual periodicity. This species reproduces only once a year, regularly, in the spring, and advanced preparation for this reproduction is under endogenous control (Kenagy, 1981).

The hibernation cycle of the golden-mantled ground squirrel (*Spermophilus lateralis*) is also endogenous and recurs at somewhat less than 365-day intervals under uniform conditions. The mechanism for an endogenous reproductive cycle of males may lie in a varying sensitivity of luteinizing hormone (LH) secretion to negative feedback. Male golden-mantled ground squirrels (*Spermophilus lateralis*) exhibit a circannual

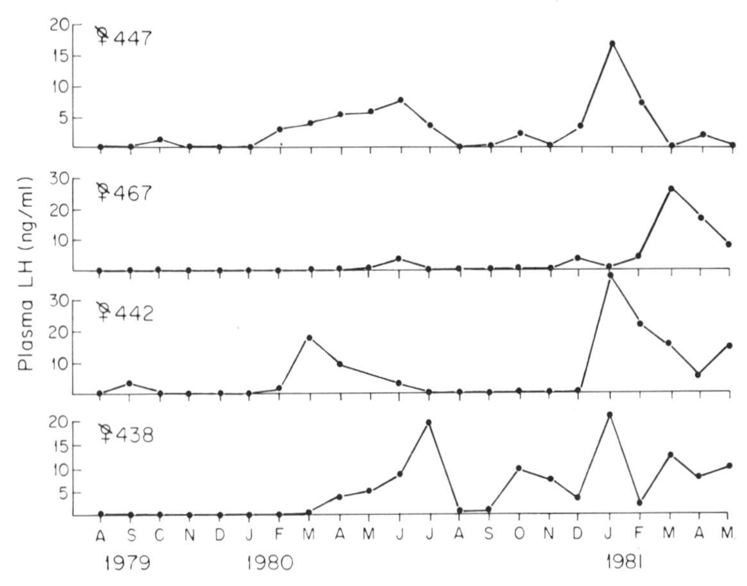

Figure 12.7 Monthly plasma LH levels of ovariectomized golden-mantled ground squirrels (*Spermophilus lateralis*). Animals were ovariectomized in August, at 70 days of age. (From Zucker and Licht, 1983a, courtesy of *Biology of Reproduction*.)

decline in negative feedback depression of LH secretion by gonadal steroids during the breeding period of the spring, which results in an elevation of serum LH in intact males. In castrated males there is no apparent seasonal rhythm to serum levels of LH (Zucker and Licht, 1983). Secretion of LH in males, then, is predicated on an annual cycle of sensitivity to the negative feedback of gonadal steroids. A different endogenous rhythm of LH release is seen in ovariectomized females of the same species, with vernal peaks of LH independent of ovarian steroid feedback (Fig. 12.7). The endogenous nature of LH release in females of this species and the seasonal (vernal) decline in sensitivity to negative feedback in males together could determine the endogenous reproductive pattern seen in these squirrels. There is nevertheless a role of the pineal gland in this species, for it seems to affect the circannual rhythms of both weight changes and onset of estrus: in pinealectomized females these circannual rhythms are shortened (Zucker, 1985).

12.7 INTERRELATIONSHIP OF EXOGENOUS FACTORS

In spite of the great amount of experimental evidence showing the effect of light on the reproduction of many diverse taxa, the role of the natural photocycle in wild populations may not be the sole major factor (e.g., Scott, 1979). The photocycle operates together with other environmental influences, including temperature, precipitation, plant growth, social factors, and both nutrition and ingested secondary plant materials (Marshall, 1970). As pointed out in this chapter, "food" is a complex of materials that have a multiplicity of effects on the reproductive system. Food should not be considered as only having a nutritional role for maintenance and growth.

12.8 REFRACTORINESS

Refractoriness probably occurs in a number of ectotherm groups and is presumably part of any endogenous circannual rhythm. In many ectotherms gonadal regression occurs when both light and temperature are stimulatory, and these patterns will become apparent in the next chapter. Refractoriness in fish occurs in a number of vernal spawning species. They may respond to changes in light, temperature, or other cues separately or in sequence, and there is frequently a postspawning gonadal regression. Refractoriness has been reported for several teleost fish (Dodd and Sumpter, 1984; Hanyu et al., 1982) and also for several squamates (Licht, 1984).

The experimental research on day length is designed to preserve uniformity in all factors but light, but even in the laboratory small events can alter reproduction. For example, irregular sound can alter levels of gonadotropins (Nalbandov, 1970); minor stress can stimulate the release of prolactin; and social factors (e.g., the Bruce effect and the Whitten effect) are known to alter sexual maturity, ovulation, and preimplantation mortality. It is important to remember that in nature the photocycle is experienced with a host of other cues and that experimental studies on captive animals are obviously not intended to replicate the conditions under which a free-living animal exists.

Although day length is easily and unequivocally controlled in a caged animal, many wild creatures, such as mice, many reptiles, most amphibians, and many small fish, very easily modify their exposure to light. That is, many species can modify their activity so that it does not coincide with the natural photocycle. Such species can decrease their exposure to the daily period of daylight or darkness. Thus although many nonavian vertebrates are well known to be photosensitive, one must not expect laboratory experiments to be exactly predictive of their behavior in nature. It is apparent that light is not a *Zeitgeber* at low latitudes, but that seasonal reproduction in the tropics is triggered by cues other than the photocycle.

Among the constellation of cues or factors, both proximate and ultimate, discussed in this chapter, some are stimulatory and others appear to be suppressive. These factors become increasingly clustered and more closely associated with the calendar at higher latitudes. Stimulatory cues occur in the spring, and suppressive factors prevail in late summer and autumn.

12.9 SUMMARY

Nonphotic exogenous factors can affect the endogenous cues of the reproductive system. Nutritional levels are usually correlated with investment in developing oocytes, and intraspecific variation in fecundity may reflect nourishment of the female.

Many secondary plant compounds either stimulate or depress one or another part of the reproductive system. These materials, whether phytoestrogens or compounds that act at a level above the gonads, may be ultimate cues and cause reproductive activity after the gonads have been brought to a subfunctional state by day length. In other species plant stimulants may be effective in the absence of the photocycle.

Finally, there are plant materials that depress or suppress reproductive activity. Some of them suppress prolactin, which is essential in the reproduction of at least some mammals. In view of the fact that most mammals do not become photorefractory, such negative cues may be important in terminating seasonal reproduction.

Among ectotherms there is frequently an interaction with light, water, and temperature, and in a few species a combination of these factors is essential to reproduction. In most ectotherms, however, temperature is much more important than any aspect of the photocycle. The reproductive system in a number of teleost fish is clearly sensitive to both day length and water temperature, and frequently both factors must be suitable for gonadal growth and reproductive activity. There is substantial evidence for a role of the pineal gland and its hormone, melatonin, in teleost reproduction: the pineal is suppressed by light, and melatonin appears to be antigonadal.

Temperature seems not to be of major importance in birds and mammals, but there are examples in which low temperatures cause the animal to draw on energy reserves and therefore reduce materials available for egg production.

Endogenous circannual rhythms are known for some vertebrates. These species, under uniform environmental conditions, experience gonadal recrudescence about once a year and are generally unresponsive to environmental cues. There is some evidence of neuroendocrine control for endogenous circannual rhythms. Refractoriness, which probably occurs in the annual periodicities of most wild species, is an endogenous phase of a reproductive cycle, and thus most reproductive cycles are at least in part endogenous.

BIBLIOGRAPHY

de Vlaming VL (1974). Environmental and endocrine control of teleost reproduction. Pp 13–83. In *Control of Sex in Fishes* (Ed: Schrenk CB). Virginia Polytechnic, Blacksburg, Virginia.

Follett BK and Follett DE (1981). *Biological Clocks in Seasonal Reproductive Cycles*. Halsted (Wiley), New York.

Lamming GE (Ed) (1984). *Marshall's Physiology of Reproduction*, 4th ed., vol 1. Churchill Livingstone, Edinburgh.

Li CP (1974). *Chinese Herbal Medicine*. U.S. Department of Health, Education and Welfare, Publ No. (NIH) 75-732.

Lincoln GA (1981). Seasonal aspects of testicular function. Pp 255–320. In *The Testis* (Eds: Burger H and de Kritser D). Raven, New York.

Lincoln GA and Short RV (1980). Seasonal breeding: Nature's contraceptive. Pp 1–52. In *Recent Progress in Hormonal Research*, vol 36. Academic, New York.

Miller PJ (Ed) (1979). *Fish Phenology: Anabolic Adaptiveness in Teleosts. Symposium of the Zoological Society of London*, No. 44. Academic, London.

Murton RK and Westwood NJ (1977). *Avian Breeding Cycles*. Clarendon, Oxford.

Potts GW and Wootton RJ (Eds) (1984). *Fish Reproduction: Strategies and Tactics*. Academic, London.

Rankin JC, Pitcher TJ, and Duggan RT (1983). *Control Processes in Fish Physiology*. Croom Helm, London, Canberra.

Richter CJJ and Goos HJT (Eds) (1982). Reproductive physiology of fish. *Proc Internat Symp Reprod Physiol Fish*, Wageningen, Netherlands, 2–6 August 1982. Centre for Agricultural Publishing and Documentation, Wageningen.

Stacey NE (1984). Control of timing and ovulation by exogenous and endogenous factors. Pp 207–222. In *Fish Reproduction* (Eds: Potts GW and Wootton RJ). Academic, London.

Wootton RJ (1982). Environmental factors in fish reproduction. Pp 210–219. In *Reproductive Physiology of Fish* (Eds: Richter CJJ and Goos HJT). Wageningen, Pudoc.

13

PATTERNS OF SEASONALITY

13.1 THE TRANSLATION OF CUES

Having considered some exogenous factors that appear to control gonadal growth and activity, we can better appreciate the rhythmicity of reproductive seasonality in nature. This chapter reviews some reproductive patterns of free-living vertebrates and attempts to identify features of the environment to which seasonality is adjusted. The proximate factors or cues that provide signals to the individual's hypothalamic-pituitary-gonadal axis initiate development of reproductive potential. Other specific environmental signals are ultimate cues and may precipitate overt reproductive behavior after gonadal maturation, but they may not necessarily have a direct effect on the gonads.

Most wild vertebrates have an annual cycle of sexual activity and circulating hormones, but this cycle has been determined for only a few species. In the little brown bat (*Myotis lucifugus*) the baseline levels of testosterone obtain during most of the year and range from 2.5 to 11.7 ng/ml, but in midsummer, before fall mating, there is a surge to 59.1 $\pm$ 9.2 ng/ml (Gustafson and Shemesh, 1976). Comparable premating increases are known for some other species that have been closely monitored. The Asiatic elephant (*Elephas maximus*) experiences a nonbreeding low of 0.7 ng/ml but a reproductive high of 44.7 during musth, the state of maximal sexual aggressiveness in male proboscideans (Jainudeen et al., 1972).

In females, gonadal steroids increase concurrently with seasonal activity. In fish, steroid hormones stimulate reproductive activity in those species with internal fertilization, but in some species with external fertilization, prostaglandins govern breeding (Stacey, 1987).

This chapter is mostly descriptive. Some of the observed patterns in nature may suggest the factors that serve as *Zeitgebers*, such as those described in Chapters 15 and 16. The patterns observed in nature can be divided into *strategies* and *tactics*. These terms have been widely and loosely applied by field biologists. Here, a reproductive strategy is the totality of traits that maintain a population in its natural range. A tactic is a modification of the strategy so as to adjust it to environmental changes from year to year or to minor or local differences (Wootton, 1984).

Discussions of the adaptive significance of life history strategies and of fitness frequently assume that observed events strictly follow genetic programming. These assumptions usually neglect the highly variable Pleistocene and Holocene environments from which modern species have emerged. If, indeed, reproductive strategies were rigidly fixed and represented "finely tuned" adaptations, these modern species probably would not have survived the Pleistocene. We shall see many reproductive patterns that are remarkably well adapted to existing conditions, but are also variable and apparently better described as phenotypic manifestations of a genotype that is flexible in its expression.

Ectotherms are generally temperature sensitive and frequently respond to combinations of factors that include temperature. Among most of the trouts, charrs, and salmon (Salmonidae), spawning is frequenlty cued by low or declining temperatures, and these fish tend to be autumnal breeders. Most minnows (Cyprinidae), on the other hand, are triggered by rising temperatures, and they usually spawn in the spring and summer. Almost all amphibians require a humid atmosphere, if not standing water, before they spawn. Although most species are found in permanently humid habitats, the occurrence of water itself may not necessarily constitute a distinct variable about which seasonality is scheduled. In many narrative accounts of amphibian reproduction, however, there is the implication that water or rain functions as a cue. Also, amphibians in temperate latitudes spawn in a broad spectrum of ambient temperatures, and clearly there is not only one season for amphibian reproduction. Instead, in some regions different species of anurans spawn in waves, so that a minimal number of species occupy a given pond at any given period. Also, totally terrestrial amphibians usually differ in reproductive seasonality from those that spawn in water.

Reptiles are diverse in their seasonality. Within a restricted region, some mate in the spring and others in the autumn. We have already seen that gametogenesis occurs in the spring in some reptiles and at the end of summer in others, and this pattern is complicated by sperm storage, in males and/or females. In addition, among related species oviparous forms may have a reproductive seasonality very different from closely related live-bearing forms in the same region.

Among birds there is the strongest tendency for uniformity in seasonality. In the mid- and high latitudes most species adhere to a pattern of vernal nesting, and photoperiodic stimuli initiate gonadal recrudescence. At lower latitudes important cues are less rigidly coupled to the annual photocycle, for nesting in some tropical birds is entirely dissociated from the Gregorian calendar. In tropical regions factors other than light seem to cue the nesting of many species.

In midlatitudes larger species of mammals (which usually have a long gestation) tend to reflect more clearly environmental changes in their breeding seasons, but many small species sometimes depart from a strict dependence on meteorological cues. Marine mammals and some bats exhibit the greatest regularity in mating and parturition, for many of them breed in synchrony with migration. Tropical mammals follow a diversity of cues.

13.2 AQUATIC REPRODUCTION

Fish and many amphibians deposit eggs in water. The aqueous environment provides protection from sudden temperature changes, and water temperature is a critical factor in the reproduction of many aquatic vertebrates. Light, especially at the red end of the spectrum, does not penetrate far through water, which limits the growth of green plants and the production of oxygen. The mass of water, approximately

840 times that of air at sea level, accounts for stratification of dissolved oxygen, minerals, and temperature, but it provides momentum to marine currents and provides force to wave and tidal action. Oxygen is increasingly soluble at lower temperatures but is nevertheless scarce in deep waters, where temperatures can be very low. Oxygen content may be highest in polar waters, where surface ocean waters may be −1.9°C. Thus there are marked oxygen gradients correlated with both depth and temperature, and spawning of fish is adjusted to these variations.

Because of the mass of water, environmental changes are much slower than in aerial habitats. Daily and seasonal variations in the physical features (oxygen, mineral content, and heat) may be rather slight in a given area but are conspicuously different from one region to another. The maximal amount of dissolved oxygen in water is approximately one-thirtieth the oxygen content of air at sea level, but carbon dioxide is approximately 28 times as soluble in water as in oxygen. In shallow, sunlit waters photosynthesis releases oxygen in the daytime, and there can be substantial daily cycles of dissolved oxygen; but where light penetration is weak, oxygen is sparse. This is important in spawning, for developing eggs and larvae have high oxygen demands.

It is ironic that the breeding seasons of the most abundant and diverse vertebrates, the fish, are the poorest known. Although many fish are major sources of food and have been carefully studied and some sport species and ornamental tropicals are virtually domestic, these examples represent a small fraction of the known species. Knowledge of the timing of reproduction of most other species accumulates from the efforts of enthusiastic students, and their attention is concentrated on freshwater and littoral fish.

13.3 FISH OF MARINE WATERS

The tremendous mass of water in the oceans accounts for extensive environmental stability in a given area as well as great diversity from place to place. The high latitudes are continuously very cold but experience drastic variations in day length, which largely account for the polar summer productivity and seasonality of reproduction.

13.3.1 Latitudinal Trends

Sharks and rays occupy a specific position apart from bony fish, because of either their unique aspects of viviparity or their hard egg cases. The egg cases allow for a prolonged fetal growth, and their presence in the sea over an extended period does not necessarily indicate a prolonged period of oviposition. Nevertheless, oviposition does appear to occur throughout much of the year in many egg-laying species. Among live-bearing elasmobranchs, some are very clearly seasonal and others lack seasonality.

At high latitudes spawning is cued by temperature and tends to be compressed into a rather brief period. Cold-adapted fish at the warmest borders of their geographic ranges breed in the coldest months. In the Pacific waters off the coast of Canada, the sablefish (*Anoplopoma fimbria*) spawns at the coldest time of the year: eggs are most abundant in January and February, and the peak for larvae is in April (Mason et al., 1983). Warm-adapted species at the coldest parts of their geographic ranges spawn in the warmest months. As part of this general pattern, warm-adapted fish tend to have follicles in various stages of development, and they may spawn repeatedly over extended periods, in contrast to cold-adapted species, which spawn only once a year (Fig. 13.1) (Qasim, 1956). In general these patterns seem to be adapted to seasons of maximal abundance of plankton. At high latitudes plankton is abundant for a relatively short season, and in warmer water plankton occurs for much longer seasons but at lower levels of abundance.

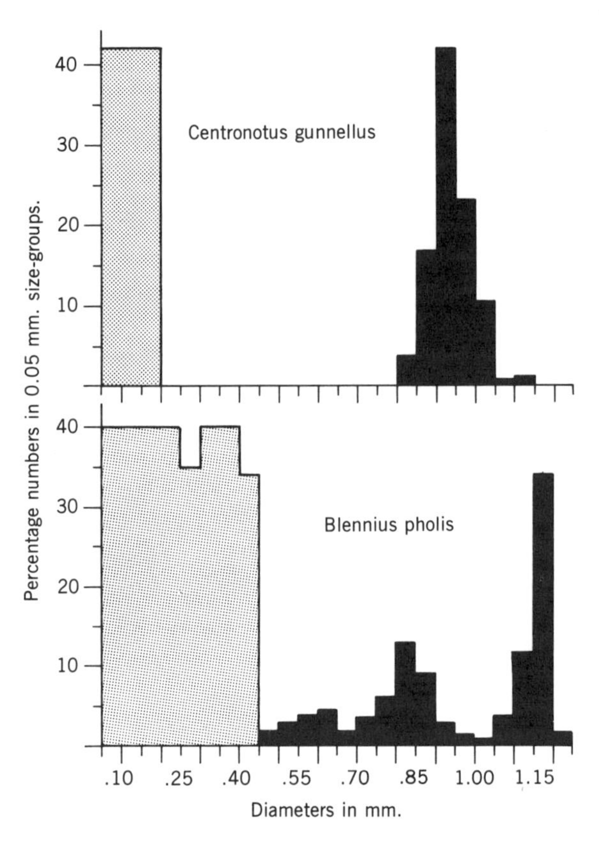

Figure 13.1 Egg formation in two shoreline species of fish. *Centronotus gunnellus*, a boreal species, spawns once a year. In contrast, *Blennius pholis*, a warm-water species, spawns several times a year. Shaded areas indicate yolkless cells, presumably follicles and oogonia, and black areas indicate maturing oocytes. (From Qasim, 1956, courtesy of the *Journal du Conseil international pour L'exploration de la Mer.)*

13.3.2 Midlatitude Species

In the open ocean, successful spawning frequently depends upon adequate plankton for pelagic larvae. Plankton levels vary in a fairly regular fashion with seasonal changes in upwelling and solar radiation, but they are subject to the more variable factor of wind intensity.

Upwellings occur along the western coasts of North and South America, and they vary seasonally. They bring nutrient-rich water to the surface, where there is an abundance of light and dissolved oxygen. Photosynthetic activity commences in the lowest photic layers. Because the upward movement is slow, the products of photosynthesis build up before the water reaches the surface. These areas of primary productivity support populations of zooplankton for fish such as sardines and anchovies. These fish spawn in plankton-rich waters off the coast of southern California, and as the season progresses, the areas of productivity move northward. This phenomenon seems to determine the migrations and spawning of the whiting or hake (*Merluccius productus*), which winters some 300 to 400 m below the surface but spawns in upwellings at 100 m or less off the coast of southern California (Bailey et al., 1982). As more northerly areas become productive, the adults move against the California Current and feed in the upwellings of the coasts of Oregon and Washington (Fig. 13.2).

Many midlatitude species spawn in shallow waters along the shore and usually lay demersal eggs. This is characteristic of many smelt (Osmeridae), gobies (Gobiidae), sculpins (Cottidae), and others. The Pacific herring (*Clupea pallasi*) spawns demersal eggs over beds of algae. These eggs adhere to the algal thalli, which hold them over the bottom so that currents and tidal movements enhance aeration of the developing embryos.

13.3.3 Low-Latitude Species

At lower latitudes plankton is much less abundant, and seasonality becomes less pronounced. Many coastal species of fish tend to produce fewer ova at one time, but they spawn over protracted periods, frequently at specific times. This seems to be true not only of coastal species but of coral reef fish as well. In the Gulf of Guinea, between 6 and 7°N, the clupeid fish *Ilisha africa* breeds throughout the year (Marcus and Kusemiju, 1984), and among snappers (Lutianidae) in marine waters of the African tropics, reproduction is essentially aseasonal (Talbot, 1960). A number of subtropical gobies (Gobiidae) appear to have rather protracted spawning periods, which have locally variable peaks (Geevarghese and John, 1983).

Among some 25 species of coastal tropical marine fish of Palau are many that spawn in spring and fall, independent of rain and temperature. Spawning aggregations are nocturnal and frequently occur when the moon is new or full and when the wind is minimal, thus providing a quiet environment for the eggs and larvae (Johannes, 1978). Among 35 species of reef fish in the Caribbean off Jamaica, most spawn between February and April, when water temperatures are minimal (Munroe et al., 1973). Presumably nocturnal spawning minimizes predation on the

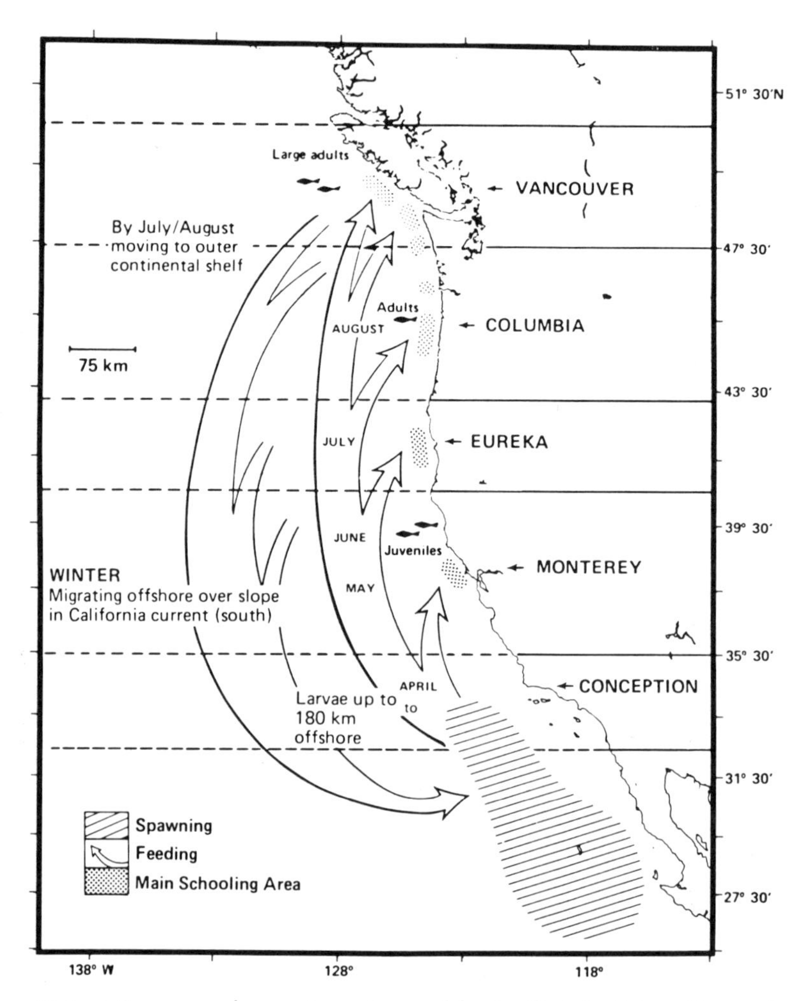

Figure 13.2 Migratory and spawning patterns of the Pacific hake (*Mercullius productus*). (From Bailey et al., 1982.)

eggs. Many of these fish have pelagic eggs and larvae that drift in gyrals, huge circular currents, which return them to the spawning ground when partly grown and better able to protect themselves from predators of coastal waters. One should remember, however, that many coral reef fish, such as carangids, groupers, and synodontids, have a prolonged larval life and assume their adult form far from their place of origin. The yellowtail damselfish (*Microspathodon chrysurus*), a coral reef pomacentrid of the Caribbean coast of Panama, spawns once a month, in a lunar rhythmicity, with maximal activity from the full to the new moon (Pressley, 1980). In these examples lunar periodicities are associated with reduced wave action and not water level per se, as in tidal-spawning species, described in the following section.

13.3.4 Tidal, Lunar, and Semilunar Rhythms

Tides are regular and constitute reliable cues to short-term changes in water levels, but the rhythms are complex. Tidal cycles, which result from combined movements of the sun and the moon, could be proximate cues in some species, whereas a high or spring tide may, by itself, be an ultimate cue. Although a complete tidal cycle occupies one lunar month, there are two spring tides monthly (with a new moon and with a full moon). Not surprisingly, tidal-spawning fish frequently have semilunar peaks of mature follicles. A number of unrelated kinds of fish are known to be rigidly tidal in their spawning periodicity, and this pattern has clearly evolved many times.

Tidal spawning can be regarded as an extreme effort to place eggs in an oxygen-rich environment and also to escape the populations of predators of the intertidal zone. The associated rhythm of oocyte development indicates that this pattern of oviposition is probably an old one.

Several species of coastal fish spawn in synchrony with the near-spring tidal cycle: spawning occurs at high tides, and two weeks later, at the next high tide, the larvae are washed to the sea. Grunion (*Leuresthes*) of the Pacific Coast of North America spawn on sandy beaches at high tide. A female partly buries herself in a vertical position in the sand. As she releases her eggs, one or several males release milt next to her, and this fluid runs down her body to carry sperm to the eggs. The fertilized eggs remain in the wet sand for two weeks. If the fry are not carried out to sea on the next high tide two weeks later, they are capable of surviving an additional two weeks (Gibson, 1978). The Japanese puffer or *fugu* (*Fugu niphobles*), on the east coast of Japan, spawns at the high tide of either the full or new moon (Fig. 13.3). Eggs remain in sand from 3 to 10 cm below the surface, but mortality is 100%. Reproduction depends upon those eggs that are promptly washed to sea (Nozaki et al., 1976). The significance of tidal spawning in the *fugu* is equivocal. On the west coast of Japan (Japan Sea), moreover, the spring tides and daily high

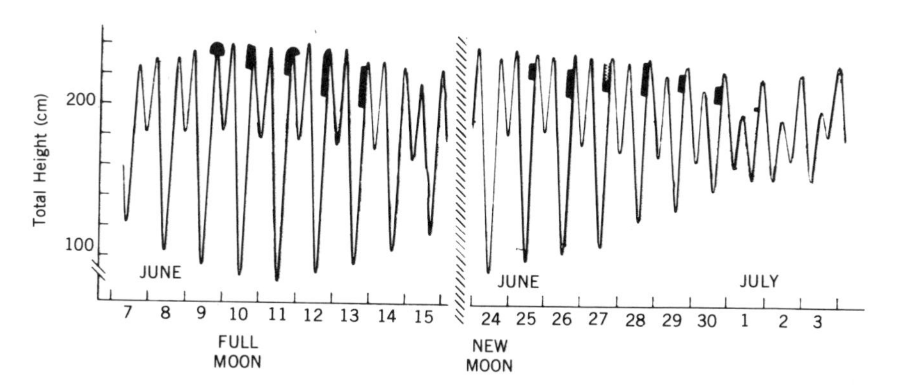

Figure 13.3 The synchrony of high tides (light line) and spawning (wide lines) of the *fugu*, or Japanese puffer (*Fugu niphobles*), at Aburatsubo Inlet, Japan. (From Nozaki et al., 1976, courtesy of *Zoological Magazine*.)

tides are minor, and there the puffer does not exhibit a tidal rhythm of spawning (Honma et al., 1980).

The mummichog (*Fundulus heteroclitus*), a topminnow, spawns at spring tides on the Atlantic Coast of North America. Synchronized with the tidal rhythm is a rhythm of ovarian activity, with eggs maturing at 14-day intervals. This ovarian cycle, although essentially lunar, seems to result from successive semilunar cycles, in which alternate peaks exceed the intervening maxima. This little fish spawns when high tides flood grassy flatlands, and the eggs lodge in the axils of a salt marsh grass (*Spartina alternifolia*) or in the empty shell of a mussel (*Geukensia demissa*). As in some other tidal-spawning species, the eggs develop in air (Taylor et al., 1979). These eggs normally hatch when submerged in water, either two or four weeks later. When experimentally submerged, the eggs begin to swell, the embryo turns within its egg some 5 minutes later, and hatching occurs 15 to 20 minutes after submergence (Taylor et al., 1977). *F. heteroclitus* has two distinct reproductive periods—one from late winter to early spring and another from summer to early fall—between which there is no spawning (Fig. 13.4). Breeding adults differ in age. The first spawning ends at the same time for all size groups, but smaller fish resume reproductive activity after the larger cohorts, perhaps because of their greater energy demand for food (Kneib, 1986). Two other cyprinodonts (*Fundulus pulvereus* and *Adinia xenica*) also have biweekly cycles of ovarian activity, with preovulatory follicles maturing synchronously with the approach of semilunar or biweekly spring tides (Greeley, 1984).

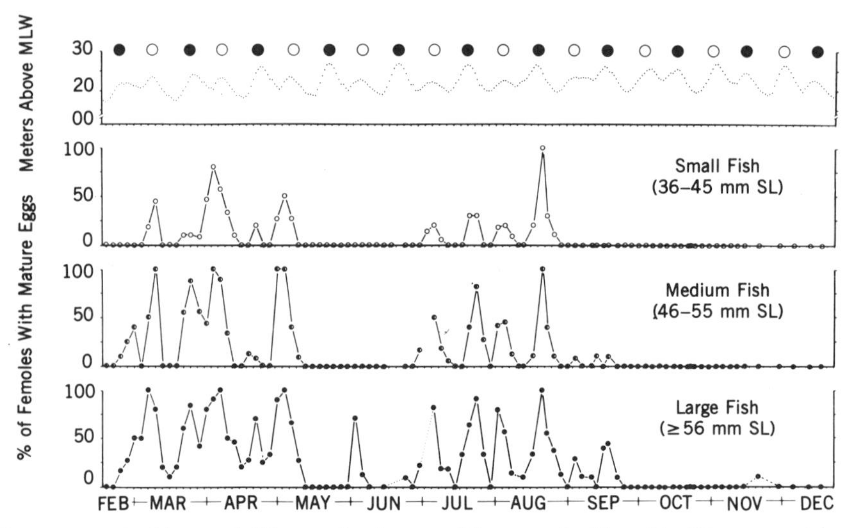

Figure 13.4 Incidence of different-sized groups (SL = standard length) of the mummichog (*Fundulus heteroclitus*) with mature ovarian eggs. Open circles at the top show the occurrence of full moons, and solid circles indicate new moons. The dotted line represents the predicted tidal level above the mean low water (MLW) of the two daily high tides. (From Kneib, 1986, courtesy of the American Society of Ichthyologists and Herpetologists.)

Tidal reproduction of the Atlantic silverside (*Menidia menidia*) has been carefully documented. These little fish move to the intertidal zone biweekly at the height of the high tides, when current velocities are minimal, perhaps so as to release sperm when there is little water movement (Middaugh and Takita, 1983). The eggs lodge in the axils of the leaves of cordgrass (*Spartina alternifolia*) and other protective sites and are exposed to air about 10 hours daily, between high tides (Middaugh et al., 1981). Fine threads that develop on the spawned eggs aid in holding them to the substrate. Hatching occurs at the next spring tide two weeks later and is restricted to the hours of darkness (Middaugh, 1981). Embryo survival is greatest in the highest intertidal zone, apparently because of minimal predation at that level (Tewksbury and Conover, 1987).

Several species of "southern trout" (Galaxiidae) of New Zealand are well known to have tidal spawning rhythms. *Galaxias attenuatus* spawns in shallow water at the peak of a spring tide, and the eggs become lodged about the bases of temporarily submerged grasses. Thus the eggs are protected from desiccation until the next spring tide two weeks later, at which time the larvae are carried out to sea. The inshore spawning migrations of this fish are synchronized with the tidal cycle (Burnett, 1965). *Galaxias maculatus* also spawns on flooded grassy flats, and the eggs develop in the axils of grasses. The eggs are ready to hatch within two weeks, and the larvae drift to open water with the next spring tide (McDowall and Whitaker, 1975).

13.3.6 Benthic Species

Deep waters are said to be virtually devoid of marked seasonal fluctuations of temperature, food supply, and oxygen, and there is no light. Among a group of benthic fish there is slight evidence of seasonal spawning (Mead et al., 1964). Snailfish, small marine species of the family Liparidae, occur from tide pools to 3500 m or deeper. Truly abyssal species living at depths of 2100 m or more spawn throughout the year in contrast to marked seasonality of shallow-water species. *Osteodiscus cascadiae*, which lives between 1900 and 2997 m, spawns aseasonally, but *Careproctus melanurus*, of shallow waters, spawns only in the summer months (Stein, 1980).

Benthic environments, however, may not be totally aseasonal. Down to depths of 2000 m some species of rattails (Macrouridae) appear to have seasonality of growth and spawning. In the northeast Atlantic Ocean (50°N, 13°W) there is a rapid fall of phytoplankton, which forms a sedimentary accretion in the spring and early summer. This detritus is derived from diatoms and dinoflagellates. Based on the date of phytoplankton bloom, the rate of fall was 100 to 150 m a day (Billett et al., 1983). Such fluctuations in the sea floor might account for synchronized reproductive cycles in four species of benthic echinoderms (Tyler et al., 1982). There is some evidence that benthic echinoderms and crustaceans are reproductively seasonal (Rokop, 1974). At depths of from 1330 to 3834 m the brittlestar (*Ophiura ljungmani*) has maximal gonadal mass in the winter months, and most young appear in the summer (Schoener, 1968). There are substantial seasonal variations in community oxygen consumption in the northern Pacific Ocean: north of Hawaii (31°N, 159°W) at a depth of 5900 m sedimentary oxygen consumption in June is almost four times higher than in November (Smith and Baldwin, 1984). Clearly, the benthos is not without seasonal environmental changes, and it appears possible that seasonality may also be found in some benthic fish.

Some species of flatnosed cod (Moridae) and the grenadiers or rattails (Macrouridae) live at 2000 m. In growth rates and spawning they seem to reflect seasonal changes at the surface, presumably a descent of energy from plankton (Gordon, 1979). *Coryphaenoides rupestris* (also called *Macrourus rupestris*) off the coast of Scotland has an autumnal peak in spawning (Gordon, 1979).

13.4 FISH OF FRESH WATERS

Seasonality of reproduction in freshwater fish usually follows one of several patterns: (1) some species develop a single generation of oocytes and spawn once a year, (2) some breed over an extended season with more than one generation of oocytes, and (3) still others have successive generations of oocytes and spawn sporadically (Qasim and Qayyum, 1961). This classification covers most reproductive strategies of freshwater fish, but it does not correlate these patterns with environmental events. The complex of patterns described in the following sections will usually fall into one of the three categories of Qasim and Qayyum, but there are many diverse environmental conditions to which these patterns are adapted.

13.4.1 Quiet Waters

Lakes and ponds vary in size but have distinctive features in common not found in rapidly moving waters. A lake has convection currents that result from seasonal temperature changes, prevailing winds, and streams entering and leaving. A deep lake will have some features of a marine sea with respect to stratification of light, temperature, and dissolved oxygen. Shallow lakes and ponds may light penetrating to the bottom, with a rapid turnover of oxygen at all depths.

13.4.1.1 Tropical Lakes. In tropical lakes environmental cues depend not only on the cycle of precipitation but also on the depth and breadth of the lake. Rains increase nutrition introduced from inflowing streams, and changes in plankton populations are also caused by fluctuations in water depths and shorelines. This suggests that reproductive seasonality varies with available nutrition and perhaps with the chemical composition of the water. Temperature and day length changes, so important in midlatitudes, are much less pronounced in the tropics and of minor importance to fish reproduction. For those species with pelagic larvae, plankton may be a major factor in their survival. In tropical lakes the growth of plankton is promoted not only by rain, but also by increased productivity from upwellings, light, and lack of wave action.

In Lake Tanganyika, species of herring (Clupeidae) have spawning peaks coinciding with high populations of plankton (Coulter, 1970). By way of contrast, African Cichlidae, especially those that carry the eggs in their buccal cavity, are rather aseasonal (Lowe-McConnell, 1958). Among a group of mouthbrooding cichlids in Lake Malawi, nine breed throughout the year, with bimodal peaks at times of plankton bloom. Food is important not only for energy in the production of eggs but also to sustain metabolism during the period of care of the young (Marsh et al., 1986). By carrying the freshly spawned eggs in the buccal cavity, a fish may create its own habitat for the eggs and small young. Some sympatric species of *Tilapia* have discrete and separate spawning seasons, timed to different phases of the annual cycle of rains (Lowe, 1953). Similarly, in Lake Victoria three of the five major species of catfish breed more or less continuously, but increases in spawning tend to occur between rainy seasons (Rinne and Wanjala, 1983). Like some of the cichlids, these catfish provide parental care. Likewise an Indian catfish (*Glyptosternum pectinopterum*) is aseasonal and spawns several times a year, and the ovaries in one fish may contain ova in several stages of maturation (Khanna and Pant, 1967). In contrast the cichlid fish *Entroplus maculatus* and *E. suratensis* spawn twice annually in Sri Lanka, during periods of maximal water clarity, and do not breed when seasonal rains render the water murky (Ward and Samarakoon, 1981).

Many oviparous cichlid fish of Africa have extended breeding seasons with local peaks in spawning, and these peaks may coincide with times of high water. Some species spawn over land that is temporarily inundated. Because hatching occurs soon after spawning and because growth is rapid, the young move to the main watercourses before the floodwaters recede (Lowe-McConnell, 1975). This same pattern also exists in tropical South America and southeast Asia.

A unique tropical pattern occurs in the cyprinodont "instant fish," which live in temporary ponds in East Africa and arid parts of South America. These fish have a rigidly defined annual life cycle, for all the adults die when ponds dry up after the rainy season. Their eggs, which remain in the dry soil, enter diapause and thus survive the seasonal drought. The eggs resume development when the ponds are flooded almost a year later (Fig 13.5). Final embryonic development is cued by submergence. These fish (species of *Nothobranchius*, *Aphyosemion*, and *Cynolebias*) have a rather restricted reproductive season and are called instant fish because of their rapid development when submerged in water. A thorough review of this group is presented by Simpson (1979).

Although the tropical Asiatic anabantid fish *Trichogaster pectoralis* breeds throughout the year, peaks of breeding coincide with seasonal rains, and there is some suggestion that rains are followed by an increase in food supply for the fry (Hails and Abdullah, 1982).

13.4.1.2 Temperate Lakes. In lakes and ponds sufficiently far from the equator so that winter ice forms, there is a kaleidoscopic array of vernal changes, any or all of which might constitute proximate cues and precipitate gonadal recrudescence.

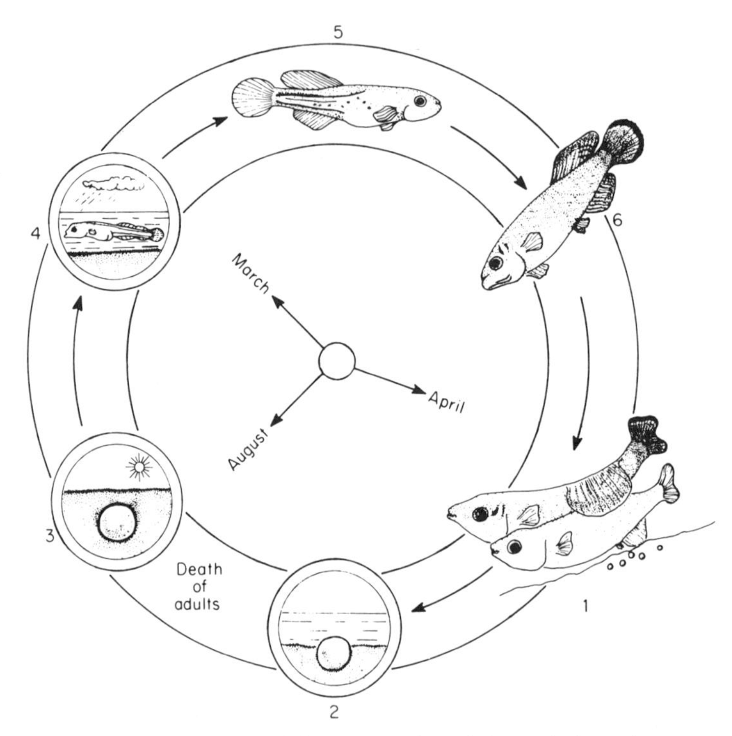

Figure 13.5 The annual cycle of the cyprinodont fish *Nothobranchius guentheri*. 1, spawning after seasonal rains; 2, eggs remain in the bottom of temporary pools; 3, water evaporates during the rainless periods, during which time the adults perish; 4, the return of the rains again floods the pools, causing some eggs to hatch; 5, the young grow rapidly; 6, spawning follows some four to eight weeks after hatching. (From Simpson, 1979, courtesy of The Zoological Society of London.)

In midlatitude lakes spawning is usually a spring or summer event, and there is much evidence that both light and temperature play major roles in timing of gonadal development. This is especially true if the water is rather shallow, for such ponds and lakes become warm by summer and their fish are sometimes referred to as *warm-water fish*.

In a warm-water reservoir on the border between North and South Dakota, a fauna of 17 species of warm-water fish adhered closely to specific patterns of spawning, but the overall season of reproduction spanned much of the summer, with little overlap between species. Year-to-year variations in peak spawning dates closely followed year-to-year variations in water temperatures. Littoral spawners seemed to be cued to specific depths of the varying shoreline (June, 1977).

13.4.1.3 Caves. Very little is available on seasonality of cave-dwelling fish. Some species appear to be reproductively active at times when there is high water and when access to caves is difficult. In the cave fish, *Amblyopsis spelaea*, reproduction is infrequent; that is, not all adults reproduce every year, but spawning is most frequent from February to April, during the annual period of high water (Poulson, 1963).

13.4.2 Moving Waters

Water levels and the flow of rivers fluctuate from season to season, depending upon patterns of precipitation and snow melt. When waters are low, surrounding banks are dry; but as rainfall saturates the land, seepage into river basins enriches the mineral and oxygen content of the rivers. In time the water levels rise and create optimal conditions for spawning of many species of fish as well as for growth of food supplies.

In tropical rivers and floodplains, there is heavy pressure on fish populations. Both predation and oxygen deprivation account for large losses, and in these environments fish either spawn large numbers of eggs or provide some sort of parental care (Lowe-McConnell, 1979). The spawning of small species is concentrated in small pools and ponds that become isolated by receding waters during dry seasons. Shallow waters are, however, vulnerable to oxygen depletion when temperatures rise, and high mortality of eggs and fry can result from overheating and suffocation (Fig. 13.6).

Among many tropical riverine fish, spawning seems to be stimulated by seasonal rains (Lake, 1967; Jhingran and Ghosh, 1978; Schwassmann, 1978). In many fish

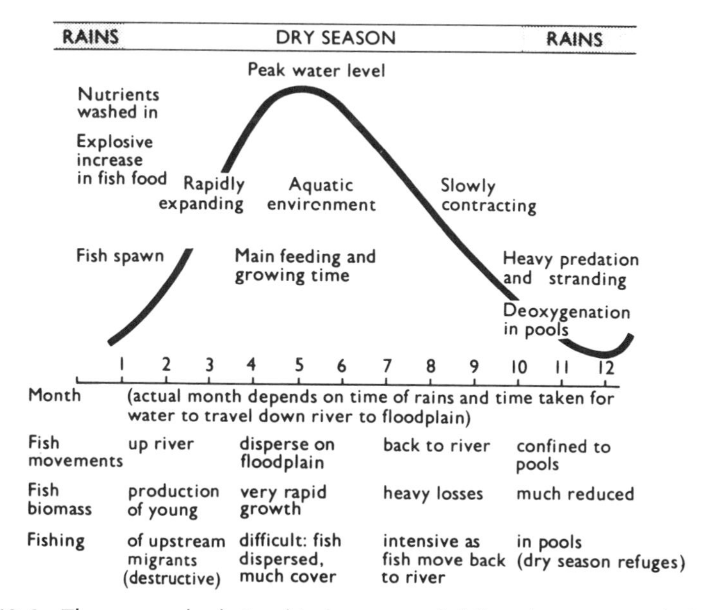

Figure 13.6 The seasonal relationship between rainfall and spawning of teleosts in a tropical river system. (From Lowe-McConnell, 1977, courtesy of *Ecology of Fishes in Tropical Waters*, Edward Arnold Publishers.)

in the Amazon Basin, spawning commences soon after flooding, indicating that gonadal ripening had occurred prior to the earliest heavy rains (Schwassmann, 1978). Wild populations of the platyfish (*Xiphophorus variatus*) breed at the onset of seasonal rains, which result in an improvement in their streams, and reproduction ceases near the start of the dry period. This cycle seems to synchronize gonadal maturation and mating within a given stream system (Borowsky and Diffley, 1981). In coastal rivers of tropical Australia the spangled perch (*Therapon unicolor*) spawns when water levels rise and turbidity increases as a result of annual rains, and gonadal preparation seems to be induced by rising temperatures and increasing day lengths (Beumer, 1979).

If there is a lack of great seasonality of flow in a tropical stream, spawning periods then resemble seasonality in temperate regions. Among six species of characoid fish in Panama (at 10°N), the spawning of two of them is restricted to one or two days of the year, while at least three have extended reproductive periods, with oocytes in several separate stages of development (Kramer, 1978).

The effect of seasonal rains is not confined to the tropics. The speckled dace (*Rhinichthys osculus*) in the Chiricahua Mountains of Arizona lives in a region where rains are seasonal but irregular. Spawning in the spring follows increased stream flow as a result of melting snowfields; there may also be a late-summer spawning, but only if flash floods again raise stream levels (John, 1963).

13.5 AMPHIBIANS

The previous chapter emphasized the importance of temperature in stimulating gonadal growth of amphibians. In temperate regions where there is no dry season, anurans and salamanders are apparently cued by seasonal fluctuations in temperature. At any latitude, however, in regions with marked wet and dry periods, amphibians seem to adapt more to soil moisture and standing water for proper reproductive conditions than to ambient temperature, and this is true whether rainfall is irregular or seasonal. Anuran reproductive seasonality in midlatitudes is suppressed by low winter temperatures and resumes as temperatures rise and snow changes to rain. In arid regions, where standing water is temporary, rains are required before amphibians can spawn. In low latitudes with continuous high humidity, mating may be more or less aseasonal, but oviposition may occur within a restricted period. Seasonality can be quite different for terrestrially ovipositing species.

At any latitude, terrestrial and arboreal amphibians may have cycles that diverge from those of species that spawn in water. For example, most species of the Nearctic genus *Ambystoma* move to water in late winter or spring, but those that lay eggs on land do so in autumn. There is a great range of spawning times of salamanders in warm climates or the tropics.

Although most amphibians in variable environments breed when there is a combination of rising temperatures, warm rains, and increasing day length, little experimental work has been done to separate and identify environmental factors. As pointed out in the previous chapter, day length, the sound produced by falling rain, and atmospheric humidity may function as cues in amphibian reproduction. Additionally, rain, as opposed to falling snow, would raise the temperature of pond water.

In regions with a fairly regular annual cycle of temperature change, there are corresponding cycles of gonadal recrudescence. In lower latitudes, where temperature changes only slightly and precipitation may be sporadic, the gonads must be in a state of readiness for prolonged periods. In these regions gonadal recrudescence may not be cyclic, although spawning usually is.

13.5.1 Temperate Species

Zoologists are concentrated in the midlatitudes, and most field studies are made in such regions. Therefore until recent years our knowledge of amphibian reproductive cycles in the wild has been largely concerned with species of the temperate regions. In such environments gonadal recrudescence often starts in late summer, and mating occurs on rising temperatures in the spring. In toads (*Bufo bufo*), for example, the ovulatory cycle begins a few weeks after egg laying, in late spring or summer, and oocytes reach near-maximal size by November, prior to hibernation. Thus the females are replete with mature eggs in late winter and early spring, when both sexes migrate to the spawning ponds (Jørgensen et al., 1978).

Some amphibians inhabit springs or lakes, in which obviously no land-to-water migration takes place. In these environments there is frequently no annual cycle of temperature change, and gametogenesis, which is heat sensitive, is continuous. In spring-inhabiting frogs of the genus *Telmatobius*, spermatogenesis is continuous throughout the year at constant water temperatures near 0°C, and ovaries also show no seasonal changes (Cei, 1949).

Northeastern United States has a rich and diverse fauna of salamanders, and the marked climatic seasonality must account for the annual regularity of reproduction in these amphibians. The seasonality of reproduction is nevertheless diverse. Almost all of these species have internal fertilization involving the transfer of a spermatophore, with the possibility of sperm storage. As a result of this situation, there can be seasons of mating and also of oviposition, and these two periods may be separated by months.

13.5.2 Tropical Species

Anurans that breed in the water, whether in the tropics or at higher latitudes, are most active during periods of high atmospheric humidity. Because tropical rains may be irregular, pond-breeding frogs and toads spawn sporadically. In an equatorial region in the upper Amazon Valley in Ecuador, Duellman (1978) recognized 11 reproductive modes, based on sites of oviposition, habitat, and type of larval development. Aseasonality is characteristic of those species that lay eggs on land, and most of the terrestrially egg-laying species have direct development, with the larval stage being passed within the egg. In pond-spawning species oviposition increases after heavy rains. Thus although there is anuran reproduction throughout the year, specific groups have their own seasonality.

Pond-spawning anurans in deserts appear when sudden rains create temporary ponds or only puddles, and such precipitation is commonly irregular. Consequently, those anurans adapted to a xeric existence must be prepared to oviposit during periods in which standing water may briefly occur. The spadefoot toads (*Scaphiopus* spp.) of the New World breed in temporary ponds and are well known for their rigid adherence to rainfall as an environmental signal. They move from their subterranean retreats to ponds during spring rains. The rapid growth and early metamorphosis of their tadpoles are adaptations to such a spawning site.

Cyclorana platycephalus, a leptodactylid frog of Australia, burrows into the soil during droughts. Females preserve ripe ova through the nonreproductive season and thus are prepared to oviposit whenever there is adequate rain. Ripe ova may be retained from one year to the next (van Beurden, 1979). In the Durban region of South Africa the microhylid frog *Brevipes adspersus* burrows deeply in the dry season, and its emergence coincides with the appearance of the alate stage of termites on which they feed (Poynton and Pritchard, 1976). The termites do not likely constitute a seasonal cue for the frogs, but it is possible that both amphibian and insect respond to the same cues.

In the subtropical rice frog (*Rana limnocharis*) of Taiwan, there are two rather distinct spawning periods: vernal spawning throughout March and April and a second spawning in July and August. Each time, adults assemble in the shallow water of freshly cultivated fields when early evening temperatures reach 20°C. This movement occurs when the fields are plowed and flooded, which together seem to constitute an immediate attractant for the rice frog: odor from the freshly disturbed soil is suspected to guide the frogs to these fields (Alexander et al., 1979).

In several well-studied species of both live-bearing and egg-laying caecilians, mating may apparently occur over an extended period or throughout the year, but the young appear near the onset of the rainy season (Wake, 1977). In French Guyana *Typhlonectes compressicaudus*, a viviparous caecilian, has an annual reproductive season, with spermatogenesis in June and July, during the rainy season. Young are born in the dry season, while water is still high (Exbrayat and Delsol, 1985).

Geographic and environmental variations in seasonality of plethodontid salamanders suggest the relative importance of ambient temperature and humidity (Fig. 13.7). In eastern United States low temperatures restrict general activity to the warm months. In the western states, where winter precipitation (in low elevations) falls as rain and where summer rain is sparse, activity is seen from autumn to spring, when the soil is moist. In both regions oviposition is mostly in the spring. In Guatemala, where there are continuously mild temperatures and a discrete dry season, activity continues throughout the year. Spermatogenesis, moreover, is uninterrupted by any seasonal quiescence. Oviposition, which is either terrestrial or arboreal, is confined to a brief period in autumn, and brooding continues through the dry period. The eggs hatch near the onset of the annual rains (Houck, 1977). Thus reproductive activity, and activity in general, is initially suppressed by low

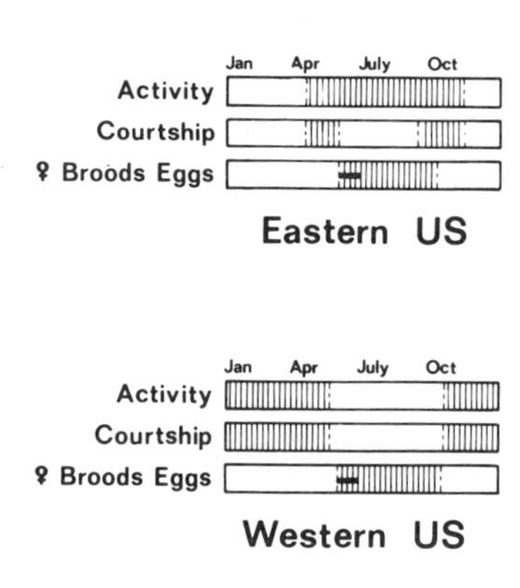

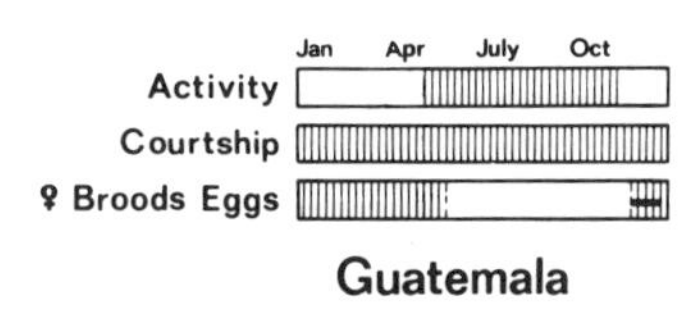

Figure 13.7 General patterns of reproductivity of three geographic groups of plethodontid salamanders. The horizontal bar indicates oviposition. Vertical lines indicate periods of surface activity. (For Guatemalan species, only terrestrial taxa were studied.) (From Houck, 1977, courtesy of Plenum Publishing Corp.)

temperatures. When temperatures are sufficiently warm to permit locomotor activity as well as gametogenesis, humidity appears to define reproductive seasonality in these amphibians.

13.5.3 Biennial Reproduction

In some amphibians, either at high latitudes or high altitudes, females tend to develop mature ova every second or third year. This characteristic is not confined to amphibians having short seasons of activity. In *Typhlonectes compressidcudus*, a Neotropical caecilian, individual females are gravid biennially, with only half the population giving birth to young in any given year (Exbrayat and Delsol, 1985). All caecilians appear to be biennial in their reproduction (Wake, 1970). Regularity of spawning in the toad (*Bufo bufo*) seems to depend upon the individual's ability to accumulate energy stores sufficient to form mature oocytes. In Switzerland (Zurich) this limitation may account for a pattern of biennial spawning in most females (Heusser, 1968). In montane populations of the fire salamander (*Salamandra salamandra quadri-virgata*), breeding is biennial (Joly, 1961). Biennial reproduction is probably more widespread among the amphibia than is generally recognized.

13.6 REPTILES

Most reptiles exhibit reproductive seasonality, whether in the temperate regions, where climatic conditions change conspicuously from month to month, or near the equator, where photoperiod and temperature are rather constant. Although not all species in a given region follow a uniform schedule of steroidogenesis, gametogenesis, mating, and birth or oviposition, most taxa have a rather regular annual pattern that varies only moderately from year to year.

Timing of reproductive cycles is probably more variable for reptiles than for other terrestrial vertebrates. Not only are there distinct patterns of gonadal recrudescence (reflecting different patterns of seasonality of gametogenesis, steroidogenesis, and circulating levels of gonadotropins), but reptilian cycles are also complicated by the occasional occurrence of viviparity and sperm storage. Unfortunately, the hormonal background of these variations has been studied in only a few species (Callard and Ho, 1980).

13.6.1 Temperate Species

Although vernal mating characterizes most species, the timing of gonadal activity varies. Some reptiles undergo spermatogenesis in the spring and mate shortly thereafter, and others experience spermatogenesis in the autumn and store spermatozoa in the epididymides for mating the following spring (Fig. 13.8). In such reptiles oviposition occurs shortly after mating, and gonadal regression follows shortly thereafter.

Two contrasting species in one area will illustrate the extent of divergence of gonadal activity. Both the cobra (*Naja naja*) and the soft-shelled turtle (*Trionyx*

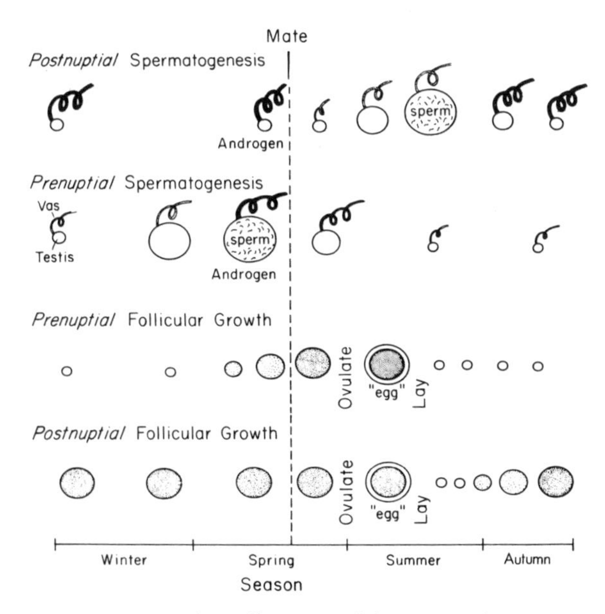

Figure 13.8 Two major types of reptillian gonadal activity relative to season of mating. Examples represent oviparous species mating in the spring, with gametogenesis postnuptial or prenuptial. (From Licht, 1984, courtesy of Longman Group Ltd.)

sinensis) hibernate in the Hong Kong area, and both mate soon after emergence from dormancy, in the spring. Although the ovulatory cycles of the two reptiles are very similar, the calendar of testicular activity differs. Spermatogenesis is prenuptial in the cobra and postnuptial in the turtle. After the production of spermatozoa by the turtle in late summer, the testes regress and the spermatozoa move to the cauda epididymides, where they remain over the winter. In the cobra, steroidogenesis is maximal in late winter, but steroidogenesis is essentially a late-summer or autumnal activity in the turtle (Lofts, 1977). In some vipers, such as the copperhead (*Agkistrodon contortrix*), spermatogenesis is postnuptial, and mating is both in the autumn and the spring (Schuett and Gillingham, 1986). In these vipers, oogenesis is prenuptial.

In the oviparous *Sceloporus undulatus*, testicular recrudescence begins in midsummer, after mating. Spermatid formation follows in the autumn; spermatozoa are stored in the epididymides over the winter, and oogenesis takes place in the spring (McKinney and Marion, 1985). Many midlatitude oviparous iguanid lizards have this pattern, in which the two sexes appear to be following different cues: spermatogenesis is initiated in the late summer, and oogenesis commences the following spring. In those species in which gametogenesis of males and females is not simultaneous, the hypothalamus would appear differentially responsive to environmental stimuli. In the majority of oviparous lizards the young emerge from their eggs at the end of the summer or in autumn, shortly before entrance into hibernation. This season would seem to be the least appropriate and is in marked contrast to the season of birth of viviparous lizards, which are born in the spring.

In some temperate reptiles the initial breeding of young adult females is slightly delayed. Emergence from hibernation may be concurrent in both young adults and those that have reproduced the previous year, but the young females tend to mate a week or more after the older females mate. In the viviparous *Lacerta vivipara*, young adult females (breeding for the first time) mate somewhat after the time at which older adult females mate (Bauwens and Verheyen, 1985).

Live-bearing in reptiles generally alters the pattern of seasonality. In temperate regions at least, viviparity seems to be accompanied by winter activity, in contrast to sympatric egg-laying species. In live-bearing iguanid lizards testicular activity commences in the spring and continues through the summer, but ovarian growth commences in mid- or late summer, at the time of testicular regression (Fig. 13.9). In *Sceloporus grammicus*, sperm are stored in the epididymides and mating occurs in September (Guillette and Casas-Andreu, 1980); in *S. jarrovi*, mating is in November (Goldberg, 1976). Similarly, in the live-bearing *Sceloporus formosus* of southern Mexico, testicular growth occurs in the early spring and regression by late spring, and the development of oocytes takes place anytime from late spring until early December. In *Sceloporus formosus* the time of mating is unknown, but it clearly follows testicular regression and probably involves storage of sperm (Guillette, 1985). These species are active in the winter and bask in sunny weather, and parturition is in the spring. In the elapid genus *Austrelaps* of Australia, however, mating occurs in the spring and young are born in the following summer (Shine, 1987).

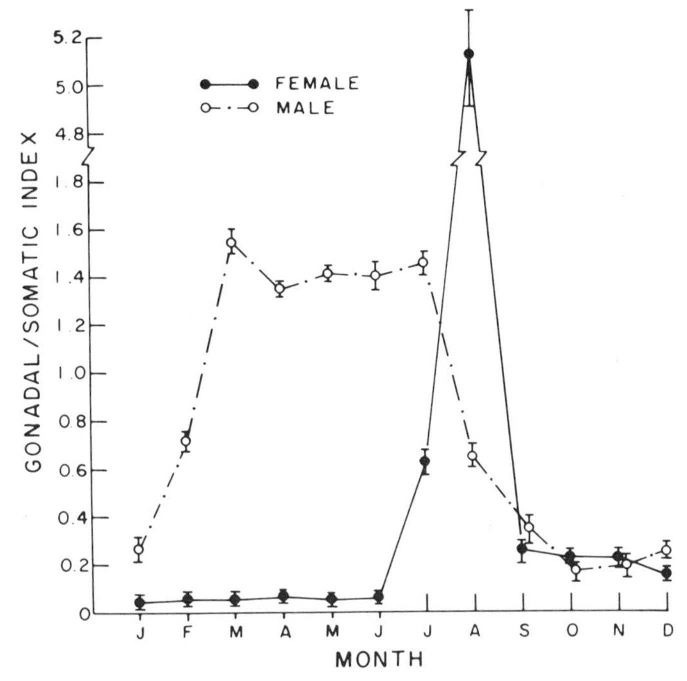

Figure 13.9 Monthly changes in gonadosomatic index of *Sceloporus grammicus microlepidotus*, a live-bearing iguanid lizard, illustrating vernal spermatogenesis and autumnal folliculogenesis. Values are means ± SE. Gonadosomatic index = gonadal mass/body mass × 100. (From Guillette and Casas-Andreu, 1980, courtesy of the *Journal of Herpetology*.)

13.6.2 Tropical Species

The concept of tropical aseasonality is based on few studies, but several tropical reptiles are known to have a discrete hiatus in reproductive activity. Clear-cut evidence for acyclic breeding is known for only a few tropical squamates. In those species in which males may have spermatozoa throughout the year, cyclicity may be apparent in the females (Licht, 1984; Vial and Stewart, 1985). In tropical rain forests reptiles are likely to breed throughout the year, but in grassland species reproductive cycles may be synchronized by seasonal rains (Barbault, 1974). Among several species of snakes in Cambodia, each has a distinct seasonality, but there is no uniformity across specific lines (Saint Girons and Pfeffer, 1971).

Among seven species of anoles (*Anolis* spp.) studied in tropical environments on Caribbean islands, reproductive activity persisted throughout the year, judging from the state of both male and female gonads. Minimal gonadal activity characterized the "winter" months, from October to January, and reproduction appears to depend upon a combination of day length and temperature cycles. In Caribbean anoles ovarian activity is closely tied to rainfall. Folliculogenesis declines with the arrival of the dry season, but rainfall has no effect on testicular activity (Licht and Gorman, 1970).

There are sexually dissimilar gonadal cycles in the viviparous boid snake, *Candoia carinata*, of the southwest Pacific. In the males testicular activity, as measured by seminiferous tubule diameter, is greatest from April to June, after which epididymal spermatozoa are preserved. There appears to be no cyclicity to ovarian activity, however (Wynn and Zug, 1985). The opposite situation obtains in the anguid lizard, *Barisia monticola*, of Central America. The males are continuously spermatogenic, but there is a tendency for reproductive seasonality in females (Vial and Stewart, 1985).

In the absence of marked seasonal changes in weather, there may be prolonged reproductive seasons in tropical squamates, which has been documented for several species. The teiid lizards, *Ameiva quadrilineata* and *A. festiva*, of Costa Rica are reproductively active at all seasons. Females with ripe follicles or oviductal eggs have small fat bodies, whereas reproductively inactive females have large fat bodies (Smith, 1968). Among a group of four species of geckos in Brazil there is a marked seasonal cycle in testis volume, with a reduction in the dry period of the year; but oogenesis and ovulation continue throughout the year. Possibly sperm storage allows this lack of synchrony between the sexes (Vitt, 1986).

In Tanzania the live-bearing skink *Mabuya striata* breeds throughout the year, but in the same area the oviparous agamid lizard *Agama cyanogaster* lays eggs only in the wet season, with young hatching only in the dry season (Barbault, 1975). Similarly, in Zambia, where there are marked seasonal variations in rainfall, the oviparous skink *Mabuya quinquetaeniata* lays eggs at the onset of the rainy period, but *Mabuya striata* breeds throughout the year (Simbotwe, 1980).

13.6.3 Chelonians

Among midlatitude freshwater turtles, egg laying is usually a vernal event, following the formation of sperm the previous year. In at least four families of New World turtles (Emydidae, Chelydridae, Kinosternidae, and Trionychidae), mating takes place in the spring, spermatogenesis is postnuptial, and sperm are stored over the winter in the cauda epididymides (Moll, 1979). Nesting of the highly aquatic snapping turtle (*Chelydra serpentina*) is closely correlated with rising temperatures in shallow water, most especially with accumulated heat units, or degree-days, which are needed for final follicle maturation (Obbard and Brooks, 1987).

In lower latitudes, where temperatures are moderate and annual variation is reduced, reproduction may be correlated with rainfall, either positively or negatively. Some species lay only in the dry season, and others lay only in the rainy period (Moll, 1979). Although testicular recrudescence occurs in late summer in the Caspian terrapin (*Mauremys caspica rivulata*), most mating activity occurs in the winter, during months of maximal precipitation and minimal temperatures, day length, and testicular mass. At this time the epididymides are swollen with spermatozoa (Gasith and Sidis, 1985).

The timing of chelonian reproduction appears to accommodate maximal survival of the eggs rather than to provide for hatching at the most favorable time for survival of the newly hatched young. Many species of freshwater and terrestrial turtles deposit one or more clutches in the summer, and the young overwinter within the egg and emerge the following spring. Delayed emergence is independent of any particular sequence of suppressive seasonal climatic changes in temperature and/or humidity, but it allows hatchlings to emerge when environmental conditions are most favorable for their survival (Gibbons and Nelson, 1978). In many chelonians the eggs are far more vulnerable than the young are, but in some turtles, such as the marine turtles, there is great predation on the young.

Some of the large marine turtles tend to lay eggs during irregularly recurring visits to isolated sandy beaches. This irregularity has been confirmed by the identification of marked individuals over a period of years. The green turtle (*Chelonia mydas*) remains at sea for several years, and females lay several times within weeks, every second or third year (Carr and Carr, 1970). On Ascension Island, green turtles experience difficulty in digging nests in the hard dry sand, which may account for nesting during the season of heaviest precipitation (Mortimer and Carr, 1987). On the Comoro Islands green turtles lay eggs sporadically, without a discrete nesting season. The loggerhead (*Caretta caretta*) appears to be irregular in its spawning periodicity. It also oviposits at intervals of from two to several years and shows little rhythmicity in its return to its nesting beaches (Hughes, 1976). The reproductive periodicity of marine turtles is probably determined partly by the time required to make a complete migration and also to accumulate sufficient fat stores for a clutch of eggs.

Some species of chelonians divide a season's production of eggs among two or several clutches. Because the nest is not subsequently guarded, it is quite vulnerable to predation, and prehatching loss may be high. Multiple nesting is of obvious advantage in reducing loss to predators. The loggerhead turtle (*Caretta caretta*) has been known to lay up to seven clutches a season (Frazer and Richardson, 1986). The leatherback turtle (*Dermochelys coriacea*) lays a mean of four clutches annually, with a maximum of ten (Eckert, 1987).

13.6.4 Biennial Reproduction

In temperate regions, with a reduced period of annual activity and lower mean temperatures, reptiles require prolonged periods to accumulate fat for embryonic growth. This environment introduces another variable in reproductive periodicity: in a number of reptiles in temperate regions, reproduction is biennial, and alternate years are devoted to accumulation of adipose tissue, most of which supports development of a subsequent generation of young. The accumulation of fat for reproduction is most critical for females, and biennial reproduction does not affect males.

In several species of vipers (Viperidae) females breed in alternate years. In northern Idaho the rattlesnake (*Crotalus viridis*) shows a strong tendency to follow a biennial pattern. The mean percentage of reproductive females in any given year varies from 54 to 77, and reproductive activity is positively correlated with fat stores (Diller and Wallace, 1984). Similarly, in northwestern Texas *Crotalus atrox* produces young every other year, and biennial breeding seems to be predicated on the amount of body fat (Tinkle, 1962). The common adder (*Vipera berus*), in the Swiss Alps, reproduces every second or third year, and birth takes place just before its entrance into hibernation (Saint Girons and Kramer, 1973). Species of *Austrelaps*, an Australian elapid snake, appear to breed in two years out of three, perhaps depending on prey availability (Shine, 1987).

In Japan the skink *Eumeces okadae* reproduces biennially, and approximately half of the females lay eggs in any year. In this little lizard, females seldom eat while they are brooding their eggs, allowing little time to restore their fat supplies (Hasegawa, 1984). The anguid lizard, *Barisia monticola*, is a tropical live-bearer and lives at 2000 m in Costa Rica, where the temperature is constant. It produces young every other year (Vial and Stewart, 1985). On the Isle of Portland (Dorset) in the British Isles (50°N), the live-bearing *Anguis fragilis* breeds every other year. The young are born just before the adults enter hibernation, and there is little time for the adults to accumulate sufficient energy for reproduction the following spring (Patterson, 1983).

Biennial egg production may be the rule among some crocodilians. In the American alligator (*Alligator mississipiensis*) nesting is biennial (Joanen and McNease, 1980).

13.6.5 Significance of Reptilian Seasonality

With the preceding live-bearing iguanid lizards the adjustment in seasonality of mating results in birth of young in the spring, when there is an abundance of small arthropods upon which they can feed. In sympatric oviparous species, however, the young emerge from eggs in the late summer or autumn. In the tropics the association of precipitation and reproduction is apparently far more important for egg-laying species than it is for live-bearers. Food may not constitute a limiting factor for young lizards in the tropics (Stamps and Tanaka, 1981), and the lack of seasonality among the viviparous squamates in the tropics suggests that seasonality among egg-laying species is an adaptation to humidity. Oviposition in *Anolis sagrei* is correlated with atmospheric humidity, suggesting that it may not require rainfall per se (Brown and Sexton, 1973).

Reptilian reproduction appears to be scheduled so that eggs are laid at the most propitious time, which may be determined at least partly by soil and/or atmospheric moisture. Oviparous species usually have clearly apparent seasons of oviposition. Live-bearing species appear to be free of this constraint, and tropical reptiles produce young over more prolonged periods.

13.8 BIRDS

Avian seasonality is far more uniform than seasonality for other classes. All birds are oviparous. Sperm storage is brief or absent, and embryonic diapause, when it occurs, is brief and postovipositional. In most birds, especially the smaller, altricial species, incubation may occupy only two weeks, and growth of prefledging young is very rapid. It is thus possible to concentrate an avian reproductive cycle into relatively discrete short periods.

Reproduction, migration, and molt are the three major activities demanding energy above that required for growth and maintenance in birds, and avian life cycles are adjusted so as to minimize concurrent demands of these activities. For example, a bird that regularly makes a long migration may have a relatively restricted reproductive period with a single brood, in contrast to a sedentary species that has a more prolonged nesting season and may produce several successive broods. Several broods do not place an unmanageable demand upon the energy of a species that does not need to store energy for a migration. The timing of the molt is adjusted so as not to conflict with nesting or migration. The schedule for these three events is a relatively rigid budget of the species' accumulation and expenditure of energy.

13.8.1 Temperate Latitude Species

The reproduction of midlatitude birds is almost always vernal. At the highest latitudes the breeding season of most species is restricted to a period when days are long and environmental productivity is intense. As pointed out in Chapter 11, Section 4.3, virtually all birds in temperate and polar regions are stimulated by day length. The critical day length varies: species that are stimulated by a relatively short day length breed earlier than do those that require a longer stimulatory period. The duration of reproduction is arrested by a refractory period in almost all birds.

In birds it is apparent that nesting begins after the most favorable date for nestling survival. This suggests that there is some physiological imbalance between the readiness to lay and the nutritional demands of the young. When provided with supplemental food (fish), magpies (*Pica pica*) laid their eggs somewhat earlier than did a control group that foraged for all its own food (Högstedt, 1981). The provision of supplementary food has been shown to advance the time of nesting for many species (Drent and Daan, 1980).

Many species show year-to-year variations in the time of nesting and egg laying. The great tit (*Parus major*) varies in the date of first laying, and approximately 40% of this variation seems to be genetically determined, so that the date of laying can be readily adjusted to changing environments (van Noordwijk et al., 1981).

This allocation of time and energy to these three events can be seen among some Old World warblers. The chiffchaff (*Phylloscopus collybita*) regularly rears two broods and may winter from southern Britain and Europe. Some other Eurasian species (e.g., *Phylloscopus trochilus* and *P. sibilatrix*) are usually single brooded and make extensive migrations to the southern half of Africa. The chiffchaff has a postnuptial molt, in contrast to the willow warbler (*P. trochilus*), which molts after arriving on the wintering ground in southern Africa (Moreau, 1972). The chiffchaff clearly expends more time and energy in reproduction than the willow warbler does, and in the latter the postnuptial molt is delayed until after the autumn migration.

In the far north, where the reproductive season is extremely abbreviated, there is little elasticity in the nesting calendar. In the white wagtail (*Motacilla alba*) in central Finland, older females, which are nesting for the second year, arrive on the nesting ground approximately a week before the younger females, which are easily distinguished in the field. Dates of egg laying do not differ between the two age groups (Leinonen, 1973).

Many species of birds cross the equator as part of their annual migration and are then exposed to another set of vernal cues, including resurgence of plant growth and a stimulatory photoperiod. Although most species of birds are refractory to these stimuli, a few species may sometimes nest on their "wintering" grounds, where it is summer. The white stork (*Ciconia ciconia*) and the house martin (*Delicon urbica*) have both been known to breed on their winter range in Africa, and the bee eater regularly breeds twice a year, once in each hemisphere (Moreau, 1966).

13.8.2 Low-Latitude Species

In tropical regions where there is pronounced seasonality of precipitation, many birds appear to use rainfall as an ultimate cue. Most species nest at the onset of the rains, and rainfall is frequently followed by a resurgence of vegetation and an increase in insect populations.

Among 117 species of insectivorous birds in the Serengeti, none breeds during the dry season, between July and October (Sinclair, 1978). Insect abundance coincides rather closely with seasonal rainfall, and peaks in nesting follow those of rainfall and insect numbers by about two months (Fig. 13.10). In *Acacia* woodlands in Tanzania, insectivorous birds breed before the rains commence, possibly because the fresh leafy growth of forest trees precedes the seasonal rains and forms the basis for the development of insect populations (Moreau, 1950).

Many species of African and Australian estrildine finches occupy open areas, and they nest well after the onset of the rainy season (Fig. 13.11). They build nests as grasses mature, and they not only construct their nests from fresh grass but also feed their young largely on developing seeds in the milk stage (Skead, 1975). In the soft condition, grass seeds consists of a rich endosperm that lacks cytoplasmic separations of cellulose. Such soft seeds are not only physically easy for the young finches to eat but probably almost totally digestible.

In the Galapagos there is a marked alternation of season, from hot and wet to cool and dry (Colvinaux, 1984). Galapagos finches (Geospininae) nest during the rainy season, closely following local and year-to-year variations in precipitation (Lack, 1947). At this time food is relatively abundant, and the feeding patterns of the different finches tend to converge (Grant, 1986).

The role of rainfall in arid regions is not restricted to low latitudes. In Arizona (32–33°N), Abert's towhee (*Pipilio aberti*) nests regularly in the spring but nests a second time in mid- or late summer in years in which there is a substantial late-summer rain (Marshall, 1963).

There are many more anecdotal references repeating the preceding correlation of nesting and rainfall in arid regions. Such patterns have clear advantages, but the

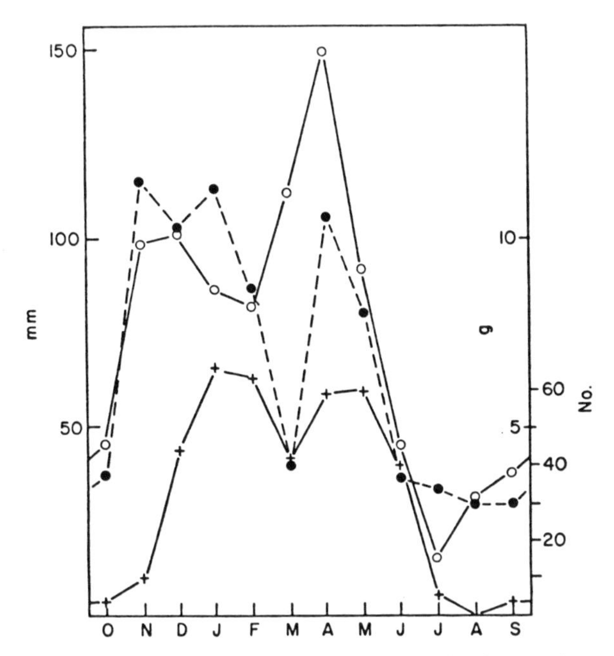

Figure 13.10 The distribution of nesting insectivorous birds (+), showing two peaks, which lag behind those of mean monthly rainfall (solid circles) and mean daily catch of insects (open circles). Breeding records and mean rainfall are for the decade 1964 to 1973. (From Sinclair, 1978, courtesy of *Ibis*, Journal of the British Ornithologists' Union.)

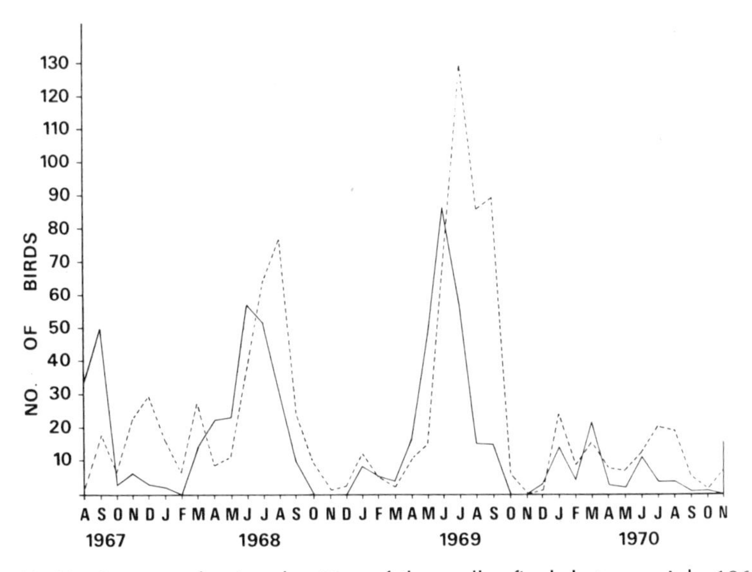

Figure 13.11 Postreproductive densities of the melba finch between July 1967 and June 1970; nesting follows maximal rainfall from November to January. Solid lines indicate number banded per month. Broken lines should number captured per month. (From Skead, 1975.)

underlying ultimate cues are not always apparent. Most experimental studies on reproductive cues or factors have been conducted with temperate species, which are exposed to marked seasonal changes in day length. In the previous chapter it was pointed out that atmospheric humidity may constitute an ultimate cue for birds.

Some species, in contrast, avoid nesting in the wet season of the year. The jungle fowl (*Gallus gallus*) in the Greater and Lesser Sunda Islands has a variable nesting season that is controlled by local rainfall patterns. Probably because of its ground-nesting habit, reproduction is confined to dry seasons (Nishida, 1980).

At the equator the rufous-crowned sparrow (*Zonotrichia capensis*) has a bimodal nesting pattern (Fig. 13.12), apparently cued by the seasonal appearance of some plants, which sprout after seasonal rains (Miller, 1962). Also, in three contiguous but contrasting habitats in Argentina (27°S), this species shows a similarity of gonadal development but nevertheless differs in nesting times. Locally, favorable nesting conditions, especially growth of ground cover, appear to stimulate nesting (King, 1973). Experimentally, this species displays gonadal growth when kept under midlatitude photocycles, indicating that the stimulatory effect of long days has not been lost by its residence in the tropics (Miller, 1959). It is likely, then, that equatorial photocycles are more or less continuously stimulatory, but that nesting is triggered by local nonphotoperiodic signals.

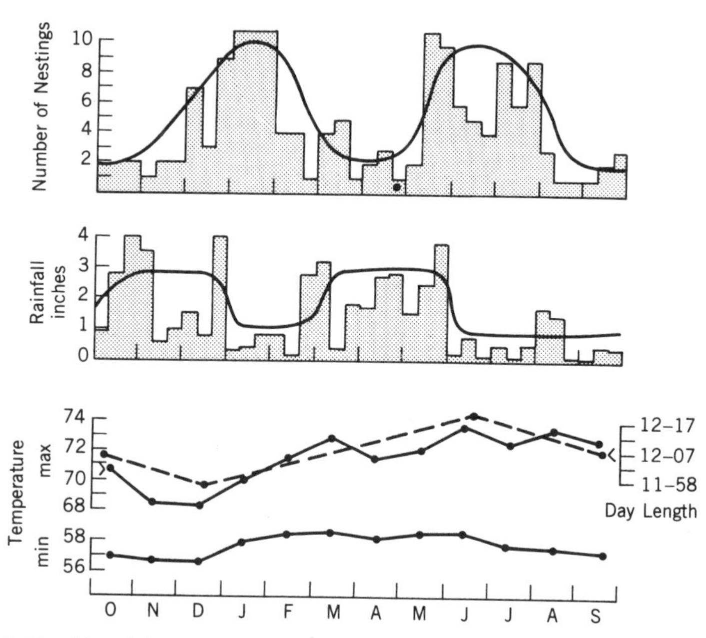

Figure 13.12 Bimodal pattern of nesting in the rufous-crowned sparrow (*Zonotrichia capensis*). Nesting follows the bimodal seasonality of precipitation and seems to relate directly to the seasonal appearance of forbs and grasses. (From Miller, 1962.)

13.8.3 Nonseasonal Species

Several nonmigratory species of birds may have semiannual breeding seasons while occupying relatively uniform environments. In the presence of an adequate food supply and nearly constant schedule of rainfall, temperature, and day length, some species breed twice a year or on a schedule not in phase with the calendar. The rufous-crowned sparrow (*Zonotrichia capensis*) of Colombia has two separate breeding periods annually (Miller, 1959), and swiftlets (*Collocalia* spp.) of southeast Asia may have three annual nesting periods (Medway, 1962; Medway and Wells, 1976). The subtropical wideawake or sooty tern (*Sterna fuscata*) has discrete nesting periods that are not necessarily correlated with environmental cues. Its breeding times are interrupted by rather brief nonreproductive phases. In colonies near the latitudinal limits of its geographic range, nesting is clearly annual; but close to the equator, nesting recurs at intervals of less than 12 months. In the West Indies (20°N) this tern nests regularly in the spring. On Ascension Island (7°S), however, the nesting cycle recurs every 9.6 months. On the Hawaiian Islands (20°N) nesting may be semiannual, with separate colonies being out of phase with one another (Ashmole, 1963).

In the absence of any *Zeitgeber*, reproduction in the sooty tern might be initiated and synchronized by mutual sexual stimulation that affects the population as a unit. In such a situation, if this suggestion is valid, a refractory period would end reproduction more or less at the same time for the entire colony. Here, rhythmicity is presumably achieved by social stimulation.

Feral pigeons (*Columbia livia*), which are semidomestic, may have irregular patterns of nesting. This pigeon breeds in the winter in lofts of vacant buildings near Tampere (Finland). Egg laying and rearing of young continue from December to mid-August, with a refractory period from September (Fig. 13.13). This population

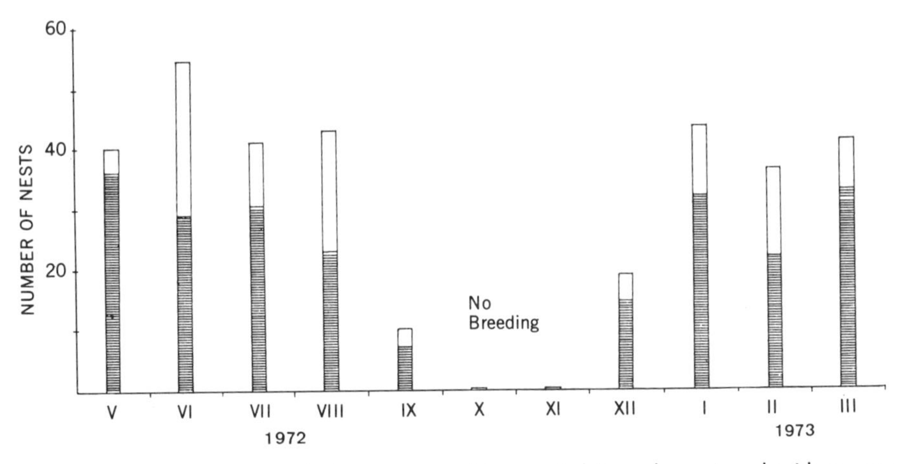

Figure 13.13 Frequency of pigeon nests with eggs (white columns) and with young (ruled columns) from May 1972 to March 1973 in Finland. Months indicated by roman numerals. (From Häkkinen et al., 1973.)

molts from August to October (Hakkinen et al., 1973). It has been suggested that pigeons (Columbidae) do not become refractory (Murton and Westwood, 1977).

13.9 MAMMALS

Discussions of seasonality in various mammals are found in a large number of widely scattered reports. Here patterns of seasonality are reviewed primarily as adaptations to habitat, climate, migration, and hibernation. Additionally, the general reproductive cycles are discussed for some of the more common, widespread, or representative taxa.

In many respects mammalian reproductive seasonality resembles that of birds. In spite of the avian pattern of diurnal activity and the mammalian pattern of nocturnality, both groups contain many photosensitive species. The majority of mid- and high-latitude mammals are markedly seasonal and usually breed in the spring. Some mammals are very sensitive to variations in food supply. Forest-dwelling and meadow-dwelling rodents, for example, may have an abbreviated reproductive season in years of sparse food production, whereas in years of plenty they breed through the summer or sometimes into the following winter. A few species, especially among the artiodactyls, mate in late summer or autumn, under the influence of short days. Mammalian seasonality, although free in most groups from the complication of migration, is frequently modified by hibernation, embryonic diapause (or delayed implantation), postpartum estrus, and sperm storage.

13.9.1 Tropical Environments

A tropical climate suggests a warm, lush, and constant milieu, free from fluctuations. Although true with respect to photoperiod and more or less true for temperature, tropical environments are not equable for rainfall, and many low-latitude mammals are conspicuously seasonal in their gonadal activity and timing of parturition.

In the New World the genus *Didelphis* is represented by three species, one or more of which may be found from southern Canada to cool temperate South America. Breeding seems to commence with the annual increase in day length; reproductive seasonality is conspicuous but highly variable with latitude and altitude. In low latitudes, where the photocycle does not offer pronounced changes in day length, reproduction typically occurs in rainy periods (Hunsaker, 1977).

Tropical crocidurine shrews have been little studied, but most appear to have rather extended reproductive seasons. The subtropical house shrew or musk shrew (*Suncus murinus*) appears to breed for very prolonged periods and may well be aseasonal (Dryden, 1969). It is commensal about human dwellings and may be somewhat removed from natural cues.

Although one would expect that tropical bats, being free from the constraints of hibernation, would be aseasonal in their reproductive cycles, this frequently is not the case. A few examples will serve to illustrate the breadth of variation in breeding of low-latitude bats.

Among tropical bats there is a tendency for reproduction to be associated with seasonal precipitation in regions where rainfall is regularly seasonal. Among a group of insectivorous bats in northeast Brazil, most timed their reproduction so that the young became independent during times of maximal rainfall. The flat-headed bat (*Neoplatymops mattogrossensis*) breeds near the close of the annual wet season, and lactation continues through the dry period (Fig. 13.14). There are a few exceptions to this pattern. *Glossophaga soricina* produces one brood in the dry season and a second during the rainy season (Willig, 1985a). The yellow-winged bat (*Lavia frons*), of East Africa, lives in a region where the rains are sporadic but concentrated. It is closely associated with *Acacia tortilis* and seems to subsist largely on insects that abound near that tree. Development of leaves and/or blossoms follows the rains and is, in turn, followed by an increase of insects, which provide a positive energy flow to the economy of the bats (Fig. 13.15). Parturition occurs at the onset of the more intense rainy period, from April to May (Vaughan and Vaughan, 1986).

In Colombia *Artibeus jamaicensis* and *A. lituratus* appear to reproduce throughout the year; they are fruit eaters and their food is available at all seasons (Tamsitt and Valdivieso, 1963). In Panama, however, *A. jamaicensis* has two distinct periods of parturition, with births in the rainy season (Fleming, 1971). The vampire bats (Desmodontidae) are found only in the New World tropics, where they feed on the blood of larger mammals. Their food is continuously available, and *Desmodus*

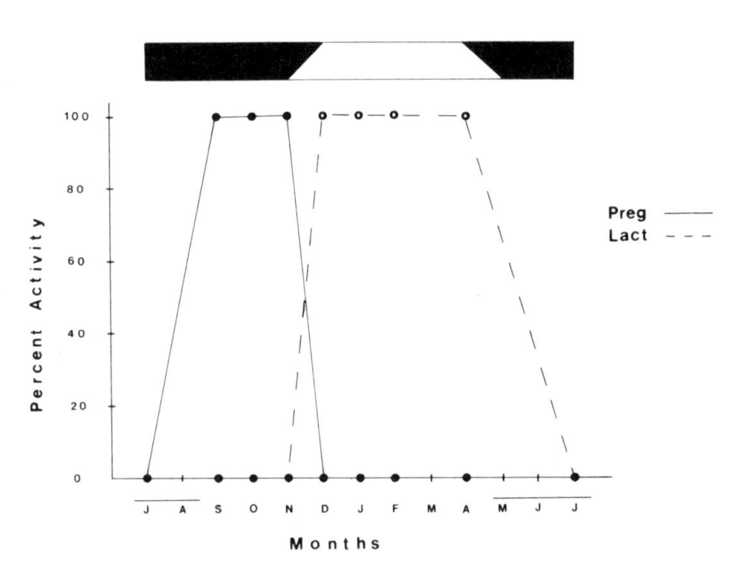

Figure 13.14 Seasonal reproduction and environmental water balance of the Neotropical flat-headed bat (*Neoplatymops mattogrossensis*). The dark parts of the bar at the top indicate periods of water deficit, and the unshaded part is the period of water surplus. The solid line above the letters for the months indicates periods when testicular activity was apparent. (From Willig, 1985b, courtesy of the *Journal of Mammalogy*.)

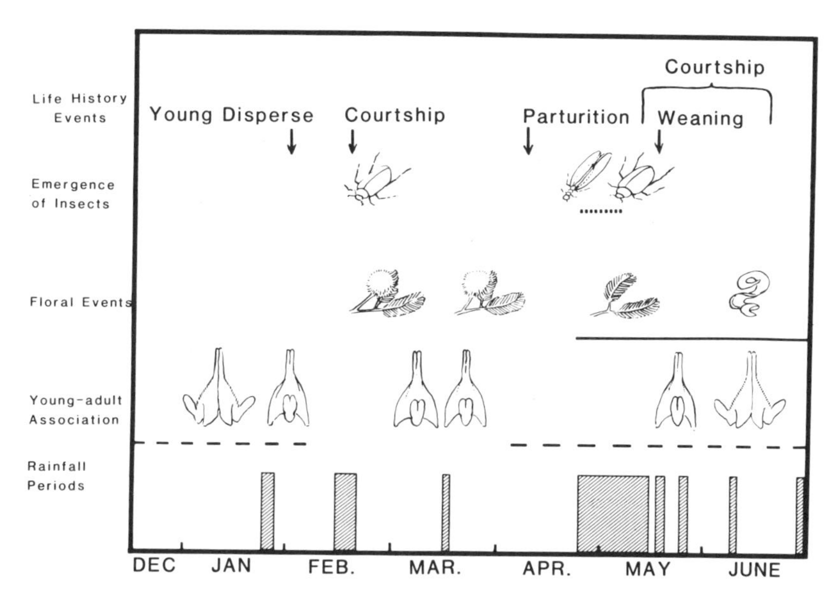

Figure 13.15 Temporal relationships of reproduction in the yellow-winged bat (*Lavia frons*) of Africa, including rainfall, stages of *Acacia tortilis*, and emergence of insects. The dashed line indicates the period of association of the young with the mother; the solid line is the period when all acacias are in leaf; the dotted line is the time of nuptial flights of termites. (From Vaughan and Vaughan, 1986, courtesy of the *Journal of Mammalogy*.)

rotundus, the most abundant species, breeds throughout the year (Wimsatt and Trapido, 1952).

Among a group of 10 species of bats in East Africa, reproductive seasonality is expressed in several patterns. These patterns include a distinct and restricted annual or semiannual reproductive period among some insectivorous species and a more extended, but synchronized, reproduction in fruit bats (Okia, 1987). In Uganda the fruit bat *Epomophorus anurus* has a semiannual breeding pattern, which includes a five- or six-month gestation followed by a postpartum estrus. Births occur in late March and late September. Males also show a reproductive seasonality, with maximal testicular development just before birth of the young. It has been suggested that this semiannual reproductive cycle is synchronized with the maturation of the fruit on which these bats feed and that rainfall is the *Zeitgeber* (Okia, 1974).

In the marked seasonal climate of Nagpur, India, the sheath-tailed bat (*Taphozous longimanus*) breeds continuously. Follicular growth commences in early pregnancy. There is a postpartum estrus and pregnancy during lactation. The cycle seems to be unaffected by exogenous cues (Gopalkrishna, 1955).

The Australian *Myotis adversus* is polyestrous. Populations at 27°S, in southern Queensland, have a bimodal reproductive seasonality, but at 22°S, in central Queensland, the same species produces at least three litters. The litter size is one in both regions (Dwyer, 1970). Similarly, *M. nigricans* in Panama is polyestrous and appears to breed almost continuously (Wilson and Findley, 1970).

In temperate regions both sperm storage and embryonic diapause are logical means of accelerating the onset of embryonic development in the spring, for subsequent weaning of young while insect food still abounds. These patterns persist, however, in some tropical and subtropical bats, which neither hibernate nor migrate. Delayed implantation is seen in tropical species of *Rhinolophus* and *Hipposideros* (Ramakrishna and Rao, 1977; Menzies, 1973). Two species of bent-winged bats (*Miniopterus*) exhibit embryonic diapause both in the temperate climates of France, Japan, and southern Australia and in the tropical climate of southeast Asia. Sperm storage is well known for some tropical species of vesper bats (Vespertilionidae).

The cane rat (*Thyronomys swinderianus*) of tropical Africa is fairly aseasonal in its sexual activity. In Ghana most adult females are gravid at any given time, although there is a concentration of pregnancies at times of greatest rainfall (Asibey, 1974). The mole rat (*Cryptomys hottentotus*) of Zimbabwe is nonreproductive for most of the year, and breeding is probably confined to the period of precipitation (Genelly, 1965).

Many of the distinctive Neotropical caviomorph rodents have been discussed by Weir (1974), but most of her data are from captive colonies. Pearson (1949) studied wild populations of the viscacha (*Lagidium peruanum*) at 5000 ft in Peru. Breeding there occurs in the austral spring, and despite a long gestation, two broods are produced annually.

A discrete seasonality might seem to be complicated by a rather long gestation, from 21 to 22 months for the African elephant. Other large tropical mammals, which may be more or less aseasonal, also have long periods of pregnancy: gestation is some 13 months for tapirs and 15 months for rhinoceroses. A long gestation, however, does not preclude a favored season for mating.

The two living species of elephants are said to exhibit little consistent seasonality. The African species (*Loxodonta africana*) may have annual peaks of pregnancy (Perry, 1953), with seasonality altered by population density (Laws, 1969). Thus seasonality would appear not to be associated with optimal conditions for the young. Separate populations of elephants, however, have fairly distinct periods of conception, such times following seasonal rains. This pattern probably reflects the stimulating effect of forage, which grows after the rains (Fig. 13.16). There is, however, no seasonal fluctuation in gonadotropins, gonadal steroids, or prolactin in the male African elephant or in nonpregnant females (McNeilly et al., 1983).

The two species of hippopotamuses are truly amphibious and do not stray from water, but their mating and parturition are associated with the seasonal cycle of rainfall. The large hippopotamus (*Hippopotamus amphibius*) mates at the close of the rainy season. After a gestation of eight months, the single young is born at the time of maximal precipitation, ground water, and pond depth.

13.9.2 Temperate and Semiarid Environments

In higher latitutdes the photocycle remains uniform from one year to the next, but other environmental features, especially rainfall, may change. Meteorological changes

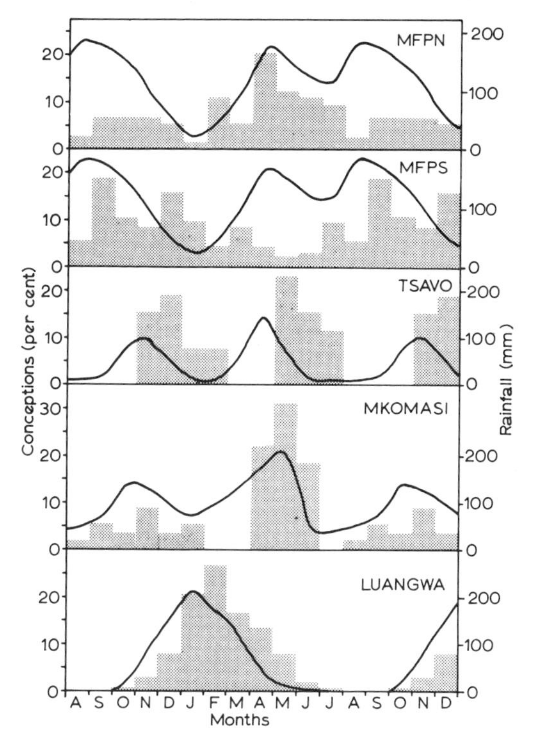

Figure 13.16 Estimated monthly conception in percent (histograms) for five populations of elephants, plotted with mean monthly rainfall (solid line). (From Laws et al., 1975, courtesy of Clarendon Press.)

in themselves may alter reproductive seasonality and additionally alter vegetative characteristics from year to year. Although the onset of breeding is rather consistent from year to year, the end of reproduction may be extended in years of heavy mast crops.

13.9.2.1 Autumnal and Winter Breeding. Numerous examples of small mice (*Apodemus* and *Peromyscus*) and voles (*Clethrionomys*) breed in late fall and winter. Such events demonstrate a partial independence from the annual photocycle. When autumnal breeding occurs, it involves most of the adult population and is frequently correlated with an unusually heavy crop of such favored foods as pine seeds, acorns, and beechnuts (Jameson, 1953; Smyth, 1966). The bank vole (*Clethrionomys glareolus*) breeds through the spring and summer in the British Isles, and sexually active animals can be found in autumn or even into winter in years when mast is abundant (Coutts and Rowlands, 1969). In a nearly pure stand of beech (*Fagus sylvaticus*) in Denmark, winter breeding of both the bank vole and the yellow-necked mouse (*Apodemus flavicollis*) followed the production of a very heavy crop of beechnuts (Jensen, 1982). In southern Sweden the bank vole (*C. glareolus*), the wood mouse (*Apodemus sylvaticus*), and the yellow-necked mouse all breed in the winters in which there is a heavy crop of beechnuts (Eriksson, 1984). Similar patterns have been reported for many mice and voles.

In temperate regions where rainfall is both sporadic and sparse, breeding of mammals may closely follow precipitation, which is similar to reproductive patterns in the tropics. Generally, reproduction in many desert rodents is maximal during periods of rainfall, at which time there is almost always a flush of desert annuals. Seasonal fecundity tends to be about as regular and predictable as the precipitation in any given region (Breed, 1981; Smith and Jørgensen, 1975). Anecdotal accounts are numerous, and a few will illustrate this relationship.

In southwestern United States the round-tailed ground squirrel (*Spermophilus tereticaudus*) commences to breed after the winter rains have initiated the growth of annual plants. The rains vary from year to year, and for each increase of 13 mm over the mean, the breeding season is advanced about nine days (Reynolds and Turkowski, 1972).

Reproductive seasonality of hares (*Lepus* spp.) is rather variable and subject to temporary changes in weather and habitat. In California the black-tailed jackrabbit (*Lepus californicus*) commences reproduction with the first appearance of green vegetation, and the duration of breeding varies with the growing season of forbs. The season of fecundity, moreover, is prolonged on irrigated land and rather short on dry cropland (Lechleitner, 1959). Similar patterns have been noted for other species of leporids, and in desert-dwelling species breeding closely follows periods of substantial rainfall. In Arizona, for example, both *Lepus californicus* and *L. alleni* have semiannual breeding seasons that occur during the two annual rainy periods of the deserts (Vorhies and Taylor, 1933).

The breeding season of the European rabbit (*Oryctolagus cuniculus*), introduced into Australia, closely parallels the pattern of sporadic rainfall. After rain, following an 18-month drought, pregnancies occurred within two weeks, before the growth of fresh vegetation (Poole, 1960). This suggests that perhaps humidity was a proximate cue, as is the case, experimentally, for the zebra finch (see Chapter 12 Section 12.5.1).

Pocket gophers (Geomyidae) are abundant over much of western North America, especially in dry, sandy areas. They feed mostly on roots of herbaceous plants and are frequently pests on cultivated land. Not surprisingly, they have been studied by many zoologists, and their reproductive cyclicity varies with the amount of water available for plant growth. The breeding season of the Botta pocket gopher (*Thomomys bottae*) is greatly prolonged on irrigated alfalfa fields in central California. In spring (March to May) pregnant females are equally numerous in both irrigated and non-irrigated alfalfa fields. By early summer (June), breeding ceases on nonirrigated fields but continues where irrigated alfalfa is green and lush (Miller, 1946).

In the deserts of southwestern United States, germination and growth of winter annuals depend on adequate autumnal precipitation. In years with sparse rainfall, the subsequent development of annual forbs and grasses is poor. For five years the reproduction of desert rodents in southern Nevada paralleled the germination and growth of annual plants. With ample fall germination and sufficient subsequent rainfall, the annuals retained green leaves until February and March, by which time the rodents had begun to breed. Green leaves constitute an important vernal food for some species of heteromyid rodents (*Perognathus* and *Dipodomys*). In years when autumnal and winter rains are absent or short lived, these desert rodents fail to breed, and populations decline (Beatley, 1969). In the Mohave Desert heteromyid rodents breed only in years of ample winter rain (Beatley, 1976). Also, in the Sonoran Desert, where rainfall is notoriously irregular, reproduction of the Merriam kangaroo rat (*Dipodomys merriami*) closely follows the rains. Within-year variations (in 1971) in pregnancies in two areas of contrasting precipitation were striking and highly significant (van de Graaf and Balda, 1973). In the hopping mouse (*Notomys alexis*), a murid rodent of Australian deserts, breeding is facultative, apparently cued by irregular precipitation (Breed, 1975, 1981).

There is evidence that water per se may promote fecundity in desert rodents. Two species of gerbils (*Gerbillurus paeba* and *Rhabdomys pumilio*) of the Namib Desert (at 25°S) breed during periods of precipitation. When rains fail, reproduction is reduced. Artificial availability of drinking water was followed both by an increase in fecundity during the period of rains and by an extension of the reproductive period into the dry season, in contrast to the reproductive performance of these species in a control area (Christian, 1980). Sexual activity in the mole rat (*Heterocephalus glaber*), a fossorial rodent of the arid regions of southern Africa, is restricted to the season of rainfall. This rodent stores food, and rain provides for greater growth of plants that can be gathered. Increased body fat and fecundity follow a period of heavy rain, and breeding may not occur when rains fail (Jarvis, 1969).

There is distinct but different seasonality in the reproduction of two hyraxes in Kenya. In *Heterohyrax brucei* the young are born shortly prior to the period of heavy rains, at which time there is still some residual foliage on woody plants. This species is a browser and finds adequate food in the branches of trees and shrubs, even during the dry season. Lactation continues throughout the rainy period, when all vegetative growth resumes. In contrast, the young of the more terrestrial *Procavia habessinica* are born after the long rainy season, when there is a fresh growth of forbs and grasses upon which the lactating females subsist. Thus reproduction in both hyraxes is tied to the seasonal precipitation but clearly in different ways (Sale, 1969).

In Rwenzori National Park (formerly Queen Elizabeth National Park), which lies on the equator, there are two distinct rainy seasons, and native rodents breed only after the onset of the rains. It has been suggested that reproduction in the mice *Lemniscomys striatus* and *Praomys (Mastomys) natalensis* is stimulated by an increase of protein intake resulting from the inclusion of termites in their diet during the rains. At this time *P. natalensis* also feeds heavily on the fresh fruits of the shrub *Erythrococcus bongensis*. Both rodents experience a sharp rise in body fat after the onset of the rains, and thus the rains and the protein-rich food seem to relate to their seasonal pattern of reproduction (Field, 1975).

Some ungulates in Africa are both gregarious and nomadic. They have long been a food source for aborigines, and consequently their movements and reproductive patterns are well known. In the seasonally arid environments in which they occur,

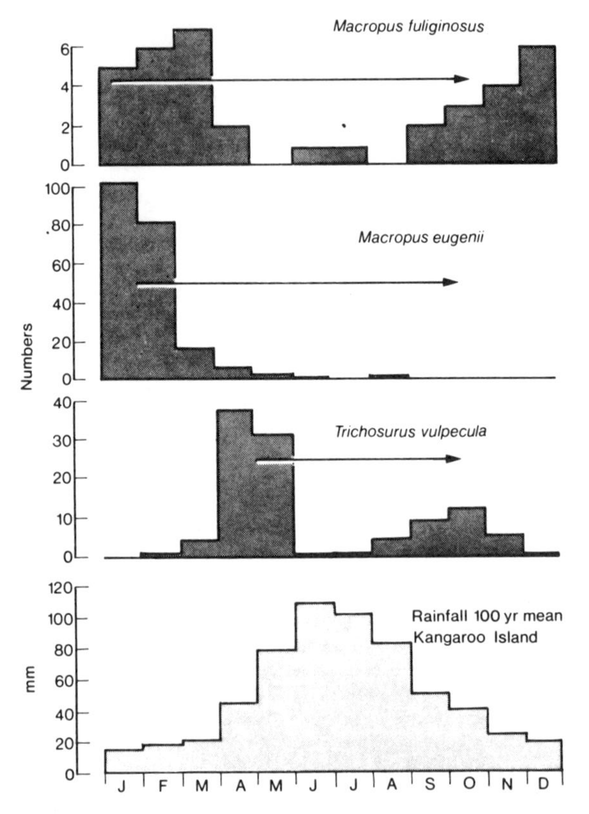

Figure 13.17 Monthly occurrence (numbers) of births of macropod marsupials and the mean monthly rainfall (mm) on Kangaroo Island, Australia. The horizontal line with the arrow indicates the duration of pouch life of the joey. (From Tyndale-Biscoe, 1984; data from various sources.)

parturition is usually scheduled so that the young start life at a time when fresh forage is growing, which can be either annually or semiannually. The dik-dik (*Rhynchotragus kirki thomasi*) conceives at both the beginning and the end of the annual rainy season, and gestation lasts about six months. There is a postpartum conception, so that most females produce two young per year, and the young tend to be born when the grass is tall and fresh (Kellas, 1954). Similarly, the impala (*Aepyceros melampus*) has two breeding periods annually, coinciding with the semiannual rainy periods (Anderson, 1975). The springbok (*Atidorcas marsupialis*), in contrast, lives in an arid region of sporadic rainfall in southwest Africa and migrates in search of fresh herbaceous growth that develops after the nonseasonal rains. Reproduction, being dependent upon an extensive flush of plant growth, is also sporadic, although males are fecund throughout the year. Breeding can therefore occur in any month under favorable environmental conditions.

Many of the Australian marsupials dwell in arid regions where the rainfall is either seasonal and restricted or sporadic. Most Australian marsupials are seasonal in reproduction, responding to seasonal fluctuations in rainfall. In some species of macropod marsupials (such as kangaroos and wallabies), breeding commences with the emergence of young from the pouch, which occurs during the period of annual precipitation (Fig. 13.17). Some dasyurids breed in late winter, apparently as a result of photostimulation. *Isoodon obesulus*, a peramelid, exhibits a precise seasonality, suggesting photostimulation (Fig. 13.18).

13.9.3 Migratory Marine Species

Among migratory seals, sea lions, and whales, parturition must be synchronized with the migratory schedule and the requirements of the newborn young. In some high arctic species these events must fit into a narrow growing season, and the environment does not allow much flexibility. Seasonal diapause is almost universal among these mammals, and there is either a postpartum or a postlactational estrus (see Chapter 7, Section 7.2.5).

In the earless, or true, seals (Phocidae), lactation is rather brief. In most species mating occurs at the end of lactation, at the time the females abandon their young. In the eared seals or sealions (Otariidae), mating is postpartum, and lactation, even in polar regions, is more prolonged than in the phocids. Thus in both families mating must occur within the brief time of their terrestriality, at the end of their spring migration and just prior to their return to lower latitudes.

The walrus (*Odobenus rosmarus*) makes only local movements in the Arctic. Mating follows the birth of the young by several weeks, and gestation is about 11 months. Thus the single young is born at the onset of the brief period of arctic productivity; lactation is prolonged.

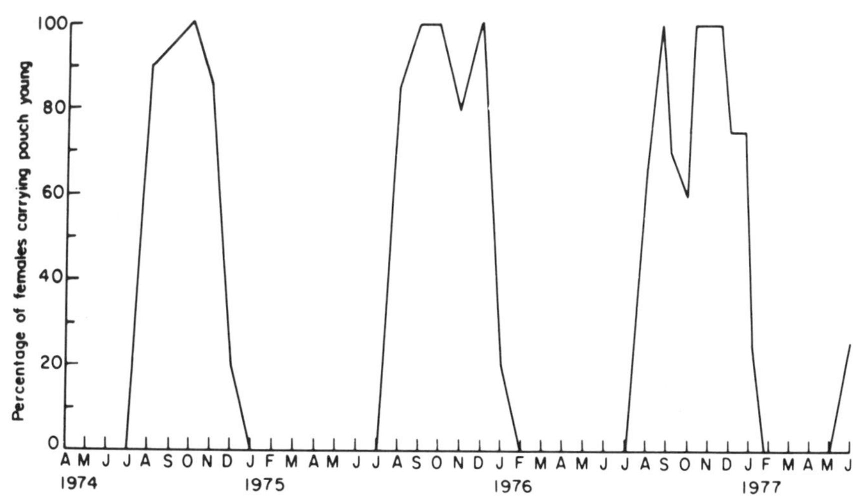

Figure 13.18 Percentage of females of the short-nosed bandicoot (*Isoodon obesulus*) with pouch young, illustrating its relatively clear-cut reproductive seasonality. (From Stoddart and Braithwaite, 1979, courtesy of the British Ecological Society.)

Births in a colony of the northern fur seal (*Callorhinus ursinus*) occur within a 36-hour period. The timing is sufficiently late that low temperatures do not endanger the newborn pups and early enough to allow them to develop to the point of migrating some four months later (Peterson, 1968).

The finback whales (Balaenopteridae) perform annual migrations to the high latitudes, where they accumulate thick layers of subcutaneous fat, and return to low latitudes as the day length shortens and the productivity of the polar seas declines. Birth of young (calving) takes place in shallow and relatively warm waters in the winter. Because of the alternation of the seasons globally, when the population of the Northern Hemisphere is calving, that of the Southern Hemisphere is in the high latitudes engorging on plankton. Seasonality is scheduled so that parturition occurs under the most auspicious environment for birth of the young, and lactation takes place at a time when energy-rich food is most abundant.

13.9.4 Hibernating Species

Insectivorous bats (Microchiroptera) are numerous and diverse in the tropics, and a few species occur at the higher latitudes, even dwelling in the subarctic. In the mid- and high latitudes hibernation is the rule and constitutes an important factor in the annual reproductive period of many bats. Additionally, many species are also migratory, and some species both migrate from their summer range and hibernate in their winter range.

In many Holarctic bats hibernation divides the reproductive season. Mating occurs in the autumn, after which both sexes become torpid for several months. Hibernation is broken by periodic arousals throughout the winter, and active individuals may copulate also at those times. Such matings, however, are not seen in all species, for in some bats the sexes separate at the time of hibernation, and others have separate migratory patterns. In most bats of the temperate regions spermatozoa are stored in the oviducts, but in several kinds fertilization immediately follows mating, and there is either an embryonic diapause or a very slow rate of embryonic growth. In species that store spermatozoa, ovulation follows emergence from hibernation. In taxa that ovulate in the autumn, embryonic growth accelerates after hibernation.

Among the various groups of hibernating mammals, the timing of mating and parturition is closely bound to the period of dormancy. Hibernation is found in many species of ground-dwelling squirrels, including the marmots (*Marmota*), the prairie dogs (*Cynomys*), most of the ground squirrels (*Spermophilus* and *Ammospermophilus*), and most of the chipmunks (*Tamias*). One of the most thoroughly studied is the golden-mantled ground squirrel (*Spermophilus lateralis*) of western North America. In wild populations vitellogenesis commences in January, in the middle of the period of hibernation, and adult females mate immediately after emergence, or at least as soon as they appear aboveground (McKeever, 1966). There is only a single brood, for the young must grow rapidly and accumulate an adequate supply of body fat before entrance into winter dormancy. Similar patterns are known for other species of ground squirrels (Asdell, 1964). Reproductive cues are not apparent for these hibernators, and gonadal recrudescence may be bound, physiologically, to the cycle of hibernation. Among dormice (Gliridae), pregnancy follows soon after emergence from hibernation, and usually there is only a single brood.

Several hibernating rodents are known to produce two or more broods a season, but they are the exception. Jumping mice (Zapodidae) may have two or even three litters annually. The gestation is only about 17 to 18 days, and there is a postpartum estrus. The eastern chipmunk (*Tamias striatus*), a hibernator, has at least two litters per year (Allen, 1938).

Hibernating bears, such as the North American black bear (*Ursus americanus*), mate in late summer, and the blastocyst experiences a diapause for several weeks. The young are born in midwinter, long before actual emergence from hibernation.

Reproduction of hibernating mammals seems to be closely tied to the time of arousal from hibernation, for pregnant females are found shortly after emergence, and emergence is partly influenced by ambient temperature. Richardson's ground squirrel (*Spermophilus richardsonii*) breeds earlier in the year at lower latitudes, and in Saskatchewan (49°N) this squirrel breeds approximately 10 days earlier in years with conspicuously warmer-than-usual days in late winter (Michener, 1973). It has also been suggested that the lemming (*Lemmus trimucronatus*) reproduces as minimum temperatures increase (Mullen, 1968).

13.9.5 Primates

For a long time there was a prevailing notion that primates were characterized by year-long sexual activity, which seemed to provide a conceptual basis for their social cohesiveness (Lancaster and Lee, 1965). However, many accounts document alternating periods of mating and sexual inactivity in free-living populations of monkeys, baboons, and other primates. It is now widely accepted that wild populations of many species of primates are, indeed, quite seasonal in sexual activity.

Because of the comparatively tight social structure of many primate communities, some techniques that might be suitable for studying some kinds of mammals, such as wild mice, cannot be employed for primates. Most primates, however, are diurnal, and inasmuch as infants are frequently carried by a parent, times of birth can be estimated from visual observations.

As social activity becomes more complex, it assumes some control over sexual activity, and the latter seems to be less under environmental control. Many early studies of primate reproduction were made on captive colonies and sometimes under controlled conditions, so that environmental cues were not apparent.

In some of the rhesus monkeys or macaques (*Macaca* spp.), there is apparent seasonality of both copulation and birth. In the Japanese macaque (*M. fuscata*) kept under a 12L:12D schedule with controlled temperature, menstrual cycles recur four to five times, commencing in December (Aso et al., 1977), but there is no clue as to a possible *Zeitgeber*. In a free-ranging population of the same species in Kyushu, mating occurs from October to April and births from April to October (Mizuhara *fide* Lancaster and Lee, 1965). When kept in large outdoor enclosures exposed to natural cues, a population of *Macaca mulatta* exhibited a clear-cut seasonality in mating. Copulation was most frequent in October and November, at which time serum testosterone reached maximal levels (Gordon et al., 1976). In other species of macaques, the schedule is slightly different, but there are generally discrete periods of copulation and parturition. Copulation usually occurs under a short or declining day length.

In contrast to macaques, baboons tend to be less seasonal in periodicity of mating and births. Some species have peaks, but there usually seems to be an extended period during which births occur.

The common langur (*Presbytis entellus*) in northern India does not show an obvious seasonality, and females in estrus can be found in any month of the year. Nevertheless, most births are in April and May. This may be an adaptation to food for the young, for fresh vegetation, on which the young feed, follows in a few weeks, when monsoons begin in June and July (Jay, 1965). On the other hand, the vervet monkey (*Cercopithecus aethiops*) of tropical western Cameroon, parturition in one population was synchronized with seasonal rains, while another group exhibited seasonality with the reverse pattern. Seasonality in this example is attributed to social, not environmental, factors (Gartlan, 1968). The lesser mouse lemur (*Microcebus murinus*) breeds seasonally both in the laboratory and in nature: mating is in the spring, and females are anestrous from September to February. The length of the estrous cycle, however, is increased by social contacts (olfactory, visual, and tactile) (Perret, 1986).

The ring-tailed lemur (*Lemur catta*) breeds in the winter in its native Madagascar, and its seasonality is under photoperiodic control (van Horn, 1975).

Seasonal patterns in human conception, but probably not in coitus, have been reviewed by Cowgill (1969) and MacFarlane (1970). Although peaks in conception are not conspicuous, they do occur and reflect social rather than environmental factors. In some northern European societies, for example, conception rises at about the time of Christmas, perhaps as a consequence of the celebration of Teutonic fertility rites. In Utah, a region in which the prevailing religion proscribes the recreational use of alcoholic beverages, there is a conspicuous decline in conception rate from December to February (MacFarlane, 1970). Another socially induced pattern has been noted among aborigines in western Abelam, New Guinea. In this equable climate, births reach a peak between October and March. The cultivation of yams is a major activity, and the social status of a man is elevated by his production of yams. Also, sexual activity is supposed to inhibit a man's capacity to grow yams, and there is therefore a strong incentive toward sexual abstinence during the yam-growing season, providing the basis for increased copulation and conception after the yams are harvested (Scaglion, 1978).

Humans may not be entirely free from environmental influences. In upper socioeconomic strata of the eastern United States, seasonal rates of conception are independent of seasonal incidence of coitus (Udry and Morris, 1967), suggesting an exogenous influence on ovulation rate. The best case for exogenous influence on human conception exists for Eskimo villages in Labrador. Data from 100 or more years in each of two communities reveal a peak of conceptions in June and a peak of births in March (Fig. 13.19). The amplititude of the changes in monthly

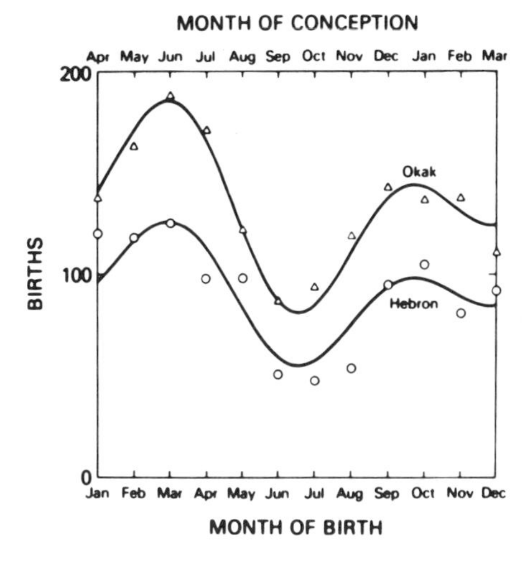

Figure 13.19 Analysis of dates of conceptions and births from two Eskimo communities in Labrador. Data from Okak (1778–1918) and Hebron (1831–1940) are fitted by the summation of two sinusoidal functions with a 12- and 6-month periodicity. (From Ehrenkranz, 1983, courtesy of The American Fertility Society.)

means in birth dates is 75% and 84% from annual means. The factor inducing this seasonality is unidentified, but regular changes in food, photoperiod, and temperature are possibilities (Ehrenkranz, 1983).

13.9.6 Temperate and Cold Environments

At the highest latitudes breeding seasons are short, and a high variance characterizes the modifying factors of the environmental variables (Fig. 13.20). The annual

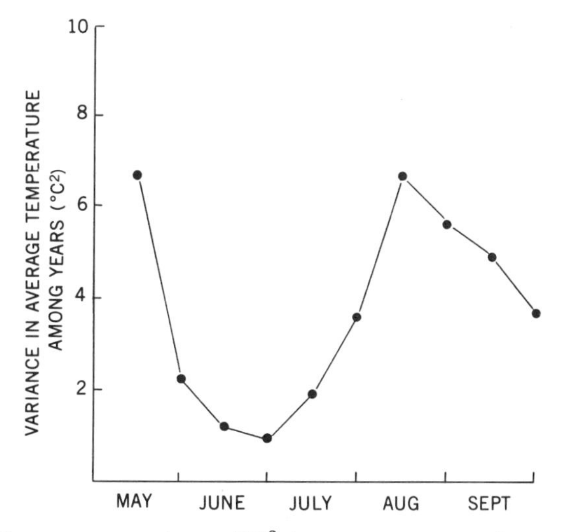

Figure 13.20 Year-to-year variance (°C)^2 in temperatures at Barrow, Alaska. (From Myers and Pitelka, 1979, courtesy of the Regents of the University of Colorado.)

photocycle, the changes in temperature, and the resurgence in plant growth are all of major magnitude and rather closely synchronized. Consequently, mammalian seasonality has little room for flexibility. In the Canadian arctic (Northwest Territories) the redback vole (*Clethrionomys rutulis*) breeds at the time of snowmelt, which varies by about 11 days from one year to the next. In years when the appearance of the first litter is delayed, sexual maturity of the young females in this litter is also delayed. If the initial litter is born earlier in the season, then those females may themselves reproduce in that summer (Martell and Fuller, 1979).

In soricid shrews, which comprise the largest group of insectivores, breeding is usually vernal, and some species have postpartum estrus. In late winter and early spring, with the approach of maximal gonadal activity, there is a conspicuous increase in weight. There is also an annual turnover in population, so that no individual experiences more than a single breeding season. Thus the breeding season is delimited by sexual maturity in the spring and death in the summer.

Artiodactyles that dwell in seasonally cold regions mate so that the developing young make maximal use of the limited season of mild weather and abundant forage. Thus most species mate in the autumn, with parturition in late winter or spring, depending upon the species and the latitude. In the fallow deer (*Dama dama*), for example, by late winter and early spring the testes contain no spermatozoa, but sudden testicular growth occurs in midsummer, after the summer solstice, with concurrent spermatogenesis and growth of a fresh set of antlers (Fig. 13.21). Similar

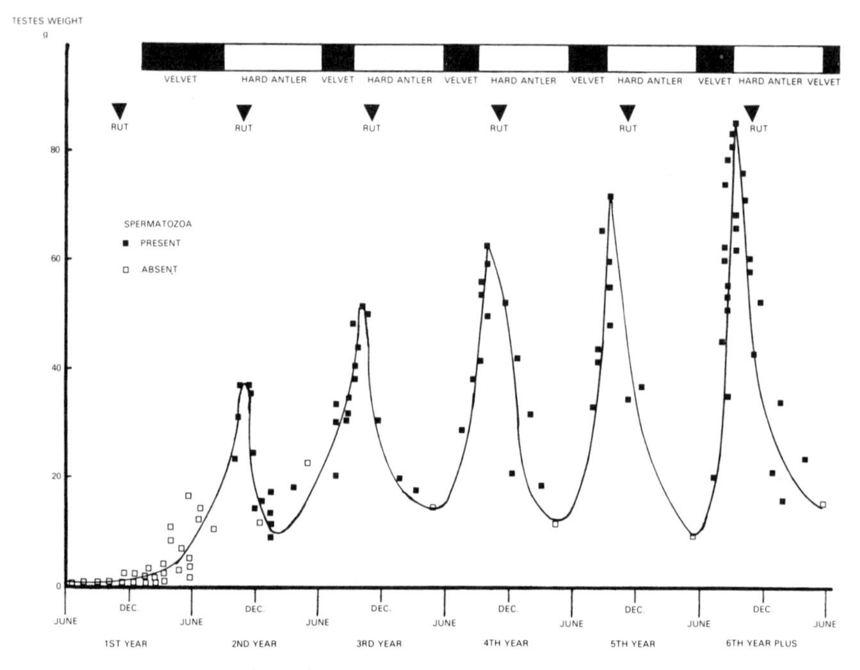

Figure 13.21 The annual testicular cycle of the fallow deer (*Dama dama*). Maximal testis size occurs in the autumn, at the time of the rut, and declines to a minimum in the spring, at which time the antlers are lost. (From Chapman and Chapman, 1975.)

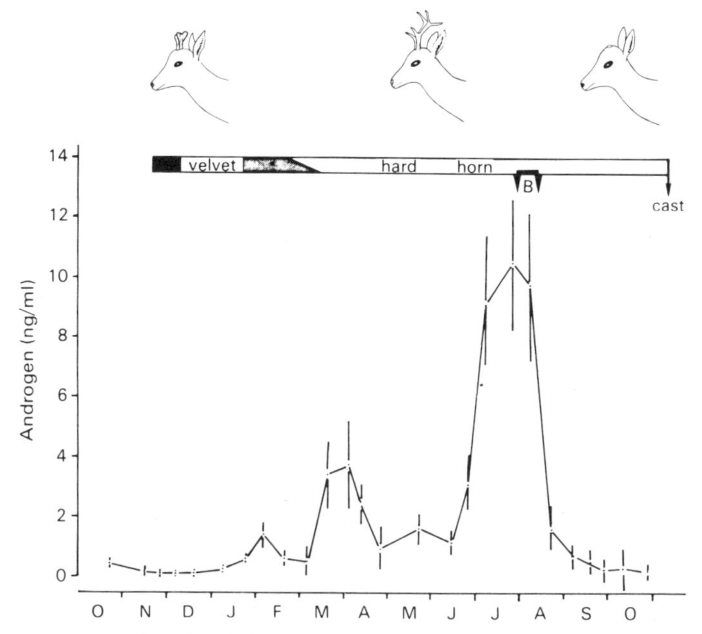

Figure 13.22 Annual cycle of plasma androgen in the roe deer (*Capreolus capreolus*), with a peak at the time of the rut (B). (From Sempéré and Boissin, 1981, courtesy of the *Journal of Reproduction and Fertility*.)

patterns of seasonality occur in the most boreal artiodactyls, the caribou (*Rangifer tarandus*), the moose (*Alces alces*), and the muskox (*Ovibos moschatus*). The autumnal rut and the annual development of antlers of deer are under the seasonal output of androgens (Fig. 13.22). There is direct development in all deer but the roe deer, with no embryonic diapause.

13.10 ADAPTIVE SIGNIFICANCE OF REPRODUCTIVE SEASONALITY

Reproductive seasonality is the association of (1) the series of physiological events from gonadal growth, steroidogenesis, gametogenesis, mating, and finally the appearance of the next generation with (2) a particular season. Although some species are now known to have strong endogenous elements to their annual reproductive cycles, environmental factors are generally responsible for synchronizing the physiological changes with annual climatic cycles.

Although the annual photocycle is regular from one year to the next, the onset of nonphotoperiodic events varies slightly. In many vertebrates this results in year-to-year differences in the timing and duration of reproduction. Birds, which are *stimulated* by both the increase in day length and the occurrence of long days, are closely bound to the annual photocycle. Mammals, on the other hand, in which the hypothalamic-pituitary-gonadal axis is *repressed* by increasing darkness, or short days, and in which the gonads eventually become scotorefractory (and not photorefractory), sometimes respond to nonphotic factors of the environment. Thus seasonality is more predictable in birds than in mammals.

There are annual variations in the length of "summer" as measured in terms of heating degree days. Year-to-year variations in precipitation occur, and such variations result in both qualitative and quantitative changes in plant growth. These nonphotic factors may alter the cyclicity of mammalian reproduction. Also, year-to-year differences in rainfall cause minor irregularities in stream flow and the formation and duration of temporary ponds, and such factors can be critical to the timing of spawning of amphibians and fish. Fluctuations in vernal warming can alter reproductive events in some polar vertebrates, especially birds.

Climatic variations that we observe in a lifetime are relatively slight. Within the brief period of human history, however there have been fluctuations of greater magnitude and duration. The Little Ice Age (1600–1850) was characterized by significantly lowered temperatures and shortened growing seasons for both North America and Europe (Ladurie, 1971). Prior to that time, however, and up to about 1000 A.D., warmer and overall milder conditions permitted agricultural communities to survive in Greenland. Such climatic changes undoubtedly affected reproductive phenology of small mammals in the mid- and high latitudes. In addition, the survival of young lizards, some turtles, and voles, mice, and lemmings strongly suggests that reproductive seasonality is not characterized by a delicate, fine-tuned adjustment so that young are born precisely at the most optimal time of the year. Rather, reproductive seasonality is somewhat flexible, with the result that modern species have the capacity to make rapid adjustments in times of mating, oviposition, birth, and other such events within a given genotypic frame.

The result of these year-to-year variations is variability in the onset and termination of the reproductive calendars of many vertebrates. Whereas environmental factors determine seasons of gonadal activity for many forms, there are very few in which hatching of eggs or birth of young are set by exogenous factors. Some flexibility occurs in developmental rates of ectotherms, but in most groups the onset and duration of mating of adults accounts for variation in the start and duration of natality. These variations are usually so minor that the young are, nevertheless, produced when environmental conditions are usually optimal for their survival.

Whereas the *onset* of reproduction is subject to some year-to-year variation due to meteorological variations, there are frequently greater variations in *duration* of reproductive activity brought about by slight deviations in rainfall, temperature, and food supply. Consequently young born at the end of summer or in autumn do not face the same environment as young born in the spring do.

Ornithologists developed the concept that reproductive seasonality is adjusted so that young are produced at the time most favorable for their survival, which is certainly applicable to birds. If this concept is modified to be the time most favorable for maximal survival of the most vulnerable stage, it is more applicable to other vertebrate groups. For example, in many species of fish and reptiles the reproductive calendar provides for maximal survival of the eggs. In oviparous fish the oxygen dissolved in water appears to be more critical than conditions for the young, which have developed gill cover movement and pursue their own food. In some congeneric and sympatric species of lizards, oviparous species have a narrowly restricted breeding season in contrast to viviparous species, which may reproduce continuously. It was pointed out previously that oviposition in some tropical reptiles is cued to the humidity of the environment in the site for oviposition and that there is ample food for young throughout the year. The more-or-less continuous availability of food for neonate lizards at low latitudes is confirmed by the restricted seasonality of some oviparous species while viviparous species have continuous production of young.

The painted turtle (*Chrysemys picta*) lays its eggs in the late spring, and these eggs usually hatch in late summer. However, some eggs may overwinter and hatch the following spring, with the result that young may appear either in the autumn or in the spring. The same phenomenon has been reported for the snapping turtle (*Chelydra serpentina*) in New York (Hamilton, 1940) and is also known for the whiptail (*Cnemidophorus sexlineatus*) in Arkansas (Trauth, 1977). The overwintering of reptilian eggs may be common, and survival of the young is probably as great in the spring as in the late fall.

Among the several species of small mammals for which reproductive phenology has been well studied, both vernal and autumnal production of young are not uncommon. It occurs in some species of Sciuridae, Muridae, and Cricetidae. Because mammals do not usually become photorefractory, autumnal reproduction has no endogenous restriction. If it were deleterious for small mammals to produce young at the onset of winter, it would surely be very rare. It is significant that while wild mice breed in the autumn when food is abundant, autumnal breeding seems to be a regular occurrence in some species of midlatitude squirrels.

With birds the timing of nesting, incubation, and hatching is certainly very critical, and the reproductive chronology of birds is widely acknowledged to be adjusted so that young appear at the time most favorable for their rapid growth and maximal survival. Because the period of embryonic development is fairly constant for a given avian species, the schedule of oviposition would seem to be the mechanism that provides for hatching at the most advantageous time. The environmental conditions that provide the adults with the stimuli and energy for nesting are frequently the same as the environmental requirements of the young; this is especially true of ultimate cues. Moreover, inasmuch as incubation of most small birds is rather brief, major environmental (including nutritional) changes between egg laying and hatching are rare. In many passerine birds incubation requires only two weeks or less, and the first broods appear at times of maximal plant growth and insect populations.

13.11 SUMMARY

The seasonality of fish is clearly associated with several environmental factors. Various strategies account for successful reproduction, but all seem to ensure adequate oxygen for the eggs and small fry as well as food for postlarval young. In fish of temperate climates, temperature is critical, but thermal needs vary among taxa. Boreal groups, such as salmonids, tend to spawn at times of annual low temperatures, whereas warm-water groups, such as cyprinids and centrarchids, spawn at the end of a vernal warming period.

In tropical fish there are seasonal differences between fish in ponds, in lakes, and in moving waters, but in each group the reproductive strategy provides for oxygenation of eggs and larvae. In pond-dwelling species, which breed in water typically low in dissolved oxygen, spawning adults provide oxygen to the nest, sometimes by placing eggs in bubble nests at the surface and in one species by attaching aerial eggs to the undersides of leaves overhanging calm water. In stream-dwelling and lake-dwelling species, spawning is frequently associated with rising waters and a concomitant increase in dissolved oxygen and food.

Spawning in marine fish invariably includes a strategy to maximize the supply of oxygen, regardless of the pattern of seasonality per se. This may include parental care, pelagic eggs, and migration. A number of species of coastal fish spawn in shallow water in a spring tide, leaving the eggs in a humid aerial site until the next high tide, either two or four weeks later. Tactics have evolved to provide for a return to the sea after either two or four weeks of prelarval growth.

Although nesting in birds is considered to be adjusted so that the eggs hatch when food for the nestlings is most abundant, in reality some species nest later than the period of maximal food availability. Perrins (1970) suggested that food needed for egg formation may determine the onset of nesting, and this theory is supported by early nesting in birds that have been provided with supplemental food. As the avian season progresses, deteriorating food supplies reduce the food available for egg formation. In some species young from nests established late in the season tend to be small and experience greater mortality than those from earlier nests (Perrins, 1970; Toft et al., 1984).

As a general rule most mammals of the low latitudes adhere to a reproductive pattern that involves synchrony to precipitation, and breeding usually occurs during the period of rain. This pattern is also true of large ungulates, some of which migrate to areas of maximal production of forage. Most large mammals of the tropics, including such forms as proboscideans, scavengers, and large carnivores, tend to be somewhat less seasonal, but careful studies frequently reveal patterns of environmentally scheduled breeding.

The majority of pinniped species mate and produce their young in the high latitudes, even though this may require extensive migration from the polar regions in hostile seasons. The time of parturition is necessarily also the time for mating, for the sexes are separate during the other seasons of the year. Because the periods of parturition and lactation are rather restricted in polar and subpolar regions, mating is also confined to a period of a few days, either postpartum or postlactational. Embryonic diapause adjusts parturition to occur approximately one year later.

In virtually all temperate latitude vertebrates reproduction is seasonal. A great many low-latitude species are also seasonal. Reproduction is adjusted so that there is maximal survival of each generation to a reproductive age. It has frequently been said that the breeding schedule ensures maximal survival of newly hatched young.

This concept developed from the study of birds, in which it is usually true, but it is somewhat less applicable to mammals. In ectotherms reproduction may be timed so as to ensure maximal survival and development of the eggs. In oviparous ectotherms, at least, eggs are extremely vulnerable to marked environmental changes, and their proper development seems to be predicated on favorable temperature and humidity and, in aquatic spawning forms, sufficient amounts of dissolved oxygen.

BIBLIOGRAPHY

Bronson FH and Perrigo G (1987). Seasonal regulation of reproduction in muroid rodents. *American Zoologist*, **27**: 929–940.

Cushing DH (1983). *Climate and Fisheries*. Wiley, New York.

Duvall D, Guillette Jr LJ, and Jones RE (1982). Environmental control of reptilian reproductive cycles. Pp 201–231. In *Biology of the Reptilia*, vol 13 (Eds: Gans C and Pough FH). Academic, London, New York.

Fitch HS (1970). *Reproductive Cycles in Lizards and Snakes*. University of Kansas Museum Natural History Miscellaneous Publication 52. Univ Kansas, Lawrence.

Ladurie EL (1971). *Times of Feast, Times of Famine: A History of Climate Since the Year 1000*. Doubleday, Garden City, New York.

Lamming GE (Ed) (1984). Reproductive cycles of vertebrates. In *Marshall's Physiology of Reproduction*, vol 1, 4th Ed. Churchill Livingstone, Edinburgh, London.

Mitchell JC (1979). The concept of phenology and its application to the study of amphibian and reptile life histories. *Herpetological Review*, **10**: 51–54.

Perrins CM (1970). The timing of birds' breeding seasons. *Ibis*, **112**: 242–255.

Potts GW and Wootton RJ (Eds) (1984). *Fish Reproduction: Strategies and Tactics*. Academic, London.

Sadleier RMFS (1969). *The Ecology of Reproduction in Wild and Domestic Mammals*. Methuen, London.

Taylor DH and Guttman SI (Eds) (1977). *The Reproductive Biology of Amphibians*. Plenum, New York.

Wake, MH (1982). Diversity within a framework of constraints. Amphibian reproductive modes. Pp 87–106. In *Environmental Adaptation and Evolution* (Eds: Mossakowski D and Roth G). Gustav Fischer, Stuttgart.

14

FECUNDITY

14.1 NUMBER AND SIZE OF EGGS

We have seen how endogenous rhythms can be stimulated and depressed by exogenous factors. The hypothalamic-hypophyseal-gonadal axis responds to annual and irregular environmental factors. It will become apparent that both regular and irregular aspects of the environment determine the degree to which a species will express its reproductive capacity.

With respect to reproductive parameters, it is useful to remember that the hormonal background that leads to gonadal recrudescence is part of an integrated process that leads to the development of gametes. Ecologists understandably divide this process into two parts: one is the timing, or seasonality, and the other is fecundity, or the number of gametes (especially ova) produced while the process is active. Physiologically, both seasonality and fecundity appear under the same hormonal support, which is to say, they are integrated processes. In many vertebrates the number of ova that mature in any given cycle appears to be a function to the nutritive resources that have accumulated in the liver, and sometimes elsewhere, and also of the resources that can be added during the seasonal reproductive cycle.

Although fitness, or the number of offspring living to reproduce, is a function of fecundity, the two are not the same. Some species experience both great fecundity and very high early mortality. These species have low survival to the time of sexual maturity and would therefore have low fitness. The reverse also obtains in other species.

The limitations on the minimal and maximal number of ova are assumed to be under genetic as well as physical constraint. That is, there is a maximal and minimal number of ova that may mature during the course of any single ovulatory cycle and there is a spatial limit to the number of yolk-filled eggs that may develop at any one time. For many species, however, the mean number of ova and the resulting clutch or litter size appear not to be under rigid genetic control.

In Chapters 11 and 12 some of the factors that alter fecundity experimentally are considered. It is now appropriate to examine variations of reproductive performance in nature. Density-dependent effects on fecundity are discussed in Chapter 15. It is important to enquire which responses appear to be genetically fixed and which are phenotypic responses that lie within a genetically fixed range of responses.

14.1.1 Measuring Fecundity

Fecundity is usually expressed as ova produced per individual per unit time, and the time unit is a calendar year or some part thereof. Different species of vertebrates have greatly different life spans, and a day or a month in the life of an elephant is not comparable, biologically, to an equal period in the life of an annual cyprinodont fish. The very interesting topic of biological time and its important consequences are beyond the scope of this volume (see Sutherland et al., 1986).

Fecundity usually can be easily compared between closely related species, which are similar in their major life history parameters. The difficulty arises in assessing the reproductive output of a guppy, an elephant, and a sparrow. For some species reproduction can be quite irregular: the short-nosed sturgeon (*Acipenser brevirostrum*) spawns for the first time between 8 and 14 years of age, and for the second time some 8 years later (Taubert, 1980). In order to establish significant differences between the fecundity of distantly related taxa, it is necessary to consider the age of first reproduction, the frequency of reproduction, the longevity, and the percentage of the entire lifetime during which a female is reproductively active. Most life history studies fail to include these very relevant statistics. A useful step in that direction is the comparison of lizard reproduction by Tinkle et al. (1970), but lizards are a relatively uniform group.

Environmental and ethological factors are best seen between closely related species, or within a single species under different conditions. For this reason this chapter emphasizes patterns of fecundity within a species and between closely related species.

Reproductive potential in ectotherms is frequently expressed as a gonadosomatic index (GSI), which is

$$\text{gonadal mass/body mass} \times 100$$

This index is widely employed, and its popularity seems to perpetuate its acceptance. The GSI, which is intended to indicate gonadal size adjusted for age (and mass), is affected by (1) seasonal changes in body mass, which include marked fluctuations in body fat, and (2) the nonlinear relationship of gonadal mass to body mass. The GSI is not an accurate or reliable expression of the changes in gonadal mass (de Vlaming and Olcese, 1981; Iles, 1984). Some students have employed body length in a ratio to indicate seasonal changes in gonadal mass (GM). Iles (1984) denoted fluctuations in GM for fish with the formula $K = \text{GM}/L^3$.

Metabolic rates among vertebrates of different sizes and of different major taxonomic groups vary widely, and conversion of food to gametes varies accordingly. The relative fecundity of a codfish, releasing millions of eggs annually over a period of years, and of a shrew, with perhaps a lifetime production of fewer than 10 ova, reflects a host of factors. In addition to longevity and metabolic rate, some of these factors include iteroparity versus semelparity, oviparity versus viviparity, altricial versus precocial births, the relationship between egg size and egg number, parental care, and density-dependent factors such as food (Bagenal, 1978).

14.1.2 The Role of Heritability in Fecundity

Heritability is the proportion of phenotypic variation in a trait due to genetic causes, and it varies from zero to one. All phenotypic variance results from three additive factors: genetic variance, exogenous variance, and the interaction between the two. The proportion of phenotypic variation due to exogenous causes is complementary to heritability. For example, when heritability of litter size is 0.1, 90% of all phenotypic variation is due to environmental factors.

Studies of laboratory mice selected for large litter size indicate that heritability is usually quite low, mostly within the range of 0.10 to 0.22. Selection for ovulation rate may be effective in species with large litters (such as mice and swine), but prenatal mortality is strongly correlated with ovulation rate, so that selection produces only slow and small increases in litter size (Zimmerman, 1979).

Few wild species have been investigated for heritability of litter size. A laboratory colony of one type of vole (*Microtus californicus*) had a heritability of 0.17 for litter size (Krohne, 1981). The role of fertility may have a stabilizing effect on litter size. In a colony of *Peromyscus leucopus*, maximal fertility occurred in individuals from litters of four young, and fertility decreased in individuals from larger and smaller litters (Fleming and Rausch, 1978).

There appear to be few data on the heritability of clutch size in birds. In many species clutch size is constant, and apparently heritability is 1.0 or nearly so. Some birds, however, lay clutches that vary from place to place, from year to year, and with age; in such species heritability of clutch size is clearly less than 1.0. In the great tit (*Parus major*), heritability for clutch size was determined to be a high 0.51, which did not include substantial year-to-year variation (Perrins and Jones, 1974). Adjusted for year-to-year variations, heritability should be considerably less than 0.51. In the magpie (*Pica pica*) from 81 to 86% of within-year variation in clutch size is "linked to territory" (Högstedt, 1980), indicating a level of heritability comparable to that reported for mice and voles.

Low heritability of litter size provides for substantial reproductive elasticity of response to exogenous factors. Data presented in this chapter suggest that upper and lower limits of fecundity may represent genetic limitations and that fluctuations of the means reflect exogenous factors.

14.2 TELEOST FISH

Data on fecundity in teleost fish are generally obtained by estimating numbers of ova in a gonad, and in many species of fish, larger individuals have larger gonads (Fig. 14.1). The same relationship is seen among unrelated species (Fig. 14.2). The strategy for maximal fecundity is the ancestral pattern, but this strategy is modified through a variety of means. The mass-fecundity relationship is not always a straight-line relationship. Older fish may be less fecund than ovary size would suggest, because much of the ovarian tissue may consist of connective tissue.

14.2.1 Variations with Parental Care

Pelagic spawners release the most eggs at any one time and provide no parental care. Parental care must have evolved with the appearance of demersal eggs, for egg size and parental care concurrently increase with the reduction in the number of eggs that settle on the substrate (Hislop, 1984; Potts, 1984). Many fish that care for their young, even trivially, lay fewer and larger eggs. Most species of trout and

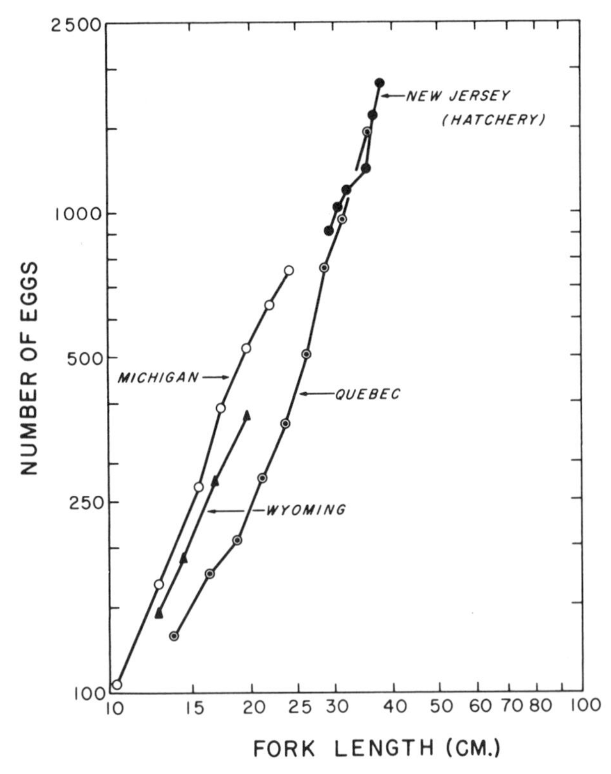

Figure 14.1 The relationship of egg number to body length in the brook trout (*Salvelinus fontinalis*) in different regions. (From Rounsefell, 1957.)

salmon (Salmonidae) and minnows (Cyprinidae) lay relatively few eggs for their size, but do exercise minimal parental care by constructing a pebble nest that offers some protection from predators. The marine catfish (Ariidae), however, carry eggs in their mouth and lay a small number of eggs measuring up to 18 mm in diameter. Among African cichlid fish the number of eggs varies with the nature of parental care: those species that carry their eggs in their buccal cavity produce fewer eggs than substrate spawners do (Lowe-McConnell, 1977).

14.2.2 Fecundity and Frequency of Spawning

Egg number tends to vary inversely with spawning frequency. Among several benthic species of liparid fish, some reproduce more or less continuously and have small complements of eggs: *Acantholiparis opercularis* produces a mean of 6 eggs and *Paraliparis megalopus* spawns a mean of 32. On the other hand, shoreline species, with restricted spawning periods, produce large clutches: *Paraliparis mento* lays about 100 eggs and *P. rosaceus* may spawn up to 1300 (Stein, 1980). Many tropical species are highly prolific, contrary to some theoretical predictions, and spawn large numbers of eggs, but over protracted periods (Johannes, 1978; Sale, 1977).

In contrast, the Norway pout (*Trisopterus esmarkii*) has a rather short spawning season during which a relatively small number of eggs are synthesized, presumably from nutrients accumulated and stored prior to the breeding season. Nutritional conditions appear to account for year-to-year fluctuations in fecundity, in contrast to more consistent fecundities seen in other gadoids (cod, haddock, and whiting). Survival rate is apparently very high in the pout (Hislop, 1984).

The Salmonidae (salmon, trout, graylings, and whitefish), which have been intensively studied, have developed several distinct reproductive patterns. Species of some genera are semelparous (*Oncorhynchus* spp.), spawning once and dying, or iteroparous (*Salmo* and *Salvelinus*), spawning several times in successive years. The semelparous forms are anadromous and also achieve the greatest size among salmonids prior to breeding. Conceptually, semelparity should produce the most eggs for a given body mass, but, in fact, just the opposite is true. The variation of reproductive parameters within the Salmonidae exhibits a high degree of flexibility. Migration affects covariation of size and age of maturity with fecundity. Anadromy and semelparity promote increased growth, a large size at sexual maturity, and a small number of large, rapidly developing eggs. The small, iteroparous species, which may live just as long, display the opposite set of traits (Hutchings and Morris, 1985).

Seasonal trends in the size of individual eggs and the relationship of egg size to growth are discussed in Chapter 5, Section 5.2.2.

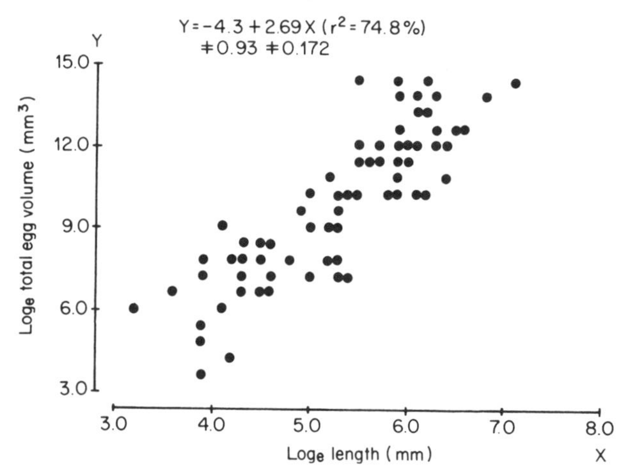

Figure 14.2 The relationship between total egg volume and fish length for a number of Canadian freshwater fish. Both axes are scaled logarithmically. Parameters (logs of X and Y) of the least-squares regression are shown. (From Wootton, 1984, courtesy of Academic Press.)

14.3 AMPHIBIANS

Amphibian fecundity varies tremendously with (1) oviparity and viviparity, (2) body mass of female (3) oviposition site (aquatic, terrestrial, or arboreal), and other environmental factors.

Within a given taxon and mode of oviposition, both clutch size and egg size vary positively with the size of the female. Within a given species larger females lay not only more but larger eggs than is typical of a smaller, and presumably younger, individual (Salthe and Duellman, 1973). Fecundity in the Eurasian common frog (*Rana temporaria*) is positively correlated with body size, but varies consistently among populations. Egg mass varies among populations and also from year to year, suggesting physiological adjustments to exogenous factors (Cummins, 1986).

An exception to this positive correlation may be viviparous caecilians. Ovarian clutch size in the live-bearing *Dermophis mexicanus* varies from 13 to 36 and is not correlated with age or size of the female (Wake, 1980). Across an assemblage of taxa, however, the trend is for clutch size and egg size to vary inversely.

14.3.1 Terrestriality and Clutch Size

With an increase in terrestriality there is an increase in egg size and yolk content and a reduction in egg number. This relationship reflects the relatively advanced state in which the terrestrial young emerges. The eggs of the totally terrestrial young amphibian egg must contain sufficient nutrients to provide for metamorphosis into a miniature adult.

Terrestrial breeding is especially characteristic of tropical anurans and may well represent an effort of a species to spawn in an oxygen-rich environment. Many quiet tropical ponds are very low in dissolved oxygen, and movement to air places anuran eggs in an environment 30 times richer in oxygen. Aerial oviposition permits the eggs to be larger, assuming that an oxygen-deficient pond would require small eggs because of their greater surface-to-mass ratio. Eggs of intermediate size are spawned in streams. This trend is well documented (Salthe, 1969; Salthe and Duellman, 1973). The ultimate in this trend is seen in the small leptodactylid frog, *Sminthillus limbatus*, of Cuba. It deposits on land a single large egg from which a fully formed frog emerges.

Salamanders exhibit the same trend between clutch size and egg size as seen in anurans, and this trend parallels the movement toward breeding on land. The family of lungless salamanders (Plethodontidae), with a variety of ovipositional patterns, illustrates this trend. The two-lined salamander (*Eurycea bislineata*) breeds in quiet streams; its clutch consists of approximately 30 eggs from 2.5 to 3 mm in diameter. The much larger purple salamander (*Gryinophilus porphyriticus*), which is permanently aquatic and breeds in cold streams or springs, lays 50 to 130 eggs approximately 3.5 mm in diameter. The genus *Desmognathus*, in which the species are smaller than the purple salamander, generally lays eggs on land, from which gilled larvae emerge: *D. fuscus* lays 15 to 30 eggs and *D. ochrophaeus* lays 10 to 15 eggs. In both species the eggs are 4 mm in diameter. *Plethodon cinereus*, of comparable size, is entirely terrestrial and lays from 5 to 13 eggs from 4 to 5 mm in diameter. The somewhat larger *Plethodon glutinosus* is also terrestrial and lays a clutch of 10 to 18 eggs 5.5 mm in diameter. These data are all from Bishop (1941).

There may be an underlying (evolutionary) tendency for larger eggs, and the terrestrial environment pattern, with its greatly increased supply of oxygen, simply allows for the expression of this tendency. The increased oxygen in the atmosphere does not *produce* a larger egg but simply *permits* an increase in egg size with a relative reduction in surface area. In addition to the greatly increased access to oxygen, terrestrial oviposition carries the increased risk of desiccation; a decrease in surface (relative to mass) of the larger ovum decreases the rate of water loss. Perhaps these two factors account for the greater size of terrestrial eggs. In the preceding examples of both anurans and salamanders, parental care increases with the tendency toward terrestriality (see Chapter 9, Section 9.5.4).

14.4 REPTILES

Clutch size in reptiles is influenced by several unrelated factors. Egg number may vary from a clutch of one or two for geckos and some species of *Anolis* to more than 200 for some species of marine turtles. The relationship of egg number to body mass is well known. Short-term (year-to-year) environmental variations have been reported for many species in temperate latitudes, and larger clutches generally follow years of greater than average rainfall. Arboreal oviposition appears to favor a reduced egg number. The relationship of latitude to clutch size is inconsistent and equivocal.

14.4.1 Environmental Variations

It is well established that reduced rainfall is followed by reduced plant growth, fewer insects, and a decline in fecundity in iguanid lizards (Hoddenbach and Turner, 1968; Mayhew, 1966; Ballinger, 1977, 1983; Pianka, 1986). For example, in the lizard *Urosaurus ornatus* a reduction in clutch size was correlated with both a previous decrease in precipitation and a decline in insect populations (Ballinger, 1977). Drought reduces clutch size in the slow worm (*Anguis fragilis*), a lizard that feeds largely on slugs and earthworms (Patterson, 1983). Year-to-year variation in brood size of the viviparous *Sceloporus malachiticus* in Costa Rica suggests that fecundity is tied to fluctuations in available resources (Reznick and Sexton, 1986). In an insular population of the adder, *Viper bereus*, the main food, voles, fluctuates from year to year. During periods of vole abundance, the body mass of female adders increases, at which times more young are produced than in years of vole scarcities (Andrén and Nilson, 1983).

The same relationship between fecundity and rainfall was noted in five populations of four species of snakes. In data taken over periods of from 2 to 15 years, productivity in three populations was higher in years of above normal rainfall, and in one species clutch size increased in years following above normal autumnal rainfall (Seigel and Fitch, 1985).

Similar patterns have been recorded for chelonians. Clutch size in turtles is more influenced by exogenous factors than by genetic factors (Gibbons, 1982). There is a positive correlation between rainfall, plant growth, and environmental productivity; and increased energy intake is followed by either larger clutches or greater numbers of clutches. The desert tortoise (*Gopherus agassizii*) responds to greater than mean rainfall by producing larger clutches (Turner et al., 1984, 1986).

14.4.2 Arboreal Reptiles

The large, arboreal agamid lizard, *Goniocephalus* spp., typically has small eggs from 2 to 4 mm in diameter. Also, in geckos (Gekkonidae) two eggs constitute a clutch, except in the sphaerodactyline geckos, which lay one egg at a time. Geckos lay their eggs under the bark of trees, and such a site might not accommodate a large clutch. Anoles (*Anolis* spp.), which include more than 150 mostly arboreal species of the New World, lay clutches of one or two eggs, which are frequently deposited under bark. Chameleons (Chamaeleontidae) are also arboreal, but they lay their eggs in soil, in large clutches of from 10 to 50. The highly arboreal agamid, *Japalura swinhonis*, lays its clutch of at least four or five eggs on the ground (Lin, 1979). Arboreal habit apparently does not relate to clutch size, but arboreal oviposition is correlated with clutches of one or two eggs. Single clutches are laid at intervals over prolonged periods.

14.4.3 Variation in Egg Size

Among the well-studied species of reptiles, there is not a tendency for larger eggs to occur in smaller clutches, as noted for amphibians. Two factors may relate to egg size in reptiles: (1) a larger egg requires a longer incubation period, and (2) few oviparous reptiles provide parental care. A prolonged incubation extends the period of vulnerability of unguarded eggs, and without parental care a smaller number of larger, unprotected eggs would not enhance survival to hatching time.

Variation in egg mass may change with the seasons. As the breeding season advances, the side-blotched lizard (*Uta stansburiana*) lays fewer but larger eggs. The shift from number to mass may be a tendency to make maximal nutritional provisions for the embryo as the season becomes increasingly unpredictable (Nussbaum, 1981).

14.4.4 Clutch Size and Latitude

Unlike prevailing trends in other vertebrates, clutch size in reptiles seems not to have a regular relationship to latitude. In some groups species at high latitudes have larger clutches, and in others the trend is reversed. Clutch size in lizards tends to *decrease* with an increase in latitude. *Uta stanburiana*, a small iguanid lizard of western United States, lays slightly larger clutches in the more southerly part of its range, and the same trend is seen in several other species of reptiles. In some species, however, greater fecundity is seen at higher latitudes (Fitch, 1970). Published data are not suitable for meticulous comparison because body mass is usually not considered in relation to latitudinal clines in egg number (Turner, 1977).

14.4.5 Mass-Related Variations

In most reptiles egg number appears to increase consistently with body mass. The positive correlation between clutch size and body mass in oviparous lizards is well known. The concept is generally valid both within species and among species (Fitch, 1970; Tinkle et al., 1970). The exceptions are all species that have very small clutches.

Clutch size in *Sceloporus graciosus*, a Nearctic iguanid lizard, increases with the mass of the female. Initial ovogenesis occurs at the end of the second year, with small primiparous females laying clutches appreciably smaller than those produced by larger and older females (Degenhardt and Jones, 1972; Jameson, 1974; Tinkle, 1973). This species ranges from Washington to Baja California and appears to lay a single clutch annually, with slightly larger clutches in the southern part of its range (Fitch, 1970). Among some other species of this genus, there are positive correlations between body mass and number of eggs per clutch, but some egg-laying taxa produce more than one clutch of eggs per season. Because of the many species over such a broad geographic range, geographic trends in clutch size are not readily apparent.

Among the many species of *Sceloporus* both oviparous and live-bearing species are represented. All live-bearing species produce young once a year. In the viviparous *Sceloporus grammicus* (Ortega and Barbault, 1984) and *Sceloporus jarrovi* (Ballinger, 1979) there is a close positive correlation between body length and litter size.

Clutch size is also related to body form, which is related to method of locomotion: more rotund species tend to develop larger clutches than more slender body shapes do (Vitt and Congdon, 1978). Body form becomes especially important when comparing clutch sizes of lizards and snakes, for the latter are generally more fecund (Fitch, 1970). There is no clear mechanism by which larger squamates produce larger numbers of eggs, but the greater body cavity would obviously be capable of containing more eggs if egg mass is not increased.

Among chelonians larger species tend to lay larger clutches, and clutch size seems to bear an inverse relation to egg size. Moll (1979) characterized two patterns: (1) large clutches of relatively small eggs laid seasonally in well-constructed and covered nests, and (2) small clutches of relatively large eggs laid over an extended period in poorly made nests or no nest at all. Both patterns occur commonly; those of the more fecund pattern (1) are potentially more vulnerable because they provide a concentration of nutrition to predators.

Turtles occur in a broad gradient of body sizes and, within a species, produce eggs in numbers proportionate to body mass. An example is the snapping turtle (*Chelydra serpentina*) in New York State (Petokas and Alexander, 1980) and Tennessee

(White and Murphy, 1973). In the painted turtle (*Chrysemys picta*) both clutch size and egg width increase with female body mass (Mitchell, 1985). Interspecific comparisons of clutch size, however, do not always reflect body mass. The giant tortoise (*Geochelone gigantea*) on Aldabra Atoll lays relatively few eggs for a large species, with clutches varying from 3.9 to 6.4, depending upon the habitat (Swingland and Lessels, 1979).

Marine turtles lay several large clutches over a period of a few weeks every several years. These large clutches decrease in size within a single season. Clutch sizes in the loggerhead turtle (*Caretta caretta*) at Little Cumberland Island, Georgia (United States), vary from 114.40 to 127.54 (Frazer and Richardson, 1986).

As in other reptiles, clutch size in crocodilians is positively correlated with maximal body size. Most data on clutch size of these large reptiles are not accompanied by actual measurements of the individuals producing the clutch, and the correlation is among known maximal sizes.

There is an intraspecific, and a less apparent interspecific, trend toward larger clutches in larger reptiles. Within a species, larger individuals are usually older. The inference is that increased clutch size follows an increase in lipid stores. Data suggest that lipid stores increase with age, possibly with more efficient foraging and with a decline in demand of energy for growth. There is also a trend for clutch size to decrease in successive clutches within a single season.

14.5 BIRDS

Avian clutch size follows some of the same patterns seen in ectotherms, except that there are no live-bearing birds. Clutch size is fairly constant in some taxa and highly variable in others. Many plovers and sandpipers lay a clutch of four eggs, whereas hummingbirds and most pigeons and doves lay a clutch of two eggs. Many marine birds lay a single egg, and some passerine species have clutches that vary in size with habitat, food, latitude, and other exogenous factors.

The determination of clutch size in birds is controlled by mechanisms somewhat different from those limiting ovulation in other vertebrates. Clutch size in birds is affected by three primary factors: adequate food stores and nutrition to synthesize a clutch of eggs; the area a brooding bird can cover; and the ability to brood and feed the young, especially in altricial species.

Species that normally lay a clutch of more than one egg deposit the eggs over a period of days, usually either one a day or one every other day. One limitation of avian clutch size, then, is the establishment of which egg is to be the last of a given series. As in other vertebrates, avian ovulation occurs with a sharp increase in estrogen, progesterone, and luteinizing hormone; and the series of ovulations indicate a succession of peaks of circulating hormones. Estrogen induces the release of nutrients (vitellogenin) from the liver in yolk formation, and the accumulation of energy is a major determination of clutch size for some birds. Ovarian activity is stimulated by day length. Clutch size, therefore, may be influenced by both nutrition and photoperiod.

The determination of which egg is to be the final one, the egg that marks the completion of the clutch, rests upon (1) the number of eggs already in the nest and (2) the body condition (stored nutrients) of the female. In some birds there appears to be a threshold of body condition, which is governed by food supply and fecundity. Drent and Daan (1980) reviewed several examples both in birds in undisturbed (unmanipulated) nests and in birds provided with supplemental feeding. They concluded that both clutch size and seasonality are related to body condition of the female, and this condition can be improved by supplemental feeding. They and several other researchers emphasized the relationship of clutch size to quality of the territory.

The clutch may not equal the number of eggs laid, for if eggs are removed at the rate at which they are being laid, the female in many species may continue to ovulate and lay for an extended period. Such species are called indeterminate layers. Not only does the successive removal of eggs prolong egg laying, but the introduction of eggs into an incomplete clutch may cause ovulation and egg laying to stop. The tactile stimulation of eggs against the brood patch apparently increases the secretion of prolactin, which in turn depresses ovulation.

14.5.1 Concepts of Avian Clutch Size

David Lack (1954, 1968) proposed that mean clutch size in birds approximates the maximal number of young that the parents can fledge and that larger clutches result in greater mortality of nestlings, while smaller clutches fail to exploit the ability of the parents to provide for the chicks. Lack assumed that time limited the parents in providing sufficient food for the nestlings. Also, he assumed that selection preserves the mean clutch size. Lack (1968) clearly believed in the evolution of a genetically determined mean clutch size, and this concept is reiterated throughout his writings. Although he was aware of environmental effects on egg production in birds, he persisted in his belief of very strong hereditary control over a mean, or "normal," clutch size.

The determination of clutch size in birds, as in other vertebrates, is now known to be far more complex than the original proposal of Lack (Winkler and Walters, 1983). Ashmole (1963) was one of the first to point out that food during the nesting season affects nesting success and population density.

Body mass is an important factor in the formation of clutch size. Among a large number of species of passerine birds, using a diversity of both concealed (as in hole nesters) and unconcealed nesting sites, clutch size decreases with increases in the body mass, but egg mass, as well as clutch mass, increases with body mass (Saether, 1985).

14.5.2 Manipulation of Brood Size

Many field experiments have involved altering the original number of eggs or young in nests of wild birds, with the intended result that the parents must care for broods that are either larger or smaller than expected from the number of eggs laid. Under these circumstances artificially increased broods generally result in poorer growth and greater mortality of nestlings.

A few examples will illustrate modified growth and survival of young in manipulated broods. When the original brood size of the pied flycatcher (*Ficedula hypoleuca*) was altered, survival of young declined with the increase in brood size. Weight loss during this period represented a normal negative energy balance, and in the experimental situation weight loss increased with the initial size of the brood (Askenmo, 1982). An analogous experiment with the magpie (*Pica pica*) produced similar results, suggesting that in this case the individual female modified its clutch size and that the clutch naturally laid is of the optimal size for that territory (Högstedt, 1980). In manipulated broods of the starling (*Sturnus vulgaris*), larger broods were given supplemental food, and they attained the same weights as did smaller broods that were not provided with food by the experimenter (Crossner, 1977).

Artificially increased broods do not always result in lower survival. In altered broods of the snow bunting (*Plectrophenax nivalis*), more young were fledged from larger broods (up to seven young to a brood), although the young in artificially enlarged broods weighed less when fledged (Hussell, 1972). When broods of 3, 6, 9, 12, and 15 young of the blue tit (*P. caeruleus*) were artificially created, survival was greatest in broods of 12 to 15 young, and it was concluded that clutch size reflected differences in "the ability to lay and/or incubate the clutch" (Nur, 1986).

In view of the single clutches of some sea birds, they may not be able to provide food for more than a single chick, and there is some experimental evidence to support this thesis. Chicks of the puffin (*Fratercula arctica*) grow faster when fed additional food than do control chicks reared only on food provided by their parents. Also, young that are deprived experimentally of one parent showed reduced growth and survival (Harris, 1978). The short-tailed shearwater (*Puffinus tenuirostris*) normally lays a single large egg; the addition of a second egg resulted in a reduction of eggs hatched (Norman and Gottsch, 1969).

Among the species of boobies and gannets (*Sula* spp.), some lay a clutch of one egg, several lay two eggs (and rarely fledge more than one young), and the piquero (*S. variegata*) lays four eggs. These small clutches appear to be under genetic control, but with poor correlation between clutch size and number of young fledged. When an egg or young chick of the gannet (*S. bassana*) was added to nests with a single egg or chick, growth of such "twinned" broods did not differ significantly from nests with one young. Moreover, there was no difference in hatching or fledgling success in the two kinds of nests (Nelson, 1964).

In contrast to these studies, manipulated clutches in waterfowl may not affect survival. "Experimental" nests of the blue-winged teal (*Anas discors*), which had clutches artificially adjusted to means of 4.9, 9.9, and 16.9 eggs, did not differ in nest success (72.5%) or hatchability (94.8%) from natural clutches, and duckling survival was unrelated to brood size (Rohwer, 1985). In this precocial species, females are capable of rearing many more young than the mean clutch size, and mean clutch size may be determined by the condition of the female during the period of egg laying. Similarly, artificially enlarged broods of Canada geese (*Branta canadensis*) did not suffer in individual growth and survival (Lessells, 1986).

These experiments all involve unnaturally altered food supply, either indirectly by changing the brood size or directly by feeding the young. By this maneuver one separates (1) the amount and quality of food available to the female prior to ovulation from (2) the amount and quality of food available to an adult foraging for its nestlings. In nature, however, the quality of food available to the female at the time of egg formation may be essentially the same as that available for the young, at least during the initial part of the nesting season. This is true for many passerine species, which have rather short incubation periods and rapid growth of altricial young. In high latitudes, with a brief nesting season, food supply may vary over short periods. In most experiments of brood size manipulation, the nests involved broods of varying original sizes. These experimentally altered broods, therefore, could have placed some large broods in nests where the original broods had been small. From these sorts of experiments some students concluded that the mean clutch size is the optimal clutch size and that this size is genetically determined.

This interpretation has two flaws. First, in species with clutches of various sizes, there is no real evidence of a rigid species-specific genetic determination of clutch size (Klomp, 1970). The determined 0.51 heritability of clutch size for *Parus major* (Perrins and Jones, 1974) may be too high. They tried to correct for years, habitat, and age, but the value of such corrections may be suspect. The sometimes highly variable (year-to-year) clutch size in a given region confirms the relatively minor role of genetic control of clutch size. Second, in manipulated broods there is confusion between (1) the parents' rather limited ability to make a given number of foraging flights during a given period of time and (2) the variable amount of food available in the birds' habitat. Inasmuch as the second factor is vastly more variable than the first, it would appear to be the more important factor in determining survival of the young and, indeed, the formation of the clutch.

14.5.3 Clutch Size and Survival in Unmanipulated Broods

In addition to the manipulated clutches mentioned previously, the ornithological literature contains many studies that compare clutch size with fledgling number in undisturbed nests and sometimes with fledgling survival for the first summer. These studies present a picture of increased survival in young with greater egg number.

In nature, among some well-studied birds, an increase in brood size is accompanied by an overall increase in nestling survival (Klomp, 1970). In many birds of diverse taxa, fecundity varies from year to year as well as from habitat to habitat. It is most plausible to relate both clutch size and survival to the factors or cues reflecting productivity of the territory. Year-to-year differences in clutch size can be related to (1) the mean age of the individuals, (2) the laying date, and (3) the food supplies.

The role of food in clutch size and survival has long been known. In 1975 a major outbreak of the oak caterpillar (*Operophtera fagata*) west of Helsinki appeared to account for an increase in clutch size as well as in survival in two tits (*Parus major* and *P. caeruleus*) (Hilden, 1977). As an example, the white wagtail (*Motacilla alba*) lays clutches of from four to six eggs in central Finland, but the percentage of young fledged does not differ significantly from brood size (Leinonen, 1973).

Productivity in the tree swallow (*Iridoprocne bicolor*) also increases with clutch size. As the egg number varies from three to seven, the number of fledglings increases from 3 to 6.7 (de Steven, 1978). Also, the crow (*Corvus corax*) in southern Sweden experiences lower nestling mortality with larger broods (Loman, 1977). In the last example, larger clutches and higher survival rates were characteristic of more productive habitats.

In Finnish populations of the great tit (*P. major*), an important factor in determining clutch size is the abundance of the oak caterpillar (*O. fagata*), a major food item of the tit during the nesting season. A high density of the caterpillar is correlated with unusual fecundity of the great tit and other hole-nesting birds. In 1975, a year of caterpillar abundance, not only did clutch size and brood size increase but fledglings per pair greatly exceeded those stastics over the previous four years. In 1975, 81 pairs of tits fledged a mean of 7.95 young in contrast to a mean of 4.20 fledged by 281 pairs from 1971 to 1974 (Hilden, 1977). In the Siberian tit (*P. cinctus*) in Finnish Lapland, clutch sizes range from 4 to 11 and nesting success (proportion of young fledged) increases with clutch size (Jarvinen, 1982).

In northern Finland, mean clutch size of the great tit appears to be density dependent, with larger clutches in more scattered populations (Orell and Ojanen, 1983). In British woodlands, clutch size is negatively correlated with population density, and more than 60% of the variation in clutch size is attributed to density (O'Connor, 1980). With a doubling in density of nesting adults, clutches averaged two eggs fewer (Perrins, 1965).

Both the blue tit (*P. caeruleus*) and the coal tit (*P. ater*) on the island of Corsica consistently lay smaller clutches than do the two species in Ventoux, in adjacent southern France (Provence), and elsewhere on the continent. Moreover, on Corsica these two tits do not lay second clutches, as they frequently do on the mainland. Breeding success on the island is 2.6 young (40.6% success) for the blue tit and 3.0 (57.3% success) for the coat tit, much less than for the adjacent mainland, where the blue tit raised 6.4 young (58.9% success) and the coal tit 9.5 young (73.3% success). In Ventoux, the blue tit feeds its young largely on caterpillars of the oak moth (*Tortrix viridana*), and the coal tit provides its young with caterpillars of the cedar moth (*Epinotia cedricida*). On Corsica both species feed their young on a broad spectrum of arthropods, including many spiders. The difference between the insular and continental populations of these two species appears to be determined by food and fits Ashmole's hypothesis (Blondel, 1985).

In a Swedish population of the magpie (*Pica pica*), not only reproductive output but also first summer survival was positively correlated with clutch size, and clutch size was correlated with territorial quality (Högstedt, 1980, 1981a).

Clutch size in the red grouse (*Lagopus lagopus scoticus*) declines with a deterioration of its staple food, heather. In winters when heather is green and vigorous, clutches the following spring number for a mean of 7.8 to 8.1. When the heather is brown and dry, clutches vary from 6.1 to 6.9 (Jenkins et al., 1963). Nestling survival and nestling weights of the house sparrow (*Passer domesticus*) and the tree sparrow (*P. montanus*) were tied to food supplies, but brood failures were randomly distributed among broods of various sizes (Seel, 1970). Heavier individuals of the pied flycatcher tend to lay larger clutches when environmental conditions are favorable, and more fledglings are produced from the larger clutches (Askenmo, 1982).

Differences in mean clutch size in the California gull (*Larus californicus*) at Mono Lake (California) and Great Salt Lake (Utah) appear to be due to food limitation prior to egg formation (Winkler, 1985).

Darwin's finches, which dwell on the Equator, not only are seasonal in their nesting schedule (Chapter 13, Section 13.8.2) but are highly facultative in their fecundity and quickly respond to departures from mean rainfall patterns. In *Geospiza fortis*, clutch size increases after heavy rain and abundant food (Price, 1985). Greater than normal precipitation may be detrimental to nesting. Prior to egg laying the female common tern (*Sterna hirundo*) is fed by the male, and this nutrition is essential to egg formation. When unusually heavy rains inhibit foraging by males, clutch size is reduced (Becker et al., 1985).

Larval midges (Chironomidae) constitute the main food of four species of diving ducks and one puddle duck in Iceland. Over seven years of observation, there was a decline in both mean clutch size and production of grown young per pair in the one year when midge larvae were much less abundant (1970) than in the other six years (Bengtson, 1971, 1972). In Scandinavia, foods of animal origin, especially small invertebrates, provide the protein needed for very rapid growth of dabbling, or puddle ducks, and their nesting is timed to the appearance of such aquatic insects as chironomids and water lice (Nummi, 1985). In boreal North America, nesting puddle ducks feed heavily on animal food, which constitutes from 70 to 99% of the diet of laying ducks. In north-central United States, nesting success of the mallard is positively correlated with abundance of larval midges, which are a major food of newly hatched young (Talent et al., 1982). Snails, crustaceans, insects, and, in wet years, earthworms all provide hen ducks with the nutritional needs for egg formation. These food supplies are known to vary from year to year with meteorological fluctuations (Swanson et al., 1979).

Many marine birds depend entirely on the sea for food, and fluctuations in marine productivity account for major shifts in reproductivity of oceanic birds. Reproductive success (fledging rate) of the brown pelican (*Pelecanus occidentalis californicus*) on the southern California coast has a direct relationship to the abundance of its major prey species, the northern anchovy (*Engraulis mordax*), and the reproductive output is especially sensitive to local prey availability and abundance (Anderson et al., 1982).

Biological effects of the El Niño Southern Oscillation (ENSO) are profound and have been reported for several well-studied avian communities. The small clutch size (usually one) and long incubation account for the vulnerability of many marine birds. They need ample food for a prolonged period, not only for egg formation but also for feeding the young. Nesting in tropical sea birds is dependent on sustained environmental productivity. Total reproductive failure in sea birds nesting on Christmas Island at 2°N in the central Pacific Ocean was seen in the 1982 to 1983 ENSO (Schreiber and Schreiber, 1984). Again, in the ENSO warming of February and

March, 1986, there was a near total nesting failure of blue-footed and red-footed boobies on the Galapagos Islands (Bowman, 1986). In October of 1985 not only was there virtually no reproduction of sea birds but there was mass starvation on Christmas Island; this was attributed to a decline in food (small fish and squid), a result of ENSO warming (Schreiber and Schreiber, 1986).

Ironically, Cody (1966) had postulated that sea birds frequently lay a clutch of one because of the stability of their environment. A more plausible explanation is that many sea birds lay a single egg because (1) during the breeding period they are concentrated on islands and (2) their food is sparse and widely scattered. They may well be unable to accumulate reserves to allow them to lay more than one egg, and if they could lay more than one, the adjacent sea might not have sufficient food to permit them to rear more than one young.

Terrestrial predators are also very sensitive to densities of prey populations. With the increase in latitude there is a decrease in numbers of species, and predators become greatly dependent on populations of a few species of prey. In Uusimaa, southwestern Finland, the goshawk (*Accipiter gentilis*) preys largely on grouse. When grouse are scarce, the brood size of the goshawk declines. In 1974, a year of grouse abundance, the mean brood size was 2.4, but in the same region in 1977, a year of grouse scarcity, the mean brood size was 0.5 (Wikman and Tarsa, 1980). Nomadic species such as the snowy owl (*Nyctea scandiaca*) and the short-eared owl (*Asio flammeus*) lay large clutches, sometimes of 10 or more eggs, when prey abounds. In years of prey scarcity they might not even nest (Pitelka et al., 1955). Among these species large clutches seem to be correlated with nomadism as well as prey abundance (Andersson, 1980). Breeding success in the golden eagle (*Aquila chrysaetos*) in Sweden is predicated on the abundance of prey, especially hares and grouse (Tjernberg, 1983).

In years of rodent abundance the rather sedentary long-tailed skua or jeager (*Stercorarius longicaudus*) lays two eggs; when rodent densities are moderate, the skua lays a single egg. When rodents are scarce, the skuas do not nest (Andersson, 1981). In the well-studied tawny owl, reproduction may not be attempted in years of prey scarcity (Southern, 1969, 1970).

When provided with fresh quail and pigeons as supplemental food, territorial sparrowhawks (*Accipiter nisus*) in southern Scotland laid larger clutches (4–7 eggs) than unfed birds did (0–2 eggs) (Newton and Marquiss, 1981). The effect of food on clutch size of predatory birds would appear to be unequivocal.

These data on undisturbed populations in nature suggest that a species has a typical range of eggs that may constitute a full clutch, but the mean clutch size varies from place to place and time to time. In addition, there is a positive correlation between number of eggs in a nest and nesting survival. From this information the range in clutch size appears to be flexible and adaptive, and the mean clutch size reflects environmental productivity. The ability of the parents to develop gametes and nest successfully is either enhanced or limited by natural fluctuations in food. The optimal clutch size is that which the bird lays, whether it be large or small. That is, there is not one optimal clutch size but many, and the optimal strategy is for the bird to be "highly flexible" in this parameter (Högstedt, 1980).

14.5.4 Condition of Young at Hatching

As previously pointed out (Chapter 9, Section 9.7), hatchlings may be altricial or precocial or hatched in some intermediate condition. The altricial young is nude and blind and lacks thermoregulatory ability. Eggs producing precocial young require additional yolk to sustain prehatching growth, and for an equal amount of material investment in the clutch, altricial birds can lay more eggs. There is not, however, a discrete dichotomy between altricial and precocial clutch sizes, and other factors affect both groups.

Precocial young, being advanced at hatching, generally require less parental care, and many species of gallinaceous birds and waterfowl lay rather large clutches. Among species that produce precocial hatchlings, some feed their young whereas others only provide protection from weather and predators. Between these two groups, the feeders and the nonfeeders, the latter tend to have larger clutches (Winkler and Walters, 1983).

14.5.5 Trends in Egg and Clutch Size with Season

In addition to the clutch size, or range in clutch size, inherent to a given species, there are correlations between time of laying with clutch and egg size. Sometimes egg size and egg number are negatively correlated.

Large clutch size appears to be related to early nesting in some birds. In the crow (*Corvus cornix*) in south Sweden, nests started earlier in the year have larger clutches and less starvation than those started later in the season. Also, predation on nests of smaller clutches (2–4 eggs) exceeded that of larger clutches (Loman, 1977).

In the great tit (*Parus major*), a species with a prolonged nesting season, egg size increases and clutch size decreases as the season advances (Perrins, 1970). In England the earlier and larger broods of the great tit have a higher degree of young fledged (94%) than the smaller, later broods do (37%) (Lack, 1955). Moreover, the young that are fledged earlier attain a higher survival by midsummer (Perrins, 1965). In northern Finland (65°N) the great tit lays an initial clutch of 9.86 eggs and a second clutch of 7.51 eggs (Orell and Ojanen, 1983). The bullfinch (*Pyrrhula pyrrhula*) rears two broods in the central Netherlands, with the first clutch being the larger, and the number of fledglings increases with clutch size (Bijlsma, 1982).

The red-billed gull (*Larus novaehollandiae*) lays larger eggs with increasing age (Fig. 14.3): egg laying commences in the second year, and 7-year-old females lay the largest eggs (Mills, 1979). The same phenomenon is seen in the willow grouse (*Lagopus lagopus*) (Myrberget, 1977).

The well-known trends of decrease in clutch size and increase in egg mass with increase of body mass also obtains in waterfowl (Anatidae). Within waterfowl the number of eggs gradually decreases as one examines data from larger species, and the egg size relative to body mass also diminishes with the increase in body mass (Lack, 1968). Apparently the larger egg is produced at the expense of a larger number of eggs. Young of larger species have greater lipid reserves at birth and thus greater chances for survival.

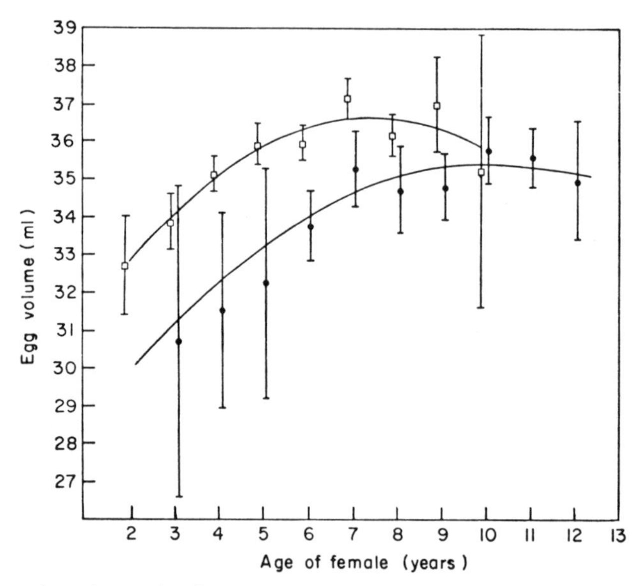

Figure 14.3 The relationship between age of female and egg volume in the red-billed gull (*Larus novaehollandiae*). The data for the years 1965 to 1966, 1967 to 1968, and 1968 to 1969 are not statistically different and are grouped together and represented by the upper series of open squares. Mean volumes in 1971 to 1972 were smaller and are represented by solid circles. (From Mills, 1979, courtesy of *Ibis*, Journal of the British Ornithologists' Union.)

The typically smaller second and third clutches of the great tit have been attributed to a deterioration in the food of the adults, from caterpillars to less nutritious spiders and adult insects (Klomp, 1970). It might also reflect a decline in gonadotropins associated with the approach of refractoriness. The phenomenon of within-season decline in clutch size is so widespread among many diverse species that there may be some basic physiological explanation. The poorer survival of late broods is also rather common and occurs in both nidicolous and nidifugous species. It is plausible that the same environmental events—high quality of food and stimulatory photoperiod—that enhance folliculogenesis and possibly account for larger clutches in the early part of the nesting season also promote an abundant supply of high-quality food for the young.

Klomp (1970) suggested that the smaller clutches of later nestings may result from "a proximate effect without having any ultimate adaptive significance." This theory is thoroughly plausible in view of the recent climatic history of northern Europe, the region where many of the studies on clutch size have been made.

Egg size in the long-billed curlew (*Numenius americanus*) varies with both body mass and vegetative condition of the nest site before laying. When plant growth was short (1977 and 1978), females did not move extensively before egg laying, and the heavier females laid larger eggs. Before cattle had trampled the tall herbaceous vegetation that covered the nesting grounds in 1979, curlew foraged some 10 km distant. Not only did this increased movement delay nesting, but lighter females laid larger eggs than heavier females did, suggesting that the latter may have had to expend more energy in the additional flying required in 1979 (Redmond, 1986).

A few birds regularly lay two dissimilar eggs, both of which are fertile. The white-fronted tern (*Sterna striata*) in New Zealand lays a clutch of one or two eggs. If there are two, the second is significantly smaller than the first (Mills and Shaw, 1980). In the five species of crested penguins (*Eudyptes* spp.), the clutch usually consists of two eggs, the first of which is conspicuously smaller than the second (Warham, 1975). Survivorship of these eggs does not seem to differ significantly.

14.5.6 Latitudinal Gradients

There tends to be a positive correlation between increasing latitude and ovulatory rates, which has been noted for many species. Although midlatitude species lay larger clutches than their tropical allies do, clutch size may be slightly reduced from midlatitudes to very high latitudes (Lack, 1968). Among 15 species of passerine birds in Finnish Lapland, only one, *Anthus pratensis*, laid larger clutches than did populations of the same species farther south. Apparently in extremely severe and unpredictable boreal environments, females must limit reproductive expenditures (Jarvinen, 1986).

Larger clutch sizes at higher latitudes are to be expected from Ashmole's hypothesis. Ashmole's thesis states that during winter, or some other resource-limiting period, populations are limited and the subsequent increase in resources accounts for an increase in clutch size. Thus clutch size is adjusted to the ratio between maximal and minimal levels of resources, and it also depends upon population density (Ricklefs, 1980). This concept is interesting because it is compatible with life history patterns for many mid- and high-latitude birds. For a given species of passerine, conditions are modified by migration, competition, and other variations that might occcur. To indicate seasonal variation in plant production, Ricklefs expressed resource changes as actual evapotranspiration (Fig. 14.4). Thus differences in actual evapotranspiration (derived from the relationship between monthly temperatures and precipitation) are proportional to primary productivity.

Purely theoretical discussions of the relationship of latitude to clutch size usually fail to consider the mechanics by which increased fecundity is achieved (e.g., Owen, 1977). The greater photosynthetic productivity, resulting from longer day lengths at high latitudes, provides for increased nourishment during folliculogenesis. This phenomenon supports the role of food as proposed by Ashmole, inasmuch as longer light periods promote the growth of photosynthetic plants.

Also, since ovulatory rates are stimulated by increasing day lengths and the stimulation is responsive to dosage, the increase of clutch size with latitude is to be expected. Surprisingly, students of clutch size variation in nature consistently fail to consider the effect of increasingly stimulatory day lengths at higher latitudes in the spring.

The generality that clutch size increases with latitudes is valid for some groups of birds, such as woodpeckers (Koenig, 1986), but not for others. Many species

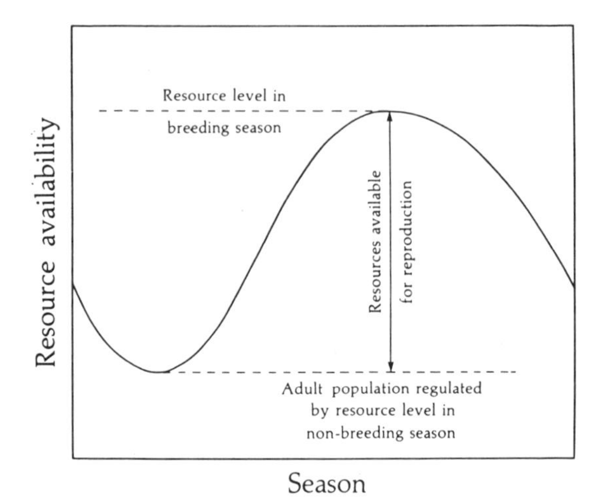

Figure 14.4 A graphic expression of Ashmole's hypothesis. Resources available for reproduction are the difference between that required in the nonbreeding season and that available in the breeding season. Clutch size is proportional to the ratio of the breeding season surplus to the adult population. (From Ricklefs, 1980, courtesy of the American Ornithologists' Union.)

of shorebirds (Charadriiformes), for example, usually lay four eggs, and departures from this pattern are not correlated with latitude (MacLean, 1972).

The partridge (*Perdix perdix*) typically lays larger clutches at higher latitudes (Pulliainen, 1971). This is true not only in Europe, where the partridge is native, but also in North America, where the species has had a rather brief history. There is no indication that introductions into the New World came from the same latitudes in Europe, and the relationship between fecundity and latitude appears to be an immediate response to day length, without overwhelming genetic control. Similarly, among the many species of birds introduced into New Zealand, most have clutch sizes that differ from those in their original homes (Klomp, 1970).

14.5.7 Strife among Nestlings

Among most species of large eagles, there is frequently sibling strife with mortality among nestlings. Among the larger eagles, productivity does not seem dependent on whether a nest contains one egg or two. For many species in Africa the annual production of young per pair is about 0.52 for species laying two eggs and 0.54 for those laying one egg. Among 205 nests of the Verreaux's eagle (*Aquila verreauxi*), 157 (76%) had two eggs. Of the two-egg clutches 14% failed completely, whereas 36.7% of the single-egg clutches failed (Brown et al., 1977). In nests with two eggs, the second chick usually succumbs to attacks by the older sibling, even when food is abundant. The cause of the difference is equivocal, for single-egg clutches characterize young females, which might lack the body condition as well as the hunting skills of an older female.

Intersibling strife seems not to occur among the smaller raptors (*Accipiter* spp., *Circus* spp., and *Falco* spp.) which may lay from four to seven eggs. Among siblings of the osprey (*Pandion haliaetus*), ranking weights vary from week to week. They eat without strife or competition, and there is little prefledgling mortality (Stinson, 1977).

14.5.8 Seemingly Imperfectly Adapted Patterns of Clutch Size and Incubation

In certain species the clutch size appears to be less than perfectly adapted. These examples will remind the reader that not all features and traits, including clutch size, are totally advantageous or maximally adaptive.

Along the Zambezi River the African skimmer (*Rhynchops flavirostris*) lays a clutch of two to four eggs. The first chick that breaks out leaves the nest, and the parents follow it, abandoning the remaining unhatched eggs (Roberts, 1976). This behavior could be adaptive if the first egg to hatch is destined to produce the most competitive individual, but there is no evidence that incubation length constitutes such selection.

The hadedah ibis (*Bostrychia hagedash*) commonly lays a clutch of three eggs but fledges only a single young. Mortality results from eggs and young falling out of the nest (Raseroka, 1975). One could suggest that a large clutch size provides for accidents to the infants or that the strongest chick ejects its weaker siblings. Although these explanations are thoroughly plausible, they do not seem satisfactory in view of the many species of birds that sustain substantial populations with clutches of one egg.

The crested penguins (*Eudyptes* spp.) usually lay a clutch of two eggs of unequal size. In 30 to 50% of the nests both eggs hatch, but invariably only a single young is fledged. Mortality of the chick is independent of the size of the egg or the size of the chick at hatching (Warham, 1975), suggesting that mortality in the nest may be random. The chinstrap penguin (*Pygoscelis antarctica*), which also lays a clutch of two eggs, has a hatching success of about 90% (Conroy et al., 1975). The Adelie penguin (*P. adeliae*) at Cape Crozier (Antarctica) rears a mean of less than one chick from a clutch of two eggs (Oelke, 1975). At Cape Hallet this species lays two eggs of uneven size, the first being slightly larger. The larger egg requires a mean incubation time of 35.5 days in contrast to 34.1 days for the smaller second egg. At hatching, about half of the chick's original yolk is unused (Reid, 1965).

All of these apparently maladaptive patterns could easily be explained as being advantageous for one reason or another. It is not difficult to imagine a scenario that would justify apparently wasteful patterns, but such explanations are speculative and unsupported by evidence. Although these examples are trivial and are not basic to concepts of evolutionary biology, these patterns do exist. Perhaps at an earlier date in the history of these species they did not suffer wastage, and genetic modifications have not kept pace with environmental changes.

Several species of birds place ovoid objects in their nests, which they then incubate. Some gulls are known to incubate stones along with their eggs. The ring-billed gull (*Larus delawarensis*) may incubate stones that bear a crude similarity to their eggs and will persist in covering these foreign objects even when their own eggs are removed. Apparently the incubating gull cannot distinguish these stones from her own eggs (Conover, 1985). Perhaps the most bizarre example is the emu (*Dromaius novaehollandiae*), which sometimes incubates paddy melons (*Cucumis myriocarpus*), fruits about the size and shape of emu eggs (Schrader, 1975). Presumably these are examples of extremely conservative behavior.

14.6 MAMMALS

Litter size or brood size in mammals is usually measured in terms of implanted embryos, and the mean number of embryos is often slightly greater than the mean number of neonates. In most species the number of implanted embryos approximates the number of ripe ova discharged at one time. In a few mammals the number of ova greatly exceeds the number of embryos (see Chapter 3, The Ovary, Section 3.2.2). Rarely are there implanted embryos of different ages, a phenomenon called superfetation.

In contrast to the positive correlation between adult size and number of ova in ectotherms, larger species of mammals tend to have smaller broods than smaller species do. In addition to this relationship, eutherian mammals exhibit a positive correlation between maternal and neonatal size and also duration of gestation. In many large ungulates the large size and advanced development of the neonate may be in response to its lack of a nest and its need to run to escape enemies. Also, many large herbivores are nomadic or migratory, and the young must be able to travel long before they are full grown. This relationship is not seen in carnivores, in which the young are altricial, commonly born in a den, and may number from two to four or more. In marsupials the young are very small, regardless of the size of the mother, but larger species tend to have smaller litters. The monotremes, the platypus and the two species of echidnas, lay one or two eggs and reproduce once a year (Griffiths, 1978).

Ova of marsupial and placental mammals are uniformly small and devoid of yolk, and the number of ova does not relate to the size of ova. In both wild and domestic species, nutrition clearly affects ovulatory rates, perhaps through the hypothalamus (see Chapter 12, Section 12.2).

14.6.1 Condition of Young at Birth

As in birds, infant mammals enter the world either in an altricial, or relatively helpless, condition or a precocial, or relatively advanced, condition, or in some intermediate state of development. The degree of in utero development relates to the number of young born in a single litter. Altricial mammals are devoid of hair and cannot thermoregulate. Their eyelids are unopened, their ear pinnae are laid tightly over the auditory meatus, and they are unable to walk. Precocial young, at the other extreme, have a normal covering of fur, and their eyes and ears are functional. They may take solid food, sometimes within minutes after birth. The development of precocity requires a longer gestation, which tends to reduce the number of young that develop within the uterus at one time. Thus precocial young usually occur in species with litters of one or a few, whereas altricial young are usually produced in larger numbers. Nearly embryonic young are produced by the marsupials, some of which produce only one young at a time, and also by the monotremes.

The caviomorph rodents, such as the porcupine and the guinea pig, exemplify the precocial state. Young are born in a very advanced condition, and they can walk and eat leafy material before the embryonic fluids have dried from their fur. Litters tend to be small in caviomorphs. Most mice and squirrels exhibit the altricial condition. The young are nude and blind and they cannot walk; their litters tend to be larger than those of caviomorphs of comparable size.

14.6.2 Litter Size and Parental Care

One might assume that altricial young receive more parental care than precocial young do and therefore species with altricial young should have smaller litters, but this is not always true. When litter size is small, parental care may be prolonged. While it is true that a young porcupine or guinea pig may become independent at an early age, parental care may be prolonged in the case of a whale, a horse, or an elephant. Extensive parental care with reduced fecundity characterizes long-lived species.

14.6.3 Fixed Litter Size

A few species produce litters of a nonvariable size. Among the many species of mammals that produce a single young at a time, the number can vary from single births to twins, which are uncommon. Single births characterize macropod marsupials, cetaceans, pinnipeds, most primates, most bats, most artiodactyls, most perissodactyls, and a few other major and minor groups. In the majority of these groups the young are precocial, primates being a conspicuous exception.

Litter size in species that bear a single young may not always be limited by ovulatory rate. In the pronghorn (*Antilocapra americana*) there may be from four to seven embryos, most of which die before implantation. The uterus may hold one or two large, implanted embryos, and twins are more common than single births (O'Gara, 1978). Although cetaceans normally have a single young in one pregnancy, there are occasional conspicuous departures from this pattern. Six and seven embryos have been found in the blue whale (*Balaenoptera musculus*), and six embryos have been found in the fin whale (*B. physalus*) (Arvy, 1974). In these finback whales the embryos were of different sizes, suggesting that ovulations were not simultaneous.

Among armadillos (Dasypodidae) polyembryony results in litters of from two to twelve monozygous young from a single ovum. In the nine-banded armadillo (*Dasypus novemcinctus*) the single ovum divides twice, and from these four cells develop four separate and identical individuals (Storrs and Williams, 1968). In the mulita armadillo (*D. hybridus*) as many as 12 embryos result from a single ovum (Hamlett, 1935).

In some marsupials litter size is fixed not by ovulatory rate but by the number of nipples of the mother. Many of these ova die in utero, but there may be more young born than there are nipples to provide milk for them. In the dasyurid *Dasyuroides byrnei*, litters of 20 or more embryos are reduced at birth to 6, the normal number of nipples for that species (Hill and O'Donoghue, 1913). *Antechinus stuartii*, another dasyurid, regularly produces more young than it has nipples, and the surplus young die (Lee and Cockburn, 1985; Selwood, 1983). This reduction of litter size at birth also occurs in the marsupial cat (*Dasyurus vivierrinus*), the Tasmanian devil (*Sarcophilus harrisi*), and the pigmy glider (*Acrobates pygmaeus*) Troughton, 1947).

14.6.4 Variation with Age

In many species of mammals the age structure of the population undergoes seasonal changes, alternating periods in which young or mature individuals predominate. These changes contribute to variations in fecundity, for age-dependent differences in ovulatory rates may account for seasonal changes in litter size. This is most clearly seen in those mammals having successive litters in a single season, with individuals born early in the year becoming sexually active in the same year, but it also characterizes long-lived species.

Litter size in voles (Arvicolidae) exhibits general patterns of change with season and age, and the two factors are not always clearly separable. Most voles attain sexual maturity at an early age, with the result that young born in the spring produce young a few weeks later, in the summer. Postpartum pregnancy is common in these rodents.

This age-dependent difference in fecundity has been known for a long time and was pointed out by Brambell and Rowlands in 1936 in the bank vole (*Clethrionomys glareolus*). Small (and presumably primiparous) females of less than 17.9 g were found with a mean of 3.7 corpora lutea, and this statistic increased consistently to 5.2 for females weighing more than 26 g (Fig. 14.5). The same trend has been subsequently confirmed for a great many species. The meadow vole (*Microtus pennsylvanicus*) in northern Minnesota has a mean litter size of 4.3 at a body size of 90 to 99 mm, which increased to a mean litter of 7.2 at 130 to 139 mm (Beer et al., 1957). In southern Indiana litter size of both *Microtus pennsylvanicus* and *M. ochrogaster* are positively correlated with body size (Keller and Krebs, 1970). In a captive population of *Microtus montanus*, fecundity increased regularly with age. First litters of subadult (not fully grown) females averaged 3.6, but the initial litter of adult females was 4.2. The mean size of the fifth litter for subadults was

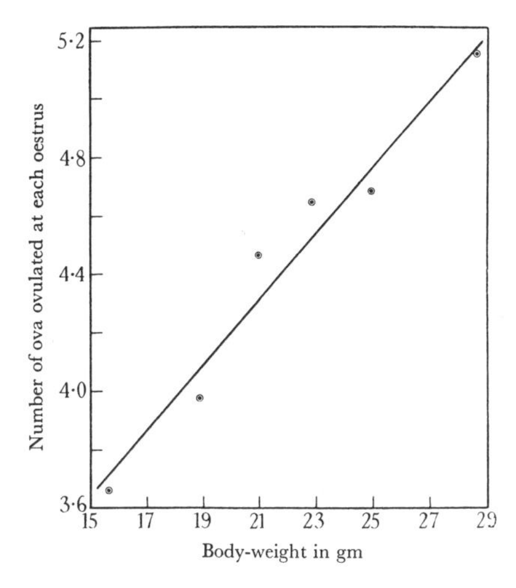

Figure 14.5 The relationship between number of ova ovulated and body mass in the bank vole (*Clethrionomys glareolus*). (From Brambell and Rowlands, 1936, courtesy of *Philosophical Transactions*, The Royal Society.)

4.9 and for adults was 5.8, the differences between age groups being statistically significant (Negus and Pinter, 1965). In a laboratory colony of the prairie deer mouse (*Peromyscus maniculatus bairdii*), litters varied from two to nine, and litter size was positively correlated with body mass of the mother at conception (Myers and Master, 1983).

In southern Finland mean litter size of the vole *Microtus agrestis* varies consistently with both season and age of female (Fig. 14.6). Similarly, in *Microtus californicus* and *M. montanus* in central California, the largest litters occur in the spring, when

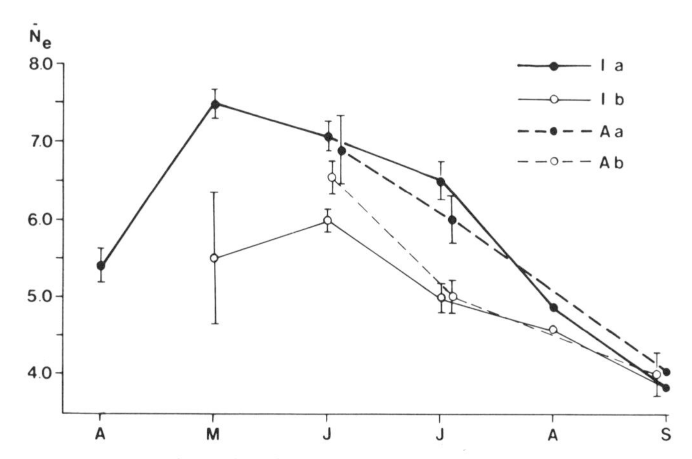

Figure 14.6 Mean number of embryos (N_e) with 95% confidence limits for *Microtus agrestis*. Solid lines (a) indicate data from overwintering females, and broken lines (b) represent litter sizes for females born in the spring of the year in which data are taken. Data are from two localities and year: I represents Ingles (1961), and A represents Ahtiala (1969). (From Myllymäki, 1977, courtesy of *Oikos*.)

most females are full grown (Hoffman, 1958; Lidicker, 1973; Krohne, 1981). The same pattern is seen in the red-backed vole (*Clethrionomys rufocanus*) in Finland (Kalela, 1957). The summer decline in litter size reflects the increase in reproductive activity by primiparous females, but it may also be caused by the effect of plant growth on mammary tissue and fecundity: as forbs mature and go to seed, there is a parallel involution of mammary tissue (Myllymaki, 1977). Ripening and drying culms of grasses may contain fungi that suppress the release of prolactin (see Chapter 12). A late-season decline in fecundity may also result from the approach of refractoriness (see Chapter 11). Thus secular changes in fecundity are caused by a number of exogenous variables.

Such secular variation in fecundity is not peculiar to voles. Litter size in the thirteen-lined ground squirrel (*Spermophilus tridecemlineatus*) increases with age of the mother. In northern Texas, primiparous females have mean litters of 4.9, in contrast to litters of 7.0 for multiparous females (McCarley, 1966).

The same relationship between body mass and fertility is seen in some domestic mammals that produce more than a single young. In sheep the percentage of multiple ovulations increased markedly with the body mass of the ewe at the time of mating (Fig. 14.7). In the red deer hinds (*Cervus elaphus*) fertility drops in the year following lactation. *Milk hinds*, those that have nursed a calf for the previous year, experience a 73.7% fertility in contrast to 90.1% in *yeld hinds*, those that were barren the previous year. Lactation results in a loss of both fat and total body mass in the next mating period (Clutton-Brock et al., 1982).

Contrary to most other mammals, in the dasyurid *Antechinus sturatii* fecundity declines significantly from the first year to the second (Selwood, 1983). In the brown bandicoot (*Isoodon obesulus*), however, litter size increases with body mass, and the oldest females produce the largest litters (Stoddart and Braithwaite, 1979).

14.6.5 Latitudinal Gradients

As with birds, most mammals within a given taxon (a genus or a family) produce larger litters at higher latitudes, but, again, the underlying mechanism is not established.

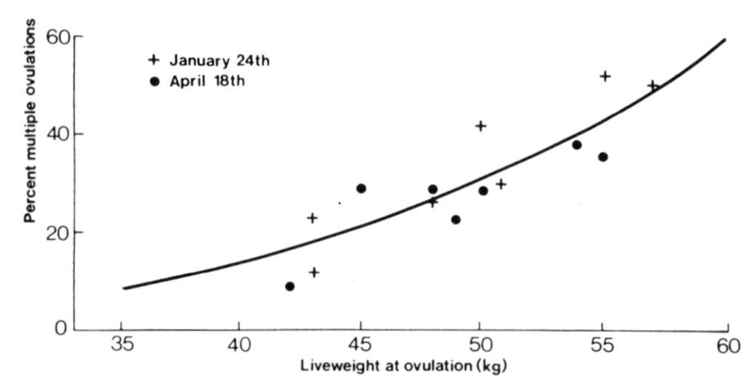

Figure 14.7 Percent of multiple ovulations of Merino ewes correlated with body mass (liveweight). (From Morley et al., 1978, courtesy of Agricultural Systems.)

This trend is assumed to be genetically determined and adaptive. Although it certainly appears to be adaptive, the genetic control allows for some intraspecific variation. Productivity of photosynthetic organisms increases with longer day length, resulting in a surge in plant growth on which many mammals feed. This sudden increase in photosynthetic activity, together with the decrease in the daily period of darkness at high latitudes, promotes fecundity independent of genetic influences.

A comparison of a large body of data on litter size in New World voles illustrates the trend, but some species have large year-to-year differences. Smaller mean litter size of low-latitude voles reflects, in part, the younger age at which primiparous females reproduce, for they produce smaller litters than do the more mature over-wintered females that dominate the more boreal populations (Innes, 1978).

The latitudinal trend for various species of *Peromyscus* indicates substantial variation of in utero litter size. The white-footed mouse (*Peromyscus leucopus*) under the same laboratory conditions of light, temperature, and food indicated only slight latitudinal differences, with larger litters at the lower latitude. Females from Campeche, southeastern Mexico, had a mean litter size of 4.67 (SE 0.11), slightly more than females from Michigan, with a mean litter size of 4.28 (SE 0.11) (Lackey, 1978). Among populations of deer mice (*Peromyscus maniculatus*) from contrasting latitudes, temperature has no role in reproductive performance (Bronson and Pryor, 1983).

Litters of the snowshoe hare (*Lepus americanus*) are usually larger at higher latitudes, and this trend tends to remain in captive populations. Hares from central Alberta (43°N) and northern Wisconsin (46°N), when reared at 43°, had essentially the same litter sizes that the parent stock produced in the wild (Keith et al., 1966).

Among species of rock hyrax (*Procavia* spp.) there is a pattern of both (1) larger litters at higher latitudes and (2) year-to-year variation in litter size within populations. In tropical Africa, mean litters range from 2.0 to 2.2; in Cape Province, means range from 3.2 to 3.3. In the relatively high latitude of Cape Province, however, mean litters of *Procavia capensis* vary from 1.5 to 3.5, the larger litters from regions of greater rainfall and heavier vegetative growth. Larger litters also occur in years of greatest rainfall. Among a number of xeric species, mean litter sizes vary with year-to-year changes in rainfall (Skinner et al., 1977). Fecundity is low in the Australian tropical murid rodents, the Hydromyinae, which entered Australia some five to ten million years ago. It is speculated that smaller litters have resulted from climatic unpredictability (Yom-Tov, 1985), which is contrary to some hypotheses on fecundity.

14.6.6 Variation with Elevation

Although many mammalian species occur over a wide range of elevations, there are few studies of a single species showing elevational ranges and trends in litter sizes.

In ground squirrels (*Spermophilus* spp.) reproduction is affected by hibernation, which is more prolonged at higher elevations. Inasmuch as hibernation draws upon stores of body fat, populations at higher elevations have a shorter summer period in which to raise young and prepare for winter.

In southwestern Alberta the Columbian ground squirrel (*Spermophilus columbianus*) produces smaller litters at 2170 m than it does at 1500 m. Seven litters from females captured at 2170 m averaged 3.0 young in contrast to an average of 3.6 young from females captured at 1500 m. Time of emergence from hibernation is always later at the higher elevation than at the lower elevation. In these populations, at the same latitude, the growing season for plants is considerably shorter at the higher elevation, and presumably this provides less energy for reproduction (Murie et al., 1980). The golden-mantled ground squirrel (*Spermophilus lateralis*) similarly produces smaller litters at higher elevations (Bronson, 1980). Between two populations of the golden-mantled ground squirrel at 2650 m in California, one (at Castle Peak) lived where the snow cover was complete for about seven months annually, and another (near Bodie) experienced about five and a half months of intermittent snow cover. Of females that were captured in the wild and produced young in captivity, those from Castle Peak had mean litters of 4.5 (range 4–5, $N = 13$), and those from Bodie averaged litters of 6.0 (range 4–9, $N = 25$) (Phillips, 1981).

In nonhibernating mammals, in which there is not such an urgency to accumulate body stores for winter, litter size appears to increase with altitude. The deer mouse (*Peromyscus maniculatus*) in an altitudinal transect in central California showed an increase in live embryos from 4.00 $\pm$ 1.00 at 4500 ft to 4.77 $\pm$ .19 at 12,400 ft (Dunmire, 1960).

Fecundity in mammals reflects a complex of factors, both phylogenetic, such as degree of embryonic development, hibernation, and parental care, and exogenous, such as latitude and elevation. Mammalian fecundity is constrained by the capacity of the uterus and the condition of the neonate in eutherians. Precocial young grow larger and tend to be fewer in number at birth than altricial young do. Young of marsupials are virtually embryos at birth, but the number of young is extremely variable. Many small marsupials have large litters, but most larger species have a few or only a single young. Monotremes lay only one or two eggs, which produce altricial young.

14.6.7 Short-Term Environmental Changes

As with birds fecundity in mammals that normally have several young may vary with environmental resources. At times of food abundance a female may accumulate sufficient energy reserves to provide for a larger than normal litter.

Litter size of the round-tailed ground squirrel (*Spermophilus tereticaudus*) varies with the winter rains. Near Phoenix (Arizona) the 1960 winter (October–February) rainfall was 225 mm, and the mean litter size was 9.0 ($N = 10$, range 8–12), but in the relatively dry winter of 1970, when 72 mm of rain fell, mean litter size was 3.3 ($N = 15$, range 2–5) (Reynolds and Turkowski, 1972). Captive roe deer (*Capreolus capreolus*) maintained in outdoor enclosures in Sweden become increasingly fecund when given supplemental food. With additional nutrition triplets are produced by young does, but, without supplemental food single births are the rule (Tegner, 1964). A correlation exists between unusually warm ambient temperatures and larger than normal litters in the prairie deer mouse (*Peromyscus maniculatus bairdii*) in nature, and heavy rains and lower than normal temperatures are followed by small litters (Myers et al., 1985).

14.7 SUMMARY

An animal makes maximal use of its resources. Fecundity is essentially driven by nutrition in the majority of vertebrates, but is modified by specific morphological restraints as well as exogenous factors. Modifying patterns, such as parental care and viviparity, tend to reduce fecundity and increase survival. With increasing latitude, fecundity tends to increase; within a given species, fecundity tends to increase with body mass.

The common pattern among many species of teleost fish is the ancestral one of producing a large number of floating eggs, which are subject to heavy loss. Generally egg mass is positively correlated with body mass. A reduction in number of eggs occurs with the development of demersal eggs, which are usually attended to by one or both parents, whose function is to provide oxygen and protection.

Spawning in amphibians follows some of the patterns seen in fish. Some anurans deposit large numbers of eggs that float on the surface, but eggs that are submerged tend to be larger and fewer and are usually guarded. Several groups of anurans, salamanders, and caecilians transit to terrestrial egg laying. This transition places the eggs in an oxygen-rich environment, which carries with it a risk of desiccation. Terrestrial eggs are always guarded, and they are larger and fewer than those laid in water.

Reptilian eggs, usually relatively few in number, are always deposited on land. Fecundity varies widely and declines with parental care, increased body size, and deposition of eggs under bark in trees. There is no clear trend of varying fecundity with latitude.

Fecundity becomes increasingly complex with a variety of reproductive strategies in birds. Many species are altricial, and for equal amounts of egg mass, altricial birds lay more eggs than precocial birds do. Fecundity tends to increase at higher latitudes. This increase could result from (1) the stimulatory effect of longer day lengths on the avian reproductive system, (2) greater environmental productivity (photosynthetic activity) and enhanced nutrition with longer day lengths, and (3) additional foraging by the parents during longer periods of day length. There is strong evidence that in birds, as in other vertebrates, fecundity is modified by nutrition, with many adaptations to different modes of living.

Fecundity, expressed as clutch or litter size, is affected by a diversity of exogenous factors. Each given species appears to have upper and lower limits. For those species that have single births or clutches of one or two eggs, variability is limited, and fecundity may be almost entirely under genetic control. For species with several or many eggs at one time, upper and lower limits appear to be genetically controlled. For a given population, means may change from one habitat to another or one locality to another.

Among the factors that tend to be correlated with reduced fecundity are viviparity, parental care, and hibernation. Associated with greater than average fecundity are nutrition, long days (which occur at high latitudes), age, and mass.

The productivity of mammals varies with the body mass of the adult female, with larger species having fewer young, but larger individuals within a species having more young. Fecundity tends to be greater in herbivorous mammals. The litters are not necessarily larger than in carnivorous species, but they are usually more frequent. During times of food abundance, mammals are usually more prolific. Mammalian fecundity also varies with the interaction of such exogenous influences as hibernation, elevation, and altitude.

If a species is using its reproductive resources most efficiently, the mean clutch size, which is usually close to the most frequent clutch size, should be the most productive. At the same time, natural selection should favor the genotype leaving the largest number of individuals. Because mean clutch size or mean litter size or a given species varies from time to time and from place to place, the most frequent clutch or litter size need not be, and often is not, the most productive. Less frequent but larger broods may leave more survivors.

Some have suggested that increased fecundity at high latitudes is a response to increased predation (e.g., Lord, 1960) or that clutch size responds to predation (Perrins, 1977). This notion is appealing but lacks a mechanism by which predators can induce fecundity in their prey (Cody, 1966). Slobodkin (1961) pointed out that predation should produce no biological changes in a prey population if it simply replaces another cause of mortality. On the other hand, if the age structure of the prey population is altered by predation, then compensatory changes may occur in the response of the prey. There seems to be no simple way for a predator to alter litter size and clutch size. In reality, predators, like most causes of mortality (disease, starvation, sudden temperature changes), exert their greatest toll on young individuals. In this situation which is the most common, predation has no special effect on prey populations. Mortality, regardless of its cause, tends to decrease the various deleterious effects of crowding and eventually tends to increase the available food for survivors.

The popular and widely held concept that fecundity in birds is genetically determined by the number of young a pair can care for (Lack, 1954, 1968) fails to fit some sets of field data (Ashmole, 1963; Cody, 1966; Högstedt, 1980; and others). Food supply may affect seasonality, fecundity, and mortality. Several major groups of sea birds are characterized by small clutches in addition to late maturity. As populations increase, reproductive rate declines concurrently with declines in food supply, which limits density without a change in mortality rate. This effect appears most likely in tropical seas, which are relatively unproductive at best, and periodic restrictions in breeding have been noted in brown boobies, black noddies, wideawakes, and others (Ashmole, 1963).

For a species to maintain itself in nature, the totality of reproduction must be adequate, but it need not be perfect. If any part of the life cycle is sufficiently weak, a species will eventually disappear; the fossil record testifies to the frequency of extinction. One assumes that extinction is caused by a climatic change to which a species cannot adapt or to competition from one or more recently evolved and better-adapted species.

We should remember that natural selection is a crude process that changes species slowly, a result of selective forces of the physical and biotic parts of the environment. Within the lifetime of a species, environments can change rapidly. If genetic constraints cannot provide a species with sufficient flexibility to maintain its numbers in the face of environmental change, extinction will follow.

The faunistically devastating effects of the Pleistocene were relatively very recent. Changes accompanying the Pleistocene were not only concomitant with the disappearance of many vertebrate species—a fact widely acknowledged—but have left us with a fauna that is, in one way or another, sufficiently flexible and tolerant of change. Not only have they survived the Pleistocene but they persist under varying modern conditions as well. A large number of avian field studies have been made in northern Europe. During the Pleistocene the entire avifauna was totally evicted from its current range in northern Europe, and the Little Ice Age must have had a devastating effect on avian reproduction in that region. It is reasonable to expect that the avian genetic control of reproduction, which was exposed to a series of varying environmental regimens during the Pleistocene, Holocene, and Little Ice Age, is capable of expressing a broad spectrum of responses to environmental changes.

BIBLIOGRAPHY

Gould SJ and Lewontin RC (1979). The spandrels of San Marco and the Panglossian paradigm: A critique of the adaptionist programme. *Proceedings of the Royal Society*, **205**: 581–598.

Hussell DJT (1972). Factors affecting clutch size in arctic passerines. *Ecological Monographs*, **42**:317–364.

Klomp H (1970). The determination of clutch-size in birds. *Ardea*, **58**: 1–124.

Lack D (1956). Evolutionary ecology. *Journal of Animal Ecology*, **34**: 223–231.

Lack D (1968). *Ecological Adaptations for Breeding in Birds*. Methuen, London.

Stearns SC (1976). Life-history tactics: A review of the ideas. *Quarterly Review of Biology*, **51**: 3–47.

Stonehouse B and Perrins C (Eds) (1977). *Evolutionary Ecology*. Macmillan, London.

Watson A (Ed) (1970). *Animal populations in Relation to Their Food Resources*. Blackwell, Oxford.

Winkler DW and Walters JR (1983). The determination of clutch size in precocial birds. In *Current Ornithology* **1**: 33–68. Plenum, New York.

15

FACTORS MODIFYING POPULATIONS

15.1 THE NATURE OF POPULATIONS

A population is an assemblage of animals unified by genetic similarity and spatial continuity and confined by ecological boundaries. This aggregation of individuals tends to act as a unit in many of its changes over time. The contrast between an individual and a population assists in understanding each. Whereas an individual experiences birth, growth, reproduction, and death, a population has a range of individuals of different ages (or an age structure) and may contract or expand its size, sometimes 100-fold or more and sometimes with regularity.

This chapter explores some of the factors altering the size and composition of populations. The study of populations is fraught with the illusive nature of the population as a unit and the need to express its characteristics as indices and estimates. You can hold a mouse in your hand, measure it, dissect it, and describe it with considerable confidence. You cannot grasp a population and usually can only see a small part of it at one time. Not only must its density be estimated from a small sample, but the essential aspects, such as age structure, birth rate, and death rate, are all estimated. Immigration and emigration are usually educated guesses. Data on populations cannot be considered to be "hard data."

Most populations of vertebrates tend to maintain densities about which they fluctuate. There is a dynamic balance altered by changes in recruitment and mortality; actual densities vary throughout the year and from one year to another. In some species fluctuations are major and recur with well-known regularity, whereas in others population changes are major but irregular and unpredictable. In most species fluctuations may be trivial and may go unnoticed by all except the most careful observer.

Demographically, populations are modified by both extrinsic (or exogenous) and intrinsic (or endogenous) factors; the two can interact, and neither one should be neglected in a study of population changes (Lidicker, 1978). Population density itself may influence the direction and degree of changes. Such changes are called density dependent, the effects of which may be direct or indirect. A departure from mean conditions of food, cover, temperature, or other variable exogenous factors, immigration and emigration, which may be either independent of density or associated with density, may also change birth rate and/or mortality rate. Certain aspects of a population, such as age structure, influence rates of increase and decrease. This chapter documents some well-known causes of alterations in both natality and mortality and reviews some carefully studied examples of major population fluctuations.

The vast number of research reports on wild vertebrate populations illustrates the extreme diversity and complexity of reproductive patterns, demographic changes, and the apparent mechanisms of density control. Lidicker (1978) reviewed many of the major difficulties in designing population research and reporting results. Densities are the consequence of many aspects of a species' life history, of environmental patterns (as well as deviations from usual environmental patterns), and of interactions between life history and environment.

Unrestricted or exponential growth proceeds geometrically and is eventually limited by carrying capacity (*K*) or environmental restraint, or by some intrinsic, density-dependent factor, such as territoriality. Generally, regulating influences become increasingly effective as a population trend continues, and they will eventually reverse the direction of change. This relationship acts to limit the swings of the dynamic equilibrium.

15.2 DENSITY-DEPENDENT PHENOMENA

Certain inherent aspects of natural populations act to stabilize densities. These factors retard population expansion when there is a great increase in number, but they may also promote reproduction and growth when numbers are low. Some density-dependent phenomena have been mentioned elsewhere in this book, especially in Chapter 5, Section 5.4.1.

Density-dependent phenomena in insects were discussed in some detail by Fujita and Utida (1953). They considered the effects of density on population growth and reproduction. Among insects there may be several distinct types of curves of density-dependent reproduction. Any one of these curves is characteristic of a given species and, moreover, may change with changes in density over time. Mortality rate also changes with density. Mortality and natality do not necessarily balance one another so as to maintain a static density, and the optimal density varies with the species (Fujita, 1954). This concept applies to vertebrates.

Unnaturally high crowding in some species, especially several well-studied rodents, stimulates growth of the adrenal cortex, which has been taken as symptomatic of stress (Christian, 1961; Christian and Davis, 1956). The stress syndrome may well lie at the basis of some intrinsic density-dependent population responses, but most of the evidence comes from captive populations and is not known to operate in cycles in wild vertebrates.

Although density-dependent mechanisms appear to be effective at *limiting* population growth at high densities, the recovery of growth at low densities is frequently a slow process. Those species with parental care and with a low reproductive potential cannot recover rapidly from low densities. A high reproductive potential accounts for large numbers of young or larvae, most of which perish at high densities. At low densities more of the young or larvae survive; density dependency provides for a population recovery. Survival of larval fish or amphibians at any population level may be density independent if survival should be reduced by low temperature or some other density-independent factor. In some fish with relatively high fecundity, however, there is little real evidence that density dependency alters survival of eggs and larvae; but recovery may be related to spatial and temporal heterogeneity in age of maturity (Garrod and Horwood, 1984).

Density-dependent relationships pertain directly to an individual's demand for space as well as the resources produced by that space. The minimal demand is the space an individual occupies. For a filter-feeding invertebrate, such as some spiders, this might be adequate, but a filter-feeding vertebrate, such as a baleen whale, forages for food and demands a considerable volume of space sufficient to produce food needed for sustenance. As the density of a population declines, the per capita resources available for use by the remaining numbers increase. The relationship is seldom a simple ratio of population/resource, for not only does the resource promote the population, but the population increases may temporarily depress the productivity of the resource if not the resource itself. Frequently predators of several species of prey may vary in abundance and change their diet with changes in population densities of their prey.

One could imagine that such a density-dependent relationship would preserve a specific level of density, but there is usually a lag in the suppressive effect of high density, so that a population may exceed the level that can be supported by its habitat. At other times the population may fall far below the density that the habitat can support. If an adult group of a grazing mammal exploits its food supply at exactly the same rate as the plant replenishes itself, the population might be in equilibrium with its food source, even though the numbers of grazers may not be constant. If, on the other hand, the individuals within the population are youthful and growing rapidly, the cumulative mass of the population continues to increase even without an increase in the number of individuals. Thus essentially unchanged densities can reduce the habitat's capacity to support their numbers. Carrying capacity can be reduced by excessive densities, such as the destruction of vegetation by a large number of deer or voles. Consequently, when the carrying capacity of the environment deteriorates and becomes inadequate, a drastic suppression of the population follows, either through a reduction in recruitment or an increase in mortality or both. The decline in population may continue until the population is substantially below the original carrying capacity. Thus equilibrium of the population density may be preserved over the long term, but the species may be characterized by short-term fluctuations.

15.2.1 Density-Dependent Reproduction

High densities may either increase or retard recruitment, according to the particular habits of a species, the degree to which competition is critical, and the presence of predators. In a territorial species, an increase in density commonly tends to depress reproductive activity, unless densities are so low that heterosexual encounters are infrequent. On the other hand, in a colonial species, an increase in density appears to promote reproduction and enhance survivorship. Even in colonial species, however, very high densities may reduce rates of reproduction. As with insects, the pattern of density-dependent behavior, or type of curve, varies from one species to another.

15.2.1.1 Density-Dependent Limitations. In aquaria, where fish are confined to a finite and specific body of water, crowding may suppress fecundity and eventually mediate population control. When maintained at constant temperature in aerated water under a 12L:12D photocycle, guppies (*Poecilia reticulata*) showed no difference in growth rate or body size, but females in low populations had more embryos and greater ovarian weights. At higher densities females had a higher number of vitellogenic oocytes, perhaps because of a prolonged period of vitellogenesis (Dahlgren, 1985).

Experimental manipulation of populations frequently provides insight into the mechanics of density-dependent events. Following a reduction (through culling) in a colony of herring gulls (*Larus argentatus*) from 1972 to 1981, young gulls began to nest one year earlier, and the egg size, usually small for young gulls, increased over that which obtained in years prior to 1972 (Coulson et al., 1982).

For some colonial species excessive density may reduce the chances for survival, and survivorship declines with increases in density. This may be due to a scarcity of nest sites or a reduction of food at high densities. This pattern is seen for species of voles, rabbits, and hares, very large colonies of sea birds, and many other vertebrates.

On islands off the coast of Northumberland (United Kingdom) the shag (*Phalacrocorax aristotelis*) competes for nesting sites. As densities increase, the mean number of fledglings declines. Following a high mortality of adults from red tide (caused by *Gonyaulax tamarensis*), productivity per pair markedly increased (Potts et al., 1980). A proper nest site is essential for successful rearing of young, and in this example nest sites constituted the mechanism of density-dependent control.

Productivity of the mallard (*Anas platyrhynchos*) sometimes varies with density of breeding pairs (Fig. 15.1). Duckling mortality is also density dependent: as duck populations increase, there is increased predation on nests by carrion crows, and removal of crows was followed by an increase in nesting success (Hill, 1984).

15.2.1.2 Density-Dependent Enhancement. Among colonial species of birds, high densities tend to protect the group from losses. Some colonial species may also obtain foraging information from colony members, thereby reducing time spent

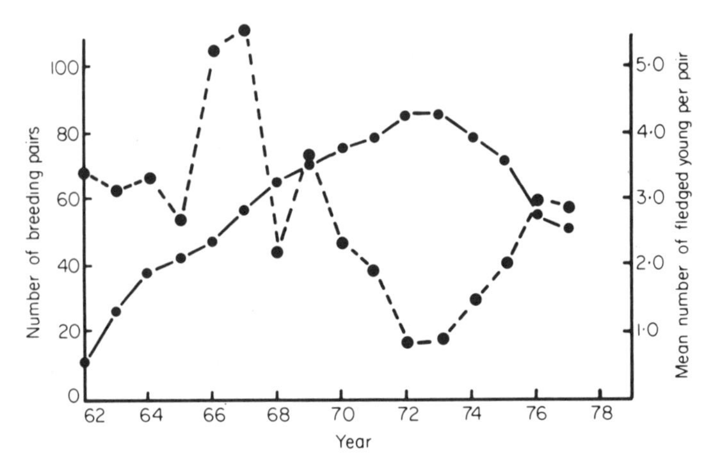

Figure 15.1 Number of breeding pairs of mallards (*Anas platyrhynchos*) (continuous line) and mean number of young produced per pair (broken line) at Sevenoaks, Kent (England) from 1962 to 1977. (From Hill, 1984, courtesy of the British Ecological Society.)

in searching for food (Greene, 1987). The guillemot (*Uria aalge*) enjoys greater reproductive success when nesting in high densities, apparently because greater numbers of adults concentrated in a colony reduce predation by gulls (Birkhead, 1977). Nesting success of the marabou stork in Uganda is greatest in colonies of medium size (40–70 nests) and declines above and below that range (Pomeroy, 1978), although the causes are not apparent. In some tern colonies loss of young appears to be independent of food supply.

The lapwing (*Vanellus vanellus*) may nest in loose aggregations, with territories in contact. Their proximity provides a nesting area with protection from predation by the carrion crow (*Corvus corone*), and predation increases with greater distances between lapwing territories (Elliot, 1985).

Colonial nesting is not confined to birds. The bluegill sunfish (*Lepomis macrochirus*), a small freshwater teleost, nests in colonies of up to 500 nests, and the congregation of adults seems to deter predators (Dominey, 1981).

15.2.2 Density-Dependent Mortality

The limitation of populations by their own densities may include both the inhibition of reproduction and the acceleration of mortality. These functions are separate, and their combined effect is the reduction in the rate of increase or an actual reduction in numbers. To say that the mortality rate of a species is dependent on its own density fails to specify the mechanism by which population control is effected. Density-dependent control is, nevertheless, an old concept, even though the mechanism may not always be understood.

Many authors have pointed out that herbivores at high densities have a tendency to consume vegetation more rapidly than it is replaced and, in effect, "eat into their capital" (Lidicker, 1975). This has been an underlying theme of much research on the cyclicity of game and pest species, and there has long been much evidence to support the concept that density dependency is frequently based on the density-influenced limitation of vegetation (Koskimies, 1955).

In voles, which have what amounts to an annual life cycle, adult survival may be seasonal due to age or density dependent, with greatest mortality in winter and spring (Myllymäki, 1977; Wolff and Lidicker, 1980; and many others). In two four-year cycles of voles (*Clethrionomys rufocanus*, *C. rutilus*, and *Microtus oeconomomus*) in Lapland (northern Sweden), seasonal breeding ceased early in peak years, while preferred food was abundant (Andersson and Jonasson, 1986).

Voles tend to reproduce on a seasonal schedule, and heavy mortality of adults follows, mostly at the close of reproduction and shortly thereafter. This decline appears to have a strong density-dependent element. Experimental removal of 50% of the adult female bank voles (*Clethrionomys glareolus*) prior to reproduction in the spring was followed by a decline in the postreproductive mortality of both sexes. The experimenters suggested that the elimination of half of the females reduced male fighting, thus lowering the decline in males (Gipps et al., 1985).

Winter mortality of reindeer (*Rangifer tarandus*) reduces the subadult population, primarily through food shortage and severity of weather, and such reductions are greater at higher densities. There is, moreover, a decline in fecundity, especially in younger does, with increases in density. The effect of density is mainly through the does, and indirectly through the vigor of their calves. During winters of food shortage, the does will abandon their calves in order to preserve their own integrity (Skogland, 1985).

15.2.3 Supplemental Feeding Experiments

Provisioning wild populations with additional food tends to confirm density-dependent events and aids in appraising the role of food in the control of natural populations. Results of such experiments are not without problems, for the supplemental foods are sometimes commercially prepared and, although nutritious, may not be as well accepted by wild vertebrates as their familiar foods.

Food is implicated in the fluctuations of many voles. When supplemental food (commercial mouse chow) was provided to a wild population of *Microtus pennsylvanicus*, they grew faster, became more fecund, enjoyed greater juvenile survival, and reached greater densities than in a control group. Nevertheless, a population decline followed at the same time as in the control group (Fig. 15.2). The decline occurred with lowered survival and emigration, suggesting that food is not the only critical element in vole cycles. Spacing and lowered survival may be density-dependent aspects independent of food (Desy and Thompson, 1983).

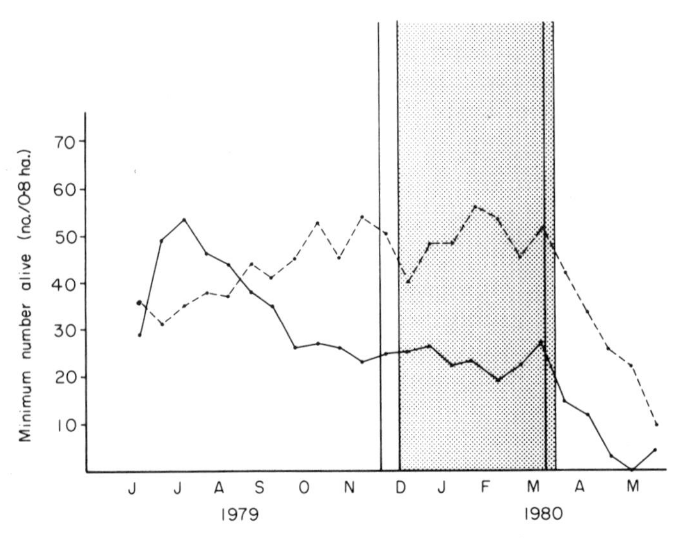

Figure 15.2 Minimum number of voles (*Microtus pennsylvanicus*) on a grid with supplemental food (broken line) and on a control grid (continuous line). Bold vertical lines indicate the nonreproductive period (November 20 to March 8), and the stippled area indicates winter months (December 1 to March 15). (From Desy and Thompson, 1983, courtesy of the British Ecological Society.)

Supplemental feeding of wild populations has sometimes shown equivocal results. A twofold or threefold increase in densities of *Peromyscus maniculatus* and *Clethrionomys rutilus* was observed after feeding them sunflower seeds, and it was postulated that this increase was caused by enhanced survival of young and by immigration. The artificial addition of food did not extend the reproductive season or increase the number of litters; there were no data on litter size (Gilbert and Krebs, 1981). Similarly, when food was provided to a wild population of *Peromyscus leucopus*, mortality was reduced, especially during the period of population growth, perhaps reflecting the better nutrition of lactating females (Bendell, 1959). When a population of Douglas squirrels (*Tamiasciurus douglasii*) was provided with supplemental food (whole oats and sunflower seeds), the population increased five- to tenfold over a control population. Pregnancy increased but the breeding season of females was not extended. Population increase in this case was attributed to greater incidence of reproduction of females and immigration (Sullivan and Sullivan, 1982); there were no data on litter size.

In a field experiment designed to compare the roles of density and nutrition of house mice in Australia, high-protein food (oats and sunflower seeds) was contrasted with low-protein food (rice). In addition, juvenile mice were culled from two plots, so that there were four kinds of study plots, with four plots of each kind. Pregnancies increased only on the plots with high-protein food, even though food was abundant on plots with rice. Culling had no effect, and it was assumed that removal of juvenile mice may have simulated normal dispersal of young individuals (Bomford and Redhead, 1987).

15.3 DENSITY-INDEPENDENT FACTORS

Population changes may be responsive to certain reproductive factors that relate to population structure but are relatively independent of population density. In many short-lived species populations fluctuate not only in density but also in the relative numbers of young and mature or aging individuals, which differ in breeding performance. These changes in survival rate of young may be independent of mortality of old individuals, and thus the age structure of a population becomes fundamental to fluctuations in density.

15.3.1 Age-Dependent Reproduction and Survival

In a great many vertebrates in all classes, reproductive efficiency, the successful production of offspring increases with age. This relationship is not simply due to an increase in fecundity, which is important in itself (see Chapter 14), but also to the success in rearing offspring, or perinatal survival. It is not clear whether greater reproductive success results from breeding experience, which one might expect in species that care for their young, or some physical or physiological aspect that changes with age.

Survivorship of captive white-footed mice (*Peromyscus leucopus*) from birth to day 28 tends to increase with increased parity of the mother. Litter size also increases with parity. Thus productivity, through an increase of both fecundity and survivorship of young, is positively correlated with parity (Fleming and Rauscher, 1978). Mortality of red deer (*Cervus elaphus*) calves is appreciably greater in those born of very young as well as aged doe, with maximal survival in doe seven to ten years of age (Guiness et al., 1978). This would tend to preclude experience as a factor in the greater success in older mothers.

In some long-lived birds, especially marine species, success in reproduction increases with age. In the arctic skua or jaeger (*Stercorarius parasiticus*), pairs that had bred together for the previous two years fledged a mean of 1.51 ± 0.034 young, in contrast to a mean of 0.65 ± 0.066 fledglings from pairs with no previous nesting experience (Davis, 1976). Likewise, the Manx shearwater (*Puffinus puffinus*) has greater breeding success when pairs have nested together for two or more years: newly mated pairs fledged 66% of their chicks in contrast to a 79% success of established pairs (Brooke, 1978). Success in fledging young increases with age and reproductive experience, up to 12 years of breeding history in the fulmer (*Fulmarius glacialis*), and breeding success increases with the duration of the pair bond (Ollason and Dunnet, 1978). Initial breeding of the goldeneye (*Bucephala clangula*) results in small clutches relative to those laid by older, experienced females (Dow and Fredga, 1984).

The relationship between breeding success and parity is widespread and probably prevails among most multiparous vertebrates.

15.3.2 Age-Dependent Mortality

Rates of mortality vary widely among major groups of vertebrates. Within a given generation, mortality may be very heavy in the early stages, relatively constant throughout life, or concentrated near the end of reproduction. Probably most species have a characteristic pattern of mortality. Extremely fecund fish, spawning millions of eggs at one time, usually suffer high prereproductive mortality, and in other species massive mortality may occur after the major reproductive activity.

Very young fish may face a *critical period*, a time of high mortality, presumably when the yolk sac is absorbed and the fry begins to feed. Winds or changes in temperature or current can cause early mortality. The difficulty in identifying and evaluating the critical period lies in the methods of collecting small fry and in estimating very early growth rates. A critical period may not characterize all pelagic spawners and may depend upon the time of spawning.

Mortality in early life characterizes many species, and greatest losses are seen in the most fecund species. The greatest variation in mortality is seen in the earliest stages, especially in the egg and larva. Fluctuations in recruitment of new individuals into a population result from changes in (1) fecundity and (2) very early mortality. In pelagic fish, fecundity is usually high, and survival of eggs and larvae depends upon satisfactory conditions. There are frequently substantial year-to-year variations in environmental conditions, especially in temperature and food, and survival of a single year class may be either very high or very low (Nikolskii, 1969).

High success in survival of one year class may result in its dominating the entire population for several years; total failure will leave a hiatus in the age structure of the population. If there is high mortality in the spawning during one year, the adults of the previous year usually continue to spawn large numbers of eggs the next year. If the species normally lives for several years, a large age class will gradually lose its prominent position, and there is an inherent tendency for a return to the original distribution of the several age classes.

The prey of larval anchovy (*Engraulis mordax*) in the ocean is concentrated in patches. A high rate of survival of larval anchovy depends upon relatively calm surface winds, which seem to allow large concentrations of plankton on which the larval fish feed (Peterman and Bradford, 1987).

In most highly prolific amphibians, early losses depend upon unpredictable environmental conditions. In the spotted salamander (*Ambystoma maculatum*), spawning may extend over two or more months, and embryonic mortality may be high if a freeze follows oviposition (Fig. 15.3). Early development of amphibians also depends upon adequate water and adequate amounts of dissolved oxygen.

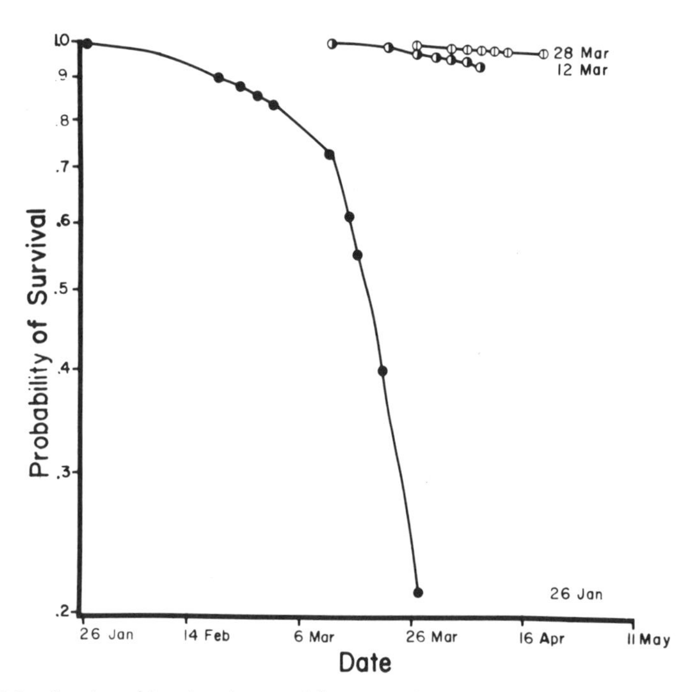

Figure 15.3 Survivorship of embryos of the spotted salamander (*Ambystoma maculatum*) from eggs laid on three different dates. Early breeding provides the advantage of rapid growth prior to drying on the pond, but it entails a risk of freezing. (From Harris, 1980, courtesy of the American Society of Ichthyologists and Herpetologists.)

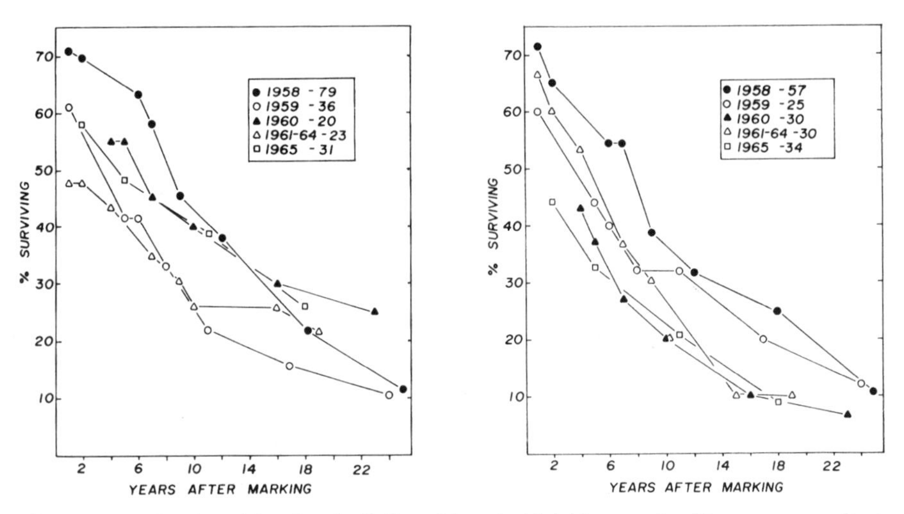

Figure 15.4 Survivorship of male (*left*) and female (*right*) box turtles (*Terrapene carolina*), plotted for the year first marked, with sample sizes shown after the year marked. (From Williams and Parker, 1987, courtesy of *Herpetologica*.)

As pointed out in Chapter 14, Section 14.4.1, fecundity of reptiles may vary markedly from one year to the next, depending upon rainfall and resultant fluctuations in food supplies. In a short-lived species, variations in fecundity quickly result in changes in population density. Under severe drought in Texas the rusty lizard (*Sceloporus olivaceous*) declined not only in numbers but also in older (three years or older) females. Because of the lesser fecundity of young females, the drought also lowered the reproductive potential of the entire population (Blair, 1952). Among some reptiles, especially the turtles, annual mortality may be extremely low. In the box turtle (*Terrapene carolina*), of 136 individuals marked in 1958, 11.4% of the males and 10.5% of the females survived at least 25 more years (Fig. 15.4).

Some species of vertebrates have a pattern of a rather constant attrition in adults, but there are few data on wild species to document rates of adult mortality. Survival of the kittiwake (*Rissa tridactyla*), a small boreal gull, was observed in a colony in which all individuals were color banded and thereby easily observable. In this particular colony, survivorship was greater in females than in males but decreased gradually with age in both sexes (Carlson and Wooler, 1976). In some long-lived marine birds, mortality increases very gradually with age (Botkin and Miller, 1974).

15.3.3 Senescence

In species in which most individuals die soon after reproduction, one is tempted to invoke death due to old age. Senescence has been viewed as a physical deterioration or physiological exhaustion of an individual. Although it is generally simplistic to regard senescence as a wearing down of physical parts, in some species this does happen. Shrews (Soricidae), for example, which feed on insects, many of them hard bodied, have greatly worn teeth before they attain one year of age, and few individuals live long after their first reproduction. Tooth wear may limit longevity in many mammals, including elephants, many large ungulates, hippos, and rhinos (Laws et al., 1975).

The poor quality of data on longevity, perhaps as much as any single measurement, limits our ability to evaluate life history theories. Fecundity may be more closely related to longevity, when longevity is known, but there is not any good information on comparative longevity for most closely related species.

In some vertebrates there appears to be a genetically programmed senescence of males. This mortality becomes effective after mating, so that the rapid elimination of males increases resources for females. Males of the mouselike dasyurid marsupial *Antechinus stuartii* die after their initial mating season; they suffer internal deterioration and disappear before the birth of their own young (Braithwaite, 1974; Braithwaite and Lee, 1978). The African skink *Mabuya beuttneri* is an annual species, and males experience earlier mortality than females do (Fig. 15.5). The increased mortality of males follows mating, but females remain in good health until eggs are laid. This sexual disparity in mortality obtains under ideal conditions in captivity. It is suggested that early mortality of males could either be (1) a side effect of other traits that have survival value or (2) a result of a positive selective pressure for early male senescence, although it is difficult to imagine a gene for early death (Barbault, 1986). There is an obvious advantage to postmating mortality of males

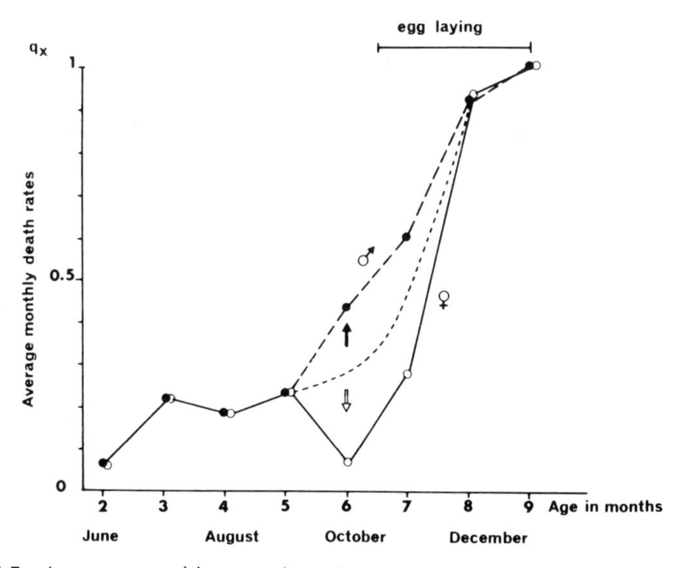

Figure 15.5 Average monthly mortality of males and females of the skink *Mabuya beuttneri*, showing the increase in deaths of males after mating. Short dash equals mean monthly mortality. Long dash (solid circles) equals actual monthly mortality of males. Solid line (open circles) shows actual monthly mortality of females. (From Barbault, 1986, courtesy of *Oikos*.)

if it increases resources for the surviving females and their young. Greater survival of these young would promote the increase of early paternal mortality.

Most data on maximal longevity relate to domestic species and do not reflect the roles of predation and competition that deplete populations of wild individuals. These data do, however, suggest that for domestic animals reproductivity continues to an age far beyond their normal life span in the wild.

15.4 THE SNOWSHOE HARE; A DENSITY-RELATED CYCLE

Professor Lloyd B. Keith and his associates have presented a meticulous account of the cyclical changes in populations of the snowshoe hare (*Lepus americanus*), together with its major predator, the lynx (*Lynx canadensis*), as well as several common avian predators. Keith and his colleagues followed a complete 10-year cycle, providing many previously unknown aspects of the biology of the prey, its food supply, and its predators.

The fluctuations of the lynx were originally documented by the Hudson's Bay Company, which recorded the cyclical frequency with which pelts were brought in by fur trappers. Elton and Nicholson (1942) introduced this record into the literature on vertebrate population cycles. Subsequently, Butler (1953) reviewed government data on the number of furbearers taken annually in Canada and compared them with responses to annual questionnaires concerning the abundance of furbearers. He was aware that the number of pelts, such as lynx, sold annually might not represent the same proportion of the total population every year, but the data did reveal a discrete 10-year cyclicity. Peaks in populations are widespread in the arctic but are not uniform throughout Canada. High densities radiate from the central provinces, suggesting movement of individuals away from the area of highest density to regions that subsequently show a population increase. Similar and nearly simultaneous patterns of peaks and radiation were shown for the snowshoe hare.

The snowshoe hare cycle is not strictly density dependent, but it is strongly density influenced. It constitutes one of the most completely studied examples of a cycle in which the main actor, the snowshoe hare, generates some of the major influences of its own population densities.

15.4.1 Reproductive Features

For 16 years the reproductive patterns of the hare were followed near Rochester, Alberta (Cary and Keith, 1979). During the 10-year cycle, the major reproductive parameters exhibited marked changes characteristic of specified parts of the cycle. In addition to seasonal changes in reproductive rates, there are conspicuous year-to-year differences in duration of breeding and ovulatory and pregnancy rates for the successive litters in any single season. These differences relate to specific phases of the population cycle.

In most years first pregnancies occur in early April, and within a year each female produces four litters, which are closely synchronized within a given population.

Within a given year, both ovulatory rates and pregnancy rates are lowest in the initial litters, following a pattern well known to occur among other leporids. There are, however, marked changes in fecundity from year to year. As populations increase to a 10-year peak, reproductive rates have already begun to decrease about 3 years earlier: that is, the decline in population follows the initial decrease in reproduction by about three years. This decline in reproductive activity is caused by a lowered ovulation rate, a shortened annual breeding season, and a reduction in the pregnancy rate in the third and fourth litters. Finally, the return to maximal reproduction occurs when the population density is minimal, five years after the peak. Thus there is a clear sequence of reproductive decline while population density is still increasing, and the peak in density follows the deterioration in reproduction by three years.

In contrast to the relationship that obtains in many species of mammals, fecundity in the hare is inversely related to age. This correlation is not due to a direct relationship of age to fecundity, but rather to the mean age of the breeding cohort resulting from the cyclical patterns in juvenile survival. Changes in male body mass from midwinter to spring correlate well with fecundity the following spring: reproduction declines with weight loss, which relates to a decline in food during late winter.

15.4.2 Demographic Features

Demography of the snowshoe hare was discussed in great detail by Keith and Windberg (1978) and Keith et al. (1976). At their peak, hare populations attained densities 20 to 40 times those of the population lows. During the four years of the declining phase, adult survival was 0.11 annually. During the four years of population growth, it was 0.34 annually; at population peaks survival was 0.53. Winter survival of adults was least for lightweight individuals (< 1400 g). Maximal winter to spring weight loss was observed in years just prior to and immediately after peak densities. The rate of winter weight losses fell at the time of the cyclic low. Similarly, survival of the mountain hare (*Lepus timidus*) of boreal Eurasia increases when population densities are low (Hewson, 1976).

Juvenile survival was less than that for adults. Survival of young was less for the third and fourth litters than for the first litters. Also, winter weights of juveniles were greatest one year prior to the population peak, and maximal juvenile weights occurred in individuals from the first litters. Juvenile mortality was greatest in years of population decline, and survival increased three- or fourfold in years of population increase. In a later cycle (1976–1984, at Kluane, Yukon) the decline was attributed to a drop in recruitment, due mostly to poor juvenile survival (Krebs et al., 1986).

Experimental removal of adult snowshoe hares in the Yukon Territory was followed by increased survival of juveniles, but removal of first-litter young did not seem to enhance survival of second-litter individuals (Boutin, 1984). This suggests that the young face competition from their parents.

The reduction in fecundity and the increase in winter weight loss are both related to the decline in population density. Nutritional deprivation had affected the hare before peak densities: reproduction declined and growth of young born in the summer deteriorated. Reduction in weight from winter to spring was associated with lower survival. Prior to the point of lowest hare densities, reproduction increases and winter mortality declines.

15.4.3 The Role of Nutrition

Implicit in the relationship between the hare cycle and reproduction is the interaction of the hare with its food supply (Pease et al., 1979). The snowshoe hare is a browser and selects terminal parts (4 mm in diameter or less) and buds, thus depleting its food supply while eating. When populations are low, the hares do not have a long-term effect on the recovery of the browse. Before the time populations of hares reach a peak, however, the amount and quality of browse is inadequate to sustain them in good health. They are thus malnourished in the winter and at times of lowest temperatures suffer increased weight loss from winter to spring as a result of browse deterioration.

15.4.4 The Role of Predators

In the area of study the snowshoe hare has several major predators, most of which are affected by the cycle of the hare. At times the lynx (*Lynx canadensis*) preys heavily on hares, and at times hares provide the most common prey of the lynx. Several raptors also play a role in the cyclicity of the hare.

Populations of the lynx are very sensitive to the abundance and scarcity of the hare (Brand and Keith, 1979). During the decline of hare populations, lynx are less fecund: litters are smaller and pregnancy rates lower. Adult lynx suffer directly during the decline in hare densities, and body fat of the lynx drops. Postpartum mortality of kittens is high (from 65 to 95%), perhaps accounting for most of the decline of the lynx. As hare populations expand, the lynx also increases its numbers, but population growth of the predator lags somewhat behind that of the hare. Cyclical peaks of other predators (e.g., the fox and fisher) occur with high densities of the lynx (Butler, 1953).

The great-horned owl (*Bubo virginianus*) increases its capture of hares during the cyclical increase. As hare populations decline, owl populations also decline, but to a lesser extent. The goshawk (*Accipiter gentilis*), another important predator of the hare, also declines with the decrease in hare numbers. Predation on hares by the red-tailed hawk (*Buteo jamaicensis*) reflects hare abundance, but populations of this raptor remain fairly stable and do not fluctuate with the hare.

With the exception of the red-tailed hawk, the abundance of predators follows the rise in hare numbers. The decline in hares subsequently increases the relative abundance of predators, and partly as a consequence rates of predation on hares increase as their densities decline. Thus hare populations drive the change in predator populations. During maximal hare populations, they provide abundant food for the lynx, horned owl, goshawk, and red-tailed hawk, but with the decline of their prey, prey/predator ratios become unfavorable for maintaining high populations of predators.

Although the decline in hare populations is not induced by predation, predation during the decline in hares apparently increases the amplitude of the cycle.

15.4.5 Relationship of Ruffed Grouse to Hare Cycles

It has long been realized that the cycles of the ruffed grouse (*Bonasa umbellus*) and the snowshoe hare tend to be synchronized (Keith, 1963). Although there is no direct relationship between the populations of the two prey species, their populations are linked through their predators. The pressure on grouse is attributed to predation by horned owls, goshawks, and red-tailed hawks during the decline of hare populations, for there is no year-to-year change in reproduction of the grouse.

Thus hare abundance allows the increase of predators, and as hare densities decline, predation increases not only on hares but also on grouse. Hare abundance, which drives the cyclicity in populations of its major predators, indirectly induces the cyclicity in the grouse.

15.5 MAMMAL POPULATIONS IN ARID LANDS

Probably the clearest examples of environmental effects on density are in arid and semiarid environments, which are characteristically subjected to wide variations in precipitation. These habitats are marginal at best, and extended dry periods can depress populations of desert vertebrates to extremely low levels. There is actually very little information on the life histories of desert species when they exist at very low levels, for at such times they are difficult to capture. Most commonly, desert species attract attention when their populations increase.

In the Namib Desert prolonged and extreme aridity depresses populations of native rodents (*Aethomys namaquensis*, *Gerbillurus paeba*, and *Desmodillus auricularis*). In 1973 rainfall dropped to a minimum of 12 mm in some areas, but the following year precipitation in the same areas increased to 165 mm or more. In 1975 rainfall varied from 60 to 120 mm. The increase in precipitation promoted a tremendous growth of desert plants. Rodent populations also increased very rapidly in 1974 and 1975. One population of *Desmodillus auricularis* expanded at a rate amounting to a 100% monthly increase; others increased more slowly until March or April 1975. Survival rates were inversely correlated with the most rapid population growth and were highest during a decline phase, indicating that the decline was due to reduced recruitment and not to increased mortality (Christian, 1980).

Changes in density of rodents of Australian deserts have been documented in a series of papers by Newsome, summarized by Newsome and Corbett (1975). A major part of that island nation is desert, with meager annual rainfall. The driest parts have a mean annual precipitation of 120 mm, but rainless periods may last several years. The highest temperatures may range from 35 to 40°C, and small rodents escape heat in subterranean burrow systems. During rainless periods, the mice move away from wheat fields, where the hard soil prevents burrowing. Homesites are available only during the rains. Successful reproduction follows heavy midsummer precipitation: mice then breed for three of four months, after which the populations "crash" (Newsome, 1969a). The mice are permanent dwellers of adjacent reed beds, where they breed annually in the summer (Newsome, 1969b).

Newsome's study followed house mice from a low density, following a 6½-year drought, to explosive increases, after unusual and extensive rains and flooding. This precipitation brought out a lush growth of grasses and forbs, which provided abundant food and cover for small rodents.

Predators [the introduced "native" dingo, the fox (*Vulpes vulpes*), and the house cat] may have assisted in suppressing rodent populations during drought periods, but predator populations were also low during drought periods. These carnivores and the barn owl (*Tyto alba*) preyed upon native murid rodents (*Rattus villosissimus* and *Notomys alexis*) as well as the alien house mouse (*Mus musculus*). Populations of predators followed the classical pattern of increase subsequent to the rise in prey abundance. After the rodent numbers declined, the carnivores were much in evidence and in poor health and were noted to die of starvation.

The Australian rodent plagues are entirely dependent upon erratic combinations of meteorological events that result in a phenomenal resurgence of vascular plants. A decline of rodents occurs when the plant-supporting rains cease and high evaporation rates return the land to its usual parched state. Predators seem to play no decisive role in the decline of rodent populations.

Changes in populations of Australian desert rodents are unpredictable, noncyclic, nondensity dependent, and based on sporadic periods of heavy precipitation. In semiarid regions, with regular but sometimes scanty rainfall, reproductive activity is usually cued by precipitation (see Chapter 13, Patterns of Seasonality, Section 13.9.2).

Rabbit populations in Australia are also very sensitive to the erratic fluctuations in precipitation. Following a water shortage in arid northeastern Australia, rabbits (*Oryctolagus cuniculus*) declined and became scarce. In the absence of fresh plants, they climbed trees and browsed and ate woody twigs and bark. In so doing they destroyed part of the source of their own food supply. The decline in rabbit densities was attributed to a failure of reproduction resulting from overgrazing by rabbits (Cooke, 1982).

15.6 MICROTINE CYCLES

Microtine rodents, or Arvicolidae (Honacki et al., 1982) or voles and lemmings, comprise a rather discrete group of small Holarctic rodents that are united by several distinctive morphological features (Anderson and Jones, 1984). Most species are terrestrial and feed mainly on forbs and grasses, both the vegetative growth and seeds and sometimes the roots. During periods of high densities, voles may eat the bark and roots of woody plants, and the muskrat (*Ondatra zibethicus*), although preferring plant materials, will sometimes eat freshwater clams, crayfish, and other aquatic comestibles. In meadows most species frequently form runways and tunnels; their nests are usually below the ground but may be on the surface during periods of heavy snow. Voles are partly concealed by plant cover and are active both day and night.

15.6.1 Reproductive Potential

The very high reproductive potential of the microtine rodents provides the mechanism for rapid growth of populations. Most species become sexually mature between three and four weeks of age under very favorable conditions, and postpartum pregnancies enable one female to produce many offspring in one year. Induced ovulation increases the likelihood that heterosexual meetings will produce fertile matings. Most species have the capability of breeding for extended periods of the year, which does vary with the region, food availability, and population density.

In the red-backed vole (*Clethrionomys rufocanus bedfordiae*) of Hokkaido, there is a high loss in early life, with a 50% mortality in the first month and an additional 20% mortality in the following two months. In this vole, population fluctuations depend on changes in mortality in early life as well as on varying rates of fecundity (Maeda, 1963, 1969). In a cyclic population of the bank vole (*Clethrionomys glareolus*), however, winter mortality of adults is high, and in peak years reproduction is restricted to a relatively brief period (Wiger, 1979). In cyclic populations of voles (*Clethrionomys rufocanus*, *C. rutilus*, and *Microtus oeconomus*) in Lapland, juveniles born late in peak years suffer especially high mortality (Andersson and Jonasson, 1986).

15.6.2 Predation on Voles

A cyclic population of *Microtus californicus* suffered heavy predation by small carnivores during the decline and at the low part of the cycle. As voles declined, the number of carnivores (mostly feral house cats) on a 35-acre (14.2-hectare) study area dropped from 4.4 to 0.4, but during the vole crash they ate from 20 to 33% of the existing population per month. As voles became fewer, small carnivores took an increasing number of other small rodents, mostly pocket gophers. The presence of gophers apparently sustained a predator population sufficient to exert continued pressure on the already depressed vole numbers. Eventually a decline in predators allowed an increased reproduction of the few remaining voles. During this interaction, the small carnivores actively depressed vole numbers during their decline and in this way increased the amplitude of the vole cycle (Pearson, 1971). The role of the small carnivores and buffer species in this vole cycle resembles that of the lynx in the cycles of the snowshoe hare. In a nearby area, availability of preferred foods changes with seasons and also with population densities, and the recovery of vole forage may delay reproduction during high densities (Batzli and Pitelka, 1971).

An insular population of *Microtus californicus* on an island in San Francisco Bay lives in essentially the same climatic regimen as the populations discussed by Pearson (1971) and Batzli and Pitelka (1971), but these insular voles have no mammalian predators and cannot disperse. This population is affected by both density-dependent and density-independent factors, which produce annual high and low population densities. Autumnal and winter rainfall, which produces a rich growth of forbs and grasses, is followed by intense reproduction one or two months later. The highest pregnancy rates and litter sizes both occur in the spring, and reproductive activity and litter size declines in late summer. Peaks in the population occur in the summers, as a result of breeding by young born early in the spring. During the dry summer, growth in the young born late in the spring is poor, and they appear to have been stressed by the summer drought (Lidicker, 1973).

15.6.3 Patterns of Cyclicity

The study of microtine cycles is replete with hypotheses that purport to explain regular population changes in terms of rather simple causes. Although it appears that a kaleidoscope of factors, both endogenous and exogenous, produce oscillations, no discrete set of explanations can account for the phenomenon in general. There appear to be sufficient biological differences among the many species of voles and lemmings to justify the assumption that among the various microtine species cyclicity is governed by somewhat different sets of factors.

The following comments associate several reproductive patterns reported for fluctuating vole populations. One will note some similarities between the three- to four-year cycles of voles and the ten-year cycles of the snowshoe hare. During low densities, there is rather high fecundity, early puberty, and a rapid succession of pregnancies (except at high latitudes). The plants in the vole habitat are recovering from heavy overgrazing. There tends to be rather low densities of predators, except immediately after a drastic decline in numbers of voles.

As voles increase, more obvious signs of their presence become noticeable: runways become well worn, accumulations of scats are larger, and piles of cuttings—usually short lengths of culms of grasses clipped by voles in their effort to reach the ripening seeds—are numerous and conspicuous. After three to four years, when densities are much higher, there are clear signs of decline in health (body mass and fecundity) of the voles and a visible deterioration of their habitat. The forbs and grasses that supported a sparse population of voles two to three years perviously have become overgrazed. At this point survival of the young has declined, and there is sometimes evidence of dispersal.

In a series of papers, Dennis Chitty proposed that microtine populations are limited by spacing behavior and that the behavior resulting from crowding limits reproduction and further population increases. The Chitty hypothesis additionally states that the element of *self-regulation*, which emphasizes individual behavioral differences, results from short-term (year-to-year) genetic changes (Chitty, 1967). This proposal has been explored subsequently by Krebs and other researchers. It appears that short-term changes in heterozygosity may occur in microtine rodents, but it is not known if such changes alter self-regulation (Krebs, 1978). Voles and lemmings do appear to be conspicuously poor at regulating their own populations, for their density fluctuations greatly exceed those of species of *Apodemus* and *Peromyscus*.

Most of the many proposals designed to account for vole cyclicity are embraced by one of two hypotheses offered by Rosenzweig and Abramsky (1980). The first, the predation hypothesis, depends upon the relationship of voles, as predators, to their plant foods, implying coevolutionary equilibria, or the stabilizing properties of the systems' dynamics. This is predicated on an assumed heterogeneity of the voles' habitats, which are occupied by voles adapted to the median habitat of medium quality or productivity. Vole fluctuations are greatest in the most productive patch of the environment, and in the poor environments fluctuations are low or absent (Fig. 15.6*a*). Rosenzweig and Abramsky assume that this relationship will promote the uniformity of the genotype because of dispersal from patches of greatest productivity to patches of poorest plant productivity and vole densities. This produces a degree of variation in stability from one habitat to another.

The second hypothesis, the phenological hypothesis, considers that optimal reproductive effort varies with time and habitat, but because the habitats vary, the voles' adaptation (or sensitivity to plant cues) is less than perfect. This small degree of maladaptation is ameliorated by dispersal from one habitat or patch to another (Fig. 15.6*b*). If one population in a patch of low productivity is only slightly sensitive to the stimulatory cues of plants, it will be adjusted to the low plant growth, and stability will obtain. Immigrants with a greater sensitivity, from a patch of greater productivity, will eventually immigrate into the poorer habitat and increase the sensitivity of that population. Subsequently, the population will become more fecund and place an excess burden on that habitat.

Both the predation hypothesis and the phenological hypothesis make three assumptions: (1) that habitats are heterogeneous, which they frequently are but at other times may be uniform over extensive areas; (2) that there is a rapid change in genotype, which is not definitely known to occur within the short span of one cycle; and (3) that dispersal is an inherent part of microtine cyclicity, which is generally assumed to be true, although it is difficult to demonstrate. It is now well accepted that cyclicity is a feature of populations in the most productive habitats and that fluctuations are modest or slight in marginal environments. These two hypotheses neither consider the role of small predators, such as weasels and raptors, on declining populations of voles nor the variation in food quality and quantity. These two hypotheses are valuable in emphasizing the importance of environmental heterogeneity.

15.7 POPULATION STUDIES ON TITS (PARUS SPP.)

In northern Europe one or more species of tits occur commonly in wooded and partly open habitats and nest in holes in trees. These little passerines have been studied in great detail for many years, and there is a wealth of information that documents the complexity of factors involved in their fluctuations in density. The great tit (*Parus major*) and the blue tit (*P. caeruleus*) are especially well studied and are the subjects of numerous reports.

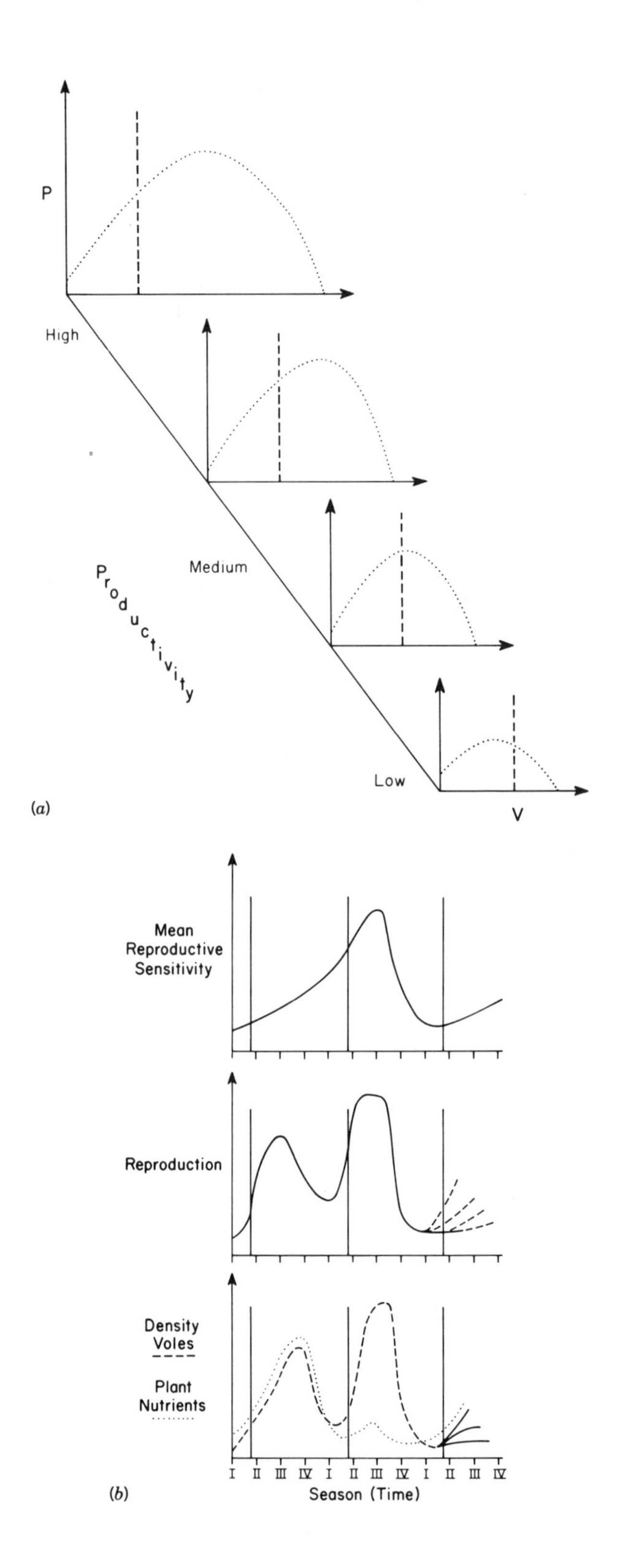

P
High
Medium
Productivity
Low
V
(a)
Mean
Reproductive
Sensitivity
Reproduction
Density
Voles
Plant
Nutrients
I II III IV I II III IV I II III IV
Season (Time)
(b)

15.7.1 Density-Dependent Effects

Several aspects of the demography of the great tit are related to densities, especially in the winter and again in the nesting period. The possibility of clutch size being density-dependent is very real, but there are not many situations in which density varies without changes in other environmental aspects. It is very likely that density per se in nature operates together with one or more other factors (such as food supply induced by a change in density) that may directly affect clutch size. (This concept is discussed in Chapter 13.)

Densities of the blue tit (*Parus caeruleus*) suffer in the presence of the great tit, with which it competes for both nesting sites and food. This relationship was demonstrated by an experimental removal of the great tit from its nesting boxes in the winter, which resulted in an increase in the breeding population of the blue tit in the spring (Dhont and Eyckerman, 1980). In the willow tit (*Parus montanus*) in southwest Sweden, mortality of the young is high in their first winter. Year-to-year variations in winter mortality depend upon mean low temperature, but they are also density-dependent. Variation in fecundity of the willow tit is also density dependent, but, in addition, clutch size relates to the structure of the population, for yearling females lay fewer eggs. A winter of high survival of first-year birds is followed by a spring with smaller than usual clutches (Ekman, 1984).

Among tits in general, social behavior, presumably through the pressure of territoriality, limits population size, and excluded individuals immigrate (Newton, 1980). In prime habitat in Britain there is a strong density-dependent control of clutch size and breeding success. At high densities subordinate individuals leave prime habitat, and females disperse greater distances than males do (O'Connor, 1980). They move to marginal areas, where the most conspicuous fluctuation in numbers occurs. In the Netherlands artificial reduction of postnesting populations reduces dispersal (Kluijver, 1951).

Figure 15.6 Graphic conceptions of two hypotheses designed to explain cyclic fluctuations in densities of voles. (*a*) The predation hypothesis: habitat heterogeneity and panmixia prevent voles from precise coadaptation; their systems' range of dynamic stabilities is expanded over a productivity gradient, and oscillation becomes a permanent feature of their consensus. P, vole density; V, food (plant) density. Dotted lines are victim (plant) isoclines, which expand with productivity. Dashed lines are the predator (vole) isocline, which, owing to the panmixia, is fixed, despite varying productivity. Intersections of the isoclines to the left of the hump produce oscillation. (*b*) The phenology hypothesis: three years of seasons (I, winter); vertical lines are for time reference. With an appropriately low reproductive sensitivity, the population successfully tracks its plant resources through the first year. Continuous immigration raises the sensitivity so that the second-year population reproduces too soon and too much. This depresses the plant resource, and a vole crash ensues. Those that are reproducing the most are in poorest health, have poorest fitness, and are heavily selected against; hence the mean reproductive sensitivity also drops suddenly. Variance in the exact condition of the vole and plant populations at the end of the crash leads to variability in vole reproduction and density for a period after the crash. (From Rosenzweig and Abramsky, 1980, courtesy of *Oikos*.)

15.7.2 The Role of Food

Beech mast constitutes an emergency winter food of the great tit near Wageningen, The Netherlands, and enhances survival of the young between fledgling time and first breeding. Densities appear to depend upon survival of the young to the first nesting season, and this survival is directly related to the winter food supply. In the absence of a good crop of beechnuts, there is a large loss of great tits in the winter and a reduction in the population the following spring (van Balen, 1980).

15.7.3 The Role of Predators

The sparrowhawk (*Accipiter nisus*) preys on tits, which may form a major item in the diet of males. A drastic reduction of sparrowhawks from organochlorine pesticides and their later recovery in numbers were not correlated with meticulously monitored densities of the great tit and the blue tit near Oxford, England (Perrins and Geer, 1980). Quite possibly the decline in the sparrowhawk was compensated by other predators. Avian predators in general fluctuate with densities of their prey species, but the extent of the change in predator populations depends upon their feeding versatility (Newton, 1980).

15.7.4 Summary of Tit Demography

Densities of these two passerine birds are affected by an array of independent influences that promote and suppress their populations. Some of these factors, such as a winter supply of beechnuts, are mostly density independent, and others, such as spacing from territoriality, are strongly density influenced. Food may enhance survival in the winter and account for a large nesting population in the spring, and territoriality appears to depress both density and fecundity in undisturbed woodlands. The number of nest sites is finite, and both intra- and interspecific competition for nest holes limits tit densities.

15.8 PREDATOR-PREY INTERACTIONS

Any study of predator-prey relationships must evaluate both sides of the hypothetical equation. That is, there is an effect of prey populations on predators and an effect of the predator on its prey species.

Effects of predators on prey populations are extremely difficult to understand, because (1) it is necessary to follow trends of both prey and predator populations and (2) there are frequently two or more species of prey and of predator, and their relationships apparently involve complex interactions. Predators rarely confine their attention to a single prey species. Changes in populations of predators usually lag behind those of the prey, in both the increase and the decrease phase of a cycle.

15.8.1 Types of Responses

Predators exhibit two types of responses to changes in prey populations. Increases in prey densities elicit a functional response—an increase of the prey species in the diet of the predator. In addition, there is a numerical response—an increase in numbers of prey individuals taken as the population of the prey increases (Solomon, 1949). Although both functional and numerical responses result in an elimination of prey individuals, the two responses may not occur simultaneously. Moreover, they depend on the levels of other prey species as well as other kinds of predators, as already described for the snowshow hare and its predators.

Among those birds and mammals that prey on vertebrates, most species lie near the generalist end of the spectrum between generalists and specialists. The osprey very rarely captures anything but fish, and small bird-hunting hawks, such as small accipiters, seldom vary their fare. When other predators *appear* to specialize, they are simply facultative specialists, or opportunists. Thus there is seldom a one-to-one relationship between most species of predators and species of prey, and this state of affairs makes it difficult to establish the effect of one upon the other.

Densities of nesting pairs of the sparrowhawk (*Accipiter nisus*) in Britain are predicated on densities of spring populations of their prey. Territories of the sparrowhawk are fairly evenly spaced, and increases in prey densities reduce the distance between hawk nests (Newton et al., 1986). This spacing is also density-dependent, stemming from competition for prime home ranges (Newton and Marquiss, 1986). The long-tailed skua or jeager (*Stercorarius longicaudus*) nests where vole and lemming populations are the greatest, and fluctuations of these raptors closely parallel those of their rodent prey (Fig. 15.7).

15.8.2 Small Rodent Populations and Their Predators

Inasmuch as populations of mice and squirrels frequently make economic inroads on human activities, biologists are paid to study them. Before biologists became concerned with the study of small mammal fluctuations, hunters had become resentful of large predators of such quarry as deer and upland game birds. It is not surprising that biologists had become conditioned to expect that predators drive population densities of their prey. The usual tendency, however, is for the welfare of predators to depend upon that of their prey.

With the decline in density of one prey species, the several kinds of resident predators turn their attention to other prey. Densities of both predators and prey are extremely difficult to establish with accuracy, and calculated rates of predation rest upon many assumptions (King, 1985).

In northern Fennoscandia synchronized population changes of voles and grouse have been explained by a commonality of predators, and as one prey population declines, predators increase their attention to the alternate prey species. This hypothesis was tested by providing supplemental food for predators on one of two 615-hectare plots. Survival of grouse was greater in the experimental area than in the control

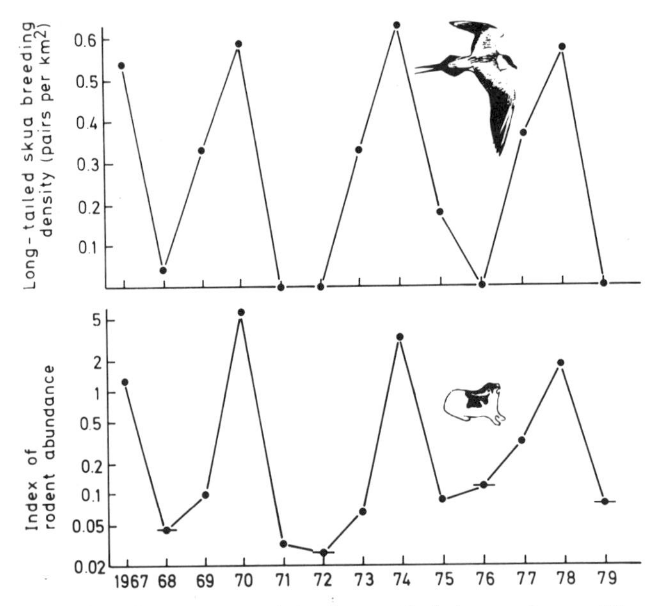

Figure 15.7 The breeding density of the long-tailed skua or jaeger (*Stercorarius longicaudus*) correlated with the abundance of lemmings (*Lemmus lemmus*). (From Andersson, 1981, courtesy of the British Ecological Society.)

area during a decline in vole populations (in 1985). It appears that predation on grouse increases as a consequence of vole decline (Lindstrom et al., 1987).

There is a general correlation between cycles of lemmings and brants (*Branta bernicula bernicula*) in the Arctic. During the cyclical increase of lemmings, predators have an abundance of food, because their densities lag about one year behind those of their prey. Following the collapse of rodent populations, the still high densities of predators concentrate on other species, including the brant (Greenwood, 1987).

Synchrony in the declines of vole and shrew cycles in Finland are attributed to predation both by (1) rather mobile avian predators and, to a lesser degree, by (2) small mustelids. The small mustelids greatly favor voles over shrews, but the decline in densities of voles shifts predation pressure by hawks and owls on shrews (Korpimaki, 1986).

In Britain densities of small rodents affect the reproductive performance of the tawny owl (*Strix aluco*). Low populations of prey (small rodents) cause a decline in fat stores of the owls, and the ensuing hunger drives incubating females to neglect their eggs, resulting in a decline in hatching. In that manner the prey densities affect incubation. Alternate prey provided food for the young that did hatch, and survival of those young was not reduced. Although rodent scarcity reduced productivity of the owls, the long-term effect on owl populations was negligible, because survival of the nestlings was density-dependent, being greater in years of low reproduction. The owls are relatively long lived and occupy territories for some six to ten years (King, 1985).

The stoat (*Mustela erminea*), introduced to New Zealand, is one of a small number of terrestrial predators there, and the introduced rat (*Rattus rattus*) and the house mouse (*Mus musculus*) comprise the bulk of the stoat's diet. The stoat mates in summer, and embryonic diapause postpones parturition until the following spring, some 10 months later. Prey abundance during the time of folliculogenesis may increase pregnancies, but a drop in prey density prior to birth of the young, some 10 months later, may restrict food for young stoats (King, 1983a, 1983b). In southern Sweden the stoat has a short life span, with a mean life expectancy (at the age of independence) of 1.4 years for males and 1.1 years for females, with a high adult mortality in the second winter. Year-to-year densities reflect reproduction, which is correlated with variations in prey (*Microtus* and *Apodemus*). Interspecific competition for small rodents increases with a decline in rabbits, which constitute a staple fare of larger predators (Erlinge, 1983). Thus fluctuations in the stoat can be food driven but are complicated by a diverse community of both prey and predators. This type of situation is similar to the previously discussed shifting of prey by the lynx during snowshoe hare fluctuations.

Densities of the weasel (*Mustela nivalis*) respond very rapidly to increases in densities of voles. Not only is the adult female weasel capable of producing two broods in one summer, but young females may mate and produce young in the summer of their birth (King, 1980).

15.9 THE ROLE OF DISPERSAL

The movement of individuals in and out of areas with fluctuating populations seems to be commonly involved in changes in densities, and this is most characteristic of smaller species. It appears to be a psychological effect of crowding, and not directly dependent upon food.

Dispersal accounts for changes in density and composition of populations; changes which are separate from natality and mortality. Dispersal from one area (emigration) results in invasion (immigration) of another area, usually of a somewhat less desirable habitat, referred to as a "sink." The effects of this process are quite different, and can be thought of as the opposite sides of the same coin.

It has been postulated that the pressure of high density may instill the drive to disperse. This has been referred to as "saturation dispersal," and movements during population increases may be considered to be "presaturation dispersal" (Lidicker, 1975). The two types may not be discretely separate, but one can imagine that saturation dispersal involves individuals under some duress, and perhaps in a deteriorating physical condition, while presaturation dispersal involves the individuals' movement away from the place of birth, usually prior to their first reproduction, and does not result from population pressures.

In the grey squirrel (*Sciurus carolinensis*), dispersal results from aggression of adults toward maturing young, and the populations are regulated by the depression of recruitment beyond losses due to loss of adults (Thompson, 1978). Beaver families are regulated by emigration of subadults prior to initial breeding (Bradt,

1938). Such examples are typical of presaturation dispersal in mammals that are more or less acyclic.

Emigration can alleviate the pressure of a population on its environment and immediately or eventually provide more resources for remaining individuals. In this sense, emigration functions as does predation or any other factor that removes individuals. Although conceptually emigration may retard population growth before it exceeds the habitat's carrying capacity (Lidicker, 1975), in many cyclic species growth may continue through the rise of densities *above* the carrying capacity. There are numerous examples, such as the snowshoe hare, of environmental deterioration caused by extremely high densities.

Dispersal plays a conspicuous role in the fluctuations of voles. The annual decline in *Microtus townsendii* is due either to emigration in slight declines or to both mortality and emigration when declines are drastic (Beacham, 1980). Spacing of adult females restricts their densities. Population trends of *Microtus pennsylvanicus* on two plots, one from which adult males and another from which adult females were continuously removed, indicated that adult females restrict the ingress of adult females from neighboring areas, regardless of the presence or absence of males. Such restriction is less pronounced in "male only" plots (Boonstra and Rodd, 1983).

A slightly different intrasexual relationship is seen in female *Microtus californicus*. In an area of contrasting habitats—high quality (with high frequency of perennial grasses) and lower quality (*Conium maculatus* and annual grasses)—females were more common in high-quality areas and were more likely to occur with other females. Males, on the other hand, distributed themselves more evenly and randomly (Ostfeld et al., 1985).

Local movements also play a role in the population changes in the red grouse (*Lagopus lagopus scoticus*). During a fluctuation of red grouse in the 1970s, there was a major decline at a time when food supplies were adequate. As densities increased, younger males were less successful in obtaining adequate territories and in finding a mate; at high densities males became more abundant than females. Following the decline in numbers, there appeared to be a relative increase in aggressive cock grouse (Watson and Moss, 1980). In an eight-year cycle there was a high rate of emigration following the peak, attributed to behavioral changes occurring at the highest densities (Moss et al., 1984).

15.10 SUMMARY

Mortality and recruitment tend to compensate each other to produce a dynamic balance—a continuous change in densities. In many, perhaps most, species fluctuations appear to be minor, but a few vertebrates experience wide swings in densities, which are either regular (or cyclic) or irregular.

Both endogenous and exogenous factors alter birth rates and density. Students have explored population fluctuations for many decades, but there is no consensus on possible general factors governing cyclical changes in vertebrate numbers. Density-dependent and density-influenced trends are usually social or nutritional in nature and may promote or inhibit reproduction. High densities of vertebrates may lower the carrying capacity of the environment and thus lower both reproduction and survival. Some herbivores consume plants more rapidly than the plants can grow or revegetate. Density-dependent effects have been confirmed by experimental removal of parts of populations.

The presumably obvious role of food in fluctuations is equivocal. Supplements of food in nature enhances individual growth and survival but, in voles, does not prevent a decline in population density. Innate features, such as age and parental experience, enhance the likelihood of reproductive success. In colonial species an increase in density may increase survival up to the point that food and homesites may be limiting.

Populations of the snowshoe hare (*Lepus americanus*) increase to high densities and then decline rapidly, about every 10 years. This is a density-influenced cycle, part of which appears to be affected by a deterioration of forage during the latter part of the increase phase. During the latter increase phase of the cycle, fecundity declines, and the peak in population density follows three years later. The peak is some 20 to 40 times greater than the low. Associated with the rather rapid decline in number of hares is increased winter mortality of young hares.

Predators, especially the lynx (*Lynx canadensis*), increase after the hare begins to recover from its rapid decline in population, but the lynx population lags some three years behind the increase of the hares. By the time the hares enter the rapid decline in density, the lynx have increased substantially, and their predation on the hares appears to accelerate the decline of the prey. Also, as hares decrease in abundance, not only the lynx but also the goshawk (*Accipiter gentilis*), which is a predator of hares as well, take increasing numbers of the ruffed grouse (*Bonasa umbellus*). The grouse population then decreases in the face of this greater predation. Thus the cyclicity of the hare drives the changes in density of the lynx, and the sudden increased predation of the lynx and the goshawk appears to induce cyclicity in the grouse. This seems to be the most logical explanation for the near synchrony of cycles of the snowshoe hare and the ruffed grouse.

In arid lands the welfare of vertebrates is closely dependent upon rains, which are commonly seasonal in timing but irregular in amounts. Increases in precipitation promote lush growth of forbs and grasses as well as the insects that feed upon them. Birds and mammals breed for longer periods and experience increased fecundity following periods of heavy rainfall, and populations increase accordingly. Contrariwise, prolonged drought restricts recruitment and is followed by extremely low densities of desert vertebrates.

Voles and lemmings (Arvicolidae) frequently experience major changes in densities, and at higher latitudes population cycles of from three to four years are common. These little rodents (at least some species) have induced ovulation, which increases the likelihood of a heterosexual meeting that results in a fertile mating. Sexual maturity may be early, so that young born in the spring may themselves produce offspring the same year.

Voles are frequently abundant and pose many questions for ecologists. To date no simple or general model explains the fluctuations of all species. Like the snowshoe hare, several species of *Microtus* tend to induce their own cycles. Factors altering reproduction and mortality are known for some species. Food promotes rapid individual growth and survivorship but apparently does not provide immunity to the inevitable decline in cyclic species. Most species of *Microtus* feed on forbs and grasses (and their seeds) and frequently increase to the degree that their feeding destroys the plants upon which they depend. At least some species of *Clethrionomys* feed heavily on nuts, and variations in forest mast alter their reproduction and numbers.

The several species of tits (*Parus* spp.) are small, hole-nesting passerine birds that have been intensively studied in northwestern Europe. They compete for nest sites, and reproduction is density-dependent: clutch size and survivorship decline with increased density but increase with food supply. Tit populations undergo major changes in density, but their fluctuations are not cyclic.

Predators frequently increase in density following population growth of their favored prey, and when prey animals decline in number, the predators are still common. This may result in (1) a relative increase in predator pressure on the favored prey, (2) increased predation on a secondary prey species, or (3) a combination of both tactics.

Dispersal is the movement of individuals away from their birthplace. Presaturation dispersal is an event common to all animals and usually occurs before reproduction. Saturation dispersal occurs as a result of aggressiveness or territoriality (or both) as densities increase above the carrying capacity of the habitat. Individuals that disperse are effectively removed from the population. Saturation dispersal is difficult to distinguish from mortality but appears to have a role in the decline of populations.

BIBLIOGRAPHY

Andrewartha HG and Birch LC (1954). *The Distribution and Abundance of Animals*. Univ of Chicago, Chicago.

Berryman AA (1981). *Population Systems. A General Introduction*. Plenum, New York.

Bronson FH (1979). The reproductive ecology of the house mouse. *Quarterly Review of Biology*, **54**: 265–299.

Dempster JP (1975). *Animal Population Ecology*. Academic, London.

Drickamer LC (1987). Behavioral aspects of rodent population biology. *American Zoologist*, **27**: 253–269.

Errington PL (1946). Predation and vertebrate population. *Quarterly Review of Biology*, **21**: 144–177, 221–245.

Finerty JP (Ed) (1980). *The Population Ecology of Cycles in Small Mammals. Mathematical Theory and Biological Fact*. Yale Univ, New Haven.

Flowerdew JR, Gurnell J, and Gipps JHW (Eds) (1985). The ecology of woodland rodents. Bank voles and wood mice. *Symposium of the Zoological Society of London*, No. 55. Academic, London.

Fowler CW and Smith TD (1981). *Dynamics of Large Mammal Populations*. Wiley, New York.

Golley FB, Petrusewicz K, and Ryszkowski L (Eds) (1975). *Small Mammals: Their Productivity and Population Dynamics*. Cambridge Univ, Cambridge.

Jewell PA and Holt S (Eds) (1981). *Problems in Management of Locally Abundant Wild Mammals*. Academic, New York.

Snyder DP (Ed) (1978). Populations of small mammals under natural conditions. *Pymatuning Symposia in Ecology* 5, Univ of Pittsburgh, Pittsburgh.

Turner FG (1977). The dynamics of populations of squamates, crocodilians and rhynchocephalians. Pp 157–264. In *Biology of the Reptilia* (Ed: Gans C), vol 7, Ecology and Behavior. Academic, New York.

Watt KEF (1971). Dynamics of populations: a synthesis. Pp 568–580. In *Dynamics of Populations* (Eds: den Boer PJ and Gradwell GR). Centre for Agricultural Publishing and Documentation, Wageningen.

REFERENCES

Adams NR, Ritar AJ, and Saunders MR (1984). Measurement of permanent clover infertility in ewes. Pp 362–364. In *Reproduction in Sheep* (Eds: Lindsay DR and Pearce DT) Australian Academy of Science, Canberra.

Adkins E and Schlesinger L (1979). Androgens and the social behavior of male and female lizards (*Anolis carolinensis*). *Hormones and Behavior* **13**:139–152.

Adkins EK (1976). Embryonic exposure to an antiestrogen masculinizes behavior of female quail. *Physiology and Behavior* **17**:357–359.

Adkins-Regan E (1981). Early organizational effects of hormones: An evolutionary perspective. In *Neuroendocrinology of Reproduction* (Ed: Adler NT) Plenum, New York:159–221.

Adler K (1976). Extraocular photoreception in amphibians. *Photochemistry and Photobiology* **23**:275–298.

Ainley DG (1975). Development of reproductive maturity in Adelie penguins. In *The Biology of Penguins* (Ed: Stonehouse B) Macmillan, London:139–157.

Ainsworth L, Tsang BK, Downey BR, Baker RA, Marcus GJ, and Armstrong DT (1979). Effects of indomethacin on ovulation and luteal function in gilts. *Biology of Reproduction* **21**:401–411.

Aitken RJ (1974). Delayed implantation in roe deer (*Capreolus capreolus*). *Journal of Reproduction and Fertility* **39**:225–233.

Alatalo RV, Gustafsson L, and Lundbers A (1984). High frequency of cuckoldry in pied and collared flycatchers. *Oikos* **42**:41–47.

Alatalo RV, Lundberg A, and Glynn C (1986). Female pied flycatchers choose territory quality and not male characteristics. *Nature* **323**:152–153.

Alcorn GT and Robinson ES (1983). Germ cell development in female pouch young of the tammar wallaby (*Macropus eugenii*). *Journal of Reproduction and Fertility* **67**:319–325.

Aleksiuk M and Gregory P (1974). Regulation of seasonal mating behavior in *Thamnophis sirtalis parietalis*. *Copeia*: 681–689.

Alerstam T and Högstedt G (1981). Evolution of hole-nesting in birds. *Ornis Scandinavica* **12**:188–193.

Alerstam T and Högstedt G (1984). How important is clutch size dependent adult mortality? *Oikos* **43**:253–254.

Alexander PS, Alcala AC, and Wu DY (1979). Annual reproductive pattern in the rice frog *Rana l. limnocharis* in Taiwan. *Journal of Asian Ecology* **1**:68–78.

Alibhai SK (1985). Effects of diet on reproductive performance of the bank vole (*Clethrionomys *glareolus*). *Journal of Zoology (A)* **205**:445–452.

Alibhai SK and Gipps JWH (1985). The population dynamics of bank voles. In *The Ecology of Woodland Rodents, Bank Voles and Wood Mice* (Eds: Flowerdew JR, Gurnell J, and Gipps JHW) Zoological Society of London Symposium 55:277–313.

Allen EG (1938). The habits and life history of the eastern chipmunk, *Tamias striatus lysteri*. New York State Museum Bulletin 314.

Allen JRM and Wootton RJ (1982a). The effect of ration and temperature on the growth of the three-spined stickleback, *Gasterosteus aculeatus* L. *Journal of Fish Biology* **20**: 409–422.

Allen JRM and Wootton RJ (1982b). Effect of food on the growth of carcase, liver and ovary in female *Gasterosteus aculeatus* L. *Journal of Fish Biology* **21**:537–547.

Almeida OFX and Lincoln GA (1984). Reproductive refractoriness in rams and accompanying changes in the pattern of melatonin and prolactin secretion. *Biology of Reproduction* **30**:143–158.

Altenburg W, Daan S, Starkenburg J, and Zijlstra A (1982). Polygamy in the marsh harrier, *Circus aeruginosus*: Individual variation in hunting performance and number of mates. *Behaviour* **79**:272–312.

Altona RE and Tilley TJ (1963). Oestrogenically active steroid extracts from grasslands. *South Africa Journal of Science* **59**:561–563.

Amman RP (1970). Sperm production rates. In *The Testis*, vol I (Eds: Johnson AD, Gomes WR, and van Demark NL) Academic, New York:433–482.

Amoroso EC (1952). Placentation. In *Marshall's Physiology of Reproduction* (Ed: Parkes AS) Longman's Green, London:127–311.

Amoroso EC (1960). Viviparity in fishes. Zoological Society of London Symposium **1**: 153–181.

Amoroso EC, Goffin A, Halley G, Matthews LH, and Mathews DJ (1951). Lactation in the grey seal. *Journal of Physiology* **113**:4P–5P.

Anderson DW, Gress F, and Mais KF (1982). Brown pelicans: Influence of food supply on reproduction. *Oikos* **39**:23–31.

Anderson HL and Lent PC (1977). Reproduction and growth of the tundra hare (*Lepus othus*). *Journal of Mammalogy* **58**:53–57.

Anderson JL (1975). The occurrence of a secondary breeding peak in the southern impala. *East African Wildlife Journal* **13**:149–151.

Anderson S and Jones JK Jr (1984). *Orders and Families of Recent Mammals of the World*. Wiley, New York.

Andersson M (1976). Population ecology of the long-tailed skua (*Stercorarius longicaudus* Vieill). *Journal of Animal Ecology* **45**:537–559.

Andersson M (1980). Nomadism and site-tenacity as alternative reproduction tactics in birds. *Journal of Animal Ecology* **49**:175–184.

Andersson M (1981). Reproductive tactics of the long-tailed skua, *Stercorarius longicaudus*. *Oikos* **37**:287–294.

Andersson M (1982). Female choice selects for extreme tail length in a widowbird. *Nature* **299**:818–820.

Andersson M and Jonasson S (1986). Rodent cycles is relation to food resources on an alpine heath. *Oikos* **46**:93–106.

Andrén C (1982). The role of the vomeronasal organs in the reproductive behavior of the adder *Vipera berus*. *Copeia*:148–152.

Andrén C and Nilson G (1983). Reproductive tactics in an island population of adders, *Vipera berus* (L.), with a fluctuating food resource. *Amphibia-Reptilia* **4**:63–79.

Andrew RJ and Rogers LJ (1972). Testosterone in search behaviour and persistence. *Nature* **237**:343–345.

Andrewartha HG and Barker S (1969). Introduction to a study of the ecology of the Kangaroo Island wallaby, *Protemnodon eugenii* (Desmarest) within Flinders Chase, Kangaroo Island, South Australia. *Transactions of the Royal Society of South Australia* **93**:127–132.

Andrews RM (1985). Mate choice by females of the lizard, *Anolis carolinensis*. *Journal of Herpetology* **19**:284–289.

Annan O (1963). Experiments on photoperiodic regulation of the testis cycle in two species of the thrush genus *Hylocichla*. *Auk* **80**:166–174.

Ar A and Yom-Tov Y (1978). The evolution of parental care in birds. *Evolution* **32**:655–669.

Aragona C and Friesen HG (1975). Specific prolactin binding sites in the prostate and testis of rats. *Endocrinology* **97**:677–684.

Arak A (1983). Sexual selection by male-male competition in natterjack choruses. *Nature* **306**:261–262.

Arendt J (1979). Radioimmunoassayable melatonin: Circulating patterns in man and sheep. In *The Pineal Gland of Vertebrates Including Man* (Eds: Kappers JA and Pevet P). *Progress in Brain Research*. Elsevier, Amsterdam:249–258.

Arendt J, Land CA, Symons AM, and Pryde SJ (1983). Plasma melatonin in ewes after ovariectomy. *Journal of Reproduction and Fertility* **68**:213–218.

Arman P, Kay RNB, Goodall ED, and Sharman GAM (1974). The composition and yield of milk from captive red deer (*Cervus elaphas* L.). *Journal of Reproduction and Fertility* **37**:67–84.

Armstrong DT, Goff AK, and Dorrington JH (1979). Regulation of follicular estrogen biosynthesis. In *Ovarian Follicular Development and Function* (Eds: Midgley AR and Sadler WA) Raven, New York:169–181.

Arnaud RS, Llker P, Kelly PA, and Labrie F (1983). Hormonal regulation of ovarian epidermal growth factor (EGF) receptors in the rat. In *Factors Regulating Ovarian Function* (Eds: Greenwald GC and Terranova PF) Raven, New York:209–213.

Arvy L (1974). Mammary glands, milk and lactation in cetaceans. *Investigations on Cetacea* **5**:157–200.

Asdell SA (1964). *Patterns of Mammalian Reproduction*. Cornell University, Ithaca, New York.

Ashley M (1979). *Reproductive Development of the Female*. PSG, Littleton.

Ashmole NP (1961). Regulation of numbers of oceanic birds. *Ibis* **103a**:491–492.

Ashmole NP (1963a). The regulation of numbers of tropical oceanic birds. *Ibis* **103b**:458–473.

Ashmole NP (1963b). Food during breeding season controls population density. *Ibis* **103b**:458–473.

Ashworth US, Ramaiah GD, and Keyes MC (1966). Species difference in the composition of milk with special reference to the northern fur seal. *Journal of Dairy Science* **49**: 1206–1211.

Asiby EOA (1974). Reproduction in the grasscutter (*Thyronomys swinderianus*) in Ghana. *Zoological Society of London Symposium* **34**:251–263.

Askenmo C (1982). Clutch size flexibility in the pied flycatcher *Ficedula hypoleuca*. *Ardea* **70**:189–196.

Aso T, Tominaga, T, Oshima K, and Matsubayashi K (1977). Seasonal changes in plasma estradiol and progesterone in the Japanese Monkey (*Macaca fuscata fuscata*). *Endocrinology* **100**:745–750.

Assenmacher I and Jallageas M (1978). Annual endocrine cycles and environment in birds with special reference to male ducks. In *Environmental Endocrinology* (Eds: Assenmacher I and Farner DS) Springer-Verlag, Berlin:52–60.

Atz JW (1964). Intersexuality in fishes. In *Intersexuality in Vertebrates Including Man* (Eds: Armstrong CN and Marshall AJ) Academic, London:145–232.

Auffenberg W (1966). On the courtship of *Gopherus polyphemus*. *Herpetologica* **22**:113–117.

Aulerich RJ, Holcomb L, Ringer RK, and Schaible JP (1963). Influence of photoperiod on reproduction in mink. *Bulletin of the Michigan State University Agricultural Research Station* **46**:132–138.

Avery M and Sherwood G (1982). The lekking behaviour of the great snipe. *Ornis Scandinavica* **13**:72–78.

Axelrod DL (1979). Age and origin of Sonoran desert vegetation. Occasional Papers of the California Academy of Sciences, no. 132.

Axelrod J (1974). The pineal gland: A neurochemical transducer. *Science* **184**:1341–1348.

Bagenal TB (1970). Fish eggs and the next generation. *Journal of Fish Biology* **2**:383.

Bagenal TB (1971). The interrelations of the size of fish eggs, date of spawning and the production cycle. *Journal of Fish Biology* **3**:207–219.

Bagenal TB (1978). Aspects of fish fecundity. In *Ecology of Freshwater Fish Production* (Ed: Gerking SD) Wiley, New York:75–105.

Baggerman B (1980). Photoperiodic and endogenous control of the annual reproductive cycle in teleost fishes. In *Environmental Physiology of Fishes* (Ed: Ali MA) Plenum, New York:533–567.

Bahr JM, Wang S-C, Huang MY, and Calvo FO (1983). Steroid concentrations in isolated theca and granulosa layers of preovulatory follicles during the ovulatory cycle of the domestic hen. *Biology of Reproduction* **29**:326–334.

Bailey KM, Francis RC, and Stevens PR (1982). The life history and the fishery of the Pacific whiting, *Merluccius productus*. *California Cooperative Oceanic Fisheries* **23**:81–89.

Baker CL, Baker LC, and Caldwell MF (1947). Observations of copulation in *Amphiuma tridactylum*. *Journal of The Tennessee Academy of Science* **22**:87–88.

Baker JR (1938). The relationship between latitude and breeding season in birds. *Proceedings of the Zoological Society of London* **108**:557–582.

Baker JR and Bird TF (1936). The seasons in a tropical rain forest (New Hebrides). 4. Insectivorous bats (Vespertilionidae and Rhinolophidae). *Zoological Journal of the Linnaean Society* **40**:143–161.

Bakst MR (1981). Sperm recovery from oviducts of turkeys at known intervals after insemination and oviposition. *Journal of Reproduction and Fertility* **62**:159–164.

Baldwin J, Temple-Smith P, and Tidemann C (1974). Changes in testis specific lactate dehydrogenase isoenzymes during the seasonal spermatogenic cycle of the marsupial *Schoinobates volans* (Petauridae). *Biology of Reproduction,* **11**:377–384.

Balfour E and Cadbury CJ (1979). Polygyny, spacing and sex ratio among hen harriers *Circus cyaneus* in Orkney, Scotland. *Ornis Scandinavica* **10**:133–141.

Ball JN (1969). Prolactin (fish prolactin or paralactin) and growth hormone. In *Fish Physiology* (Eds: Hoar WS and Randall DJ) Academic, New York:207–240.

Ballinger R (1978). Variation in and evolution of clutch and litter size. In *The Vertebrate Ovary* (Ed: Jones RE) Plenum, New York:789–825.

Ballinger RE (1977). Reproductive strategies: Food availability as a source of proximal variation in a lizard. *Ecology* **58**:628–635.

Ballinger RE (1979). Intraspecific variation in demography and life history of the lizard, *Sceloporus jarrovi*, along an altitudinal gradient in southern eastern Arizona. *Ecology* **60**:901–909.

Ballinger RE (1983). Life-history variations. In *Lizard Ecology* (Eds: Huey RB, Pianka ER, and Schoener T). Harvard University Press, Cambridge:241–260.

Balon EK (1975). Reproductive guilds of fishes: A proposal and a definition. *Journal of the Fisheries Research Board of Canada* **32**:821–864.

Balon EK (1981). Additions and amendments to the classification of reproductive styles in fishes. *Environmental Biology of Fishes* **6**:377–389.

Balon EK (1984). Patterns in the evolution of reproductive styles of fishes. In *Fish Reproduction. Strategies and Tactics* (Eds: Potts GW and Wootton RJ) Academic, London:35–53.

Balon EK (1986). Saltatory ontogeny and evolution. *Rivista di Biologia* **79**:151–190.

Balsano JS, Darnell RM, and Abramoff P (1972). Electrophoretic evidence of triploidy associated with populations of the gynogenetic teleost *Poecilia formosa*. *Copeia*:292–297.

Balthazart J, Prove E, and Gilles R (Eds) (1983). *Hormones and Behaviour in Higher Vertebrates*. Springer-Verlag, Berlin, Heidelberg.

Balthazart J and Schumacher M (1983). Testosterone metabolism and sexual differentiation in the brain. In *Hormones and Behavior in Higher Vertebrates* (Eds: Balthazart J and Schumacher M) Springer-Verlag, Berlin:237–260.

Barbault R (1974). Observations ecologique sur la reproduction des lezards tropicaux: Les strategies de ponte en foret et en savane. *Bulletin Societie Zoologie de France* **100**: 153–160.

Barbault R (1975). Dynamique des populations de lezards. *Bulletin d'ecologie* **6**:1–22.

Barbault R (1986). Rapid-aging in males, a way to increase fitness in a short-lived tropical lizard? *Oikos* **46**:258–260.

Barkley MS, Bradford GE, and Geschwind II (1978). The pattern of plasma prolactin concentration during the first half of mouse gestation. *Biology of Reproduction* **19**:291–296.

Barlow EK (1981). Patterns of parental investment, dispersal and size among coral reef fishes. *Environmental Biology of Fishes* **6**:65–85.

Barlow GW (1986). A comparison of monogamy among freshwater and coral-reef fishes. In *Indo-Pacific Fish Biology: Proceedings of the Second International Conference on Indo-Pacific Fishes*. (Eds: Uyeno T, Taniuchi T and Matsumura K) Ichthyological Society of Japan, Tokyo:767–775.

Barnes MB, Kretzman M, Licht P, and Zucker I (1986). The influence of hibernation on testis growth and spermatogenesis in the golden-mantled ground squirrel, *Spermophilus lateralis*. *Biology of Reproduction* **35**:1289–1297.

Barnett SA and Dickson RB (1985). A paternal influence on survival of wild mice in the nest. *Nature* **317**:617–618.

Barr WA (1963). The endocrine control of the sexual cycle in the plaice, *Pleuronectes platessa* (L.). 1. *General and Comparative Endocrinology* **3**:197–204.

Barr WA (1968). Patterns of ovarian activity. In *Perspectives in Endocrinology* (Eds: Barrington EJW and Jorgensen CB) Academic, New York:164–238.

Bartholomew GA (1952). Reproductive and social behavior of the northern elephant seal. *University of California Publications in Zoology* **47**:369–472.

Bartholomew GA (1970). A model for the evolution of pinniped polygyny. *Evolution* **24**:546–559.

Bartholomew GA and Hoel PG (1953). Reproductive behavior of the Alaska fur seal. *Journal of Mammalogy* **34**:417–436.

Bartke A, Craft BT, and Dalterio S (1975). Prolactin restores plasma testosterone levels and stimulates testicular growth in hamsters exposed to short day-length. *Endocrinology* **97**:1601–1604.

Batzli GO and Pitelka FA (1971). Condition and diet of cycling populations of the California vole, *Microtus californicus*. *Journal of Mammalogy* **52**:141–163.

Bauchot R (1965). La placentation chez les reptiles. *Anneé Biologie* **4**:547–575.

Baum MJ (1987). Hormonal control of sex differences in the brain and behavior of mammals. In *Psychobiology of Reproductive Behavior* (Ed: Crews D) Prentice-Hall, Englewood Cliffs, New Jersey:231–257.

Bauwens D and Verheyen RF (1985). The timing of reproduction in the lizard *Lacerta vivipara*: Differences between individual females. *Journal of Herpetology* **19**:353–364.

Bay EC (1966). Adaptation studies with the Argentine pearl fish, *Cynolebias bellottii*, for its introduction into California. *Copeia*:839–846.

Baylis JR (1981). The evolution of parental care in fishes, with reference to Darwin's rule of sexual selection. *Environmental Biology of Fishes* **6**:223–251.

Bazer FW, Chen TT, Knight JW, Schlosnagle D, Baldwin NJ, and Roberts RM (1975). Presence of a progesterone-induced, uterine specific, acid phosphatase in allantoic fluid of gilts. *Journal of Animal Science* **41**:1112–1119.

Beach F (1968). Factors involved in the control of mounting behavior by female mammals. In *Reproduction and Sexual Behavior* (Ed: Diamond M). Johns Hopkins University, Baltimore, 83–101.

Beach FA (1975). Hormonal modification of sexually dimorphic behavior. *Psychoneuroendocrinology* **1**:3–23.

Beacham TD (1980). Demography of declining populations of the vole, *Microtus townsendii*. *Journal of Animal Ecology* **49**:453–464.

Beasely LJ, Pelz KM, and Zucker I (1984). Circannual rhythms of body weight in pallid bats. *American Journal of Physiology* **246**:R955–R958.

Beasley LJ and Zucker I (1984). Photoperiod influences the annual reproductive cycle of the male pallid bat (*Antrozous pallidus*). *Journal of Reproduction and Fertility* **70**:567–573.

Beatley JC (1969). Dependence of desert rodents on winter annuals and precipitation. *Ecology* **50**:721–724.

Beatley JC (1976). Rainfall and fluctuating plant populations in relation to distributions and numbers of desert rodents in southern Nevada. *Oecologia* **24**:21–42.

Beauchamp GK, Doty RL, Mounton DG, and Mugford RA (1976). The pheromone concept in mammalian chemical communication: A critique. In *Mammalian Olfaction, Reproductive Processes and Behavior* (Ed: Doty RL) Academic, New York:143–160.

Becak W (1983). Evolution and differentiation of sex chromosomes in lower vertebrates. *Differentiation 23* (Supplement):S3–S12.

Beck AB (1964). The oestrogen isoflavones of subterranean clover. *Australian Journal of Agricultural Research* **15**:223–230.

Becker PH, Finck P, and Anlauf A (1985). Rainfall preceding egg-laying—a factor in breeding success in common terns (*Sterna hirundo*). *Oecologia* **65**:431–436.

Bedford JM (1973). Limitations of the uterus in the development of the fertilizing ability (capacitation) of spermatozoa. In *Sperm Capacitation* (Ed: Hamner CE) MSS Information Corporation, New York:40–47.

Bedford JM (1977). Evolution of the scrotum: The epididymis as the prime mover. In *Reproduction and Evolution* (Eds: Calaby JH and Tyndale-Biscoe CH) *Australian Academy of Science*, Canberra:171–182.

Beer JR, MacLeod CF, and Frenzel LD (1957). Prenatal survival and loss in some cricetid rodents. *Journal of Mammalogy* **38**:392–402.

Begg GW and MacLean GL (1976). Belly soaking in the white-crowned plover. *Ostrich* **47**:65.

Behrman HR (1979). Prostaglandins in hypothalamo-pituitary and ovarian function. *Annual Review of Physiology* **41**:685–700.

Belonje PC and van Niekerk CH (1975). A review of the influence of nutrition upon the oestrous cycle and early pregnancy in the mare. *Journal of Reproduction and Fertility*, Supplement **23**:167–169.

Bendell JF (1959). Food as a control of population of white-footed mice, *Peromyscus leucopus noveboracensis* (Fisher). *Canadian Journal of Zoology* **37**:173–209.

Beneson IE (1983). On the maintenance of the unique system of sex determination of lemmings. *Oikos* **41**:211–218.

Bengston SA (1971). Variations in clutch size in ducks in relation to food supply. *Ibis* **113**:523–526.

Bengston SA (1972). Breeding ecology of the harlequin duck *Histrionicus histrionicus*. *Ornis Scandinavicus* **3**:1–9.

Bennett RD and Heftmann E (1965). Progesterone biosynthesis from pregnenolone in *Holarrhena floribunda*. *Science* **149**:652–653.

Bennett RD, Heftmann E, and Winter BJ (1969). Conversion of sitosterol to progesterone by *Digitalis lanata*. *Naturwissenschaften* **56**:463.

Bennett RD, Ko S-T, and Heftmann E (1966). Isolation of estrone and cholesterol from the date palm, *Phoenix dactylifera* L. *Phytochemistry* **5**:231–235.

Benoit J (1970). Étude de l'action des radiations visibles dur la gonadostimulation et leur pénétration intra-cranienne chez le oiseaux et les mammifères. In *La Photoregulation de la Reproduction Chez les Oiseaux et les Mammifères* (Eds: Benoit J and Assenmacher I) Centre National de la Recherche Scientifique, Paris: 121–149.

Berger J (1986). *Wild Horses of the Great Basin*. University of Chicago, Chicago.

Berger PJ and Negus NC (1974). Influence of dietary supplements of fresh lettuce on ovariectomized *Microtus montanus*. *Journal of Mammalogy* **55**:747–750.

Berger PJ and Negus NC (1982). Stud male maintenance of pregnancy in *Microtus montanus*. *Journal of Mammalogy* **63**:148–151.

Berger PJ, Negus NC, Sanders EH, and Gardner PD (1981). Chemical triggering of reproduction in *Microtus montanus*. *Science* **214**:69–70.

Berger PJ, Sanders EH, Gardner PD, and Negus NC (1977). Phenolic plant compounds functioning as reproductive inhibitors in *Microtus montanus*. *Science* **195**:575–577.

Bermant G and Davidson JM (1974). *Biological Basis of Sexual Behavior*. Harper & Row, New York.

Bernard RTF (1986). Seasonal changes in plasma testosterone concentrations and Leydig cell and accessory gland activity in the Cape horseshoe bat (*Rhinolophus capensis*). *Journal of Reproduction and Fertility* **78**:413–422.

Bernstein IS (1976). Dominance, aggression and reproduction in primate societies. *Journal of Theoretical Biology* **60**:493–497.

Beumer JP (1979). Reproductive cycles of two Australian freshwater fishes: The spangled perch, *Therapon unicolor* Gunther, 1859 and the East Queensland rainbowfish, *Nematocentris splendida* Peters, 1866. *Journal of Fish Biology* **15**:111–134.

Bex FJ, Bartke A, Goldman BD, and Dalterio S (1978). Prolactin, growth hormone, luteinizing hormone receptors, and seasonal changes in testicular activity in the golden hamster. *Endocrinology,* **103**:2069–2080.

Bickoff EM (1963). Estrogen-like substances in plants. In *Physiology of Reproduction* (Ed: Hisaw HI) Oregon State College, Corvallis.

Bickoff EM, Booth AN, Lyman RL, Livingston AL, Thompson CR, and De Eds F (1957). Coumestrol, a new estrogen isolated from forage crops. *Science* **126**:969–970.

Bieniarz K, Epler P, Breton B, and Thuy LN (1978). The annual reproductive cycle in adult carp in Poland: Ovarian state and serum gonadotropin. *Annales de Biologie Animale, Biochemie et Biophysique* **18**:917–921.

Bijlsma RG (1982). Breeding season, clutch size and breeding success in the bullfinch *Pyrrhula pyrrhula*. *Ardea* **70**:25–30.

Billard R (1978). Testicular feedback on the hypothalamus-pituitary axis in rainbow trout (*Salmo gairdnerii* R). *Annales de Biologie Animale, Biochimie, Biophysique* **18**:813–818.

Billard R and Breton B (1978). Rhythms of reproduction in teleost fish. In *Rhythmic Activity in Fishes* (Ed: Thorpe JE). Academic, London:31–53.

Billard R, Breton B, Fostier A, Jalabert B, and Weil C (1978). Endocrine control of the teleost reproductive cycle and its relation to external factors: Salmonid and cyprinid models. In *Comparative Endocrinology* (Eds: Gaillard PJ and Boer HH) Elsevier/North Holland, Amsterdam: 37–48.

Billett DSM, Lampitt RS, Rice AL, and Mantoura RFC (1983). Seasonal sedimentation of phytoplankton to the deep-sea benthos. *Nature* **302**:520–522.

Binkley S (1974). Pineal and melatonin: Circadian rhythms and body temperature of sparrows. In *Chronobiology* (Eds: Scheving LE, Halberg F, and Pauly JE). Igaku Shoin, Tokyo: 582–585.

Binkley S (1976). Pineal gland biorhythms: N-acetyltransferase in chickens and rats. *Federation*

Proceedings **35**:2347–2352.

Binkley, SA, MacBride SE, Klein DC, and Ralph C (1974). Pineal enzymes: Regulation of avian melatonin synthesis. *Science* **181**:273–275.

Binkley SA, MacBride SE, Klein DC, and Ralph C (1975). Regulation of pineal rhythms in chickens: Refractory period and nonvisual light perception. *Endocrinology* **96**:848–853.

Binkley SA, Riebman JN, and Reilly KB (1978). The pineal gland: A biological clock in vitro. *Science* **202**:1198–1201.

Birkhead TR (1977). The effect of habitat and density on breeding success in the common guillemot (*Uria aalge*). *Journal of Animal Ecology* **47**:751–764.

Birnbaumer L and Kirchick HJ (1983). Regulation of gonadotropic action: The molecular mechanisms of gonadotropin-induced activation of ovarian adenylyl cyclases. In *Factors Regulating Ovarian Function* (Eds: Greenwald GS and Terranova PF) Raven, New York:287–310.

Bishop SC (1941). The salamanders of New York. *New York State Museum Bulletin* 324.

Bittman EL (1978a). Photoperiodic influences on testicular regression in the golden hamster: Termination of refractoriness. *Biology of Reproduction* **18**:871–877.

Bittman EL (1978b). Hamster refractoriness: The role of insensitivity of pineal target tissues. *Science* **202**:648–650.

Bittman EL and Karsch FJ (1984). Nightly duration of pineal melatonin secretion determines the reproductive response to inhibitory daylength in the ewe. *Biology of Reproduction* **30**:585–593.

Blackburn DG (1981). An evolutionary analysis of vertebrate viviparity. *American Zoologist* **21**:120.

Blackburn DG (1982). Evolutionary origins of viviparity in the Reptilia. I. Sauria. *Amphibia-Reptilia* **3**:185–205.

Blackburn DG (1985). Evolutionary origins of viviparity in the Reptilia. II. Serpentes, Amphisbaenia and Ichthyosauria. *Amphibia-Reptilia* **6**:259–291.

Blackburn DG, Vitt LJ, and Beuchat CA (1984). Eutherian-like reproductive specializations in a viviparous reptile. *Proceedings of the National Academy of Science* **81**:4860–4863.

Blair WF (1952). *The Rusty Lizard*. University of Texas, Austin.

Blank JL and Desjardins C (1984). Spermatogenesis is modified by food intake in mice. *Biology of Reproduction* **30**:410–415.

Blask DE, Reiter RJ, Vaughan MK, and Johnson LY (1979). Differential effects of the pineal gland on LH-RH and FSH-RH activity in the medial basal hypothalamus of the male golden hamster. *Neuroendocrinology* **28**:36–45.

Blaxter KL (1961). Lactation and growth of the young. In *Milk: The Mammary Gland and Its Secretion*, vol 1 (Eds: Kon SK and Cowie AT) Academic, New York:305–361.

Bleier WJ and Ehtesham M (1981). Ovulation following unilateral ovariectomy in the California leaf-nosed bat (*Macrotus californicus*). *Journal of Reproduction and Fertility* **63**:181–183.

Blondel J (1985). Breeding strategies of the blue tit and coal tit (*Parus*) in mainland and island habitats: A comparison. *Journal of Animal Ecology* **54**:531–566.

Blum V and Fiedler K (1965). Hormonal control of reproductive behavior in some cichlid fishes. *General and Comparative Endocrinology* **5**:186–196.

Blumer LS (1979). Male parental care in bony fishes. *Quarterly Review of Biology* **54**: 149–161.

Bodenheimer FS (1949). Problems of vole populations in the Middle East. *The Research Council of Israel, Jerusalem.*

Bodenheimer FS and Sulman F (1946). The estrous cycle of *Microtus guentheri* D and A and its ecological implications. *Ecology* **27**:255–256.

Boisseau JP (1965). Action de quelques hormones sur l'incubation Hippocampus males normaux, or castres ou hypophysectomises. *Comptes Rendes* **260**:313–314.

Bomford M and Redhead T (1987). A field experiment to examine the effects of food quality and population density on reproduction of wild house mice. *Oikos* **48**:304–311.

Bondrup-Nielsen S and Ims RA (1986). Reproduction and spacing behaviour of females in a peak density population of *Clethrionomys glareolus*. *Holarctic Ecology* **9**:109–112.

Bonner WN (1984). Lactation strategies in pinnipeds: Problems for a marine mammal group. In *Physiological Strategies in Lactation* (Eds: Peaker M, Vernon RG, and Knight CH) Zoological Society of London Symposium **51**:253–272.

Boonstra R and Rodd FH (1983). Regulation of breeding density in *Microtus pennsylvanicus*. *Journal of Animal Ecology* **52**:757–780.

Borowsky R and Diffley J (1981). Synchronized maturation and breeding in natural populations of *Xiphophorus variatus* (Poeciliidae). *Environmental Biology of Fish* **6**:49–58.

Borowsky RL (1978). Social inhibition of maturation in natural populations of *Xiphophorus* (Pisces: Poeciliidae). *Science* **201**:933–935.

Bosc MJ, Dujiane J, Durrand P, Pellettie J, and Ravault JR (1982). Influence of season on mean plasma levels of prolactin, placental lactogen hormone and luteinizing hormone during the 2nd half of gestation in the ewe. *Reproduction and Nutrition* **22**:485–493.

Boutin S (1984). The effect of conspecifics on juvenile survival and recruitment of snowshoe hares. *Journal of Animal Ecology* **53**:623–637.

Bowman RI (1986). Unusual seabird activity on the Galapagos Islands. *Climate Diagnostics Bulletin* No. 86/**5**:7–8.

Bradbury JW and Gibson RM (1983). Leks and mate choice. In *Mate Choice* (Ed: Bateson P) Cambridge University, Cambridge:109–138.

Bradford GE (1979). Genetic variation in prenatal survival and litter size. *Journal of Animal Science* **49**, Supplement **II**:66–74.

Bradt GW (1938). A study of beaver colonies in Michigan. *Journal of Mammalogy* **19**:139–152.

Braithwaite RW (1974). Behavioural changes associated with the population cycle of *Antechinus stuartii* (Marsupialia). *Australian Journal of Zoology* **22**:45–62.

Braithwaite RW and Lee AK (1978). A mammalian example of semelparity. *American Naturalist* **113**:151–155.

Brambell FWR (1944). The reproduction of the wild rabbit, *Oryctolagus cuniculus* (L.). *Proceedings of the Zoological Society of London* **114**:1–45.

Brambell FWR and Rowlands IW (1936). Reproduction of the bank vole (*Evotomys glareolus*, Schreber). II. Seasonal changes in the reproductive organs of the male. *Philosophical Transactions of the Royal Society B* **226**:71–83.

Brand CJ and Keith LB (1979). Lynx demography during a snowshoe hare in Alberta. *Journal of Wildlife Management* **43**:827–849.

Bray OE, Kennelly JJ, and Guarino JL (1975). Fertility of eggs produced on territories of vasectomized red-winged blackbirds. *Wilson Bulletin* **87**:187–195.

Breder CM Jr and Rosen DE (1966). *Modes of Reproduction in Fishes*. Natural History, Garden City, New York.

Breed WG (1975). Environmental factors and reproduction in the female hopping mouse, *Notomys alexis*. *Journal of Reproduction and Fertility* **45**:273–281.

Breed WG (1981). Reproduction of the Australian hopping mouse, *Notomys alexis*, and other bipedal desert rodents. In *Environmental Factors in Mammal Reproduction* (Eds: Gilmore D and Cook B) Macmillan Ltd, London:186–198.

Breed WG (1982). Morphological variation in the testes and accessory sex organs of Australian rodents in the genera *Pseudomys* and *Notomys*. *Journal of Reproduction and Fertility* **66**:607–613.

Breneman WR, Zeller FJ, Carmack M, and Kelley CJ (1976). In vivo inhibition of gonadotropins and thyrotropin in the chick by extracts of *Lithospermum ruderale*. *General and Comparative Endocrinology* **28**:24–32.

Bresslau E (1920). *The Mammary Apparatus of the Mammalia in the Light of Ontogenesis and Phylogenesis*. Methuen, London.

Breton B and Billard R (1977). Effects of photoperiod and temperature on plasma gonadotropin and spermatogenesis in the rainbow trout, *Salmo gairdnerii* Richardson. *Annales de Biologie Animale, Biochimie Biophysique* **17**:331–340.

Bretschneider LH and de Wit JJD (1947). Sexual endocrinology of non-mammalian vertebrates. *Monographs on the Progress of Research in Holland During the War*, no. 11. Elsevier, Amsterdam.

Bridges EL (1951). *Uttermost Part of the Earth*. Hodder and Stroughten, London.

Brockelman WK (1975). Competition, the fitness of offspring, and optimal clutch size. *American Naturalist* **109**:677–699.

Bronson FH (1971). Rodent pheromones. *Biology of Reproduction* **4**:344–357.

Bronson FH (1976). Urine marking in mice: Causes and effects. In *Mammalian Olfaction, Reproductive Processes and Behavior* (Ed: Doty RL) Academic, New York:119–141.

Bronson FH (1979). The role of priming pheromones in mammalian reproductive strategies. In *Chemical Ecology: Odour Communication in Animals* (Ed: Ritter FJ) Elsevier/North Holland, Amsterdam:97–115.

Bronson FH and Dezell HE (1968). Studies on the estrus-inducing (pheromonal) action of male deermice urine. *General and Comparative Endocrinology* **10**:339–492.

Bronson FH and Marsden HM (1964). Male-induced synchrony of estrus in deermice. *General and Comparative Endocrinology* **4**:634–637.

Bronson FH and Pryor S (1983). Ambient temperature and reproductive success in rodents living at different latitudes. *Biology of Reproduction* **29**:72–80.

Bronson MT (1979). Altitudinal variation in the life history of the golden-mantled ground squirrel (*Spermophilus lateralis*). *Ecology* **60**:272–279.

Bronson MT (1980). Altitudinal variation in the emergence time of golden-mantled ground squirrels (*Spermophilus lateralis*). *Journal of Mammalogy* **61**:124–126.

Brooke ML (1978). Some factors affecting the laying date, incubation, and breeding success of the Manx shearwater, *Puffinus puffinus*. *Journal of Animal Ecology* **47**:477–495.

Brooks DE (1973). Epididymal and testicular temperatures in the unrestrained conscious rat. *Journal of Reproduction and Fertility* **35**:157–160.

Brooksbank BWL, Brown R, and Gufstasson JA (1974). The detection of 5a-androst-16-en-3a-ol in human male axillary sweat. *Experientia* **30**:864–865.

Brown JL and Brown ER (1981). Extended family system in a communal bird. *Science* **211**:959–960.

Brown KM and Sexton OJ (1973). Stimulation of reproductive activity of female *Anolis sagrei* by moisture. *Physiological Zoology* **45**:168–172.

Brown LH, Gargett V, and Steyn P (1977). Breeding success in some African eagles related to theories about sibling aggression and its effects. *Ostrich* **48**:65–71.

Bruce HM (1959). An exteroceptive block to pregnancy in the mouse. *Nature* **182**:105.

Bruce HM (1966). Smell as an exteroceptive factor. In *Environmental Influences on Reproductive Processes* (Eds: Hansel W and Dutt RH) *Journal of Animal Science* **25** (Supplement): 83–87.

Bruder RH and Lehrman DS (1967). Role of the mate in elicitation of hormone-induced incubation behavior in the ring dove. *Journal of Comparative and Physiological Psychology* **63**:382–384.

Truslé S (1982). Ultrastructure of testis organization and resting cells in *Liza aurata* Risso, 1810 (Teleostei, Mugulidae). In *Proceedings of the International Symposium of Reproductive Physiology of Fish* (Comp: Richter CJJ and Goos HJT) Central Agricultural Publications and Documents, Wageningen:193–194.

Truslé S and Truslé J (1978). Early sex determination in *Mugil* (*Liza*) *auratus* Risso, 1810. An ultrastructural study. *Annales de Biologie Animale, Biochemie et Biophysique* **18**: 871–875.

Bryden MM (1968). Lactation and suckling in relation to early growth of the southern elephant seal, *Mirounga leonina* (L.). *Australian Journal of Zoology* **16**:739–748.

Buckley FG (1966). Egg placement in the teleost *Badis badis*. *Copeia*:605–606.

Buechner HK, Morrison AJ, and Leuthold W (1966). Reproduction in the Uganda kob with special reference to behaviour. *Zoological Society of London Symposium* **15**:69–76.

Bull JJ (1980). Sex determination in reptiles. *Quarterly Review of Biology* **55**:3–21.

Burgoyne P (1984). The origins of men with two X chromosomes. *Nature* **307**:109.

Burke WH, Ogasawara FX, and Fuqua CL (1972). A study of the ultrastructure of the uterovaginal sperm-storage glands on the hen, *Gallus domesticus*, in relation to a mechanism for the release of spermatozoa. *Journal of Reproduction and Fertility* **29**:29–36.

Burnett AMR (1965). Observations on the spawning migration of *Galaxias attenuatus* (Jenyns). *New Zealand Journal of Science* **8**:79–87.

Burns JM (1981). Aspects of endocrine control of delay phenomena in bats with special emphasis on delayed development. In *Embryonic Diapause in Mammals* (Eds: Flint APF, Renfree MB, and Weir MJ) *Journal of Reproduction and Fertility*, Supplement **29**:61–66.

Burns JR (1985). The effect of low latitude photoperiods on the reproduction of female and male *Poeciliopsis gracilis* and *Poecilia sphenops*. *Copeia*:961–965.

Burns JR and Flores JA (1981). Reproductive biology of the cuatro ojos, *Anableps dowi* (Pisces: Anablepidae), from El Salvador and its seasonal variations. *Copeia*:25–32.

Busse C and Hamilton, WJ III (1980). Infant carrying by male chacma baboons. *Science* **212**:1281–1283.

Bustard HR (1965). Observations on the life history and behavior of *Chamaeleo hohnelii* (Steindachner). *Copeia*:401–410.

Butterstein GM, Schadler MH, Lysogorski E, Robin L, and Sipperly S (1985). A naturally occurring plant compound, 6-methoxybenzoaxazolinone, stimulates reproductive responses in rats. *Biology of Reproduction* **32**:1018–1023.

Butler L (1953). The nature of cycles in populations of Canadian mammals. *Canadian Journal of Zoology* **31**:242–262.

Bye VJ (1984). The role of environmental factors in timing of reproductive cycles. In *Fish Reproduction: Strategies and Tactics* (Eds: Potts GW and Wootton RJ) Academic, London:187–205.

Byers JA and Bekoff M (1981). Social, spacing and cooperative behavior of the collared peccary, *Tayassu tajacu*. *Journal of Mammalogy* **60**:767–785.

Byskov AG (1978). Follicular atresia. In *The Vertebrate Ovary* (Ed: Jones RE) Plenum, New York:533–562.

Byskov AG (1979). Atresia. In *Ovarian Follicular Development and Function* (Eds: Midgley AR and Sadler WA) Raven, New York:41–57.

Callard IP and Ho MH (1980). Seasonal reproductive cycles in reptiles. In *Seasonal Reproduction in Higher Vertebrates* (Eds: Reiter RJ and Follett BK) S Karger, Basel:5–38.

Callard IP, Chan SWC, and Potta MA (1972). The control of the reptilian gonad. *American Zoologist* **12**:273–287.

Callard IP and Lance V (1977). The control of reptilian follicular cycles. In *Reproduction and Evolution* (Eds: Calaby JH and Tyndale-Biscoe CH) Australian Academy of Science, Canberra:199–209.

Callard IP, Lance V, Salhanick AR, and Barad D (1978). The annual ovarian cycle of *Chrysemys picta*: Correlated changes in plasma steroids and parameters of vitellogenesis. *General and Comparative Endocrinology* **35**:245–257.

Callard IP, McChesney I, Scanes C, and Callard GV (1976). The influence of mammalian and avian gonadotropins on in vitro steroid synthesis in the turtle (*Chrysemys picta*). *General and Comparative Endocrinology* **28**:2–9.

Cameron GN (1973). Effect of litter size on postnatal growth and survival in the desert woodrat. *Journal of Mammalogy* **54**:489–493.

Canivenc R and Bonnin M (1981). Environmental control of delayed implantation in the European badger (*Meles meles*). In *Embryonic Diapause in Mammals* (Eds: Flint APF, Renfree MB and Weir BJ) *Journal of Reproduction and Fertility*, Supplement **29**:25–53.

Cardinali DP (1981). Melatonin: A mammalian pineal hormone. *Endocrine Reviews* **2**: 327–346.

Carey C (1983). Structure and function of avian eggs. *Current Ornithology* **3**:69–103.

Carr A and Carr MH (1970). Modulated reproductive periodicity in Chelonia. *Ecology* **51**:335–337.

Carr WJ, Solberg B, and Pfaffmann C (1962). The olfactory threshold for estrous female urine in normal and castrated male rats. *Journal of Comparative Physiology and Psychology* **55**:415–417.

Carrick FN and Setchell BP (1977). The evolution of the scrotum. In *Reproduction and Evolution* (Eds: Calaby JH and Tyndale-Biscoe CH) Australian Academy of Science, Canberra: 165–170.

Carrick R, Csordas SE, and Ingham SE (1962). Studies on the southern elephant seal, *Mirounga leonina* L. IV. Breeding and development. C.S.I.R.O., *Wildlife Branch* **7**:161–197.

Cary JR and Keith LB (1979). Reproduction change in the 10-year cycle of snowshoe hares. *Canadian Journal of Zoology* **57**:375–390.

Case TJ (1978). On the evolutionary and adaptive significance of postnatal growth rates in the terrestrial vertebrates. *Quarterly Review of Biology* **53**:243–282.

Cei JM (1949). Sobre la biologie sexual de un batracio de grande altura de la region andina (*Telmatobius schreiteri* Vellard). *Acta Zoologica Lilloana* **7**:467–488.

Challis JRG, Kendall JZ, Robinson JS, and Thorburn GD (1977). The regulation of corticosteroids during late pregnancy and their role in parturition. *Biology of Reproduction* **16**:57–69.

Champbell AGM, Dawes GS, Fishman AP, Hyman AI, and James GB (1966). The oxygen consumption of the placenta and fetal membranes of the sheep. *Journal of Physiology* **182**:439–464.

Chan STH and Yeung WSB (1983). Sex control and sex reversals in fish under natural conditions. In *Fish Physiology*, vol IXB (Eds: Hoar WS, Randall DJ, and Donaldson EM) Academic, New York:171–222.

Chandola A, Singh R, and Thapliyal JP (1976). Evidence for a circadian oscillation in the gonadal response of the tropical weaverbird to photoperiod. *Chronobiologia* **3**:219–227.

Chandola A and Thapliyal JP (1978). Regulation of reproductive cycles of the tropical spotted munia and weaverbird. In *Environmental Endocrinology* (Eds: Assenmacher I and Farner DS) Springer-Verlag, Berlin:61–63.

Chapman DI and Chapman N (1975). *Fallow deer: Their History, Distribution and Biology.* Dalton, Lavenham.

Chappel SC and Selker F (1979). Relation between the secretion of FSH during the next preovulatory period and ovulation during the next cycle. *Biology of Reproduction* **21**:347–352.

Charnov EL (1982). *The Theory of Sex Allocation.* Princeton University, Princeton, New Jersey.

Chaturvedi CM and Thapliyal JP (1983). Thyroid, photoperiod and gonadal regression in the common myna *Acridotheres tristis. General and Comparative Endocrinology* **52**: 279–282.

Chen HJ, Brainard GC and Reiter RJ (1980). Melatonin given in the morning prevents the suppressive action on the reproductive system of melatonin given in the afternoon. *Neuroendocrinology* **31**:129–137.

Cheng KM, Burns JT, and McKinney F (1983). Forced copulation in captive mallards III. Sperm competition. *The Auk* **100**:320–310.

Chieffi G (1962). Endocrine aspects of reproduction in elasmobranch fishes. *General and Comparative Endocrinology*, Supplement **1**:275–285.

Chitty D (1967). The natural selection of self-regulatory behaviour in animal populations. *Proceedings of the Ecological Society of Australia* **2**:51–78.

Christian DP (1980). Patterns of recovery from low numbers in Namib Desert rodents. *Acta Theriologica* **25**:431–450.

Christian JJ (1961). Phenomena associated with population density. *Proceedings of the National Academy of Science* **47**:428–449.

Christian JJ and Davis DE (1956). The relationship between adrenal weight and population status of urban Norway rats. *Journal of Mammalogy* **37**:475–486.

Clark H, Florio B, and Hurowitz R (1955). Embryonic growth of *Thamnophis s. sirtalis* in relation to fertilization date and placental function. *Copeia*:9–13.

Clark H and Sisken BF (1956). Nitrogenous excretion by embryos of the viviparous snake, *Thamnophis s. sirtalis* (L.). *Journal of Experimental Biology* **33**:384–393.

Clark JH and Zarrow MX (1971). Influence of copulation on time of ovulation in women. *American Journal of Obstetrics and Gynecology* **109**:1083–1085.

Clark MJ (1967). Pregnancy in the lactating pigmy possum, *Cercartetus concinnus*. *Australian Journal of Zoology* **15**:673–683.

Clarke JR (1977). Long and short day changes in gonadal activity of field voles and bank voles. *Oikos* **29**:457–467.

Clattenburg RE, Montemurro DE, and Bruni JE (1975). Neurosecretory activity within the suprachiasmatic neurons of the female rabbit following castration. *Neuroendocrinology* **17**:211–224.

Clattenburg RE, Singh RP, and Montemurro DE (1972). Postcoital ultrastructural change in neurons of the suprachiasmatic nucleus of the rabbit. *Zeitschrift für Zellforschung* **125**: 448–459.

Claus R, Hoppen HO, and Karg H (1981). The secret of truffles: A steroid pheromone? *Experientia* **37**:1178–1179.

Clawson RC and Domm LV (1963). Developmental changes in glycogen content of primordial germ cells in chick embryos. *Proceedings of the Society for Experimental Biology and Medicine* **112**:533–537.

Clee MD, Humphreys EM, and Russell JA (1975). The suppression of ovarian cyclical activity in groups of mice, and its dependence on ovarian hormones. *Journal of Reproduction and Fertility* **45**:395–398.

Clements JA, Reyes FI, Winter JSD, and Faiman C (1976). Studies on human sexual development: III. Fetal pituitary, serum and amniotic fluid concentrations of LH, CG and FSH. *Journal of Clinical Endocrinology and Metabolism* **42**:9–19.

Clermont Y and Harvey SC (1967). Effects of hormones on spermatogenesis in the rat. *Ciba Colloquium on Endocrinology* **16**:173–189.

Clulow J and Jones RC (1982). Production, transport, storage and survival of spermatozoa in the male Japanese quail, *Coturnix coturnix*. *Journal of Reproduction and Fertility* **64**:259–266.

Clutton-Brock TH, Guiness FE, and Albon SD (1982). *Red Deer. Behavior and Ecology of Two Sexes*. University of Chicago, Chicago.

Cockburn A, Lee AK, and Martin RW (1983). Macrogeographic variation in litter size in *Antechinus* (Marsupialia, Dasyuridae). *Evolution* **37**:86–95.

Cockrem JF and Follett BK (1985). Circadian rhythm of melatonin in the pineal gland of the Japanese quail (*Coturnix coturnix japonica*). *Journal of Endocrinology* **107**:317–324.

Cody ML (1966). A general theory of clutch size. *Evolution* **20**:174–184.

Cohen DM and Wourms JP (1976). *Microbrotula randalli*, a new viviparous ophidioid fish from Samoa and New Hebrides, whose embryos bear trophotaeniae. *Proceedings of the Biological Society of Washington* **89**:81–98.

Cohen F and Zuelzer WW (1965). The transplacental passage of maternal erythrocytes into the fetus. *American Journal of Obstetrics and Gynecology* **93**:566–569.

Cohen F, Zuelzer WW, Gufstafson DC, and Evans MM (1964). Mechanics of isoimmunization. I. The transplacental passage of fetal erythrocytes in homospecific pregnancies. *Blood* **23**:621–646.

Cohen J and Cheng MF (1979). Role of vocalization in the reproductive cycle of ring doves (*Streptopelia risoria*): Effects of hypoglossal nerve section on the reproductive behavior and physiology of the female. *Hormones and Behavior* **13**:113–127.

Cole HH, Howell CE, and Hart GH (1931). The changes occurring in the ovary of the mare during pregnancy. *Anatomical Record* **49**:199–209.

Colombo L, Belvedere PC, Marconato A, and Bentivegna F (1982). Pheromones in teleost fish. In *Reproductive Physiology of Fish* (Eds: Richter CJJ and Goos HJT) Pudoc, Wageningen:84–94.

Colombo L and Yaron Z (1976). Steroid 21-hydrolase activity in the ovary of a snake *Storeria dekayi* during pregnancy. *General and Comparative Endocrinology* **28**:403–412.

Colvinaux PA (1984). The Galapagos climate: Past and present. In *Galapagos* (Ed: Perry R) Pergamon, Oxford:55–69.

Colwell RK (1974). Predictability, constancy, and contingency of periodic phenomena. *Ecology* **55**:1148–1153.

Conaway CH and Fleming WR (1960). Placental transmission of Na22 and I131 in *Natrix*. *Copeia*:53–55.

Concannon PW, Hansel PW, and Visek WJ (1975). The ovarian cycle of the bitch: Plasma estrogen, LH and progesterone. *Biology of Reproduction* **13**:112–121.

Congdon JD, Tinkle DW, Breitenbach GL, and van Loben Sels, RC (1983). Nesting ecology and hatching success in the turtle *Emydoidea blandingi*. *Herpetologica* **39**:417–429.

Conover CD and Kynard BE (1981). Environmental sex determination: Interaction of temperature and genotype in a fish. *Science* **213**:577–579.

Conover DO and Heins SW (1987). Adaptive variation in environmental and genetic sex determination in a fish. *Nature* **326**:496–498.

Conover DO and Kynard BE (1981). Environmental sex determination: Interaction of temperature and genotype in a fish. *Science* **213**:577–579.

Conover MR (1985). Foreign objects in bird nests. *Auk* **102**:696–700.

Conroy JWH, Darling OHS, and Smith HG (1975). The annual cycle of the chinstrap penguin Pygoscelis antarctica on Signy Island, South Orkney Islands. In *The Biology of Penguins* (Ed: Stonehouse B). Macmillan, London:353–362.

Cooke BD (1982). A shortage of water in natural pastures limiting a population of rabbits, *Oryctolagus cuniculus* (L.) in arid northeastern Australia. *Australian Wildlife Research* **9**:465–476.

Cooper WE Jr and Vitt LJ (1984). Conspecific odor detection by the male broad-headed skink, *Eumeces laticeps*: Effects of sex and site of odor source and male reproductive condition. *Journal of Experimental Zoology* **230**:199–209.

Copp AJ (1979). Interaction between inner cell mass and trophectoderm of the mouse blastocyst. I. A study of cellular proliferation. *Journal of Embryology and Experimental Morphology* **48**:109–125.

Coppola DM and Vandenberg JG (1987). Induction of a puberty-regulating chemosignal in wild mouse populations. *Journal of Mammalogy* **68**:86–91.

Costa DP, Le Boeuf BJ, Huntley AC, and Ortiz CL (1986). The energetics of lactation in the northern elephant seal, *Mirounga angustirostris*. *Journal of Zoology*, London **209**: 21–33.

Coulson JC, Duncan N, and Thomas C (1982). Changes in the breeding biology of the herring gull (*Larus argentatus*) induced by the reduction in size and density of the colony.

Journal of Animal Ecology **51**:739–756.

Coulson JC and Wooler RD (1976). Differential survival rates among breeding kittiwake gulls *Rissa tridactyla* (L.). *Journal of Animal Ecology* **45**:205–213.

Coulter GW (1970). Population changes with a group of fish species in Lake Tanganyika following their exploitation. *Journal of Fish Biology* **2**:329–353.

Coutts RR and Rowlands IW (1969). The reproductive cycle of the Skomer vole (*Clethrionomys glareolus skomerensis*). *Journal of Zoology* **158**:1–25.

Cowgill UM (1969). The season of birth and its biological implications. *Journal of Reproduction and Fertility*, Supplement **6**:89–103.

Cowie AT (1969). Variation in the yield and composition of the milk during lactation of the rabbit and the galactopoietic effect of prolactin. *Journal of Endocrinology* **44**:437–450.

Cowie AT (1972). Lactation and its hormonal control. In *Reproduction in Mammals: Hormones in Reproduction* (Eds: Short RV and Austin CR) Cambridge University, Cambridge: 106–143.

Cowie AT, Forsyth IA, and Hart IC (1980). *Hormonal Control of Lactation*. Springer-Verlag, Berlin.

Cowie AT and Tindal JS (1971). *The Physiology of Lactation*. Arnold, London.

Cowie AT, Tindal JS and Yokoyama A (1966). The induction of mammary growth in the hypophysectomized goat. *Journal of Endocrinology* **34**:185–195.

Cowles RB (1944). Parturition in the yucca lizard. *Copeia*:98–100.

Cox RI (1978). Plant estrogens affecting livestock in Australia. In *Effects of Poisonous Plants on Livestock* (Eds: Keeler RF, van Kampen KR and James LF) Academic, New York: 451–464.

Craig J and Shine R (1985). The seasonal timing of reproduction: A tropical-temperate comparison in Australian lizards. *Oecologia* **67**:464–474.

Craig JF (1977). The body composition of adult perch, *Perca fluviatilis*, in Windermere, with reference to seasonal changes in reproduction. *Journal of Animal Ecology* **46**:617–632.

Crews D (1975). Psychobiology of reptilian reproduction. *Science* **189**:1059–1065.

Crews D (1979). Endocrine control of reptilian reproductive behavior. In *Endocrine Control of Sexual Behavior* (Ed: Beyer C) Raven, New York:167–222.

Crews D and Fitzgerald K (1980). "Sexual" behavior in the parthenogenetic lizards (*Cnemidophorus*). *Proceedings of the National Academy of Science* **77**:499–502.

Crews DP and Licht P (1975). Stimulation of in vitro steroid production in turtle ovarian tissue by reptilian, amphibian and mammalian gonadotropins. *General and Comparative Endocrinology* **27**:71–83.

Crim LW, Peter RE, and Billard R (1976). Stimulation of gonadotropin secretion by intraventricular injection of hypothalamic extracts in the goldfish, *Carassius auratus*. *General and Comparative Endocrinology* **30**:77–82.

Croft DB (1982). Communication in the Dasyuridae (Marsupiala): A review. In *Carnivorous Marsupials* (Ed: Archer M). Royal Zoological Society of New South Wales, Sydney **1**:291–309.

Crompton AW, Taylor CR, and Jagger JA (1978). Evolution of homeothermy in mammals. *Nature* **272**:333–336.

Cross BA (1961). Neural control of lactation. In *Milk: The Mammary Gland and Its Secretion*, vol 1 (Eds: Kon SK and Cowie AT) Academic, New York:229–277.

Crossner KA (1977). Natural selection and clutch size in the European starling. *Ecology* **58**:885–892.

Cuellar HS, Roth JJ, and Fawcett JD (1972). Evidence for sperm sustenance by secretions of the renal sexual segment of male lizards, *Anolis carolinensis*. *Herpetologica* **28**:53–57.

Cuellar O (1970). Reproduction and the mechanisms of meiotic restitution in the parthenogenetic lizard *Cnemidophorus uniparens*. *Journal of Morphology* **133**:139–166.

Cuellar O (1977). Animal parthenogenesis. *Science* **197**:837–843.

Cuellar O (1981). Long term analysis of reproductive periodicity in the lizard *Cnemidophorus uniparens*. *American Midland Naturalist* **105**:93–101.

Cummins CP (1986). Temporal and spatial variation in egg size and fecundity in *Rana temporaria*. *Journal of Animal Ecology* **55**:303–316.

Currie C and Taylor HL (1970). A histochemical study of the circumtesticular Leydig cells of a teiid lizard, *Cnemidophorus tigris*. *Journal of Morphology* **132**:101–108.

D'Sousa F and Martin RD (1974). Maternal behavior and the effects of stress in tree shrews. *Nature* **251**:309–311.

Dahlgren BT (1980). The influence of three dietary non-protein levels on fecundity and fertility in the guppy *Poecilia reticulata*. *Oikos* **34**:337–346.

Dahlgren BT (1980). The effects of three different dietary protein levels on fecundity in the guppy, *Poecilia reticulata* (Peters). *Journal of Fish Biology* **16**:83–97.

Dahlgren BT (1985). Relationships between population density and reproductive success in the female guppy, *Poecilia reticulata* (Peters). *Ekologia Polska* **33**:677–703.

Daketse M-J and Martinet L (1977). Effect of temperature on the growth and fertility on the field-vole, *Microtus arvalis*, raised in different daylength and feeding conditions. *Annales de Biologie Animale, Biochimie Biophysique* **17**:713–721.

Daly M (1979). Why don't male mammals lactate? *Journal of Theoretical Biology* **78**:325–345.

Daniel JC Jr (1981). Delayed implantation in the northern fur seal (*Callorhinus ursinus*) and other pinnipeds. In *Embryonic Diapause in Mammals* (Eds: Flint APF, Renfree MB, and Weir BJ) *Journal of Reproduction and Fertility*, Supplement **29**:35–50.

Darevsky IS (1966). Natural parthenogenesis in a polymorphic group of Caucasian rock lizards related to *Lacerta saxicola* Eversmann. *Journal of the Ohio Herpetological Society* **5**:115–160.

Darevsky IS, Kupriyanova LA, and Roschin VV (1984). A new all-female triploid species of gecko and karyological data on bisexual *Hemidactylus frenatus* from Vietnam. *Journal of Herpetology* **18**:277–284.

Davies NB (1985). Cooperation and conflict among dunnocks, *Prunella modularis*, in a variable mating system. *Animal Behaviour* **33**:628–648.

Davies NB (1986). Reproductive success of dunnocks, *Prunella modularis*, in a variable mating system. I. Factors influencing provising rate, nestling weight and fledgling success. *Journal of Animal Ecology* **55**:123–138.

Davies NB and Houston N (1986). Reproductive success of dunnocks (*Prunella vulgaris*) in a variable mating system. II. Conflicts of interest among breeding adults. *Journal of Animal Ecology* **55**:139–154.

Davies NB and Lundberg A (1984). Food distribution and a variable mating system in the dunnock, *Prunella modularis*. *Journal of Animal Ecology* **53**:895–912.

Davis JWF (1976). Breeding success and experience in the arctic skua *Stercorarius parasiticus* (L.). *Journal of Animal Ecology* **45**:531–535.

Dawkin R and Carlisle TR (1976). Prenatal investment, mate desertion and a fallacy. *Nature* **262**:131–133.

Dawley EM (1984). Identification of sex through odors by male red-spotted newts, *Notophthalmus viridescens*. *Herpetologica* **40**:101–105.

Dawson A, Goldsmith AR, and Nicholls TJ (1986). Seasonal changes in testicular size and in plasma follicle-stimulating hormone and prolactin concentrations in thyroidectomized male and thyroidectomized castrated starlings (*Sturnus vulgaris*). *General and Comparative Endocrinology* **63**:38–44.

Day SL and Nalbandov AV (1977). Presence of prostaglandin F (PGF) on hen follicles and its physiological role in ovulation and oviposition. *Biology of Reproduction* **16**:486–494.

Dayanithi G, Cazalis M, and Nordmann JJ (1987). Relaxin affects the release of oxytocin and vasopressin from the neurohypophysis. *Nature* **325**:813–816.

de Steven D (1978). The influence of age on the breeding biology of the tree swallow, *Iridoprocne bicolor*. *Ibis* **120**:516–523.

de Vlaming V (1974). Experimental and endocrine control of teleost reproduction. In *Control of Sex in Fishes* (Ed: Schreck CB) Virginia Polytechnic Institute, Blacksburg, Virginia: 13–83.

de Vlaming V, Baltz D, Anderson S, Fitzgerald R, Delahunty G, and Barkeley M (1983). Aspects of embryo nutrition and excretion among viviparous embiotocid teleosts: Potential endocrine involvements. *Comparative Biochemistry and Physiology* **76A**:189–198.

de Vlaming V and Olcese J (1981). The pineal and reproduction in fish, amphibians and reptiles. In *The Pineal Gland. II. Reproductive effects* (Ed: Reiter RJ) CRC, Boca Raton: 1–29.

de Vlaming V, Sage M, and Charlton CB (1974). The effects of melatonin treatment on gonadosomatic index in the teleost, *Fundulus similis*, and the tree frog, *Hyla cinerea*. *General and Comparative Endocrinology* **22**:433–438.

de Vlaming V and Vodicnik MJ (1977). Effects of pinealectomy on pituitary gonadotrophs, pituitary gonadtropin potency and hypothalamic gonadotropin releasing activity in *Notemigonus crysoleucas*. *Journal of Fish Biology* **10**:73–86.

de Vlaming V and Vodicnik MJ (1978). Seasonal effects of pinealectomy on gonadal activity in the goldfish, *Carassius auratus*. *Biology of Reproduction* **19**:57–63.

de Vlaming VL (1971). The effects of food deprivation and salinity changes on reproductive function in the estuarine gobiid fish, *Gillichthys mirabilis*. *Biological Bulletin* **141**:458–471.

de Vlaming VL (1972). Reproductive cycling in the estuarine gobiid fish, *Gillichthys mirabilis*. *Copeia*:278–291.

de Vlaming VL (1974). Environmental and endocrine control of teleost reproduction. In *Control of Sex in Fishes* (Ed: Schreck CB) Virginia Polytechnic Institute and State University, Blacksburg, Virginia:13–83.

de Vlaming VL, Wiley HS, Delahunty G, and Wallace RA (1980). Goldfish (*Carassius auratus*) vitellogenin: Induction, isolation, properties and relationship to yolk proteins. *Comparative Biochemistry and Physiology* **67B**:613–623.

Degenhardt WG and Jones KL (1972). A new sagebrush lizard, *Sceloporus graciosus*, from New Mexico and Texas. *Herpetologica* **28**:212–217.

Deguchi T (1979). Circadian rhythm of serotonin *N*-Acetyltransferase activity in organ culture of chicken pineal gland. *Science* **203**:1245–1247.

Deguchi T (1981). Rhodopsin-like photosensitivity of isolated chicken pineal gland. *Nature* **290**:706–707.

del Pino E, Galarza ML, de Albuja CM, and Humphries Jr AA (1975). The maternal pouch development in the marsupial frog *Gastrotheca riobambae* (Fowler). *Biological Bulletin* **149**:480–491.

del Pino EM (1980). Morphology of the pouch and incubatory integument in marsupial frogs (Hylidae). *Copeia*:10–17.

del Pino EM, Galarza ML, de Albuja CM, and Humphries Jr AA (1975). The maternal pouch and development in the marsupial frog *Gastrotheca riobambae* (Fowler). *Biological Bulletin* **149**:480–491.

del Pino EM and Humphries Jr AA (1978). Multiple nuclei during early oogenesis in *Flectonotus pygmaeus* and other marsupial frogs. *Biological Bulletin* **154**:198–212.

del Pino EM and Sanchez G (1977). Ovarian structure of the marsupial frog, *Gastrotheca riobambae* (Fowler). *Journal of Morphology* **153**:153–162.

DeMartini EE, Moore TO, and Plummer KM (1983). Reproduction and growth dynamics of *Hyperprosopon argenteum* (Embiotocidae) near San Diego, California. *Environmental Biology of Fish* **8**:29–38.

Dent JN (1970). The ultrastructure of the spermatheca in the red-spotted newt. *Journal of Morphology* **132**:397–424.

Desjardins C (1979). Potential sources of variation affecting studies on pituitary-gonad function. In *Animal Models for Research on Contraception and Fertility* (Ed: Alexander NJ) Harper and Row, Hagerstown, Maryland:13–32.

Desy EA and Thompson CF (1983). Effects of supplemental food on a *Microtus pennsylvanicus* population in central Illinois. *Journal of Animal Ecology* **52**:127–140.

Devinoy E, Houdebine LM, and Delouis LM (1978). Role of prolactin and glucocorticoids in the expression of casein genes in rabbit mammary gland in organ culture. Quantification of casein mRNA. *Biochemica et Biophysica Acta* **517**:360–366.

Dewan AP and Sumpter JP (1983). The control of trout reproduction: Basic and applied research on hormones. In *Control Processes in Fish Physiology* (Eds: Rankin JC, Duggan RT, and Pitcher TJ) Praeger, Eastbourne:200–220.

Dewan EM and Rock J (1969). Phase locking of the human menstrual cycle by periodic stimulation. *Biophysical Journal Society Abstracts*:SAM-B-2.

Dewsbury DA (1982). Ejaculate cost and male choice. *American Naturalist* **119**:601–610.

Dhont AA and Eyckerman R (1979). Temperature and date of laying by tits *Parus* spp. *Ibis* **121**:329–331.

Dhont AA and Eyckerman R (1980). Competition and the regulation of numbers in the great and blue tit. *Ardea* **68**:121–132.

Diamond J (1983). The reproductive biology of mound-building birds. *Nature* **301**:288–289.

Diamond JM (1981). Birds of paradise and the theory of natural selection. *Nature* **298**: 257–258.

Didio LJA, Allen DJ, Correr S, and Motta PM (1980). Smooth musculature in the ovary. In *Biology of the Ovary* (Eds: Motta PM and Hafez ESE) Martinus Nijhoff, The Hague: 106–118.

Dierschke DJ, Weiss G, and Knobil E (1974). Sexual maturation in the female rhesus monkey and the development of estrogen-induced gonadotropin hormone release. *Endocrinology* **94**:198–206.

Diller LV and Wallace RL (1984). Reproductive biology of the northern Pacific rattlesnake (*Crotalus viridis oreganus*) in northern Idaho. *Herpetologica* **40**:182–193.

Dimmit MA and Ruibal R (1980). Environmental correlates of emergence in spadefoot toads (*Scaphiopus*). *Journal of Herpetology* **14**:21–29.

Dixson AF and George L (1982). Prolactin and parental behavior in a male New World primate. *Nature* **299**:551–553.

Dobkin DS and Gettinger RD (1985). Thermal aspects of anuran foam nests. *Journal of Herpetology* **19**:271–275.

Dobson S and Dodd JM (1977). The roles of temperature and photoperiod in the response of the testis of the dogfish, *Scyliorhinus canicula* L. to partial hypophysectomy (ventral lobectomy). *General and Comparative Endocrinology* **32**:53–71.

Dodd JM (1975). The hormones of sex and reproduction and their effects on fish and lower chordates: 20 years on. *American Zoologist*, Supplement **1**:137–171.

Dodd JM (1983). Reproduction in cartilaginous fishes (Chondrichthys). In *Fish Physiology*, vol 9, part A (Eds: Hoar WS, Randall DJ, and Donaldson EM) Academic, New York: 31–95.

Dodd JM, Evenett PJ, and Goddard KC (1960). Reproductive endocrinology in cyclostomes and elasmobranchs. In *Hormones in Fish* (Ed: Chester Jones I) Symposium of the Zoological Society of London **I**:77–103.

Dodd JM and Sumpter PJ (1984). Fishes. In *Marshall's Physiology of Reproduction*, vol 1, *Reproductive cycles of vertebrates* (Ed: Lamming GE) Churchill Livingstone, Edinburgh: 1–126.

Dominey WJ (1980). Female mimicry in male bluegill sunfish—a genetic polymorphism? *Nature* **284**:546–548.

Dominey WJ (1981). Anti-predator function of bluegill sunfish nesting colonies. *Nature* **290**:586–588.

Donaldson EM (1973). Reproductive endocrinology of fishes. *American Zoologist* **13**:909–927.

Döring GK (1963). Uber die relative Haufigkeit des anovulatorischen Zyklus im Leben der Frau. *Archiv duer Gynakologie* **199**:115.

Dorner G (1976). *Hormones and Brain Differentiation*. Elsevier, Amsterdam.

Dorrington J and Gore-Langton RE (1981). Prolactin inhibits oestrogen synthesis in the ovary. *Nature* **290**:600–602.

Doty RL (1976). A review of recent psychophysical studies examining the possibility of chemical communication of sex and reproductive state in humans. In *Chemical Signals in Vertebrates* (Eds: Muller-Schwarze D and Mozell MM) Plenum, New York:273–286.

Doty RL, Ford M, Preti G, and Huggins GR (1975). Changes in the intensity and pleasantness of human vaginal odors during the menstrual cycle. *Science* **190**:1316–1317.

Dow H and Fredga S (1984). Factors affecting reproductive output of the goldeneye *Bucephala clangula*. *Journal of Animal Ecology* **53**:679–692.

Drent R (1973). The natural history of incubation. In *Breeding Biology of Birds* (Ed: Farner DS) U.S. National Academy of Sciences, Washington:262–322.

Drent RH (1975). Incubation. In *Avian Biology* (Eds: Farner DS and King JR) Academic, New York:333–420.

Drent RH and Daan S (1980). The prudent parent: Energetic adjustments in avian breeding. *Ardea* **68**:225–252.

Drickamer LC (1974). Social rank, observability, and sexual behavior of rhesus monkeys (*Macaca mulatta*). *Journal of Reproduction and Fertility* **37**:117–120.

Drickamer LC (1981). Acceleration and delay of sexual maturation in female house mice previously selected for early and late first vaginal estrus. *Journal of Reproduction and Fertility* **63**:325–329.

Drickamer LC (1986). Effects of urine from females in oestrus on puberty in female mice. *Journal of Reproduction and Fertility* **77**:613–622.

Dryden GL (1968). Growth and development of *Suncus murinus* in captivity on Guam. *Journal of Mammalogy* **49**:51–62.

Dryden GL (1969). Reproduction in *Suncus murinus*. *Journal of Reproduction and Fertility*, Supplement **6**:377–396.

Dryden GL and Anderson JN (1977). Ovarian hormone: Lack of effect on reproductive structures of female Asian musk shrews. *Science* **197**:782–784.

Dudley D (1974). Contributions of paternal care to the growth and development of the young in *Peromyscus californicus*. *Behavioural Biology* **11**:155–166.

Duellman WE (1978). The biology of an equatorial herpetofauna. In *Amazonian Ecuador*. Miscellaneous Publications, Museum of Natural History, University of Kansas **65**:1–352.

Dujone AR, de Laborde NP, Carril LM, Cheviakoff S, Pedroza E, and Rosner JM (1976). Correlation between catecholamine content of the human fallopian tube and the uterus and plasma levels of estradiol and progesterone. *American Journal of Obstetrics and Gynecology* **124**:229–233.

Dulka JG, Stacey NE, Sorensen P, and Van Der Kraak GJ (1987). A steroid sex pheromone synchronizes male-female spawning readiness in goldfish. *Nature* **325**:251–253.

Dunmire WW (1960). An altitudinal survey of reproduction in *Peromyscus maniculatus*. *Ecology* **41**:174–182.

Dunnet GM (1964). A field study of local populations of the brush-tailed possum *Trichosurus vulpecula* in eastern Australia. *Proceedings of the Zoological Society of London* **142**: 665–695.

Dwyer PD (1970). Latitude and breeding season in a polyestrous species of *Myotis*. *Journal of Mammalogy* **51**:405–410.

Dym M and Fawcett DW (1971). Further observations on the numbers of spermatogonia, spermatocytes and spermatids connected by intercellular bridges in the mammalian testis. *Biology of Reproduction* **4**:195–215.

Easley KA, Culley Jr DD, Horseman ND, and Penkala JE (1979). Environmental influences on hormonally induced spermiation of the bullfrog, *Rana catesbeiana*. *Journal of Experimental Zoology* **207**:407–416.

Eastin Jr WC and Spaziani E (1978). On the control of calcium secretion in the avian shell gland (uterus). *Biology of Reproduction* **19**:493–504.

Eckardt MJ and Whimster IW (1971). Skin homografts in the all female gekkonid lizard,

Hemidactylus garnotii. *Copeia*:152–154.

Eckert KL (1987). Environmental unpredictability and leatherback sea turtle (*Dermochelys coriacea*) nest loss. *Herpetologica* **43**:315–323.

Ehrenkranz JRL (1983). Seasonal breeding in humans: Birth records of the Labrador eskimo. *Fertility and Sterility* **40**:485–489.

Eichler VB and Moore RY (1975). Studies on hydroxyindole-O-methyltransferase in frog brain and retina: Enzymology, regional distribution and environmental control of enzyme levels. *Comparative Biochemistry and Physiology* **50C**:89–95.

Ekman J (1984). Density-dependent seasonal mortality and population fluctuations of the temperate-zone willow tit (*Parus montanus*). *Journal of Animal Ecology* **53**:119–134.

El-Badrawi H and Hafez ESE (1980). The postmenopausal ovary and premature ovary failure. In *Biology of the Ovary* (Eds: Motta PM and Hafez ESE) Martinus Nijhoff, The Hague: 291–320.

Elliott JA (1976). Circadian rhythms and periodic time measurement in mammals. *Federation Proceedings* **35**:2339–2346.

Elliot RD (1985). The exclusion of avian predators from aggregations of nesting lapwings (*Vanellus vanellus*). *Animal Behaviour* **33**:308–314.

Ellis LC and Balph DF (1976). Age and seasonal differences in the synthesis and metabolism of testosterone by testicular tissue and pineal HIOMT activity of Uinta ground squirrels (*Spermophilus armatus*). *General and Comparative Endocrinology* **28**:42–51.

Elton C and Nicholson M (1942). The ten-year cycle in numbers of the lynx in Canada. *Journal of Animal Ecology* **11**:215–244.

Elwood RW (1983). Paternal care in rodents. In *Parental Behaviour of Rodents* (Ed: Elwood RW) Wiley, Chichester:235–257.

Emlen ST (1978). The evolution of cooperative behaviour in birds. In *Behavioural Ecology* (Eds: Krebs JR and Davies HB) Blackwell, Oxford:245–281.

Emlen ST and Oring LW (1977). Ecology, sexual selection and the evolution of mating systems. *Science* **197**:215–223.

Emlen ST and Vehrencamp SL (1983). Cooperative breeding strategies among birds. In *Perspectives in Ornithology* (Eds: Brush AH and Clark GA Jr) Cambridge University, Cambridge:93–100.

Enders RK (1952). Reproduction in the mink (*Mustela vison*). *Proceedings of the American Philosophical Society* **96**:691–755.

Epple G (1974). Primate pheromones. In *Pheromones* (Ed: Birch MC) North Holland, Amsterdam:366–385.

Epple G (1975). The behavior of marmoset monkeys (Callithricidae). In *Primate Behavior*, vol. 4 (Ed: Rosenblum RA) Academic, New York:195–239.

Eriksson M (1984). Winter breeding in three rodent species, the bank vole *Clethrionomys glareolus*, the yellow-necked mouse *Apodemus flaviocollis*, and the wood mouse *A. sylvaticus*. *Holarctic Ecology* **7**:428–429.

Erikstad KE and Andersen R (1983). The effect of weather on survival, growth rate and feeding time in different sized willow grouse broods. *Ornis Scandinavica* **14**:249–252.

Erlinge S (1983). Demography and dynamics of a stoat *Mustela erminea* population in a diverse community of vertebrates. *Journal of Animal Ecology* **52**:705–726.

Eskes GA (1983). Gonadal response in gonadectomized, intact and pinealectomized male golden hamsters. *Journal of Reproduction and Fertility* **68**:86–90.

Evans HJ, Buckton KE, Spowart G, and Carothers AD (1979). Human heteromorphic X chromosomes in 46 XX males; evidence for the involvement of interchange. *Genetics* **49**:11–31.

Evans, LT (1959). A motion picture study of maternal behavior of the lizard, *Eumeces obsoletus* Baird and Girard. *Copeia*:103–110.

Ewer RF (1973). *The Carnivores*. Cornell University, Ithaca.

Exbryat J-M and Delsol M (1985). Reproduction and growth of *Typhlonectes compressicaudus*, a viviparous gymnophione. *Copeia*:960–955.

Fairchild L (1981). Mate selection and behavioral thermoregulation in Fowler's toads. *Science* **212**:950–951.

Falconer DS (1960). Genetics of litter size in mice. *Journal of Cell and Comparative Physiology* **56**, Supplement **I**:153–167.

Farner DA and Wingfield JC (1978). Environmental endocrinology and the control of annual reproductive cycles in passerine birds. In *Environmental Endocrinology* (Eds: Assenmacher I and Farner DS) Springer-Verlag, Berlin:44–51.

Farner DS (1964). The photoperiodic control of reproductive cycles in birds. *American Scientist* **52**:137–156.

Farner DS (1975). Photoperiodic controls in the secretion of gonadotropins in birds. In *Trends in Comparative Endocrinology* (Ed: Barrington FJW) *American Zoologist* **15** (Supplement **1**): 117–135.

Farner DS (1980a). Evolution of the control of reproductive cycles in birds. In *Hormones, Adaptation and Evolution* (Eds: Ishii S, Hirano T, and Wada M) Japan Scientific Societies, Tokyo, and Springer-Verlag, Berlin:185–191.

Farner DS (1980b). Endogenous periodic functions in the control of reproductive cycles. In *Biological Rhythms in Birds* (Eds: Tanabe Y, Tanaka K, and Ookawa TO) Japan Scientific Societies, Tokyo, and Springer-Verlag, Berlin: 123–138.

Farner DS, King JR, and Parkes KC (1983). *Avian Biology*, vol 3. Academic, New York.

Farner DS and Serventy DL (1960). The timing of reproduction in birds in the arid regions of Australia. *Anatomical Record* **137**:354–372.

Farnsworth NR, Bingel AS, Cordell AS, Crane FA, and Fong HFS (1975). Potential value of plants as sources of new antifertility agents. *Journal of Pharmacological Science* **65**:717–754.

Fay FH (1982). Ecology and biology of the Pacific walrus, *Odobenus rosmarus divergens* Illiger. *North American Fauna* 74.

Feder HH (1981). Hormonal actions on the sexual differentiation of the genitalia and the gonadotropin-regulating systems. In *Neuroendocrinology of Reproduction* (Ed: Adler NT) Plenum, New York:89–126.

Fenwick JC (1970a). Demonstration and effect of melatonin in fish. *General and Comparative Endocrinology* **14**:86–97.

Fenwick JC (1970b). The pineal organ: Photoperiod and reproductive cycles in the goldfish, *Carassius auratus*. *Journal of Endocrinology* **46**:101–111.

Ferguson MWJ and Joanen T (1982). Temperature of egg incubation determines sex in *Alligator mississippiensis*. *Nature* **296**:850–853.

Field AC (1975). Seasonal changes in reproduction, diet and body composition of two equatorial rodents. *East African Wildlife Journal* **13**:221–235.

Fildes V (1986). *Breasts, Bottles and Babies*. Edinburgh University, Edinburgh.

Filewood LWC, Hough K, Morris IC, and Peters DE (1978). Helpers at the nest of the rainbow bee-eater. *Emu* **78**:43–44.

Firth BT, Kennaway DJ, Rosenbelds AM (1979). Plasma melatonin in the scincid lizard, *Trachydosaurus rugosis*: Diel rhythm, seasonality, and the effect of constant light and darkness. *General and Comparative Endocrinology* **37**:493–500.

Fischer EA (1980). Speciation in the hamlets (Hypoplectrus: Serranidae)—a continuing enigma. *Copeia*:649–659.

Fitch HS (1970). Reproductive cycles in lizards and snakes. University of Kansas Museum of Natural History Miscellaneous Publication 52.

Fitzgerald KM and Zucker I (1976). Circadian organization of the estrous cycle of the golden hamster. *Proceedings of the Natural Academy of the Science USA* **73**:2923–2927.

Fitzgerald KT, Guillette Jr LJ, and Duvall D (1979). Notes on birth, development and care of *Gastrotheca riobambae* tadpoles in the laboratory (Amphibia, Anura, Hylidae). *Journal of Herpetology* **13**:457–460.

Fitzsimons VM (1943). The lizards of South Africa. *Transvaal Museum Memoir* **1**:1–528.

Fleming TH (1971). *Artibeus jamaicensis*—delayed embryonic development in a neotropical bat. *Science* **171**:402.

Fleming TH and Rauscher RJ (1978). On the evolution of litter size in *Peromyscus leucopus*. *Evolution* **32**:45–55.

Flint APF and Renfree MB (1981). Bromocryptine-induced implantation during pregnancy in the rat. *Journal of Reproduction and Fertility* **62**:181–183.

Flint APF and Sheldrick EL (1982). Ovarian secretion of oxytocin is stimulated by prostaglandin. *Nature* **297**:587–588.

Flint APF and Sheldrick EL (1983). Evidence for a systemic role for ovarian oxytocin in luteal regression in sheep. *Journal of Reproduction and Fertility* **67**:215–225.

Fluckinger E (1978). Effects of bromocryptine on the hypothalamus-pituitary axis. *Acta Endocrinologica* **88**:111–117.

Flux JEC (1987). Myths and mad March hares. *Nature* **325**:737.

Follett BK (1984). Birds. In *Marshall's Physiology of Reproduction*, vol 1, 4th ed, *Reproductive cycles of vertebrates* (Ed: Lamming GE) Churchill Livingstone, Edinburgh:283–350.

Follett BK and Davies DT (1979). The endocrine control of ovulation in birds. In *Animal Reproduction* (Ed: Hawk HA) Allenheld, Osmun, Montclair:323–344.

Follett BK and Maung SL (1978). Rate of testicular maturation, in relation to gonadotropin and testosterone levels, in quail exposed to various artificial photoperiods and to natural daylengths. *Journal of Endocrinology* **78**:267–280.

Follett BK and Nicholls TJ (1985). Influences of thyroidectomy and thyroxin replacement on photoperiodically controlled reproduction in quail. *Journal of Endocrinology* **107**:211–221.

Follett BK and Robinson JE (1980). Photoperiod and gonadotropin secretion in birds. In *Seasonal Reproduction in Higher Vertebrates* (Eds: Reiter RJ and Folett BK), Progress in Reproductive Biology 5. Karger, Basel:39–61.

Folley SJ (1956). Lactation. In *Marshall's Physiology of Reproduction*, vol 2, 3rd ed (Ed: Parkes AS) Longmans Green, London:525–649.

Fontaine YA (1975). Hormones in fishes. *Biochemistry, Biophysics and Perspectives in Marine Biology* **2**:139–212.

Foote CL (1964). Intersexuality in amphibians. In *Intersexuality in Vertebrates Including Man* (Eds: Armstrong CN and Marshall AJ) Academic, New York:233–272.

Ford CS and Beach FA (1952). *Patterns of Sexual Behavior*. Eyre and Spottiswoode, London.

Ford NL (1983). Variation in mate fidelity in monogamous birds. *Current Ornithology* **1**:329–356.

Ford SP and Christenson RK (1979). Blood flow to uteri of sows during the estrous cycle and early pregnancy: Local effect of the conceptus on the uterine blood supply. *Biology of Reproduction* **21**:619–624.

Forester DC (1984). Brooding behavior by the mountain dusky salamander: Can the female's presence reduce clutch desiccation? *Herpetologica* **40**:105–109.

Forester DC and Lykens DV (1986). Significance of satellite males in a population of spring peepers (*Hyla crucifer*). *Copeia*:719–724.

Foster RG, Follett BK, and Lythgoe JN (1985). Rhodopsin-like sensitivity of extra-retinal photoreceptors mediating the photoperiodic response in quail. *Nature* **313**:50–52.

Fox H (1977). The urogenital system of reptiles. In *Biology of the Reptilia* (Eds: Gans C and Parsons TS) Morphology E. Academic, New York:1–157.

Fox SL and Guillette Jr LJ (1987). Luteal morphology, atresia, and plasma progesterone concentrations during the reproductive cycle of two oviparous lizards, *Crotophytes collaris* and *Eumeces obsoletus*. *American Journal of Anatomy* **179**:324–332.

Franchimont P, Henderson K, Verhoeven G, Hagelstein M, Renard C, Demoulin A, Bourguignov J and Lecompte-Yerna M (1981). Inhibin: Mechanisms of action and secretion. In *Intragonadal Regulation of Reproduction* (Eds: Franchimont P and Channing CP) Academic, London:167–191.

Francis CM and Millington AJ (1971). The presence of methylated coumestans in annual Medicago species: Response to a fungal pathogen. *Australian Journal of Agricultural Research* **22**:75–80.

Frazer NB (1987). Preliminary estimates of survivorship for wild juvenile loggerhead sea turtles (*Caretta caretta*). *Journal of Herpetology* **21**:232–235.

Frazer NB and Richardson JI (1986). The relationship of clutch size and frequency to body size in loggerhead turtles, *Caretta caretta*. *Journal of Herpetology* **20**:81–84.

Fredga K (1983). Aberrant sex chromosome mechanisms in mammals. Evolutionary aspects. *Differentiation* **23** (Supplement):S23–S30.

Fredga K, Gropp A, Winking H, and Frank F (1976). Fertile XX- and XY-type females in the wood lemming *Myopus schistocolor*. *Nature* **261**:225–227.

Friedman MH and Friedman GS (1939). Gonadotrophic extracts from green leaves. *American Journal of Physiology* **125**:486–490.

Friedman MH and Friedman GS (1940). The relation of diet to the restriction of the gonadotrophic hormone content of the discharged rabbit pituitary. *American Journal of Physiology* **128**: 493–499.

Friedmann MB (1977). Interactions between visual and vocal courtship stimuli in the neuroendocrine response of female doves. *Journal of Comparative Physiology and Psychology* **91**:1408–1416.

Fuchs A-R, Fuchs F, Soloff MS, and Fernstrom MJ (1982). Oxytocin receptors and human parturition: A dual role for oxytocin in the initiation of labor. *Science* **215**:1396–1398.

Fujita H (1954). An interpretation of the changes in type of population density effect upon the oviposition rate. *Ecology* **35**:253–257.

Fujita H and Utida S (1953). The effect of population density on the growth of an animal population. *Ecology* **34**:488–498.

Fulkerson WL (1979). *Hormonal Control of Lactation* I. Eden, Westmount, Ontario.

Furness RW and Birkhead TR (1984). Seabird colony distributions suggest competition for food supplies during the breeding season. *Nature* **311**:655–656.

Gaddum-Rosse P and Blandau RJ (1976). Comparative observations on ciliary currents in mammalian oviducts. *Biology of Reproduction* **14**:605–609.

Gadgil M and Bossert WH (1970). Life historical consequences of natural selection. *American Naturalist* **104**:1–24.

Gambell R (1973). Some effects of exploitation on reproduction in whales. *Journal of Reproduction and Fertility*, Supplement **19**:533–553.

Gardner DM (1978). Utilization of extracellular glucose by spermatozoa of two viviparous fishes. *Comparative Biochemistry and Physiology* **59A**:165–168.

Garris DR and Mitchell JA (1979). Intrauterine oxygen tension during the estrous cycle in the guinea pig: its relation to uterine blood volume and plasma estrogen and progesterone levels. *Biology of Reproduction* **21**:149–159.

Garrod DJ and Horwood JW (1984). Reproductive strategies and response to exploitation. In *Fish Reproduction: Strategies and Tactics* (Eds: Potts GW and Wootton RJ) Academic, London:367–384.

Garstka WR and Crews D (1981). Female sex pheromones in the skin and circulation of a garter snake. *Science* **214**:681–683.

Garstka WR and Crews D (1982). Female control of male reproductive function in a Mexican snake. *Science* **217**:1159–1160.

Gartlan JS (1968). Ecology and behaviour of an isolated population of vervet monkeys on Lolui Island, Lake Victoria. *Human Biology* **40**:122–122.

Gasith A and Sidis I (1985). Sexual activity in the terrapin, *Mauremys caspica rivulata*, in Israel, in relation to the testicular cycle and climatic factors. *Journal of Herpetology* **19**:254–260.

Gavaud J (1977). La gametogenese du male de *Nectrophrynoides occidentalis* Angel (Amphibien Anoure vivipare) II-Etude experimental de role des facteurs externes sur la spermatogenese de l'adulte, au cours du cycle annuel. *Annales de Biologie Animale, Biochimie Biophysique* **17**:679–694.

Gavaud J (1986). Vitellogenesis in the lizard *Lacerta vivipara* Jacquin. II. Vitellogenin synthesis during the reproductive cycle and its control by ovarian steroids. *General and Comparative Endocrinology* **63**:11–23.

Geevarghese C and John PA (1983). Maturation and spawning of a gobiid fish, *Oligolepis acutipennis* (Cuv. & Val.), from the south-west coast of India. *Journal of Fish Biology* **23**:611–624.

Genelly RE (1965). Ecology of the common mole-rat (*Cryptomys hottentotus*) in Rhodesia. *Journal of Mammalogy* **46**:647–654.

Gern WA and Norris DO (1979). Plasma melatonin in the neotenic tiger salamander (*Ambystoma tigrinum*): Effects of photoperiod and pinealectomyt. *General and Comparative Endocrinology* **38**:393–398.

Gern WA, Owens DW, and Ralph CL (1978). Plasma melatonin in the trout: Day-night change demonstrated by radioimmunoassay. *General and Comparative Endocrinology* **34**:453–458.

Gerozissis K, Jouannet P, Soufir JC, and Dray F (1982). Origins of prostaglandins in human semen. *Journal of Reproduction and Fertility* **65**:401–404.

Ghiselin MT (1969). The evolution of hermaphroditism among animals. *Quarterly Review of Biology* **44**:189–208.

Gibbons JW (1982). Reproductive patterns in freshwater turtles. *Herpetologica* **38**:222–227.

Gibbons JW and Nelson DH (1978). The evolutionary significance of delayed emergence from the nest of hatchling turtles. *Evolution* **32**:297–303.

Gibbons JW, Greene JL, and Patterson KK (1982). Variation in reproductive characteristics of aquatic turtles. *Copeia*:776–784.

Gibbons MM and McCarthy TK (1984). Growth, maturation and survival of frogs *Rana temporaria*. *Holarctic Ecology* **7**:419–427.

Gibori G, Chatterton RT Jr, and Chien JL (1979). Ovarian and serum concentrations of androgen throughout pregnancy in the rat. *Biology of Reproduction* **21**:53–56.

Gibori G, Keyes PL, and Richards JS (1978). A role for intraluteal estrogen in the mediation of luteinizing hormone action on rat corpus luteum during pregnancy. *Endocrinology* **103**:162–171.

Gibori G and Richards JS (1978). Dissociation of two distinct luteotropic effects of prolactin: Regulation of luteinizing hormone receptor content and progesterone secretion during pregnancy. *Endocrinology* **102**:767–774.

Gibson RN (1978). Lunar and tidal rhythms in fish. In *Rhythmic Activities in Fish* (Ed: Thorpe JE) Academic, New York:202–213.

Gielen JT, Goos HJT, Moncourt-de Bruyn MGA, and van Oordt PGWJ (1982). Direct action of gonadal steroids on the maturation of gonadotropic cells in the rainbow trout, *Salmo gairdneri*. In *Proceedings of the International Symposium of Reproductive Physiology of Fish* (Comp: Richter CJJ and Goos HJT) Wageningen 54.

Gilbert BS and Krebs CJ (1981). Effects of extra food on *Peromyscus* and *Clethrionomys* in the southern Yukon. *Oecologia* **51**:326–331.

Gilbert BS, Krebs CJ, Talarico D, and Cichowski DB (1986). Do *Clethrionomys rutilus* females suppress maturation of juveniles? *Journal of Animal Ecology* **55**:543–552.

Gilbert PW and Heath GW (1972). Clasper-siphon mechanism in *Squalis acanthias* and *Mustelus canis*. *Comparative Biochemistry and Physiology* **3**:206–227.

Gilbert PW and Schlernitzauer DA (1966). The placenta and gravid uterus of *Carcharinus falciformis*. *Copeia*:451–457.

Gillingham JC (1979). Reproductive behavior of the rat snakes of eastern North America, Genus *Elaphe*. *Copeia*:319–331.

Gillingham JC and Chambers J (1982). Courtship and pelvis spur use in the Burmese python, *Python molurus bivittatus*. *Copeia*:193–196.

Ginther OJ (1974). Internal regulation of physiological processes through local venoarterial pathways: A review. *Journal of Animal Science* **39**:550–564.

Ginther OJ (1979). Reproductive seasonality and regulation of LH and FSH in pony mares. In *Animal Reproduction* (Ed: Hawk HW) Beltsville Symposia in Agricultural Research 3. Halsted, New York.

Gipps JHW, Flynn MP, Gurnell J, and Healing TD (1985). The spring decline in populations of the bank vole, *Clethrionomys glareolus*, and the role of female density. *Journal of Animal Ecology* **54**:351–358.

Glass JD and Lynch GR (1981). Melatonin: Identification of antigonadal action in mouse brain. *Science* **214**:821–823.

Glazier DS (1985). Energetics of litter size in five species of *Peromyscus* with generalizations for other mammals. *Journal of Mammalogy* **66**:629–642.

Godfrey GK (1969). Influence of increased photoperiod on reproduction in the dasyurid marsupial, *Sminthopsis crassicaudata*. *Journal of Mammalogy* **50**:132–133.

Goetz FW (1983). Hormonal control of oocyte final maturation and ovulation in fishes. In *Fish Physiology*, vol 9, part B (Eds: Hoar WS, Randall DJ, and Donaldson EM) Academic, New York:117–170.

Goff GP (1984). Brood care of the longnose gar (*Lepisosteus osseus*) by smallmouth bass (*Micropterus dolomieui*). *Copeia*:149–152.

Goff LJ and Stein JR (1978). Ammonia: Basis for algal symbiosis in salamander egg masses. *Life Science* **22**:1463–1468.

Goicoechea O, Garrido O, and Jorquera B (1986). Evidence for a trophic paternal-larval relationship in the frog *Rhinoderma darwinii*. *Journal of Herpetology* **20**:168–178.

Goldberg SR (1970). Seasonal ovarian histology of the ovoviviparous lizard *Sceloporus jarrovi* Cope. *Journal of Morphology* **132**:265–276.

Goldberg SR (1976). Reproduction in a mountain population of the coastal whiptail lizard, *Cnemidophorus multiscutatus*. *Copeia*:260–266.

Goldsmith AR (1982). Plasma concentrations of prolactin during incubation and parental feeding throughout repeated breeding cycles in canaries (*Serinus canarius*). *Journal of Endocrinology* **94**:51–59.

Goldsmith AR (1983). Prolactin in avian reproductive cycles. In *Hormones and Behaviour in Higher Vertebrates*. (Eds: Balthazart J, Prove E, and Gilles R) Springer-Verlag, Berlin, Heidelberg:375–387.

Goldsmith AR, Edwards C, Koprucu M, and Silver R (1981). Concentrations of prolactin and luteinizing hormone in plasma of doves in relation to incubation and development of the crop gland. *Journal of Endocrinology* **90**:437–443.

Gondos B (1978). Oogonia and oocytes in mammals. In *The Vertebrate Ovary* (Ed: Jones RE) Plenum, New York:83–120.

Goodman D (1974). Natural selection and a cost ceiling on reproductive effort. *American Naturalist* **108**:247–268.

Goodman RL and Karsch FJ (1980). Pulsatile secretion of luteinizing hormone: Differential suppression by ovarian steroids. *Endocrinology* **107**:1286–1290.

Gopalakrishna A (1955). Observations on the breeding habits and ovarian cycle in the Indian sheath-tailed bat, *Taphozous longimanus* (Hardwick). *Proceedings of the National Institute of Science of India B* **21**:29–41.

Gordon JDM (1979). Lifestyle and phenology in deep sea anacanthine teleosts. In *Fish Phenology: Anabolic Adaptiveness in Fishes* (Ed: Miller PJ) Symposia of the Zoological Society of London, no. 44. Academic, London:327–359.

Gordon TP, Rose RM, and Bernstein IS (1976). Seasonal rhythm in plasma testosterone levels in the rhesus monkey (*Macaca mulatta*): A three year study. *Hormones and Behavior* **7**:229–243.

Gorman ML (1974). Endocrine basis of pair-formation behaviour in male eider *Somateria mollissima. Ibis* **116**:451–465.

Gorman ML (1977). Sexual behaviour and plasma concentrations in the male eider duck (*Somateria mollissima*). *Journal of Reproduction and Fertility* **49**:225–230.

Gorman ML and Milne H (1972). Creche behaviour in the common eider *Somateria mollissima* L. *Ornis Scandinavica* **3**:21–25.

Gould JE, Overstreet JW, and Hanson FW (1984). Assessment of human sperm function after recovery from the female reproductive tract. *Biology of Reproduction* **31**:888–894.

Gourdji D and Tixier-Vidal A (1966). Variations du contenu hypophysaire en prolactine chex la Canard Pekin male au cours des cycles sexuels et de la photostimulation. *Comptus Rendes Academie Sciences* (Paris) **262**:1746–1749.

Gowaty PA (1981). An extension of the Orians-Verner-Willson model to account for mating systems besides polygyny. *American Naturalist* **118**:851–859.

Gowaty PA (1982). Sexual terms in sociobiology: Emotionally evocative and, paradoxically, jargon. *Animal Behaviour* **30**:630–631.

Gowaty PA and Karlin AA (1984). Multiple maternity and paternity in single broods of apparently monogamous eastern bluebirds (*Sialia sialis*). *Behaviour, Ecology and Sociobiology* **15**:91–95.

Graham RCB and Noble RL (1955). Comparison of in vitro activity of various species of *Lithospermum* and other plants to inactivate gonadotropins. *Endocrinology* **56**:239–247.

Grant GS (1982). *Avian Incubation: Egg Temperature, Nest Humidity and Behavioral Thermoregulation in a Hot Environment.* American Ornithologists' Union, Washington, DC.

Grant PR (1986). *Ecology and Evolution of Darwin's Finches.* Princeton University, Princeton.

Graul EW (1973). Adaptive aspects of the mountain plover social system. *Living Bird* **12**:69–94.

Greeley Jr MS (1984). Spawning of *Fundulus pulvereus* and *Adinia xenica* (Cyprinodontidae) along the Alabama Gulf Coast is associated with the semilunar tidal cycles. *Copeia*:797–800.

Green B (1984). Composition of milk and energetics of growth of marsupials. In *Physiological Strategies in Lactation* (Eds: Peaker M, Vernon RG, and Knight CH) Zoological Society of London Symposium 51:369–387.

Greene (1987). Individuals in an osprey colony discriminate between high and low quality information. *Nature* **329**:239.

Greenwood JJD (1987). Three-year cycles of lemmings and Arctic geese explained. *Nature* **328**:577.

Greenwood PJ (1980). Mating systems, philopatry and dispersal in birds and mammals. *Animal Behaviour* **28**:1140–1162.

Greep RO (Ed) (1983). *Reproductive Physiology IV.* International Review of Physiology 27.

Greer AE (1977). The systematics and evolutionary relationships of scincid lizard genus *Lygosoma. Journal of Natural History* **11**:515–540.

Grey RD, Wolf DP, and Hedrick JL (1974). Formation and structure of the fertilization in *Xenopus laevis. Developmental Biology* **36**:44–61.

Grier HJ (1981). Cellular organization of the testis and spermatogenesis in fishes. *American Zoologist* **21**:345–357.

Grier HJ (1984). Testis structure and formation of spermatophores in the atherinomorph teleost *Horaichthys setnai*. *Copeia*:833–839.

Griffiths M (1978). *The Biology of the Monotremes*. Academic, New York.

Griffiths M, McIntosh DL, and Leckie RM (1972). The mammary glands of the red kangaroo with observations on the fatty acid components of the milk triglycerides. *Journal of Zoology* **166**:265–275.

Grigg GC and Harlow P (1981). A fetal-maternal shift of blood oxygen affinity in an Australian viviparous lizard, *Sphenomorphus quoyii* (Reptilia, Scincidae). *Journal of Comparative Physiology* **142**:495–499.

Grimes LG (1976). The occurrence of cooperative breeding behavior in African birds. *Ostrich* **47**:1–15.

Grocock CA (1979). Testis development in the vole, *Microtus agrestis*, subjected to long or short photoperiods from birth. *Journal of Reproduction and Fertility* **55**:423–427.

Grocock CA (1981). Effect of different photoperiods on testicular weight changes in the vole, *Microtus agrestis*. *Journal of Reproduction and Fertility* **62**:25–32.

Grocock CA and Clarke JR (1974). Photoperiodic control of testes activity in the vole, *Microtus agrestis*. *Journal of Reproduction and Fertility* **39**:337–347.

Gross MR (1982). Sneakers, satellites and parentals: Polymorphic mating strategies in North American sunfishes. *Zeitschrift fuer Tierpsychologie* **60**:1–60.

Gross MR (1984). Sunfish, salmon, and the evolution of alternate reproductive strategies and tactics in fishes. In *Fish Reproduction. Strategies and Tactics* (Eds: Potts GW and Wootton RJ) Academic, London:55–75.

Gross MR and Charnov EL (1980). Alternate male life history strategies in bluegill sunfish. *Proceedings of the National Academy of Science* **77**:6937–6940.

Gross MR and Shine R (1981). Parental care and mode of fertilization in ectothermic vertebrates. *Evolution* **35**:775–793.

Grunwald C (1980). Steroids. In *Secondary Plant Products* (Eds: Bell EA and Charlwood BV) Springer-Verlag, Berlin:221–256.

Gubernick DJ (1981). Parent and infant attachment in mammals. In *Parental Care in Mammals* (Eds: Gubernick DJ and Klopfer PH) Plenum, New York:243–305.

Gubernick DJ and Klopfer PH (Eds) (1981). *Parental Care in Mammals*. Plenum, New York.

Guillette LJ (1979). Stimulation of parturition in a viviparous lizard (*Sceloporus jarrovi*) by arginine vasotocin. *General and Comparative Endocrinology* **38**:457–460.

Guillette LJ, Spielvogel S, and Moore FL (1981). Luteal development, placentation, and plasma progesterone concentration in the viviparous lizard *Sceloporus jarrovi*. *General and Comparative Endocrinology* **43**:20–29.

Guillette LJ Jr (1982). The evolution of viviparity and placentation in the high elevation Mexican lizard *Sceloporus aeneus*. *Herpetologica* **38**:94–103.

Guillette LJ Jr (1985). The reproductive and fat cycles of the lizard *Sceloporus formosus*. *Journal of Herpetology* **19**:474–480.

Guillette LJ Jr and Cacas-Andreu G (1980). Fall reproductive activity in the high altitude Mexican lizard, *Sceloporus grammicus microlepidotus*. *Journal of Herpetology* **14**:143–147.

Guillette LJ Jr and Casas-Andreu G (1987). The reproductive biology of the high elevation Mexican lizard *Barisia imbricata*. *Herpetologica* **43**:29–38.

Guillette LJ Jr, Jones RE, Fitzgerald KT, and Smith HM (1980). Evolution of viviparity in the lizard genus *Sceloporus*. *Herpetologica* **36**:201–215.

Guiness FE, Clutton-Brock TH, and Albon SD (1978). Factors affecting calf mortality in red deer (*Cervus elaphus*). *Journal of Animal Ecology* **47**:817–832.

Gurney ME and Konishi M (1980). Hormone-induced sexual differentiation of brain and behavior in zebra finches. *Science* **208**:1380–1383.

Gustafson AW and Shemesh M (1976). Changes in plasma testosterone levels during the annual reproductive cycle of the hibernating bat, *Myotis lucifugus lucifugus*, with a survey of plasma testosterone levels in adult male vertebrates. *Biology of Reproduction* **15**:9–24.

Guthrie M (1933). The reproductive cycles of some cave bats. *Journal of Mammalogy* **14**:199–216.

Gutzke WHN and Bull JJ (1986). Steroid hormones reverse sex in turtles. *General and Comparative Endocrinology* **64**:368–372.

Gwinner E (1977). Circannual rhythms in bird migration. *Annual Review of Ecology and Systematics* **8**:381–405.

Gwinner E (1981). Circannual rhythms: Their dependence on the circadian system. In *Biological Clocks in Seasonal Reproductive Cycles* (Eds: Follett BK and Follett DE) Wiley, New York:153–169.

Hafeez MA and Quay WB (1970). The role of the pineal organ in the control of phototaxis and body coloration in the rainbow trout (*Salmo gairdneri*, Richardson). *Zeitschrift fuer Vergleichende Physiologie* **68**:403–416.

Hahn DW and Turner CW (1966). Effect of corticosterone and aldosterone upon milk yield in the rat. *Proceedings of the Society for Experimental Medicine* **121**:1056–1058.

Hahn EO and Cheng SC (1967). Gonadal hormones in Wilson's phalarope (*Steganopus tricolor*) and other birds in relation to plumage and sex behavior. *General and Comparative Endocrinology* **8**:1–11.

Hails AJ and Abdullah Z (1982). Reproductive biology of the tropical fish *Trichogaster pectoralis* (Regan). *Journal of Fish Biology* **21**:157–170.

Hairston NG, Tinkle DW, and Wilbur HM (1970). Natural selection and the parameters of population growth. *Journal of Wildlife Management* **34**:681–690.

Hakkinen I, Jokinen M, and Tast J (1973). The winter breeding of the feral pigeon *Columbia livia domestica* at Tampere in 1972/1973. *Ornis Fennica* **50**:83–88.

Halawani MEEI, Silsby JL, Behnke EJ, and Fehrer SC (1984). Effect of ambient temperature on serum prolactin and luteinizing hormone levels during the reproductive life cycle of the female turkey (*Meleagris gallopava*). *Biology of Reproduction* **30**:809–815.

Halbert SA, Tam PY, and Blandau RJ (1976). Egg transport in the rabbit oviduct: The roles of cilia and muscle. *Science* **191**:1052–1053.

Haldar CM and Thapliyal JP (1977). Effect of pinealectomy on the annual testicular cycle of *Calotes versiclor*. *General and Comparative Endocrinology* **32**:395–399.

Hall MR and Goldsmith AR (1983). Factors affecting prolactin secretion during breeding and incubation in the domestic duck (*Anas platyrhynchos*). *General and Comparative Endocrinology* **49**:270–276.

Hall RJ (1969). Ecological observations on Graham's water snake (*Regina grahami* Baird and Girard). *American Midland Naturalist* **81**:156–163.

Halliday T (1980). *Sexual Strategy*. The University of Chicago, Chicago.

Halliday TR (1978). Sexual selection and mate choice. In *Behavioural Ecology* (Eds: Krebs JR and Davies NB) Blackwell, Oxford:180–213.

Halliday TR (Ed) (1983). The study of mate choice. In *Mate Choice* (Ed: Bateson P) Cambridge University, Cambridge:3–32.

Hamada Y, Schlaff S, Kobayashi Y, Santulli R, Wright KH, and Wallach EE (1980). Inhibitory effect of prolactin on ovulation in the in vitro pefused rabbit ovary. *Nature* **285**:161–163.

Hamilton A (1976). The significance of patterns of distribution shown by forest plants and animals in tropical Africa for the reconstruction of Upper Pleistocene palaeoenvironments: A review. In *Palaeoecology of Africa and of the Surrounding Islands and Antarctica* (Ed: van Zinderen Bakker EM) vol 9. Balkema, Cape Town.

Hamilton DW, Allen WR, and Moor RM (1973). The origin of equine endometrial cups. *Anatomical Record* **177**:503–518.

Hamilton WJ Jr (1940). Observations on the reproductive behavior of the snapping turtle. *Copeia*:124–126.

Hamlett GWD (1935). Delayed implantation and discontinuous development in mammals. *Quarterly Review of Biology* **10**:432–447.

Hamner CE and McLaughlin KC (1974). Capacitation of sperm: As a function of the oviduct. In *The Oviduct and its Functions* (Eds: Johnson AD and Foley CW) Academic, New York:161–191.

Hamner WH (1968). The photorefractory period of the house finch. *Ecology* **49**:212–227.

Hanks J (1977). Comparative aspects of reproduction in the male hyrax and elephant. In *Reproduction and Evolution* (Eds: Calaby JH and Tyndale-Biscoe CH) Australian Academy of Science, Canberra:155–164.

Hansson L (1984). Predation as the factor causing extended low densities in microtine cycles. *Oikos* **43**:255–256.

Hanyu I, Asahina K, and Shimizu A (1982). The roles of light and temperature in the reproductive cycles of three bitterling species: *Rhodeus ocellatus ocellatus, Acheilognatus tabira* and *Pseudoperilampus typus*. In *Reproductive Physiology of Fish* (Eds: Richter CJJ and Goos HJT) PUDOC, Wegeningen:229–232.

Hardisty MW (1978). Primordial germ cells and the vertebrate line. In *The Vertebrate Ovary* (Ed: Jones RE) Plenum, New York:1–45.

Hardisty MW, Potter IC, and Hilliard RW (1986). Gonadogenesis and sex differentiation in the Southern Hemisphere lamprey, *Geotria australis* Gray. *Journal of Zoology* **209**:477–499.

Harper F (1955). *The Barren Ground Caribou of Keewatin*. University of Kansas Museum of Natural History, Miscellaneous Publications no. 6.

Harrington RW Jr (1957). Sexual photoperiodicity of the cyprinid fish, *Notropis bifrenatus* (Cope), in relation to the phases of its annual reproductive cycle. *Journal of Experimental Zoology* **135**:529–553.

Harrington RW Jr (1971). How ecological and genetic factors interact to determine when self-fertilizing hermaphrodites of *Rivulus marmoratus* change into functional secondary males, with a reappraisal of the modes of intersexuality among fishes. *Copeia*:389–432.

Harris MP (1978). Variations within British puffin populations. *Ibis* **120**:129.

Harris RN (1980). The consequences of within-year timing of breeding in *Ambystoma maculatum*. *Copeia*:719–722.

Hart BL and Ladewig J (1979). Serum testosterone of neonatal male and female dogs. *Biology of Reproduction* **21**:289–292.

Hasegawa M (1984). Biennial reproduction in the lizard *Eumeces okadae* on Miyake-jima, Japan. *Herpetologica* **40**:194–19.

Haseltine FP and Ohno S (1981). Mechanism of gonadal differentiation. *Science* **211**:1272–1278.

Hasler JF, Buhl AE, and Banks EM (1976). The influence of photoperiod on growth and sexual function in male and female collard lemmings (*Dicrostonyx groenlandicus*). *Journal of Reproduction and Fertility* **46**:323–329.

Hasler MJ and Nalbandov AV (1978). Pregnancy maintenance and progesterone concentrations in the musk shrew, *Suncus murinus* (Order Insectivora). *Biology of Reproduction* **19**: 407–413.

Hasler MJ, Hasler JF, and Nalbandov AV (1977). Comparative breeding biology of musk shrews (*Suncus murinus*) from Guam and Madagascar. *Journal of Mammalogy* **58**:285–290.

Haugre RL, Karsch FJ, and Foster DL (1977). A new concept for control of the estrous cycle of the ewe based on the temporal relationships beween luteinizing hormone, estradiol and progesterone in peripheral serum and evidence that progesterone inhibits tonic LH secretion. *Endocrinology* **101**:807–817.

Hayman DL and Rofe RH (1977). Marsupial sex chromosomes. In *Reproduction and Evolution* (Eds: Calaby JH and Tyndale-Biscoe CH) Australian Academy of Science, Canberra City:69–79.

Heap RB and Illingworth DV (1977). The mechanism of action of estrogens and progesterone. In *The Ovary*, vol 3 (Eds: Zuckerman L and Weir BJ) Academic, New York:59–150.

Hearn JP (1974). The pituitary gland and implantation in the tammar wallaby (*Macropus eugenii*). *Journal of Reproduction and Fertility* **39**:325–341.

Hearn JP (1975). Hypophysectomy of the tammar wallaby, *Macropus eugenii*: Surgical approach and general effects. *Journal of Endocrinology* **64**:403–416.

Hearn JP, Short RV, and Baird DT (1977). Evolution of the luteotrophic control of the mammalian corpus luteum. In *Reproduction and Evolution* (Eds: Calaby JH and Tyndale-Biscoe CH) Australian Academy of Science, Canberra:255–263.

Heftmann E (1975). Steroid hormones in plants. *Lloydia* **38**:195–209.

Hegner RE, Emlen ST, and Demong NJ (1982). Spatial organization of the white-fronted bee-eater. *Nature* **298**:264–266.

Heiligenberg W (1977). *Principles of Electrocation and Jamming Avoidance in Electric Fish. Studies of Brain Function*. Springer-Verlag **1**:1–85.

Henderson KM, Willcox DL, and Bruce NW (1983). Effect of infusion of PG12, 6-keto-PGF2a on luteal function in the pregnant rat. *Journal of Reproduction and Fertility* **69**:11–16.

Herbert J (1981). The pineal gland and its photoperiodic control of the ferret's reproductive cycle. In *Biological Clocks in Seasonal Reproductive Cycles* (Eds: Follett BK and Follett DE) Wiley, New York:261–276.

Herrenkohl LR and Campbell C (1976). Mechanical stimulation of mammary gland development in virgin and pregnant rats. *Hormones and Behavior* **7**:183–197.

Hertelendy F and Bielleir HV (1978). Prostaglandin levels in avian blood and reproductive organs. *Biology of Reproduction* **18**:204–211.

Heusser H (1963). Die ovulation des Erdkrotenweibchens im Rahmen der Verhaltensorganisation von *Bufo bufo* L. *Revue Suisse de Zoologie* **70**:741–758.

Heusser H (1968). Die Lebenswise der Erdkrote, *Bufo bufo* (L.); Wanderungen and Sommerquartiere. *Revue Suisse de Zoologie* **75**:927–982.

Hweson R (1976). A population study of mountain hares (*Lepus timidus*) in north-east Scotland from 1956–1969. *Journal of Animal Ecology* **45**:395–414.

Hib J, Ponzio R, and Vilar O (1982). Contractility of the rat cauda epididymides and vas deferens during seminal emission. *Journal of Reproduction and Fertility* **66**:47–50.

Hickie JP, Lavigne DM, and Woodward WD (1982). Vitamin D and winter reproduction in the collared lemming, *Dicrostonyx groenlandicus*. *Oikos* **39**:71–76.

Highfill DR and Mead RA (1975a). Sources and levels of progesterone during pregnancy in the garter snake, *Thamnophis elegans*. *General and Comparative Endocrinology* **27**: 389–400.

Highfill DR and Mead RA (1975b). Function of corpora lutea of pregnancy in the viviparous garter snake, *Thamnophis elegans*. *General and Comparative Endocrinology* **27**:401–407.

Hilden O (1975). Breeding system of Temminck's stint *Calidris temminckii*. *Ornis Fennica* **52**:117–144.

Hilden O (1977). The nesting of birds affected by mass occurrence of caterpillars. *Ornis Fennica* **54**:36–37.

Hill DA (1984). Population regulation in the mallard *Anas platyrhynchos*. *Journal of Animal Ecology* **53**:191–202.

Hill JP and O'Donoghue CH (1913). The reproductive cycle in the marsupial *Dasyurus viverrinus*. *Quarterly Journal of Microscopic Science* **59**:133–174.

Hinde RA and Steel E (1976). The effect of male song on an estrogen-dependent behavior pattern in the female canary (*Serinus canarius*). *Hormones and Behavior* **7**:293–304.

Hinkley Jr R (1966). Effects of plant extracts in the diet of male *Microtus montanus* on cell types of the anterior pituitary. *Journal of Mammalogy* **47**:396–400.

Hirons GJM (1982). The effects of fluctuations of rodent numbers on breeding success in the tawny owl *Strix aluco*. *Mammal Review* **12**:155–157.

Hirschfeld PR and Bradley WG (1977). Growth and development of two species of chipmunks: *Eutamias panamintinus* and *E. palmeri*. *Journal of Mammalogy* **58**:44–52.

Hisano N, Cardinali DP, Rosner JM, Nagle CA, and Tramezzani JH (1972). Pineal role in the duck extraretinal photoreception. *Endocrinology* **93**:1318–1322.

Hisaw FL and Abramowitz AA (1939). Physiology of reproduction in the dogfishes, *Mustelus canis* and *Squalus acanthias*. *Reports of the Woods Hole Oceanographic Institute* **1938**:22.

Hisaw FL and Albert A (1947). Observations on the reproduction of the spiny dogfish, *Squalus acanthias*. *Biological Bulletin* **92**:187–199.

Hislop JRG (1984). A comparison of the reproductive tactics and strategies of cod, haddock, whiting and Norway pout in the North Sea. In *Fish Reproduction* (Eds: Potts GW and Wootton RS) Academic, New York:311–329.

Hislop JRG, Robb AB, and Gauld JA (1978). Observations on effects of feeding level on growth and reproduction of haddock, *Melanogrammus aeglefinis*(L.) in captivity. *Journal of Fish Biology* **13**:85–98.

Hoar WS and Nagahana Y (1978). The cellular source of sex steroids in teleost gonads. *Annales de Biologie Animale, Biochimie et Biophysique* **18**:893–898.

Hoddenbach GA and Turner FB (1968). Clutch size of the lizard *Uta stansburnaia* in southern Nevada. *American Midland Naturalist* **80**:262–265.

Hodgen GD, Goodman AL, O'Connor A, and Johnson DK (1977). Menopause in rhesus monkeys: Model for study of disorders in the human climacteric. *American Journal of Obstetrics and Gynecology* **127**:581–584.

Hoffman LH (1970). Placentation in the garter snake, *Thamnophis sirtalis*. *Journal of Morphology* **131**:57–88.

Hoffmann K (1974). Testicular involution in short photoperiods inhibited by melatonin. *Naturwissenshaften* **61**:364–365.

Hoffmann K (1981). The role of the pineal gland in the photoperiodic control of seasonal cycles in hamsters. In *Biological Clocks in Seasonal Reproductive Cycles* (Eds: Follett BK and Follett DE). Wiley, New York:237–250.

Hoffmann LH and Wimsatt WA (1972). Histochemical and electron microscope observations on the sperm receptacles in the garter snake oviduct. *American Journal of Anatomy* **134**: 71–96.

Hoffmann RS (1958). The role of reproduction and mortality in population fluctuations of voles (*Microtus*). *Ecological Monographs* **28**:80–109.

Hofmann JE, Getz LL, and Gavish L (1987). Effect of multiple short-term exposures of pregnant *Microtus ochrogaster* to strange males. *Journal of Mammalogy* **68**:166–169.

Högstedt G (1974). Length of the prelaying period in the lapwing *Vanellus vanellus* L. in relation to food resources. *Ornis Scandinavica* **5**:1–4.

Högstedt G (1980). Evolution of clutch size in birds: Adaptive variation in relation to territory quality. *Science* **210**:1148–1150.

Högstedt G (1981a). Effect of additional food on reproductive success in the magpie (*Pica pica*). *Journal of Animal Ecology* **50**:219–229.

Högstedt G (1981b). Should there be a positive or negative correlation between survival of adults in a bird population and their clutch size? *American Naturalist* 118:568–571.

Holcomb HH, Costlow ME, Boschow RA, and McGuire WL (1976). Prolactin binding in rat mammary gland during pregnancy and lactation. *Biochemica et Biophysica Acta* **428**: 104–112.

Holley AJF and Greenwood PJ (1984). The myth of the mad March hare. *Nature* **292**:549–550.

Homma K, Ohta M and Sakakibara Y (1980). Surface and deep photoreceptors in photoperiodism in birds. In *Biological Rhythms in Birds* (Eds: Tanabe Y, Tanaka K, and Ookawa T) Japan Scientific Societies, Tokyo, and Springer-Verlag, New York:149–156.

Honacki JH, Kinman KE, and Koepel JW (1982). *Mammal Species of the World*. Association of Systematics Collections and Allen Press, Lawrence, Kansas.

Honma Y, Ozawa T, and Chiba A (1980). Maturation and spawning behavior of the puffer, *Fugu niphobles*, occurring on the coast of Sado Island in the Sea of Japan (A preliminary report). *Japanese Journal of Ichthyology* **27**:129–138.

Hornfledt B, Lofgren E, and Carlsson B-G (1986). Cycles in voles and small game in relation to variations in plant production indices in Northern Sweden. *Oecologia* **68**:496–502.

Houck LD (1977). Life history patterns and reproductive biology of neotropical salamanders. In *The Reproductive Biology of Amphibia* (Eds: Taylor DH and Guttman SI) Plenum, New York:43–72.

Howard RD (1978). The evolution of mating strategies in bullfrogs, *Rana catesbeiana*. *Evolution* **32**:850–871.

Howarth B Jr (1974). Sperm storage: As a function of the female reproductive tract. In *The Oviduct and Its Functions* (Eds: Johnson AD and Foley CW) Academic, New York: 237–270.

Howe HF (1978). Initial investment, clutch size, and brood reduction in the common grackle (*Quisculus quiscala* L.). *Ecology* **59**:1109–1122.

Howell AB (1930). *Aquatic Mammals. Their Adaptations to Life in the Water*. Charles C Thomas, Springfield.

Hseuh AJW, Erickson GE, and Lu KH (1979). Changes in uterine estrogen receptor and morphology in aging female rats. *Biology of Reproduction* **21**:793–800.

Hseuh AJW, Jones PBC, Adashi EY, Wang C, Zhuang L-Z, and Welsh Jr TH (1983). Intraovarian mechanisms in the hormonal control of granulosa cell differentiation in rats. *Journal of Reproduction and Fertility* **69**:325–342.

Hsu CY and Liang HM (1970). Sex races of *Rana catesbeiana* in Taiwan. *Herpetologica* **26**:214–221.

Hubbs C (1966). Fertilization, initiation of cleavage, and developmental temperature tolerance of the cottin fish, *Clinocottus analis*. *Copeia*:29–36.

Hubbs CL (1938). Fishes from the caves of Yucatan. In *Fauna of the Caves of Yucatan*. Publications of the Carnegie Institute of Washington, no. 491, **21**:261–295.

Hubbs CL and Hubbs LC (1932). Apparent parthenogenesis in nature in a form of fish hybrid origin. *Science*, n.s. **76**:628–630.

Hubbs CL (1921). The ecology and life-history of *Amphigonopterus aurora* and other viviparous perches of California. *Biological Bulletin* **40**:181–209.

Hubert J (1968). A propos de la lignee germinale chez 2 reptiles: *Anguis fragilis* L et *Vipera aspis* L. *Comptes Rendus de l'Academie des Sciences, Paris* **226**:231–233.

Hubert J (1976). La Lignee germinale chez les reptiles au cours du developpment embryonnaire. *L'Annee Biologique* **15**:548–565.

Huey RB (1977). Egg retention in some high-altitude *Anolis* lizards. *Copeia*:373–375.

Huggins RA (1941). Egg temperatures of wild birds under natural conditions. *Ecology* **22**: 148–157.

Hughes GR (1976). Irregular reproductive cycles in the Tongaland loggerhead sea-turtle, *Caretta caretta* L. (Cryptodira: Chelonidae). *Zoologica Africana* **11**:275–291.

Hughes JP, Stabenfeldt GH, and Evans JW (1975). The oestrous cycle of the mare. *Journal of Reproduction and Fertility*, Supplement **23**:161–166.

Hughes RL (1974). Morphological studies on implantation in marsupials. *Journal of Reproduction and Fertility* **39**:173–186.

Hughes RL (1984). Structural adaptations of the eggs and the fetal membranes of monotremes and marsupials for respiration and metabolism. In *Respiration and Metabolism of Embryonic Vertebrates* (Ed: Seymour RS) Junk, Dordrecht:389–421.

Hunsaker D II (1977). Ecology of New World marsupials. In *The Biology of Marsupials*

(Ed: Hunsaker II D) Academic, New York:95–156.

Hunter JR and Hasler AD (1965). Spawning association of the redfin shiner, *Notropis umbratilis*, and the green sunfish, *Lepomis cyanellus*. *Copeia*:265–281.

Hureau J-C and Ozouf C (1977). Determination de l'age et croissance du coelacanthe *Latimeria chalumnae* Smith, 1939 (Poisson, Crossopterygien Coelacanthidae). *Cybium* **2**:129–137.

Husby M (1986). On the adaptive value of brood reduction in birds: Experiments with the magpie, *Pica pica*. *Journal of Animal Ecology* **55**:75–83.

Hussell DJT (1972). Factors affecting clutch size in Arctic passerines. *Ecological Monographs* **42**:317–364.

Hussell DJT (1985). Clutch size, daylength and seasonality of resources: Comments on Ashmole's hypothesis. *Auk* **102**:632–634.

Hutchings JA and Morris DW (1985). The influence of phylogeny, size and behaviour on patterns of covariation in salmonid life histories. *Oikos* **45**:118–124.

Idler DR and Ng TB (1983). Teleost gonadotropins: isolation, biochemistry and function. In *Fish Physiology* (Eds: Hoar WS, Randall DJ and Donaldson EM) Reproduction, Part A. Academic, New York:IX:187–221.

Iles TD (1974). The tactics and strategy of growth in fishes. In *Sea Fisheries Research* (Ed: Harden Jones FR) Paul Elek (Scientific Books), London:331–345.

Iles TD (1984). Allocation of resources to gonadal and soma in Atlantic herring *Clupea harengus* L. Pp 331–347. In *Fish Reproduction: Strategies and Tactics* (Eds: Potts GW and Wootton RJ). Academic, London.

Immelmann K (1970). Environmental factors controlling reproduction in African and Australian birds—A comparison. *Ostrich*, Supplement **8**:193–204.

Ingermann RL and Terwilliger RC (1984). Facilitation of maternal fetal oxygen transfer in fishes: Anatomical and molecular specializations. In *Respiration and Metabolism of Embryonic Vertebrates* (Ed: Seymour RS) Junk, Dordrecht:1–15.

Innes DGL (1978). A reexamination of litter size in some North American microtines. *Canadian Journal of Zoology* **56**:1488–1496.

Jacobsohn D (1961). Hormonal regulation of mammary gland growth. In *Milk: The Mammary Gland and Its Secretion*, vol 1 (Eds: Kon SK and Cowie AT) Academic, New York: 127–160.

Jainudeen MR, Katongole CB, and Short RV (1972). Plasma testosterone levels in relation to musth and sexual activity in the male Asiatic elephant, *Elephas maximus*. *Journal of Reproduction and Fertility* **29**:99–103.

Jalabert B (1976). In vitro oocyte maturation and ovulation in rainbow trout (*Salmo gairdneri*), northern pike (*Esox lucius*) and goldfish (*Carassius auratus*). *Journal of the Fishery Research Board of Canada* **33**:974–988.

Jallageas M and Assenmacher I (1986). Effects of castration and thyroidectomy on the annual biological cycles of the edible dormouse *Glis glis*. *General and Comparative Endocrinology* **63**:301–308.

Jameson EW Jr (1953). Reproduction of deer mice (*Peromyscus maniculatus* and *P. boylei*) in the Sierra Nevada. *California*. *Journal of Mammalogy* **34**:44–58.

Jameson EW Jr (1974). Fat and breeding cycles in a montane population of *Sceloporus graciosus*. *Journal of Herpetology* **8**:311–322.

Jarman M and Thompson TE (1986). The testis temperature of anaesthetized quail. *Journal of Reproduction and Fertility* **78**:307–310.

Jarman PJ (1974). The social organization of antelope in relation to their ecology. *Behaviour* **48**:215–267.

Jarvinen A (1982). Ecology of the Siberia tit *Parus cinctus* in NW Finnish Lapland. *Ornis Scandinavica* **13**:47–55.

Jarvinen A (1986). Clutch size of passerines in harsh environments. *Oikos* **46**:365–371.

Jarvis JUM (1969). The breeding season and litter size of African mole-rats. *Journal of Reproduction and Fertility*, Supplement **6**:237–248.

Jarvis JUM (1978). Energetics of survival in *Heterocephalus glaber (Ruppell), the naked mole-rat (Rodentia: Bathyergidae). Bulletin of Carnegie Museum of Natural History* **6**:81–87.

Jarvis JUM (1981). Eusociality in a mammal: Cooperative breeding in naked mole-rat colonies. *Science* **212**:571–572.

Jay P (1965). The common langur of North India. In *Primate Behavior. Field Studies of Monkeys and Apes* (Ed: de Vore I) Holt, Rinehart & Winston, New York:197–249.

Jenkins D, Watson A, and Miller GR (1963). Population studies on red grouse, *Lagopus lagopus scoticus* (Lath.) in north-east Scotland. *Journal of Animal Ecology* **32**:317–377.

Jenkins N and Dodd JM (1980). Effects of synthetic mammalian gonadotropin releasing hormone and dogfish hypothalamic extracts on levels of androgens and oestrodial in the circulation of the dogfish (*Scyliorhinus canicula* L.). *Journal of Endocrinology* **86**:171–177.

Jenness R and Stidier EH (1976). Lactation and milk. In *Biology of Bats of the New World*, Part 1 (Eds: Baker RJ, Jones Jr JK and Carter DC) Special Publication of the Museum of Texas Technological University, 10, Lubbock, Texas:1–218.

Jenni DA and Collier G (1972). Polyandry in the American jacana (*Jacana spinosa*). *Auk* **89**:743–765.

Jensen PT (1977). Isolation and immunological determination of sow colostrum trypsin inhibitor. *Acta Pathologica et Microbiologica Scandinavica (B)* **85**:441–448.

Jensen TS (1982). Seed production and outbreaks of non-cyclic rodent populations in deciduous forests. *Oecologia* **54**:184–192.

Jerrett DP (1979). Female reproductive patterns in nonhibernating bats. *Journal of Reproduction and Fertility* **56**:369–378.

Jewell PA (1977). The evolution of mating systems in mammals. In *Reproduction and Evolution* (Eds: Calaby JH and Tyndale-Biscoe CH) Australian Academy of Science, Canberra:23–32.

Jhingran AG and Ghosh KK (1978). Fisheries of Ganga River system in context of Indian aquaculture. *Aquaculture* **14**:141–162.

Joanen T and McNease L (1980). Reproductive biology of the American alligator in southwest Louisiana. In *Reproductive Biology an Diseases of Captive Reptiles*. Contributions to Herpetology, no. 1, Society for the Study of Amphibians and Reptiles:153–159.

Johannes RE (1978). Reproductive strategies of coastal marine fishes of the tropics. *Environmental Biology of Fish* **3**:65–84.

Johke T (1978). Hormone secretion at lactogenesis and during lactation in dairy farm animals. In *Physiology of Mammary Glands* (Eds: Yokoyama A, Mizuno H, and Nagasawa H) University Park, Baltimore:325–344.

John KR (1963). The effect of torrential rains on the reproductive cycle of *Rhinichthys osculus* in the Chirakahua Mountains, Arizona. *Copeia*:286–291.

Johns MA, Feder HH, Komisaruk BR, and Mayer AD (1978). Urine induced reflex ovulation in anovulatory rats may be a vomeronasal effect. *Nature* **272**:446–448.

Johnson LF and Jacob JS (1984). Pituitary activity and reproductive cycle of male *Cnemidophorus sexlineatus* in west Tennessee. *Journal of Herpetology* **18**:396–405.

Johnston PG and Zucker I (1979). Photoperiodic influences on gonadal development and maintenance in the cotton rat, *Sigmodon hispidus. Biology of Reproduction* **21**:1–8.

Johnston PG and Zucker I (1980). Antigonadal effects of melatonin in white-footed mice (*Peromyscus leucopus*). *Biology of Reproduction* **23**:1069–1074.

Johnston RE (1975). Scent marking by golden hamsters (*Mesocricetus auratus*). *Zeitschrift fuer Tierpsychologie* **37**:75–98.

Johnston RE (1977). Sex pheromones in golden hamsters. In *Chemical Signals in Vertebrates* (Eds: Muller-Schwarze and Mozell MM) Plenum, New York:225–249.

Johnstone R and Youngson AF (1982). The production of genetically all-female Atlantic salmon stocks. In *Proceedings of the International Symposium on Reproductive Physiology of FIsh* (Comp: Richter CJJ and Goos HJT) Centre Agricultural Publication Documents, Wageningen:79.

Joly J (1961). Le cycle sexuel biennial chez la femelle de Salamandra salamandra quadrivirgasta dans les hautes-Pyrenees. *Comptes Rendus Academie des Sciences, Paris* **252**:3145–3147.

Joly JMJ (1971). Les cycles sexueles de *Salamandra* (L.). I. Cycle sexuel des males. *Annales des Sciences Naturelles, Zoologie et Biologie Animale* **13**:451–504.

Jones GP (1980). Contribution to the biology of the redbanded perch, *Ellerkeldia huntii* (Hector), with a discussion of hermaphroditism. *Journal of Fish Biology* **17**:197–207.

Jones PBC and Hseuh AJW (1983). Modulation of steroidogeness enzymes by gonadotropin-releasing hormone in cultured granulosa cells. In *Factors Regulating Ovarian Function* (Eds: Greenwald GS and Terranova PF) Raven, New York:275–279.

Jones R, Althea E, Gerrard M, and Roth JJ (1973). Estrogen and brood pouch formation in the marsupial frog, *Gastrotheca riobambae. Journal of Experimental Zoology* **184**:177–184.

Jones RE (1978). Ovarian cycles in nonmammalian vertebrates. In *The Vertebrate Ovary* (Ed: Jones RE) Plenum, New York:731–762.

Jones RE and Guillette Jr LJ (1982). Hormonal control of oviposition and parturition in lizards. *Herpetologica* **38**:80–93.

Jones RE, Fitzgerald KT, and Duvall D (1978). Quantitative analysis of the ovarian cycle of the lizard *Lepidodactylus lugubris. General and Comparative Endocrinology* **35**:70–76.

Jones SM, Ballinger RE, and Porter WP (1987). Physiological and environmental sources of variation in reproduction: Prairie lizards in a food rich environment. *Oikos* **48**:325–335.

Jones WT (1985). Body size and life-history variables in heteromyids. *Journal of Mammalogy* **66**:128–132.

Jørgensen CB, Hede KE, and Larsen LO (1970). Environmental control of annual ovarian cycle in the toad, *Bufo bufo bufo* L: Role of temperature. In *Environmental Endocrinology*

(Eds: Assenmacher I and Farner DS) Springer-Verlag, New York:28–36.

Jørgensen CB, Hede K-E, and Larsen LO (1978). Environmental control of annual ovarian cycle in the toad, *Bufo bufo bufo* L.: Role of temperature. In *Environmental Endocrinology* (Eds: Assenmacher I and Farner DS) Springer-Verlag, New York:28–36.

Josimovich JB and Archer DF (1977). The role of lactogenic hormones in the pregnant woman and fetus. *American Journal of Obstetrics and Gynecology* **129**:777–780.

Joss JMP (1977). Pineal gonad relationships in the lamprey *Lampetra fluviatilis. General and Comparative Endocrinology* **21**:118–122.

Joss JMP (1977). Hydroxyindol-O-methyltransferase (HIOMT) activity and the uptake of 3H-melatonin in the lamprey, *Geotria australis* Gray. *General and Comparative Endocrinology* **31**:270–275.

Joss JMP (1978). Rhythmicity in the production of melatonin by *Lampropholas guichenoti* (scincid lizard) and *Geotria australis* (lamprey). In *Comparative Endocrinology* (Eds: Gaillard GJ and Boer HH). Elsevier/North Holland, New York:172–173.

Jouannes RE (1978). Reproductive strategies of coastal marine fishes in the tropics. *Environmental Biology of Fish* **3**:65–84.

June FC (1977). Reproductive patterns in seventeen species of warmwater fishes in a Missouri reservoir. *Environmental Biology of Fish* **2**:285–296.

Kah O, Chambolle P, Dubourg P, and Dubois MP (1984). Immunocytochemical localization of luteinizing-hormone releasing hormone in the brain of the goldfish, *Carassius auratus. General and Comparative Endocrinology* **53**:107–115.

Kalela O (1957). Regulation of reproductive rate in subarctic populations of the vole *Clethrionomys rufocanus* (Sand.). *Annales Academie Scientiarum Fennicae*, Series A, IV. Biologia 34.

Kallmann KD (1962). Gynogenesis in the teleost *Mollienesia formosa* (Girard) with a discussion of the detection of parthenogenesis by tissue transplant. *Journal of Genetics* **58**:7–24.

Kamel F, Wright WW, Mock EJ, and Frankel AI (1977). The influence of mating and related stimuli on plasma levels of luteinizing hormone, follicle stimulating hormone, prolactin and testosterone in the male rat. *Endocrinology* **101**:421–429.

Kappers JA (1967). The sensory innervation of the pineal organ of the lizard, *Lacerta viridis*, with remarks on its position in the trends of pineal phylogenetic structural and functional evolution. *Zeitschrift fuer Zellforschung und mikroskopische Anatomie* **119**:289–294.

Karg H and Schams D (1974). Prolactin release in cattle. *Journal of Reproduction and Fertility* **39**:463–472.

Karsch FJ and Foster DL (1981). Environmental control of seasonal breeding: A common final mechanism governing seasonal breeding and sexual maturation. In *Environmental Factors in Mammal Reproduction* (Eds: Gilmore D and Cook B) Macmillan, London.

Karsch FJ, Legan SJ, Hauger RL, and Foster DL (1977). Negative feedback action of progesterone on tonic luteinizing hormone secretion in the ewe: Dependence on the ovaries. *Endocrinology* **101**:800–806.

Karsch FJ, Legan SL, Ryan KD and Foster DL (1978). The feedback effects of ovarian steroids on gonadotropin secretion. In *Control of Ovulation* (Eds: Crighton DB, Haynes NB, Foxcroft GR, and Lamming GE) Butterworths, London: 29–48.

Katsurirangan LR (1951). Placentation in the sea-snake, *Enhydra schistosa* (Daudin). *Proceedings of the Indian Academy of Science, B* **34**:1–32.

Kaufman DW and Kaufman GA (1987). Reproduction of *Peromyscus polionotus*: Number, size, and survival of offspring. *Journal of Mammalogy* **68**:275–280.

Kavaliers M (1980). The pineal organ and circadian rhythms of fishes. In *Environmental Physiology of Fishes* (Ed: Ali MA) NATO Advanced Study Institute, Ser. 35, Plenum, New York:631–645.

Keenleyside MHA (1979). *Diversity and Adaptation in Fish Behaviour*. Springer-Verlag, Berlin.

Keith LB (1963). *Wildlife's Ten-Year Cycle*. University of Wisconsin, Madison.

Keith LB (1974). *Population Dynamics of Mammals*. XI International Congress of Game Biologists 17–58.

Keith LB and Windberg LA (1978). A demographic analysis of the snowshoe hare cycle. *Wildlife Monographs*, no. 58.

Keith LB, Ronstad OJ, and Meslow EC (1966). Regional differences in reproductive rates of the snowshoe hare. *Canadian Journal of Zoology* **44**:953–961.

Keith LB, Todd AW, Brand CJ, Adamcik RS, and Rusch DH (1976). *An Analysis of Predation During a Cyclic Fluctuation of Snowshoe Hares*. XIII International Congress of Game Biologists 151–175.

Kellas LM (1954). Observations on the reproductivities, measurements, and growth rate of the dikdik (*Rhynchotragus kirkii thomasi* Neumann). *Proceedings of the Zoological Society of London* **124**:751–784.

Keller BL and Krebs CJ (1970). *Microtus* population biology. III. Reproductive changes in fluctuating populations of *Microtus ochrogaster* and *M. pennsylvanicus* in southern Indiana. *Ecological Monographs* **40**:263–294.

Kenagy GJ (1981). Endogenous annual rhythm of reproductive function in the non-hibernating desert ground squirrel *Ammospermophilus leucurus*. *Journal of Comparative Physiology* **142**:251–258.

Kenagy GJ and Bartholomew GA (1979). Effects of daylength and endogenous control on the annual reproductive cycle of the antelope ground squirrel, *Ammospermophilus leucurus*. *Journal of Comparative Physiology* **130**:131–136.

Kenagy GJ and Bartholomew GA (1981). Effects of daylength, temperature and green food on testicular development in a desert pocket mouse *Perognathus formosus*. *Physiological Zoology* **54**:62–73.

Kennedy TG (1977). Evidence for a role for prostaglandins in the initiation of blastocyst implantation in the rat. *Biology of Reproduction* **16**:286–291.

Keverne EV (1983). Endocrine determinants and constraints on sexual behaviour in monkeys. In *Mate Choice* (Ed: Bateson P) Cambridge University, Cambridge:407–420.

Khanna SS and Pant MC (1967). Seasonal changes in the ovary of a sisorid catfish, *Glyptosternum pectinopterum*. *Copeia*:83–88.

Kime DE (1982). The control of androgen biosynthesis in fish. In *Proceedings of the Intenational Symposium on Reproductive Physiology of Fish* (Eds: Richter CJJ and Goos HJT). *Centre for Agricultural Publishing and Documentation*, Wageningen:95–98.

Kimura K and Uchida TA (1984). Development of the main and accessory placentae in the Japanese long-fingered bat, *Miniopterus schreibersii fuliginosus*. *Journal of Reproduction and Fertility* **71**:119–126.

King CM (1980). Population biology of the weasel *Mustela nivalis* on British game estates. *Holarctic Ecology* **3**:160–168.

King CM (1983a). The life-history of strategies of *Mustela nivalis* and *M. erminea*. *Acta*

Zoologica Fennica **174**:183–184.

King CM (1983b). The relationships between beech (*Nothofagus* sp), seedfall and populations of mice (*Mus musculus*), and the demographic and dietary responses of stoats (*Mustela erminea*), in three New Zealand forests. *Journal of Animal Ecology* **52**:141–166.

King CM (1984). The origin and adaptive advantages of delayed implantation in *Mustela erminea*. *Oikos* **42**:126–128.

King CM (1985). Interactions between woodland rodents and their predators. In *The Ecology of Woodland Rodents, Bank Voles and Woodmice* (Eds: Flowerdew JR, Gurnell J, and Gipps JHW) Zoological Society of London Symposium **55**:219–247.

King JR (1973). The annual cycle of the rufus-collared sparrow (*Zonotrichia capensis*) in three biotopes in north-western Argentina. *Journal of Zoology* **170**:163–188.

King M (1977). The evolution of sex chromosomes in lizards. In *Reproduction and Evolution* (Eds: Calaby JH and Tyndal-Biscoe CH) Australian Academy of Science, Canberra City: 55–60.

Kipling C (1983). Changes in the growth of pike (*Esox lucius*) in Windemere. *Journal of Animal Ecology* **56**:647–657.

Kirschbaum F (1975). Environmental factors control reproduction of tropical electric fish. *Experientia* **31**:1159–1160.

Kirschbaum F (1979). Reproduction of the weakly electric fish *Eigenmannia virescens* (Rhamichthyidae, Teleostei) in captivity. *Behavioral Ecology and Sociobiology* **4**:331–355.

Kirschbaum F (1984). Reproduction of weakly electric teleosts: Just another example of convergent development? *Environmental Biology of Fishes* **10**:3–14.

Kitchener DJ and Halse SA (1978). Reproduction in female *Eptesicus regulus* (Thomas) (Vespertilionidae), in south-western Australia. *Australian Journal of Zoology* **26**:257–267.

Kleiman DG (1971). The courtship and copulatory behaviour of the green acouchi, *Myoprocta pratti*. *Zeitschrift fuer Tierpsychologie* **29**:259–278.

Klein DC and Willer JL (1970). Indole metabolism in the pineal gland: A circadian rhythm of N-acetyltransferase. *Science* **169**:1093–1095.

Klingel H (1975). Social organization and reproduction in equids. *Journal of Reproduction and Fertility*, Supplement **23**:7–11.

Klomp K (1980). Fluctuations and stability in great tit populations. *Ardea* **68**:205–224.

Kluge AG and Eckardt MJ (1969). *Hemidactylus garnotii* Dumeril and Bibron, a triploid all-female species of gekkonid lizard. *Copeia*:651–664.

Kluijver HN (1951). The population ecology of the great tit, *Parus m. major*. *Ardea* **39**: 1–135.

Klump GM and Gerhardt HC (1987). Use of non-arbitrary acoustic criteria in mate choice by female gray tree frogs. *Nature* **326**:286–288.

Klun JA and Robinson JF (1969). Concentration of two, 1,4-benzoxazinones in dent corn at various stages of development of the plant and its relation to resistance of the host plant to the European corn borer. *Journal of Economic Entomology* **62**:214–220.

Knecht M, Ranta T, Naor Z, and Catt KJ (1983a). Direct effects of GnRH on the ovary. In *Factors Regulating Ovarian Function* (Eds: Greenwald GS and Terranova PF) Raven, New York:225–243.

Knecht M, Ranta T, Naor Z and Catt KJ (1983b). Direct effects of GnRH on the ovary. In *Factors Regulating Ovarian Function* (Eds: Greenwald GS and Terranova PF) Raven, New York:225–243.

Knecht M, Ranta T, and Catt KJ (1983b). Regulation of cAMP-phosphodiesterase in the rat ovary by FSH and gonadotropin-releasing hormone. In *Factors Regulating Ovarian Function* (Eds: Greenwald GS and Terranova PF) Raven, New York:269–273.

Kneib RT (1986). Size-specific patterns in the reproductive cycle of the killifish, *Fundulus heteroclitus* (Pisces: Fundulidae). *Copeia*:342–351.

Knight TW, Oldham CM, and Lindsay DR (1975). Studies on ovine infertility in agricultural regions in western Australia: The influence of a supplement of lupins (*Lupinus angustifolius* cv. *Uniwhite*) at joining on the reproductive performance of ewes. *Australian Journal of Agricultural Research* **26**:567–575.

Kodric-Brown A (1977). Reproductive success and the evolution of breeding territories in pupfish (*Cyprinodon*). *Evolution* **31**:750–766.

Koenig WD (1981). Reproductive success, group size, and the evolution of cooperative breeding in the acorn woodpecker. *American Naturalist* **117**:421–443.

Koenig WD (1986). Geographical ecology of clutch size variation in North American Woodpeckers. *The Condor* **88**:499–504.

Koenig WD and Pitelka FA (1981). Ecological factors and kin selection in the evolution of cooperative breeding in birds. In *Natural Selection and Social Behavior: Recent Research and New Theory* (Eds: Alexander RD and Tinkle DW). Chiron, New York:261–280.

Kolata GB (1974). !Kung hunter-gatherers: Feminism, diet and birth control. *Science* **185**: 932–934.

Kong YC, Hu SY, Lau FK, Che CT, Yeung HW, Cheung S, and Hwang JCC (1976). Potential antifertility plants from Chinese medicine. *American Journal of Chinese Medicine* **4**:105–128.

Kooyman GL and Drabek CM (1968). Observations on milk, blood and urine constituents of the Weddell seal. *Physiological Zoology* **41**:187–194.

Kopcewicz J (1971). Estrogens in developing bean (*Phaseolus vulgaris*) plants. *Phytochemistry* **10**:1423–1427.

Kordon C (1978). The role of neurotransmitters in the secretion of pituitary gonadotropins and prolactin. In *Control of Ovulation* (Eds: Crighton DB, Foxcroft GR, Haynes NB, and Lamming GE) Butterworths, London:21–28.

Korf H-W, Liesner L, Meissl H, and Kirk A (1981). Pineal complex of the clawed toad, *Xenopus laevis* Daud.: Structure and function. *Cell and Tissue Research* **216**:113–130.

Korpimaki E (1986). Predation causing synchronous decline phases in microtine and shrew populations in western Finland. *Oikos* **46**:124–127.

Korsrud GO and Baldwin RL (1972). Hormonal regulation of rat mammary gland enzyme activities and metabolite patterns. *Canadian Journal of Biochemistry* **50**:386–389.

Koskimies J (1955). Ultimate causes of cyclic fluctuations in numbers in animal populations. *Perhaps on Game Research*, Finnish Game Foundation 15.

Kraemer JE and Bennett SH (1981). Utilization of posthatching yolk in loggerhead sea turtles, *Caretta caretta*. *Copeia*:406–411.

Kramer DL (1978). Terrestrial group spawning of *Brycon petrosus* (Pisces: Characidae) in Panama. *Copeia*:536–537.

Kraus F (1985). Unisexual salamander lineages in northwestern Ohio and southeastern Michigan: A study of the consequences of hybridization. *Copeia*:309–324.

Krebs CJ (1966). Demographic changes in fluctuating populations of *Microtus californicus*. *Ecological Monographs* **36**:239–273.

Krebs CJ (1978). A review of the Chitty hypothesis of population regulation. *Canadian Journal of Zoology* **56**:2463–2480.

Krebs CJ, Gilbert BS, Boutin S, Sinclair ARE, and Smith JNM (1986). Population biology of snowshoe hares. I. Demography of food-supplemented populations in the southern Yukon, 1976–1984. *Journal of Animal Ecology* **55**:963–982.

Krebs JR and Davies NB (1984). *Behavioural Ecology. An Evolutionary Approach*, 2nd ed. Blackwell, Oxford.

Krekorian CO (1976). Field observations in Guyana on the reproductive biology of the spraying characid, *Copeina arnoldi* Regan. *American Midland Naturalist* **96**:88–97.

Krekorian CO and Dunham DW (1972). Preliminary observations on the reproductive and parental behavior of the spraying characid *Copeina arnoldi* Regan. *Zeitschrift fuer Tierpsychologie* **31**:419–437.

Krementz DG and Handford P (1984). Does avian clutch size increase with altitude? *Oikos* **43**:256–259.

Kristoffersson R, Broberg S, and Pekkarinen M (1973). Histology and physiology of embryotrophe formation, embryonic nutrition and growth in the eel pout, *Zoarces vivipara* (L.). *Annals Zoologicae Fennici* **10**:467–477.

Krohne DT (1980). Intraspecific litter size variation in *Microtus californicus*. II. Variation between populations. *Evolution* **34**:1174–1182.

Krohne DT (1981). Intraspecific litter size variation in *Microtus californicus*: Variation within populations. *Journal of Mammalogy* **62**:29–40.

Krutzsch PH and Wells WW (1960). Androgenic activity in the interscapular brown adipose tissue of the male hibernating bat (*Myotis lucifugus*). *Proceedings of the Society for Experimental Biology and Medicine* **105**:578–581.

Kuhn NJ (1977). Lactogenesis: The search for trigger mechanisms in different species. In *Comparative Aspects of Lactation* (Ed: Peaker M) Zoological Society of London Symposium **41**:165–192.

Kulkarni CV (1940). On the systematic position, structural modifications, bionomics and development of a remarkable new family of cyprinodont fishes from the province of Bombay. *Records of the Indian Museum* **42**:379–423.

Kunz TH (1974). Feeding ecology of a temperate insectivorous bat (*Myotis velifer*). *Ecology* **55**:693–711.

Kunz YW (1971). Histological study of greatly enlarged pericardial sac in the embryos of the viviparous teleost *Lebistes reticulatus*. *Revue Suisse Zoology* **78**:187–207.

Kurfess JF (1967). Mating, gestation and growth rate in *Lichanura r. roseofusca*. *Copeia*: 477–479.

Kushlan JA and Simon JC (1981). Egg manipulation by the American alligator. *Journal of Herpetology* **15**:451–454.

Labov JB (1977). Phytoestrogens and mammalian reproduction. *Comparative Biochemistry and Physiology* **57A**:3–9.

Lack D (1947). *Darwin's Finches*. Cambridge University Press, Cambridge.

Lack D (1947). The significance of clutch-size. *Ibis* **89**:302–352.

Lack D (1954). *The Natural Regulation of Animal Numbers*. Clarendon, Oxford.

Lack D (1955). British tits (*Parus* spp.) in nesting boxes. *Ardea* **43**;50–84.

Lack D (1965). Evolutionary ecology. *Journal of Animal Ecology* **34**:223–231.

Lack D (1966). *Population Studies of Birds*. Clarendon, Oxford.

Lack D (1968). *Ecological Adaptations for Breeding in Birds*. Methuen, London.

Lackey JA (1978). Reproduction, growth, and development in high-latitude and low-latitude populations of *Peromyscus leucopus* (Rodentia). *Journal of Mammalogy* **59**:69–83.

Ladewig J and Hart BL (1980). Flehmen and vomeronasal organ function in male goats. *Physiology and Behavior* **24**:1067–1071.

Ladewig J, Price EO and Hart BL (1980). Flehmen in male goats: Role in sexual behavior. *Behavioral and Neural Biology* **30**:312–322.

Ladurie EL (1971). *Times of Feast, Times of Famine: A History of Climate Since the Year 1000*. Doubleday, Garden City.

Lagios MD (1975). The pituitary gland of the coelacanth *Latimeria chalumnae* Smith. *General and Comparative Endocrinology* **25**:126–146.

Lake JS (1967). Rearing experiments with five species of Australian freshwater fishes. I. Inducement to spawning. *Australian Journal of Marine and Freshwater Research* **18**: 137–153.

Lam TJ (1983). Environmental influences on gonadal activity in fish. A review. *Fish Physiology* **9**:65–116.

Lam TJ and Soh CL (1975). Effects of photoperiod on gonadal maturation in the rabbitfish, *Siganus caniculatus*. *Aquaculture* **5**:407–410.

Lambe DR and Erickson CJ (1973). Ovarian activity of female ring doves (*Streptopelia risoria*) exposed to marginal stimuli from males. *Physiology and Psychology* **1**:281–283.

Lambert JGD (1970). The ovary of the guppy *Poecilia reticulata*. The granulosa cells as sites of steroid biosynthesis. *General and Comparative Endocrinology* **15**:464–476.

Lamotte M, Rey P and Vogeli M (1964). Recherches sur l'ovaire de *Nectrophrynoides occidentalis* bactracien anoure vivipare. *Archives d'Anatomie Microscopique et de Morphologie Experimental* **53**:179–224.

Lancaster JB and Lee RB (1965). The annual reproductive cycle in monkeys and apes. In *Primate Behavior: Field Studies and Apes* (Ed: DeVore R) Holt, Rinehart & Winston, New York:486–513.

Lanman JT (1977). Parturition in nonhuman primates. *Biology of Reproduction* **16**:28–38.

Laqabhsetwar AP (1975). Prostaglandins and studies related to reproduction in laboratory animals. In *Prostaglandins and Reproduction* (Ed: Karim SMM) University Park, Baltimore:241–270.

Larsen LO (1974). Effects of testosterone and oestradiol on gonadectomized and intact male and female river lampreys (*Lampetra fluviatilis* L.). *General and Comparative Endocrinology* **24**:303–313.

Larsen LO and Rothwell B (1972). Adenohypophysis. In *The Biology of Lampreys*, vol 2 (Eds: Hardesty MW and Potter IC) Academic, New York:1–67.

Larsen GJH Jr, Schroeder PC, and Waldo AE (1977). Structure and function of the amphibian follicular epithelium during ovulation. *Cell and Tissue Research* **181**:505–518.

Lavigne DM, Stewart REA, and Fletcher F (1982). Changes in composition and energy content of harp seal milk during lactation. *Physiological Ecology* **55**:1–9.

Laws RM (1969). Aspects of reproduction in the African elephant, *Loxodonta africana*. *Journal of Reproduction and Fertility*, Supplement **6**:495–531.

Laws RM, Parker ISC, and Johnstone RCB (1975). *Elephants and Their Habits*. Clarendon, Oxford.

Lawton AD and Whitsett JM (1979). Inhibition of sexual maturity by a urinary pheromone in male prairie deer mice. *Hormones and Behavior* **13**:128–138.

Leamy L (1981). The effect of litter size on fertility in *Peromyscus leucopus*. *Journal of Mammalogy* **62**:692–697.

Leathem JH (1966). Nutritional effects on hormone production. In *Environmental Influences on Reproductive Processes* (Eds: Hansel W and Dutt RH) *Journal of Animal Science* 25 (Supplement):68–78.

Lechleitner RR (1959). Sex ratio, age classes and reproduction of the black-tailed jackrabbit. *Journal of Mammalogy* **40**:63–81.

Lecyk M (1962). The effect of the length of daylight on reproduction of the field vole (*Microtus arvalis* Pall.). *Zoologica Poloniae* **12**:189–221.

Lee AK and Cockburn A (1985). *The Evolutionary Ecology of Marsupials*. Cambridge University, Cambridge.

Legan SJ and Karsch FJ (1978). Mechanism for periodic control of seasonal breeding in the ewe. *Federation Proceedings* **37**:297.

Legan SJ and Karsch FJ (1979). Neuroendocrine regulation of the estrous cycle and seasonal breeding of the ewe. *Biology of Reproduction* **20**:74–85.

Legan SJ, Karsch FJ, and Foster DL (1977). The endocrine control of seasonal reproductive function in the ewe: A marked change in response to the negative feedback action of estradiol on luteinizing hormone secretion. *Endocrinology* **101**:818–824.

Legan SJ and Winans SS (1981). The photoneuroendocrine control of seasonal breeding in the ewe. *General and Comparative Endocrinology* **45**:317–328.

Legg SP, Curnow DH, and Simpson SA (1950). The seasonal and species distribution of oestrogen in British pasture plants. *Biochemical Journal* **46**:19–20.

Lehrman DS and Brody P (1961). Does prolactin induce incubation behaviour in the ring dove? *Journal of Endocrinology* **22**:269–275.

Leider V (1959). Uber die Eientwicklung bei Mannchenlosen stammen der Silberkarausche *Carassius auratus* gibelio (Bloch) (Vertebrate, Pisces). *Biologisches Zentralblatt* **135**:212.

Leinonen M (1973). Comparison between the breeding of biology of year-old and older females of the white wagtail *Motacilla alba*. *Ornis Fennica* **50**:126–133.

Lemnell PA (1978). Social behaviour of the great snipe *Capella media* at the arena display. *Ornis Scandinavica* **9**:146–163.

Leon ML, Croskerry PG, and Smith GK (1978). Thermal control of maternal behavior. *Journal of Comparative Physiology and Psychology* **21**:793–811.

Leopold AS, Erwin M, Oh J, and Browning B (1975). Phytoestrogens: Adverse effects on reproduction in California quail. *Science* **191**:98–100.

Lepri JJ and Noden PF (1984). Reproductive function is independent of photoperiod in adult male *Microtus pinetorum*. *Journal of Mammalogy* **65**:696–697.

Lerner IM and Libby MJ (1980). *Heredity, Evolution and Society*. Freeman, San Francisco.

Lessells CM (1986). Brood size in Canada geese: A manipulative experiment. *Journal of Animal Ecology* **55**:669–689.

Lewis M and Dodd JM (1974). Thyroid function and the ovary in the spotted dogfish *Scyliorhinus canicula. Journal of Endocrinology* **63**:63P.

Liang-Tang LK and Soloff MS (1972). Characterization of a binding protein specific for 17β-estradiol and estrone in rat pregnancy plasma. *Biochemica et Biophysica Acta* **263**:753–763.

Licht LE (1976). Sexual selection in toads (*Bufo americanus*). *Canadian Journal of Zoology* **54**:1277–1284.

Licht P (1971a). Regulation of the annual testis cycle by photoperiod and temperature in the lizard *Anolis carolinensis. Ecology* **52**:240–252.

Licht P (1971b). Response of the male reproductive system to interrupted-night photoperiods in the lizard *Anolis carolinensis. Zeitschrift für vergleichende Physiologie* **73**:274–284.

Licht P (1972). Environmental physiology of reptilian breeding cycles: role of temperature. *General and Comparative Endocrinology*, Supplement **3**:477–488.

Licht P (1979). Reproductive endocrinology of amphibians and reptiles. *Annual Review of Physiology* **41**:337–351.

Licht P (1984). Reptiles. In *Marshall's Physiology of Reproduction*, vol 1, 4th ed (Ed: Lamming GE) Churchill Livingstone, Edinburgh:206–282.

Licht P and Gorman GC (1970). Reproductive and fat cycles in Caribbean anolis lizards. *University of California Publications in Zoology* **95**:1–52.

Licht P and Popkoff H (1974). Separation of two distinct gonadotropins from the pituitary gland of the bullfrog, *Rana catesbeiana. Endocrinology* **94**:1587–1594.

Lidicker WZ Jr (1973). Regulation of numbers in an island population of the California vole, a problem in community dynamics. *Ecological Monographs* **43**:271–302.

Lidicker WZ Jr (1975). The role of dispersal in the demography of small mammals. In *Small Mammals: Their Productivity and Population Dynamics* (Eds: Golley FB, Petrusewicz K, and Ryszkowski L) Cambridge University, Cambridge:103–128.

Lidicker WZ Jr (1978). Regulation of numbers in small mammal populations—historical reflections and a synthesis. In *Populations of Small Mammals Under Natural Conditions* (Ed: Snyder DP) Pymatuning Laboratory of Ecology, University of Pittsburgh, Special Publication Series, Volume 4:122–141.

Liem KF (1963). Sex reversal as a natural process in the symbranchiform fish *Monopterus albus. Copeia*:303–312.

Liggins GC (1973). The physiological role of prostaglandins in parturition. In *Research Related to the Control of Fertility* (Eds: Bindon BM, Emmens CW, and Smith MSR) *Journal of Reproduction and Fertility*, Supplement **18**:143–150.

Lightfoot RJ and Marshall T (1976). Effect of lupin grain supplementation on ovulation rate and fertility of Merino ewes. *Journal of Reproduction and Fertility* **46**:518.

Liley NR and Stacey NE (1983). Hormones, pheromones and reproductive behavior in fish. In *Fish Physiology* (Eds: Hoar WS, Randall DJ, and Donaldson EM) Academic, New York, Part B:1–63.

Lin JY (1979). Ovarian, fat body and liver cycles in the lizard *Japalura swinhonis formosensis* in Taiwan (Lacertilia: Agamidae). *Journal of Asian Ecology* **1**:29–38.

Lincold GA, Almeida OFX, Klandorf H, and Cunningham RA (1982). Hourly fluctuations in the blood levels of melatonin, prolactin, luteinizing hormone, follicle stimulating hormone, testosterone, triiodothyronine, thyroxin and cortisol in rams under artificial photoperiods and effects of cranial sympathectomy. *Journal of Endocrinology* **92**:237–250.

Lincoln DW and Renfree MB (1981). Mammary gland growth and milk ejection in the agile wallaby, *Macropus agilis*, displaying concurrent asynchronous lactation. *Journal of Reproduction and Fertility* **63**:193–203.

Lincoln G (1978). The photoperiodic control of seasonal breeding in rams. In *Comparative Endocrinology* (Eds: Gaillard PJ and Boer HH) Elsevier/North Holland, Amsterdam:149–152.

Lincoln GA (1981). Seasonal aspects of testicular function. In *The Testis* (Eds: Burger H and de Kretser D) Raven, New York:255–203.

Lincoln GA and Kay RNB (1979). Effects of season on the secretion of LH and testosterone in intact and castrated red deer stags. *Journal of Reproduction and Fertility* **55**:75–80.

Lincoln GA and MacKinnon PC (1976). Study of seasonally delayed puberty in the male hare, *Lepus europaeus. Journal of Reproduction and Fertility* **46**:123–128.

Lincoln GA and Peet MJ (1977). Photoperiodic control of gonadotropin secretion in the ram: A detailed study of the temporal changes in plasma levels of follicle-stimulating hormone, luteinizing hormone and testosterone following an abrupt switch from long to short days. *Journal of Endocrinology* **74**:355–367.

Lincoln GA, Peet MJ, and Cunningham RA (1977). Seasonal or circadian changes in the episodic release of FSH, LH and testosterone in rams exposed to artificial photoperiods. *Journal of Endocrinology* **72**:337–349.

Lindner HR, Bar-Ami S, and Tsafrini A (1983). Control of the resumption of meiosis in mammals. In *The Ovary* (Ed: Serra GB) Raven, New York:83–94.

Lindsey AA (1937). Weddell seal in the Bay of Whales, Antarctica. *Journal of Mammalogy* **18**:127–144.

Lindstrom E, Angelstam P, Widen P, and Andren H (1987). Do predators synchronize vole and grouse fluctuations? An experiment. *Oikos* **48**:121–124.

Ling N, Ying S-Y, Ueno N, Shimasaki S, Esch F, Hotta M, and Guillemin R (1986). Pituitary FSH is released by a heterodimer of the B-subunits from the two forms of inhibin. *Nature* **321**:779.

Lisk RD (1971a). Oestrogen and progesterone synergism and elicitation of maternal nest-building in the mouse (*Mus musculus*). *Hormones and Behavior* **11**:232–247.

Lisk RD (1971b). Oestrogen and progesterone synergism and elicitation of maternal nest-building in the mouse. *Animal Behaviour* **19**:606–610.

Lochmiller RL, Whelan JB, and Kirkpatrick RL (1982). Energetic cost of lactation in *Microtus pinetorum. Journal of Mammalogy* **63**:475–481.

Lodge JR, Fechheimer MS, and Jaap RG (1971). The relationship of in vivo sperm storage interval to fertility and embryonic survival in the chicken. *Biology of Reproduction* **5**:252–257.

Lofts B (1964). Evidence of an autonomous reproductive rhythm in an equatorial bird (*Quelea quelea*). *Nature* **201**:523–524.

Lofts B (1970). *Animal Photoperiodism*. Edward Arnold, London.

Lofts B (1974). Reproduction. In *Physiology of the Amphibia*, vol 2 (Ed: Lofts B). Academic, New York:107–218.

Lofts B (1977). Patterns of spermatogenesis and steroidogenesis in male reptiles. In *Reproduction and Evolution* (Eds: Calaby JH and Tyndale-Biscoe CH) Australian Academy of Science, Canberra:127–136.

Lofts B (1978). Reptilian reproductive cycles and environmental regulators. In *Environmental Endocrinology* (Eds: Assenmacher I and Farner DS) Springer-Verlag, Berlin:37–43.

Lofts B (1984). Amphibians. In *Marshall's Physiology of Reproduction*, vol 1, 4th ed (Ed: Lamming GE) Churchill Livingstone, Edinburgh:127–205.

Lofts B and Bern HA (1972). The functional morphology of steroidogenic tissues. In *Steroids in Nonmammalian Vertebrates* (Ed: Idler D) Academic, New York:35–125.

Lofts B and Murton RK (1968). Photoperiodic and physiological adaptations regulating avian breeding cycles and their ecological significance. *Journal of Zoology* **155**:327–394.

Loman J (1977). Factors affecting clutch and brood size in the crow, *Corvus cornix*. *Oikos* **29**:294–301.

Loman J and Madsen T (1986). Reproductive tactics of large and small male toads, *Bufo bufo*. *Oikos* **46**:57–61.

Lomas DE and Keverne EB (1982). Role of the vomeronasal organ and prolactin in the acceleration of puberty in female mice. *Journal of Reproduction and Fertility* **66**:101–107.

Lombardi J and Wourms JP (1985). The trophotaeniae placenta of a viviparous goodeid fish. II. Ultrastructure of trophotaeniae, the embryonic component. *Journal of Morphology* **184**:293–309.

Lord RD Jr (1960). Litter size and latitude in North American mammals. *American Midland Naturalist* **64**:488–499.

Loudan ASI, McNeilly AS, and Milne JA (1983). Nutrition and lactational control of fertility in red deer. *Nature* **302**:145–147.

Löve A and Löve D (1945). Experiments on the effects of animal sex hormones on dioecious plants. *Arkiv for Botanik* **32A** (13):1–60.

Lovell R and Rees TA (1961). Immunological aspects of colostrum. In *Milk: The Mammary Gland and Its Secretion*, vol II (Eds: Kon SK and Cowie AT) Academic, New York: 363–381.

Lowe CH, Wright JW, Cole CJ, and Bezy RL (1970). Natural hybridization between the teiid lizards *Cnemidophorus sonorae* (parthenogenetic) and *Cnemidophorus tigris* (bisexual). *Systematic Zoology* **19**:114–127.

Lowe RH (1953). Notes on the ecology and evolution of Nyassa fishes of the genus *Tilapia*, with a description of *T. saka* Lowe. *Proceedings of the Zoological Society of London* **122**:1035–1041.

Lowe-McConnell RH (1958). *Ecology of Fishes in Tropical Waters. Studies in Biology*, no. 76. Edward Arnold, London.

Lowe-McConnell (1975). *Fish Communities in Tropical Freshwaters*. Longman, London.

Lowe-McConnell RH (1977). *Ecology of Fishes in Tropical Waters*. Edward Arnold, London.

Lu KH, Hopper BR, Vargo TM, and Yen SSC (1979). Chronological changes in sex steroid, gonadotropin and prolactin secretion in aging female rats displaying different reproductive states. *Biology of Reproduction* **21**:193–203.

Luckey TD, Mende TJ, and Pleasants J (1954). The physical and chemical characterization of rats' milk. *Journal of Nutrition* **54**:345–359.

Lynch GR, Heath HW, and Johnston CM (1981b). Effect of geographical origin on the photographic control of reproduction in the white-footed mouse, *Peromyscus leucopus*. *Biology of Reproduction* **25**:475–480.

Lynch HJ, Rivest RW, Ronshein PM, and Wurtman RJ (1981a). Light intensity and the control of melatonin secretion in the rat. *Neuroendocrinology* **33**:181–193.

Lynn WG (1970). The thyroid. In *Biology of Reptilia*, vol 3 (Eds: Gans C and Parsons TS) Academic, New York:201–234.

MacFarlane JD and Taylor JM (1982). Nature of estrus and ovulation in *Microtus townsendii* (Bachman). *Journal of Mammalogy* **63**:104–109.

MacFarlane WV (1970). Seasonality of conception in human populations. *Biometeorology* **4**:167–182.

Macgregor HC and Kezer J (1970). Gene amplification in oocytes with 8 germinal vesicles from the tailed frog, *Ascaphus truei* Stejneger. *Chromosoma* **29**:189–206.

MacKinnon JC (1972). Summer storage of energy and its use for winter metabolism and gonad maturation in American plaice (*Hippoglossoides platessoides*). *Journal of the Fisheries Research Board of Canada* **29**:1749–1759.

MacLean GL (1972). Clutch size and evolution in the Charadrii. *Auk* **89**:299–324.

MacLusky N and Naftolin F (1981). Sexual differentiation of the central nervous system. *Science* **211**:1294–1303.

Macrides F, Bartke A, and Dalterio S (1974). Effects of exposure to vaginal odor and receptive females on plasma testosterone in the male hamster. *Neuroendocrinology* **15**:355–364.

Macrides F, Bartke A, Fernandez F, and D'Angelo W (1974). Effects of exposure to vaginal odor and receptive females on plasma testosterone in the male hamster. *Neuroendocrinology* **15**:355–364.

MacRoberts MH and MacRoberts BR (1976). Social organization and behavior of the acorn woodpecker in central coastal California. *Ornithological Monographs* **21**:1–115.

Maeda M (1963). Field experiment on the biology of field mice in woodland 2. Natality and mortality of the red-backed vole, *Clethrionomys rufocanus bedfordiae* (Thomas). *Bulletin of the Government Forest Experiment Station* **160**:1–18.

Maeda M (1969). Field experiments on the biology of field mice in woodlands in Hokkaido (III). Seasonal changes in food habits of the red-backed vole, *Clethrionomys rufocanus bedfordiae*. *Annual Report of the Hokkaido Branch in the Forest Experiment Station* **1968**:129–142.

Maeda M (1973). Ecology of field mice in forest lands in Hokkaido. Report 4. Food habits and nutrition of the red-backed vole. *Clethrionomys rufocanus bedfordiae* (Thomas). *Bulletin of the Government Forest Experiment Station* **258**:1–12.

Magnusson WE (1979). Maintenance of temperature of crocodile nests. *Journal of Herpetology* **13**:439–443.

Magnusson WE, Lima AP, and Sampaio RM (1985). Sources of heat for nests of *Paleosuchus trigonatus* and a review of crocodilian nest temperatures. *Journal of Herpetology* **19**:199–207.

Mahmoud IY (1967). Courtship behavior and sexual maturity in four species of kinosternid turtles. *Copeia*:314–319.

Main SJ, Davies RV, and Setchell BP (1979). The evidence that inhibin must exist. *Journal of Reproduction and Fertility*, Supplement **26**:3–14.

Malacarne G and Vellano C (1987). Behavioral evidence of a courtship pheromone in the crested newt, *Triturus cristatus carnifex* Laurenti. *Copeia*:245–247.

Malcom ADB (1981). Production and processing of prolactin. *Nature* **290**:546.

Maltz E and Shkolnik A (1984). Lactational strategies of desert ruminants: The Bedouin goat, ibex and desert gazelle. In *Physiological Strategies in Lactation* (Eds: Peaker M, Vernon RG, and Knight CH) Zoological Society of London Symposium **51**:193–213.

Mann M, Michael SD, and Svare R (1980). Ergot drugs suppress plasma prolactin and lactation but not aggression in parturient mice. *Hormones and Behavior* **14**:319–328.

Mann T (1984). Spermatophores. *Zoophysiology* **15**.

Marcus O and Kusemiju K (1984). Some aspects of the reproductive biology of the clupeid *Ilisha africana* (Bloch) off the Lagos Coast, Nigeria. *Journal of Fish Biology* **25**:679–689.

Marsh BA, Marsh AC, and Ribbink AJ (1986). Reproductive seasonality in a group of rock-frequenting cichlid fishes in Lake Malawi. *Journal of Zoology (London)* **209**:9–20.

Marshall AJ (1970). Environmental factors other than light involved in the control of sexual cycles in birds and mammals. In *La Photoregulation de la Reproduction Chez les Oiseaux et les Mammifères* (Eds: Benoit J and Assenmacher I) Centre National de la Recherche Scientifique, Paris:53–69.

Marshall AJ and Serventy DL (1958). The internal rhythm of reproduction in xerophilous birds under conditions of illumination and darkness. *Journal of Experimental Biology* **35**:666–670.

Marshall JT Jr (1963). Rainy season nesting in Arizona. *Proceedings of the XIIIth International Ornithological Congress* **2**:620–622.

Martell AM and Fuller WA (1979). Comparative demography of *Clethrionomys rutilus* in taiga and tundra in the low Arctic. *Canadian Journal of Zoology* **57**:2106–2120.

Martin JR and Battig K (1980). Exploratory behaviour of rats in oestrus. *Animal Behaviour* **28**:900–905.

Martin RD (1968). Reproduction and ontogeny in tree-shrews (*Tupaia belangeri*). *Zeitschrift für Tierpsychologie* **25**:409–495.

Martin RD (1984). Scaling effects and adaptive strategies in mammalian lactation. In *Physiological Strategies in Lactation* (Eds: Peaker M, Vernon RG, and Knight CH) Zoological Society of London Symposium **51**:87–117.

Martinet L (1970). Role du photoperiodisme sur la biologie sexuelle du campagnol des champs (*Microtus arvalis*). In *La Photoregulation de la Reproduction Chez les Oiseaux et les Mammifères* (Eds: Benoit J and Assenmacher I) Centre National de la Recherche Scientifique, Paris:435–452.

Martinet L, Allais C, and Allain D (1981). The role of prolactin and LH in luteal function and blastocyst growth in the mink (*Mustela vison*). In *Embryonic Diapause in Mammals* (Eds: Flint APF, Renfree MB, and Weir BJ) *Journal of Reproduction and Fertility*, Supplement **29**:119–130.

Maruniak JA and Bronson FH (1976). Differential effects of male stimuli on follicle-stimulating hormone, luteinizing hormone and prolactin secretion in prepubertal female mice. *Endocrinology* **98**:1101–1108.

Marynick SP (1971). Long term storage of sperm in *Desmognathus fuscus* from Louisiana. *Copeia*:345–347.

Mashaly MM, Birrenkott GP, El-Begearmi E, and Wentworth BC (1976). Plasma LH and progesterone concentrations in the turkey hen during the ovulatory cycle. *Poultry Science* **55**:1226–1234.

Maslin TP (1971). Conclusive evidence of parthenogenesis in three species of *Cnemidophorus* (Teiidae). *Copeia*:156–158.

Maslin TP (1971). Parthenogenesis in reptiles. *American Zoologist* **11**:361–380.

Mason JC, Beamish RJ, and McFarlane GA (1983). Sexual maturity, fecundity, spawning and early life history of sablefish (*Anoplopoma fimbria*) off the Pacific coast of Canada. *Canadian Journal of Fisheries and Aquatic Science* **40**:2126–2134.

Mason RT and Crews D (1985). Female mimicry in garter snakes. *Nature* **316**:59–60.

Masson GR and Guillette LJ Jr (1987). Changes in oviducal vascularity during the reproductive cycle of three oviparous lizards (*Eumeces obsoletus*, *Sceloporus undulatus* and *Crotophytus collaris*). *Journal of Reproduction and Fertility*, **80**:361–371.

Matthews LH (1937). The female sexual cycle in the British horse-shoe bat (*Rhinolophus ferrum-equinum insulanus* Barrett-Hamilton and *R. hipposideros minutus* Montagu). *Transactions of the Zoological Society of London* **23**:213–255.

Matthews LH (1941). Notes on the genitalia and reproduction of some African bats. *Proceedings of the Zoological Society of London* **111B**:289–346.

Matthews LH (1950). Reproduction in the basking shark, *Cetorhinus maximus* (Gunner). *Philosophical Transactions of the Royal Society B* **234**:247–316.

Matthews MJ, Benson B, and Richardson BA (1978). Partial maintenance of testes and accessory organs in blinded hamsters by homoplastic anterior pituitary grafts or exogenous prolactin. *Life Science* **23**:1311–1323.

Matty AJ (1978). Pineal and some pituitary hormone rhythms in fish. *Rhythmic Activity of Fishes* (Ed: Thorpe JE) Academic, New York:21–30.

Maule Walker FM (1984). Exocrine and endocrine secretions of the mammary gland: Local control in ruminants. In *Physiological Strategies in Lactation* (Eds: Peaker M, Vernon RG, and Knight CH) Zoological Society of London Symposium **51**:171–191.

Maule Walker FM and Peaker M (1978). Production of oestrodial-17β by the goat mammary gland during late pregnancy in relation to lactogenesis. *Journal of Physiology* **284**:71P–72P

Maule Walker FM and Peaker M (1981). The role of the mammary gland in late pregnancy and parturition of the goat. *Journal of Physiology* **312**:P63.

Maung SL and Follett BK (1978). The endocrine control of LH of testosterone secretion from the testis of the Japanese quail. *General and Comparative Endocrinology* **36**:78–89.

Maurel D (1978). Seasonal changes of the testicular and thyroid functions in the badger, *Meles meles* L. In *Environmental Endocrinology* (Eds: Assenmacher I and Farner DS) Springer-Verlag, Berlin:85–86.

May R and Mead RA (1986). Evidence for pineal involvement in timing implantation in the western spotted skunk. *Journal of Pineal Research* **3**:1–8.

Mayhew WW (1966). Reproduction in the arenicolous lizard *Uma notata*. *Ecology* **47**:9–18.

Mazzi V (1970). The hypothalamus as a thermodependent neuroendocrine center in urodeles. In *The Hypothalamus* (Eds: Martini L, Motta M, and Fraschini F) Academic, New York:663–676.

Mazzi V (1971). Preliminary observations on the effect of subtotal adenohypophysectomy on some endocrine glands in the crested newt. Contribution to the histophysiology. *Accademia della Scienze, Torino* **105**:639–645.

Mazzi V (1978). Effects on spermatogenesis of permanent lesions to the rostral preoptic area in the crested newt (*Triturus cristatus carnifex* Laur.). *General and Comparative Endocrinology* **24**:1–9.

McAllan BM and Dickman CR (1986). The role of photoperiod in the timing of the reproduction of the dasyurid marsupial *Antechinus stuartii*. *Oecologia* **68**:259–264.

McCarley H (1966). Annual cycle, population dynamics and adaptive behavior of *Citellus tridecimenlineatus*. *Journal of Mammalogy* **47**:294–316.

McCarrey JR and Abbott UK (1979). Mechanics of genetic sex determination, gonadal sex differentiation, and germ-cell development in animals. *Advances in Genetics* **20**:217–290.

McClenaghan LR Jr (1987). Lack of effect of 6-MBOA on reproduction of *Dipodomys merriami*. *Journal of Mammalogy* **68**:150–152.

McComb K (1987). Roaring by red deer stags advances the date of oestrus in hinds. *Nature* **330**:648–649.

McConnell SJ, Tyndale-Biscoe CH, and Hinds LA (1986). Change in duration of elevated concentrations of melatonin is the major factor in photoperiod response of the tammar, *Macropus eugenii*. *Journal of Reproduction and Fertility* **77**:623–632.

McDiarmid RW (1978). Evolution of parental care in frogs. In *The Development of Behavior: Comparative and Evolutionary Aspects* (Eds: Burghardt GM and Bekoff M) Garland STMP, New York:127–147.

McDowell RM and Whitaker AH (1975). Freshwater fishes. In *Biogeography and Ecology in New Zealand* (Ed: Kuschel G) Junk, The Hague:277–299.

McEwen BS (1983). Gonadal steroid influences on brain development and sexual differentiation. In *Reproductive Physiology IV* (Ed: Greep RO) *International Review of Physiology* **27**:99–145.

McKaye KR (1977). Defense of a predator's young by a herbivorous fish: An unusual strategy. *American Naturalist* **111**:301–315.

McKaye KR (1983). Ecology and breeding behavior of a cichlid fish, *Cyrtocara eucinostomus*, on a large lek in Lake Malawi, Africa. *Environmental Biology of Fishes* **8**:81–96.

McKaye KR and McKaye NM (1976). Communal care and kidnapping of young by parental cichlids. *Evolution* **31**:674–681.

McKeever S (1964). The biology of the golden-mantled ground squirrel. *Ecological Monographs* **34**:385–401.

McKeever S (1966). Reproduction in *Citellus lateralis* and *Citellus beldingi* in northeastern California. In *Comparative Biology of Reproduction in Mammals* (Ed: Rowlands IW) Academic, London:365–385.

McKinney RB and Marion KR (1985). Reproductive and fat body cycles in the male lizard, *Sceloporus undulatus*, with comparisons of geographic variation. *Journal of Herpetology* **19**:208–217.

McKinney TD and Desjardins C (1973). Intermale stimuli and testicular function in adult and immature house mice. *Biology of Reproduction* **9**:370–378.

McLaren A, Simpson E, Tomonari K, Chandler P, and Hogg H (1984). Male sexual differentiation in mice lacking H-Y antigen. *Nature* **321**:552–555.

McLaren IA (1956). Summary of the biology of the ringed seal in waters of southwest Baffin Island. In *The "Calanus" Expeditions in the Canadian Arctic* (Ed: Dunbar MJ) *Arctic* **9**:178–190.

McMillan JP, Underwood HA, Elliott JA, Stetson MH, and Menaker M (1975). Extraretinal light perception in the sparrow. IV. Further evidence that the eyes do not participate in photoperiodic photoreception. *Journal of Comparative Physiology A* **97**:205–213.

McMillin JM, Seal US, Rogers L, and Erickson AW (1976). Annual testosterone rhythm in the black bear (*Ursus americanus*). *Biology of Reproduction* **15**:163–167.

McNatty KP, McNeilly AS, and Sawers RS (1974). A possible role for prolactin in control of steroid secretion by the human Graafian follicle. *Nature* **240**:653–655.

McNaughton SJ and Wolf LL (1979). *General Ecology*, 2nd ed. Holt, Rinehart and Winston, New York.

McNeilly AS, Glasier A, Jonassen J, and Howie PW (1982). Evidence for direct inhibition of ovarian function by prolactin. *Journal of Reproduction and Fertility* **65**:559–569.

McNeilly AS, Martin RD, Hodges JK, and Smuts GL (1983). Blood concentrations of gonadotropins, prolactin and gonadal steroids in males and in nonpregnant female African elephants (*Loxodonta africana*). *Journal of Reproduction and Fertility* **67**:113–120.

McNulty JA (1982). The effects of constant light and constant darkness on the pineal organ of the goldfish, *Carassius auratus*. *Journal of Experimental Zoology* **219**:29–37.

McPherson RJ and Marion KR (1981). The reproduction biology of female *Sternothorus odoratus* in an Alabama population. *Journal of Herpetology* **15**:389–396.

Mead GW, Bertlesen E, and Cohen DM (1964). Reproduction among deep-sea fishes. *Deep Sea Research* **11**:569–596.

Mead RA (1968a). Reproduction in western forms of the spotted skunk (genus *Spilogale*). *Journal of Mammalogy* **49**:373–390.

Mead RA (1968b). Reproduction in eastern forms of the spotted skunk (genus *Spilogale*). *Journal of Zoology, London* **156**:119–136.

Mead RA (1971). Effects of light and blinding upon delayed implantation in the spotted skunk. *Biology of Reproduction* **5**:214–220.

Mead RA (1972). Pineal gland: Its role in controlling delayed implantation in the spotted skunk. *Journal of Reproduction and Fertility* **30**:147–150.

Mead RA (1981). Delayed implantation in mustelids, with special emphasis on the spotted skunk. In *Embryonic Diapause in Mammals* (Eds: Flint APF, Renfree MB, and Weir BJ) *Journal of Reproduction and Fertility*, Supplement **29**:11–14.

Mead RA (1986). Role of the corpus luteum in controlling implantation in mustelid carnivores. *Annals of the New York Academy of Sciences* **476**:25–35.

Mead RA (1988). The physiology and evolution of delayed implantation in carnivores. In *Carnivore Behavior, Ecology, and Evolution* (Ed: Gittleman J) Cornell University, Ithaca (in press).

Mead RA and Swannack A (1978). Effects of hysterectomy on luteal function in the western spotted skunk (*Spilogale putorius latifrons*). *Biology of Reproduction* **18**:379–383.

Mead RA and Swannack A (1980). Aromatase activity in corpora lutea of the ferret. *Biology of Reproduction* **22**:560–565.

Measroch V (1955). Growth and reproduction in the females of two species of gerbil, *Tatera brantsi* (A. Smith) and *Tatera afra* (Gray). *Proceedings of the Zoological Society of London* **124**:631–658.

Medway L (1962). The swiftlets (*Collocalia*) of Niah Cave, Sarawak. *Ibis* **104**:45–66; 228–245.

Medway L and Wells DR (1976). *The Birds of the Malay Peninsula*, vol 5. HF & G Witherby, London.

Meier AH (1975). Chronophysiology of prolactin in lower vertebrates. *American Zoologist* **15**:905–916.

Mellenberger RW and Bauman DE (1974). Metabolic adaptations during lactogenesis. Lactose synthesis in rabbit mammary tissue during pregnancy and lactation. *Biochemical Journal* **141**:659–665.

Mena F and Grosvenor CE (1971). Release of prolactin in rats by exteroceptive stimulation: Sensory stimuli involved. *Hormones and Behavior* **2**:107–116.

Mendoza G (1940). The reproductive cycle of the viviparous teleost, *Neotoca bilineata*, a member of the family Goodeidae. II. The cyclic changes in the ovarian soma during gestation. *Biological Bulletin* **78**:349–365.

Mendoza G (1972). The fine structure of an absorptive epithelium in a viviparous teleost. *Journal of Morphology* **136**:109–130.

Menzies JI (1973). A study of the leaf-nosed bats (*Hipposideros caffer* and *Rhinolophus lauderi*) in a cave in northern Nigeria. *Journal of Mammalogy* **54**:930–945.

Metter DE (1964). On breeding and sperm retention in *Ascaphus*. *Copeia*:710–711.

Michael R and Keverne EB (1970). Primate sex pheromones of vaginal origin. *Nature* **225**:84–85.

Michael RP and Bonsall RW (1970). Hormones and the sexual behavior of rhesus monkeys. In *Endocrine Control of Sexual Behavior* (Ed: Beyer C) Raven, New York:279–302.

Michael RP and Bonsall RW (1977). Chemical signals and primate behavior. In *Chemical Signals in Vertebrates* (Eds: Muller-Schwarze D and Mozell MM) Plenum, New York:251–272.

Michael RP and Zumpe D (1981). Environmental influences on the reproductive behavior of rhesus monkeys. In *Environmental Factors in Mammalian Reproduction* (Eds: Gilmore D and Cooke B) Macmillan, London:77–99.

Michael RP, Bonsall RW, and Warner P (1975). Primate sexual pheromones. In *Olfaction and Taste* (Eds: Denton DA and Coghlan JP) *Proceedings of the 5th International Symposium Melbourne*, 1974:417–424.

Michener GR (1973). Climatic conditions and breeding in Richardson's ground squirrel. *Journal of Mammalogy* **54**:499–503.

Middaugh DP (1981). Reproductive ecology and spawning periodicity of the Atlantic silverside, *menidia menidia* (Pisces: Atherinidae). *Copeia*:766–776.

Middaugh DP and Takita T (1983). Tidal and diurnal spawning cues in the Atlantic silverside, *Menidia menidia*. *Environmental Biology of Fish* **8**:97–104.

Middaugh DP, Cott GI, and Dean JM (1981). Reproductive behavior of the Atlantic silverside, *Menidia menidia* (Pisces, Atherinidae). *Environmental Biology of Fish* **6**:269–276.

Midgley AR and Sadler WA (1979). *Ovarian Follicular Development and Function*. Raven, New York.

Millar JS (1973). Evolution of litter size in the pika *Ochotona princeps* (Richardson). *Evolution* **27**:134–143.

Millar RP (1967). Oestrogenic activity in central African highveld grasses. Rhodesia, Zambesia and Walawi. *Journal of Agricultural Research* **5**:179–183.

Millar RP and Glover TD (1973). Regulation of seasonal sexual activity in an ascrotal mammal, the rock hyrax, *Procavia capensis*. *Journal of Reproduction and Fertility*, Supplement **19**:203–220.

Miller AH (1959a). Response to experimental light increments by Andean sparrows from an equatorial area. *Condor* **61**:344–347.

Miller AH (1959b). Reproductive cycles in an equatorial sparrow. *Proceedings of the National Academy of Science* **45**:1095–1100.

Miller AH (1962). Bimodal occurrence of breeding in an equatorial sparrow. *Proceedings of the National Academy of Science* **48**:396–400.

Miller CM (1944). Ecological relations and adaptations of the legless lizard of the genus *Anniella*. *Ecological Monographs* **14**:271–289.

Miller MA (1946). Reproductive rates and cycles in the pocket gopher. *Journal of Mammalogy* **27**:335–358.

Miller MR (1951). Some aspects of the life history of the yucca night lizard, *Xantusia vigilis*. *Copeia*:114–120.

Miller PJ (Ed) (1979). Fish phenology: Anabolic adaptiveness in teleosts. *Zoological Society of London Symposium* 44.

Milligan SR (1976). Pregnancy blocking in the vole, *Microtus agrestis*. I. Effects of the social environment. *Journal of Reproduction and Fertility* **46**:91–95.

Milligan SR (1981). Analysis of the LH surge induced by mating and LH-RH in the vole, *Microtus agrestis*. *Journal of Reproduction and Fertility* **63**:39–45.

Millington AJ, Francis CM, and McKeown NR (1964). Wether bioassay of annual pasture legumes 2. Oestrogenic effects of 9 strains of *Trifolium subterraneum*. *Australian Journal of Agriculture* **15**:527–336.

Mills JA (1979). Factors affecting the egg size of the red-billed gull *Larus novaehollandiae scopulinus*. *Ibis* **121**:53–67.

Mills JA and Shaw PW (1980). The influence of age on laying date, clutch size, and egg size of the white-fronted tern, *Sterna striata*. *New Zealand Journal of Zoology* **7**:147–153.

Mills MGL (1985). Related spotted hyaenas forage together but do not cooperate in rearing young. *Nature* **316**:61–62.

Milton DA and Arthington AH (1983). Reproductive biology of *Gambusia affinis holbrooki* Baird and Girard, *Xiphophorus helleri* (Gunther) and *X. maculatus* (Heckel) (Pisces; Poeciliidae) in Queensland, Australia. *Journal of Fish Biology* **23**:23–41.

Misra HC and Thapliyal JP (1980). Inverse seasonal cytological cyles in testes and pineal organs of the bloodsucker *Calotes versicolor*. *Herpetologica* **36**:116–119.

Mitchell JC (1985). Female reproductive cycle and life history attributes in a Virginia population of painted turtles, *Chrysemys picta*. *Journal of Herpetology* **19**:218–226.

Mittwoch U (1983). Heterogametic sex chromosomes and the development of the dominant gonad in vertebrates. *American Naturalist* **122**:159–180.

Mittwock U (1967). *Sex Chromosomes*. Academic, New York.

Mizuno H and Sensui N (1978). Effects of concurrent pregnancy and continuous suckling of two consecutive lactations. In *Physiology of Mammary Glands* (Eds: Yokoyama A, Mizuno H, and Nagasawa H) Japan Scientific Societies, Tokyo:287–302.

Moehlman PD (1979). Jackal helpers and pup survival. *Nature* **277**:382–383.

Mohr H (1936). Hemirhamphiden-Studien IV. Die Gattung Dermogenys van Hasselt, V. Die Gattung *Normorhamphus* Weber and de Beaufort, IV. Die Gattung *Hemirhamphidon* Bleeker. *Mitteilungen Zoologische Museum des Berlin* **21**:34–64.

Moll EO (1979). Reproductive cycles and adaptations. In *Turtles: Perspectives and Research* (Ed: Morlock M) Wiley, New York:305–331.

Moller AP (1987). Intruders and defenders on avian breeding territories: The effect of sperm competition. *Oikos* **48**:47–54.

Moller OM (1973). The progesterone concentrations in the peripheral plasma in the mink (*Mustela vison*) during pregnancy. *Journal of Endocrinology* **56**:121–132.

Moller OM (1974). Plasma progesterone before and after ovariectomy in unmated and pregnant mink, *Mustela vison*. *Journal of Reproduction and Fertility* **37**:367–372.

Montgomerie RD (1981). Why do jackals help their parents? *Nature* **289**:824–825.

Moore MC, Whittier JM, Billy AJ, and Crews D (1985). Male-like behavior in an all-female lizard: Relationship to ovarian cycle. *Animal Behaviour* **33**:284–289.

Moore RY (1977). Neural control of pineal function in birds and mammals. In *The Pineal Gland* (Eds: Niv I, Reiter RJ, and Wurtman RJ) Springer-Verlag, New York:47–58.

Moore RY (1978). The innervation of the mammalian pineal gland. In *The Pineal and Reproduction* (Ed: Reiter RJ) vol. 4:1–29, Karger, Basel.

Moore WS, Miller RR, and Schultz RJ (1970). Distribution, adaptation and probable origin of an all-female form of *Poeciliopsis* (Pesces: Poeciliidae) in northeastern Mexico. *Evolution* **24**:789–795.

Moran LP and Zucker I (1978). Photoperiodic regulation of copulatory behaviour in the male hamster. *Journal of Endocrinology* **77**:249–258.

Moreau RE (1950). The breeding seasons of African birds. 1, Land birds. *Ibis* **92**:223–267.

Moreau RE (1966). Water birds over the Sahara. *Ibis* **108**:456.

Moreau RE (1972). *The Palaearctic-African Bird Migration Systems*. Academic, London.

Môri T and Uchida TA (1981). Ultrastructural observations of fertilization in the Japanese long-fingered bat, *Miniopterus schreibersii fuliginosus*. *Journal of Reproduction and Fertility* **63**:231–235.

Môri T, Son SW, and Uchida TA (1986). Implications of prolonged sperm storage from the viewpoint of capacitation in the Japanese house-dwelling bat, *Pipistrellus abramus*. *Development, Growth and Differentiation Supplement*:**94.**

Morita Y and Bergmann G (1971). Physiologische Untersuchungen und weitere Bemerkungen zur Struktur des lichtemfindlichen Pinealorgans von *Pterophyllum scalare* Civ. et Val. (Cichlidae, Teleostei). *Zeitschrift für Zellforschung und Milroskopische Anatomie* **119**: 289–294.

Morley FH, Bennett D, Braden AWH, Turnbull KE, and Axelsen A (1968). Comparison of mice, guinea-pigs and sheep as test animals for bioassay of oestrogenic pasture legumes. *Proceedings of the New Zealand Society of Animal Production* **28**:11–21.

Morris DW (1985). Natural selection for reproductive optima. *Oikos* **45**:290–292.

Mortimer JA and Carr A (1987). Reproduction and migrations of the Ascension Island green turtle (*Chelonia mydas*). *Copeia*:103–113.

Moser HG (1967). Seasonal histological changes in the gonads of *Sebastodes paucispinis* (Ayers), an ovoviviparous teleost (Family Scorpaenidae). *Journal of Morphology* **123**: 329–354.

Moss CJ (1983). Oestrous behaviour and female choice in the African elephant. *Behaviour* **86**:167–196.

Moss R, Watson A, and Rothery P (1984). Inherent changes in the body size, viability, and behaviour in a fluctuating red grouse (*Lagopus lagopus scoticus*) population. *Journal of Animal Ecology* **53**:171–189.

Mossman HW and Duke KL (1973). *Comparative Morphology of the Mammalian Ovary*. University of Wisconsin, Madison.

Mossman HW and Judas I (1949). Accessory corpora lutea. Lutein cell origin and the ovarian cycle in the Canadian porcupine. *American Journal of Anatomy* **85**:1–40.

Moudgal NR (1981). A need for FSH in maintaining fertility of adult male subhuman primates. *Archives of Andrology* **7**:117.

Mueller CC and Sadleir RMFS (1977). Changes in the nutrient composition of milk of black-tailed deer during lactation. *Journal of Mammalogy* **58**:421–423.

Mueller RP and Sadleir RMFS (1979). Age of first conception in black-tailed deer. *Biology of Reproduction* **21**:1099–1104.

Mullen DA (1968). Reproduction in brown lemmings (*Lemmus trimucronatus*) and its relevance to their cycle of abundance. *University of California Publications in Zoology* **85**:1–24.

Muller CH (1977). Plasma 5a-dihydrotestosterone and testosterone in the bullfrog, *Rana catesbeiana*: stimulation by bullfrog LH. *General and Comparative Endocrinology* **33**:122–132.

Müller P (1973). *The Dispersal Centres of Terrestrial Vertebrates in the Neotropical Realm*. Junk, The Hague.

Muller RE (1981). Some effects of prolactin on the parental behavior of the convict cichlid, *Cichlosoma nigrofasciatum* (Gunther). *Environmental Biology of Fish* **6**:134–152.

Muller-Schwarze D (1971). Pheromones in black-tailed deer (*Odocoileus hemionus columbianus*). *Animal Behaviour* **19**:141–152.

Mumme RE, Koenig WD, and Pitelka FA (1983). Reproductive competition in the communal acorn woodpecker: Sisters destroy each other's eggs. *Nature* **306**:583–584.

Munro J and Bedard J (1977). Gull predation and creching behaviour in the common eider. *Journal of Animal Ecology* **47**:799–810.

Munro JL, Gaut VC, Thompson R, and Reeson PH (1973). The spawning seasons of Caribbean reef fishes. *Journal of Fish Biology* **5**:69–84.

Munsick RA, Sawyer WH, and Van Dyke HB (1960). Avian neurohypophysial hormones: Pharmacological properties and tentative identification. *Endocrinology* **66**:860–872.

Murie JO, Boag DA, and Kivett VK (1980). Litter size in Columbian ground squirrels (*Spermophilus columbianus*). *Journal of Mammalogy* **61**:237–244.

Murphy BD (1979). The role of prolactin in implantation and luteal maintenance in the ferret. *Biology of Reproduction* **21**:517–521.

Murphy BD, Concannon PW, Travis HF, and Hansel W (1981). Prolactin: The hypophyseal factor that terminates embryonic diapause in the mink. *Biology of Reproduction* **25**: 487–491.

Murphy MR (1973). Effects of female hamster vaginal discharge on the behavior of male hamsters. *Behavioral Biology* **9**:367–375.

Murtagh CE and Sharman BG (1977). Chromosomal sex determination in monotremes. In *Reproduction and Evolution* (Eds: Calaby JH and Tyndale-Biscoe CH) Australian Academy of Science, Canberra City.

Murton RK and Westwood NJ (1977). *Avian Breeding Cycles*. Oxford University, Oxford.

Muul I (1969). Photoperiod and reproduction in flying squirrels, *Glaucomys volans*. *Journal of Mammalogy* **50**:542–549.

Myers GS (1939). A possible method of evolution of oral habits in cichlid fishes. *Stanford Ichthyological Bulletin* **1**:85–87.

Myers JP and Pitelka FA (1979). Variations in summer temperature patterns near Barrow, Alaska: Analysis and ecological interpretation. *Arctic and Alpine Research* **11**:131–144.

Myers P and Master LL (1983). Reproduction by *Peromyscus maniculatus*: Size and compromise. *Journal of Mammalogy* **64**:1–18.

Myers P, Master LL, and Garrett RA (1985). Ambient temperature and rainfall: An effect on sex ratio and litter size in deer mice. *Journal of Mammalogy* **66**:289–298.

Myllymäki A (1977). Demographic mechanisms in the fluctuating populations of the field vole *Microtus agrestis*. *Oikos* **29**:468–493.

Myrberget A (1977). Size and shape of eggs of willow grouse *Lagopus lagopus*. *Ornis Scandinavica* **8**:39–46.

Nagahama Y (1983). The functional morphology of teleost gonads. In *Fish Physiology. Reproduction*, part A (Eds: Hoar WS, Randall DJ, and Donaldson EM). Academic, New York:223–275.

Nagasawa H and Yanai R (1971). Quantitative participation of placental mammotropic hormones in mammary development during pregnancy of mice. *Endocrinologia Japanica* **18**:507–510.

Nagasawa H and Yanai R (1976). Mammary nucleic acids and pituitary prolactin secretion during prolonged lactation in rats. *Journal of Endocrinology* **70**:389–395.

Nagle CA, Denari JH, Quiroga S, Riarte A, Merlo A, Germino NI, Gomez-Argana F, and Rosner JM (1979). The plasma pattern of ovarian steroids during the menstrual cycle in Capuchin Monkeys (*Cebus apella*). *Biology of Reproduction* **21**:979–983.

Nalbandov AV (1970). Comparative aspects of corpus luteum function. *Biology of Reproduction* **2**:7–13.

Neaves WB (1969). Adenosine deaminase phenotypes among sexual and parthenogenetic lizards in the genus *Cnemidophorus*. *Journal of Experimental Zoology* **171**:175–184.

Needham J and Lu G-D (1968). Sex hormones in the Middle Ages. *Endeavour* **27**:130–132.

Negus NC and Berger PJ (1971). Pineal weight response to a dietary variable in *Microtus montanus*. *Experientia* **27**:215–216.

Negus NC and Berger PJ (1977). Experimental triggering of reproduction in a natural population of *Microtus montanus*. *Science* **196**:1230–1231.

Negus NC, Berger PJ, and Forslund LG (1977). Reproductive strategy of *Microtus montanus*. *Journal of Mammalogy* **58**:347–353.

Negus NC and Pinter AJ (1965). Litter sizes of *Microtus montanus* in the laboratory. *Journal of Mammalogy* **46**:434–437.

Negus NC and Pinter AJ (1966). Reproductive responses of *Microtus montanus* to plants and plant extracts in the diet. *Journal of Mammalogy* **47**:596–601.

Nelson JB (1964). Factors influencing clutch-size and chick growth in the North Atlantic gannet *Sula bassana*. *Ibis* **106**:63–77.

Nelson JB (1966). Clutch size in the Sulidae. *Nature* **210**:435–436.

Nelson L (1964). Behavior and morphology in the glandulocaudine fishes (Ostariophysi,

Characidae). *University of California Publications in Zoology* **75**:59–152.

Nelson RJ and Zucker I (1981). Absence of extraocular photoreception in diurnal and nocturnal rodents exposed to direct sunlight. *Comparative Biochemistry and Physiology* **69A**: 145–148.

Nelson RJ, Dark J and Zucker I (1983). Influence of photoperiod, nutrition and water availability on reproduction of male California voles (*Microtus californicus*). *Journal of Reproduction and Fertility* **69**:473–477.

Newsome AE (1969a). A population study of house-mice temporarily inhabiting a South Australian wheatfield. *Journal of Animal Ecology* **38**:341–359.

Newsome AE (1969b). A population study of house-mice permanently inhabiting a reed-bed in South Australia. *Journal of Animal Ecology* **38**:361–377.

Newsome AE and Corbett LK (1975). Outbreaks of rodents in central Australia: Causes, preventions and evolutionary considerations. In *Rodents in Desert Environments* (Eds: Prakash I and Gosh P) Monographiae Biologicae, no. 28. Junk, The Hague:117–173.

Newton I (1976). Breeding of sparrowhawks (*Accipiter nisus*) in different environments. *Journal of Animal Ecology* **45**:831–849.

Newton I (1980). The role of food in limiting bird numbers. *Ardea* **68**:11–30.

Newton I and Marquiss M (1981). Effect of additional food on laying dates and clutch size of sparrowhawks. *Ornis Scandinavica* **12**:224–229.

Newton I and Marquiss M (1986). Population regulation in sparrowhawks. *Journal of Animal Ecology* **55**:463–480.

Newton I, Wyllie I, and Mearns R (1986). Spacing of sparrowhawks in relation to food supply. *Journal of Animal Ecology* **55**:361–370.

Nicholas KR and Tyndale-Biscoe H (1985). Prolactin-dependent accumulation of α-lactalbumin in mammary gland explants from the pregnant tammar wallaby (*Macropus eugenii*). *Journal of Endocrinology* **106**:337–342.

Nicholls TJ, Follett BK, and Evennett PS (1968). The effect of oestrogens and other steroid hormones on the ultrastructure of the liver of *Xenopus laevis* Daudin. *Zeitschrift fuer Zellforschung* **90**:19–27.

Nieuwkoop PD and Sutasurya LA (1976). Embryological evidence of a possible polyphyletic origin of the recent amphibians. *Journal of Embryology and Experimental Morphology* **35**:159–167.

Nikolics K, Mason AJ, Szonyi E, Ramachandran J, and Seeburg PH (1985). A prolactin-inhibiting factor within the precursor for human gonadotropin-releasing hormone. *Nature* **316**:511–515.

Nikolskii GV (1969). *Theory of Fish Population Dynamics as the Biological Background for Rational Exploitation and Management of Fishery Resources*. Oliver and Boyd, Edinburgh.

Nikolsky GV (1963). *The Ecology of Fishes*. Academic, New York.

Nishida T (1980). Ecological and morphological studies on the jungle fowl in southeast Asia. In *Biological Rhythms in Birds: Neural and Endocrinological Aspects* (Eds: Tanabe Y, Tanaka K, and Ookawa T) Japan Scientific Society, Tokyo, and Springer-Verlag, New York:301–313.

Norman FI and Gottsch MD (1969). Artificial twinning in the short-tailed shearwater *Puffinus tenuirostris*. *Ibis* **111**:391–393.

Norris DM and Duvall H (1981). Hormone induced ovulation in *Ambystoma tigrinum*—

influence of prolactin and thyroxin. *Journal of Experimental Zoology* **216**:175–180.

Norris DM and Platt JE (1973). Effects of pituitary hormones melatonin and thyroid inhibitors on radioiodide uptake by the thyroid glands of larval and adult tiger salamanders, *Ambystoma tigrinum*. *General and Comparative Endocrinology* **21**:386–376.

Nozaki M, Tsutsumi T, Kobayashi H, Takei Y, Ixhikawa T, Tsuneki K, Miyagawa K, Uemura H, and Tatsumi Y (1976). Spawning habit of the puffer, *Fugu niphobles* (Jordan et Snyder). I. *Zoological Magazine, Tokyo* **85**:156–168.

Nummi P (1985). The role of invertebrates in the nutrition of dabbling ducks—a review. *Suomen Riista* **32**:43–49.

Nur N (1986). Is clutch size variation in the blue tit (*Parus caeruleus*) adaptive? *Journal of Animal Ecology* **55**:983–999.

Nussbaum RA (1981). Seasonal shifts in clutch size and egg size in the side-blotched lizard, *Uta stansburiana* Baird and Girard. *Oecologia* **49**:8–13.

Nybelin O (1957). Deepsea bottom fishes. Reports of the Swedish Deep-Sea Expedition, *Zoology* **2**(20):250–345.

O'Connor RJ (1980). Pattern and process in great tit (*Parus major*) populations in Britain. *Ardea* **68**:165–183.

O'Gara BW (1978). *Antilocapra americana*. Mammalian Species. *American Society of Mammalogists* **90**:1–7.

Obbard ME and Brooks RJ (1987). Prediction of the onset of the annual nesting season of the common snapping turtle, *Chelydra serpentina*. *Herpetologica* **43**:324–328.

Oelke H (1975). Breeding behaviour and success in a colony of Adélie penguins Pygoscelis adeliae at Cape Crozier, Antarctica. In *The Biology of Penguins* (Ed: Stonehouse B). Macmillan, London:363–395.

Oftedal OT (1984). Milk composition, milk yield and energy output at peak lactation: A comparative review. In *Physiological Strategies in Lactation* (Eds: Peaker M, Vernon RG, and Knight CH) *Zoological Society of London Symposium* **51**:33–85.

Oh YK, Môri T, and Uchida TA (1983). Studies on the vaginal plug of the Japanese greater horseshoe bat, *Rhinolophus ferrumequinum nippon*. *Journal of Reproduction and Fertility* **68**:365–369.

Oh YK, Môri T, and Uchida TA (1985a). Spermiogenesis in the Japanese greater horseshoe bat, *Rhinolophus ferrumequinum nippon*. *Journal of the Faculty of Agriculture, Kyushu University* **29**:203–209.

Oh YT, Môri T, and Uchida TA (1985b). Prolonged survival of the Graffian follicle and fertilization in the Japanese greater horseshoe bat, *Rhinolophus ferrumequinum nippon*. *Journal of Reproduction and Fertility* **73**:121–126.

Ohno S (1977). Homology of X-linked genes in mammals and evolution of sex determining mechanisms. In *Reproduction and Evolution* (Eds: Calaby JH and Tyndale-Biscoe CH) Australian Academy of Science, Canberra City:49–53.

Ohno S, Stenius C, and Christian L (1966). The XO as the normal female of the creeping vole (*Microtus oregoni*). *Chromosomes Today* **1**:182–187.

Okia NO (1974). The breeding pattern of the eastern epulatted bat, *Epomophorus anurus* Heuglin, in Uganda. *Journal of Reproduction and Fertility* **37**:27–31.

Okia NO (1987). Reproductive cycles of East African bats. *Journal of Mammalogy* **68**: 138–141.

Oksche A (1965). Survey of the development and comparative morphology of the pineal organ. In *Structure and Function of the Epiphysis Cerebri* (Eds: Kappers JA and Schade JP) Elsevier, Amsterdam:3–29.

Ollason JC and Dunnet GM (1978). Age, experience and other factors affecting the breeding success of the fulmer, *Fulmarus glacialis*, in Orkney. *Journal of Animal Ecology* **47**: 961–976.

Oppenheimer JR (1970). Mouthbreeding in fishes. *Animal Behaviour* **18**:493–503.

Orell M and Ojanen M (1983). Effect of habitat, date of laying and density on clutch size of the great tit, *Parus major*, in northern Finland. *Holarctic Ecology* **6**:413–423.

Organ JA (1960). Studies on the life history of the salamander Plethodon welleri. *Copeia*:287–297.

Orgebin-Christ M and Dijiane J (1979). Properties of a prolactin receptor from the rabbit epididymis. *Biology of Reproduction* **21**:135–139.

Ortega A and Barbault R (1984). Reproductive cycles in the mesquite lizard, *Sceloporus grammicus*. *Journal of Herpetology* **18**:168–175.

Ortega A and Barbault R (1986). Reproduction in the high elevation Mexican lizard *Sceloporus scalaris*. *Journal of Herpetology* **20**:111–114.

Osborne VA (1966). An analysis of the pattern of ovulation as it occurs in the annual reproductive cycle of the mare in Australia. *Australian Veterinary Journal* **42**:149–154.

Ostermeyer MC and Elwood RW (1984). Helpers (?) at the nest in the Mongolian gerbil, *Meriones unquiculatus*. *Behaviour* **91**:61–77.

Ostfeld RS (1987). On the distinction between female defense and resource defense polygyny. *Oikos* **48**:239–240.

Ostfeld RS, Lidicker WZ Jr, and Heske EJ (1985). The relationship between habitat heterogeneity, space use, and demography in a population of California voles. *Oikos* **45**:433–442.

Overstreet JW and Cooper GW (1978). Sperm transport in the reproductive tract of the female rabbit: I. The rapid transit phase of transport. *Biology of Reproduction* **19**:101–114.

Owen DF (1977). Latitudinal gradients in clutch size: An extension of David Lack's theory. In *Evolutionary Biology* (Eds: Stonehouse B and Perrins C) Macmillan, London:171–179.

Owens DD and Owens MJ (1984). Helping behaviour in brown hyenas. *Nature* **308**:843–845.

Oxender WD, Noden PA, and Hafs HD (1977). Estrus, ovulation and serum progesterone, estradiol and LH concentrations in mares after an increased photoperiod during the winter. *American Journal of Veterinary Research* **38**:203–207.

Packard GC, Tracy CR, and Roth JJ (1977). The physiological ecology of reptilian eggs and embryos, and the evolution of viviparity within the class Reptilia. *Biological Review* **52**:71–105.

Packard MJ and Packard GC (1977). Antigonadotrophic effect of melatonin in male lizards (*Callisaurus draconoides*). *Experientia* **33**:1665.

Padykula HA and Taylor JM (1977). Uniqueness of the bandicoot chorioallantoic placenta (Marsupiala: Paramelidae). Cytological and evolutionary interpretations. In *Reproduction and Evolution* (Eds: Calaby JH and Tyndale-Biscoe CH) Australian Academy of Science, Canberra:303–324.

Panigel M (1956). Contribution à l'etude de l'ovoviparité chez reptiles: Gestation et parturition chez le lizard vivipare *Zootoca vivipara*. Annales des Sciences Naturelles. *Zoologie et Biologie Animale*, ser. 11, **18**:569–669.

Parenti LR (1986). Bilateral symmetry in phallostethid fishes (*Atherino morpha*), with description of a new species from Sarawak. *Proceedings of the California Academy of Sciences* **44**:225–236.

Parker GA (1983). Mate quality and mating decisions. In *Mate Choice* (Ed: Bateson P) Cambridge University, Cambridge:141–164.

Parker HW (1956). Viviparous caecilians and amphibian phylogeny. *Nature* **178**:250–252.

Partridge BL, Liley NR, and Stacey NE (1976). The role of pheromones in the sexual behaviour of the goldfish. *Animal Behaviour* **24**:291–299.

Partridge L and Harvey P (1986). Contentious issues in sexual selection. *Nature* **323**:580–581.

Patterson IJ, Gilboa A, and Tozar DJ (1982). Rearing others people's young: Brood-mixing in the shelduck *Tadorna tadorna*. *Animal Behaviour* **30**:199–202.

Patterson JW (1983). Frequency of reproduction, clutch size and clutch energy in the lizard *Anguis fragilis*. *Amphibia-Reptilia* **4**:195–203.

Payne RB (1973a). The breeding season of a parasitic bird, the brown-headed cowbird, in central California. *Condor* **75**:80–99.

Payne RB (1973b). Behavior, mimetic songs and song dialects and relationships of the parasitic indigobirds (*Vidua*) of Africa. *Ornithological Monographs* **11**.

Payne RB (1977). The ecology of brood parasitism in birds. *Annual Review of Ecology and Systematics* **8**:1–28.

Payne RB (1984). Sexual selection, lek and arena behavior and sexual size dimorphism in birds. *Ornithological Monographs*, no. 33, American Ornithologists' Union, Washington, DC.

Peaker M and Goode JA (1978). Milk of the fur-seal *Arctocephalus tropicalis gazella*; in particular the composition of the aqueous phase. *Journal of Zoology* **185**:469–476.

Peaker M and Maule Walker FM (1980). Mastectomy and mammary glands in reproductive control in the goat. *Nature* **284**:165–166.

Pearson AK, Tsui HW, and Licht P (1976). Effect of temperature on spermatogenesis and the production and action of androgens and on the ultrastructure of gonadotropic cells in the lizard *Anolis carolinensis*. *Journal of Experimental Zoology* **195**:291–304.

Pearson DP (1966). The prey of carnivores during one cycle of mouse abundance. *Journal of Animal Ecology* **35**:217–233.

Pearson JF (1979). Gas exchange. In *Fetal Transfer* (Eds: Chamberlain GVP and Wilkinson AW) Pitman, Kent:108–117.

Pearson OP (1944). Reproduction in the shrew (*Blarina brevicauda* Say). *American Journal of Anatomy* **75**:39–82.

Pearson OP (1949). Reproduction of a South American rodent, the mountain viscacha. *American Journal of Anatomy* **75**:39–93.

Pearson OP (1954). Habits of the lizard *Liolaemus multiformis multiformis* at high latitudes in southern Peru. *Copeia*:155–170.

Pearson OP (1971). Additional measurements of the impact of carnivores on California voles. *Journal of Mammalogy* **52**:41–49.

Pearson OP (1985). Predation. In *Biology of New World* Microtus (Ed: Tamarin RH) Special Publication of the American Society of *Mammalogists*, no. 8:535–566.

Pease JL, Vowles RH, and Keith LB (1979). Interaction of snowshoe hares and woody vegetation. *Journal of Wildlife Management* **43**:43–60.

Pedersen CA, Ascher JA, Monroe YL, and Prange AJ Jr (1982). Oxytocin induces maternal behavior in virgin female rats. *Science* **216**:648–649.

Pemble LB and Kaye PL (1986). Whole protein uptake and metabolism by mouse blastocysts. *Journal of Reproduction and Fertility* **78**:149–157.

Pengelley ET (1966). Differential developmental patterns and their adaptive value in various species of the genus *Citellus*. *Growth* **30**:137–142.

Peppler RD and Greenwald GS (1970). Effects of unilateral ovariectomy on ovulation and cycle length in 4- and 5-day cycling rats. *American Journal of Anatomy* **127**:1–8.

Perker GA (1983). Mate quality and mating decisions. In *Mate Choice* (Ed: Bateson P) Cambridge University, Cambridge:141–166.

Perret M (1986). Social influences on oestrous cycle length and plasma progesterone concentrations in the female lesser mouse lemur (*Microcebus murinus*). *Journal of Reproduction and Fertility* **77**:303–311.

Perrins CM (1965). Population fluctuations and clutch-size in the great tit, *Parus major* L. *Journal of Animal Ecology* **34**:601–647.

Perrins CM (1970). The timing of birds' breeding seasons. *Ibis* **112**:242–255.

Perrins CM (1977). The role of predation in the evolution of clutch size. In *Evolutionary Biology* (Eds: Stonehouse B and Perrins C) Macmillan, London:182–191.

Perrins CM and Geer TA (1980). The effects of sparrowhawks on tit populations. *Ardea* **68**:133–142.

Perrins CM and Jones PJ (1974). The inheritance of clutch size in the great tit (*Parus major* L.). *Condor* **76**:225–228.

Perrone M (1978). The economy of brood defense by parental cichlid fishes, *Cichlasoma maculicauda*. *Oikos* **31**:137–141.

Perry JS (1953). The reproduction of the African elephant, *Loxodonta africana*. *Philosophical Transactions of the Royal Society, Ser. B.* **237**:93–149.

Perry JS (1981). The mammalian fetal membranes. *Journal of Reproduction and Fertility* **62**:321–335.

Peter RE (1973). Neuroendocrinology of teleosts. *American Zoologist* **13**:743–755.

Peter RE (1982). Nature, localization and actions of neurohormones regulating gonadotropin secretions in teleosts. In *Proceedings of the International Symposium of the Reproductive Physiology of Fish* (Comp: Richter CJJ and Goos HJT) Centre for Agricultural Publications and Documents, Wageningen:30–39.

Peter RE (1983). The brain and neurohormones in teleost reproduction. In *Fish Physiology* (Eds: Hoar WS, Randall DJ, and Donaldson EM) Reproduction, part A. Academic, New York IX:97–135.

Peter RE and Crim LW (1979). Reproductive endocrinology of fishes: Gonadal cycles and gonadotropin in teleosts. *Annual Review of Physiology* **41**:323–335.

Peter RE and Hontela A (1978). Annual gonadal cycles in teleosts: Environmental factors and gonadotropin levels in blood. In *Environmental Endocrinology* (Eds: Assenmacher I and Farner DS) Springer-Verlag, New York:20–25.

Peterman RM and Bradford MJ (1987). Windspeed and mortality rate of a marine fish, the northern anchovy (*Engraulis mordax*). *Science* **235**:354.

Peters H (1978). Folliculogenesis in mammals. In *The Vertebrate Ovary* (Ed: Jones RE) Plenum, New York:121–144.

Peters H (1979). Some aspects of early follicular development. In *Ovarian Follicular Development and Function* (Eds: Midgley AR and Sadler WA) Raven, New York:1–13.

Peters RR, Chapin LT, Leining KB, and Tucker HA (1978). Supplemental lighting stimulates growth and lactation in cattle. *Science* **199**:911–912.

Peterson RS (1968). Social behavior in pinnipeds with particular reference to the northern fur-seal. In *The Behavior and Physiology of Pinnipeds* (Eds: Harrison RJ, Hubbard RC, Peterson RS, Rice CE, and Schusterman) Appleton-Century-Crofts, New York:3–53.

Petranka JW, Sih A, Kats LB, and Holomuzki JR (1987). Stream drift, size-specific predation and the evolution of ovum size in an amphibian. *Oecologica* **71**:624–630.

Phillips GA and Poyser NL (1981). Studies on the involvement of prostaglandins in implantation in the rat. *Journal of Reproduction and Fertility* **62**:73–81.

Phillips JA (1981). Growth and its relationship to the initial annual cycle of the golden-mantled ground squirrel, *Spermophilus lateralis*. *Canadian Journal of Zoology* **59**:865–871.

Phillips JA and Harlow HJ (1981). Elevation of upper voluntary temperatures after shielding the parietal eye of horned lizards (*Phrynosoma douglassi*). *Herpetologica* **37**:199–205.

Pianka ER (1986). *Ecology and Natural History of Desert Lizards*. Princeton University Press, Princeton.

Pickering AD (1976). Effects of gonadectomy, oestradiol and testosterone on the migrating river lamprey, *Lampetra fluviatilis* L. *General and Comparative Endocrinology* **28**: 473–480.

Pienkowski MW and Evans PR (1982). Breeding behaviour, productivity and survival of colonial and non-colonial shelducks, *Tadorna tadorna*. *Ornis Scandinavica* **13**:101–116.

Pinter AJ and Negus NC (1965). Effects of nutrition and photoperiod on reproductive physiology of *Microtus montanus*. *American Journal of Physiology* **208**:633–638.

Pitcher TJ (Ed) (1986). *The Behavior of Teleost Fishes*. Johns Hopkins University, Baltimore.

Pitelka FA, Tomich QP, and Treichel GW (1955). Ecological relations of jeagers and owls as lemming predators near Barrow, Alaska. *Ecological Monographs* **25**:85–117.

Pleszczynska WK (1978). Microgeographic prediction of polygyny in the lark bunting. *Science* **201**:935–937.

Plowman MM and Lynn WG (1973). The role of the thyroid in testicular function in the gecko, *Coleonyx variegatus*. *General and Comparative Endocrinology* **20**:342–346.

Poindron P and Le Neindre P (1980). Endocrine and sensory regulation of maternal behavior in the ewe. In *Advances in the Study of Behavior* (Eds: Rosenblatt RA, Hinde RA, Beer C, and Bushnel MC). Academic, New York **11**:75–119.

Pomeroy DE (1978). The biology of marabou storks in Uganda II. Breeding biology and general review. *Ardea* **66**:1–23.

Poole JH and Moss CJ (1981). Musth in the African elephant, *Loxodonta africana*. *Nature* **292**:830–831.

Poole WE (1960). Breeding of the wild rabbit, *Oryctolagus cuniculus* (L.) in relation to the environment. *CSIRO Wildlife Research* **5**:21–43.

Poole WE (1976). Breeding biology and current status of the grey kangaroo, *Macropus fuliginosus* (Desmarest). II. Gestation, parturition and pouch life. *Australian Journal of Zoology* **23**:333–353.

Porter DA and Fivizzani AJ (1983). Spontaneous occurrence of a synchronous hermaphrodite in the banded killifish, *Fundulus diaphanus* (Lesueur). *Journal of Fish Biology* **22**:671–675.

Porter RH and Etscorn F (1974). Olfactory imprinting resulting from brief exposure in *Acomys cahirinus*. *Nature* **250**:732–733.

Poston HA (1978). Neuroendocrine mediation of photoperiod and other experimental influences on physiological responses in salmonids. *Technical Paper, U.S. Fish and Wildlife Service* **96**:1–14.

Potts GR, Coulson JC, and Deans IR (1980). Population dynamics and breeding success of the shag, *Phalacrocorax aristotelis*, on the Farne Islands, Northumberland. *Journal of Animal Ecology* **49**:465–484.

Potts GW (1974). The colouration and its behavioural significance in the corkwing wrasse, *Crenilabrus melops*. *Journal of the Marine Biological Association of the United Kingdom* **54**:925–938.

Potts GW (1984). Parental behaviour in temperate marine teleosts with special reference to the development of nest structure. In *Fish Reproduction* (Eds: Potts GW and Wootton RJ) Academic, New York:223–244.

Poulson TL (1963). Cave adaptation in amblyopsid fishes. *American Midland Naturalist* **70**:257–290.

Poynton JC and Pritchard S (1976). Notes on the biology of *Breviceps* (Anura: Microhylidae). *Zoologica Africana* **11**:313–318.

Pressley PH (1980). Lunar periodicity in the spawning of the yellowtail damselfish, *Microspathodon chrysurus*. *Environmental Biology of Fish* **5**:153–159.

Price DJ (1984). Genetics of sex determination in fishes—a brief review. In *Fish Reproduction. Strategies and Tactics* (Eds: Potts GW and Wootton RJ) Academic, London:77–89.

Price T (1985). Reproductive responses to varying food supply in a population of Darwin's finches: Clutch size, growth rates and hatching. *Oecologia* **66**:411–416.

Priedkalns J and Bennett RK (1978). Environmental factors regulating gonadal growth in the zebra finch, *Taeniopygia guttata castanotis*. *General and Comparative Endocrinology* **34**:80–85.

Priedkalns J, Oksche A, Vleck C, and Bennett RK (1984). The response of the hypothalamo-gonadal system to environmental factors in the zebra finch, *Poephila guttata castanotis*. *Cell and Tissue Research* **238**:23–35.

Pulliainen E (1971). Clutch-size of the partridge, *Perdix perdix* L. *Ornis Scandinavica* **2**:69–73.

Purvis K and Haynes NB (1974). Short-term effects of copulation, human chorionic gonadotropin injection and non-tactile association with a female on testosterone levels in the male rat. *Journal of Endocrinology* **60**:429–439.

Qasim SZ (1956). Time and duration of the spawning season in some marine teleosts in relation to their distribution. *Journal de Conseil International pour L'exploration de la Mar* **21**:144–145.

Qasim SZ and Qayyum A (1961). Spawning frequencies and breeding seasons of some freshwater fishes with species reference to those occurring in the plains of northern India. *Indian Journal of Fish* **8**:24–43.

Qayyum A and Qasim SZ (1962). Behavior of the Indian murrel, *Ophiocephalus punctatus*, during brood care. *Copeia*:465–467.

Quay WB (1956a). Volumetric and cytological variation in the pineal body of *Peromyscus leucopus* (Rodentia) with respect to sex, captivity and daylength. *Journal of Morphology* **98**:471–495.

Quay WB (1965). Retinal and pineal hydroxindole-O-methyl transferase in vertebrates. *Life Science* **4**:983–991.

Quay WB (1969). The role of the pineal gland in environmental adaptation. In *Physiology and Pathology of Adaptation Mechanisms* (Ed: Bajusz E) Pergamon, New York:508–550.

Quay WB (1970). The significance of the pineal. In *Hormones and the Environment* (Eds: Benson GK and Phillips JG) Cambridge University, Cambridge:423–445.

Quay WB (1974). *Pineal Chemistry*. Charles C Thomas, Springfield.

Quinn TW, Quinn JS, Cooke F, and White BN (1987). DNA marker analysis detects multiple maternity and paternity in single broods of the lesser snow goose. *Nature* **326**:392–394.

Rabb GB and Rabb MS (1963). On the behavior and breeding biology of the African pipid frog, *Hymenochirus boettgeri*. *Zeitschrift fuer Tierpsychologie* **20**:215–241.

Rabb GB and Rabb MS (1963). Additional observations on breeding behavior of the Surinam toad, *Pipa pipa*. *Copeia*:636–642.

Racey PA (1972). Viability of bat spermatozoa after prolonged storage in the epididymis. *Journal of Reproduction and Fertility* **28**:309–312.

Racey PA (1979). The prolonged storage and survival of spermatozoa in Chiroptera. *Journal of Reproduction and Fertility* **56**:391–402.

Racey PA, Uchida TA, Môri T, Avery MI, and Fenton MB (1987). Sperm-epithelium relationships in relation to the time of insemination in little brown bats (*Myotis lucifugus*). *Journal of Reproduction and Fertility* **80**:445–454.

Rahn H and Ar A (1974). The avian egg: Incubation time, water loss and nest humidity. *Condor* **76**:147–152.

Rajalakshmi M and Prasad MRN (1970). Sites of formation of fructose, citric acid and sialic acid in the accessory glands of the giant fruit bat, *Pteropus giganteus giganteus* (Brunnich). *Journal of Endocrinology* **46**:413–416.

Ralph CL (1978). Pineal control of reproduction: Non-mammalian vertebrates. *Progress in Reproductive Biology* **4**:30–50.

Ralph CL, Firth BT, and Owens DW (1979). The pineal complex and thermoregulation. A review. *Biological Reviews* **54**:41–72.

Ramakrishna PA (1951). Some aspects of reproduction in the oriental vampires, *Lyroderma lyura lyra* and *Megaderma spasma* (Linn.). *Journal of Myosore University* **11**:107–118.

Ramakrishna PA and Rao KVB (1977). Reproductive adaptations in the Indian rhinolophid bat, *Rhinolophus rouxi* (Temminck). *Current Science* **41**:270–271.

Rand RW (1955). Reproduction in the female Cape fur seal. *Proceedings of the Zoological Society of London* **124**:631–658.

Raney EC (1947). Nocomis nests used by other breeding cyprinid fishes in Virginia. *Zoologica* **32**:125–132.

Raseroka BH (1975). Breeding of the Hadedah ibis. *Ostrich* **46**:208–212.

Rasweiler JJ IV (1978). Unilateral oviductal and uterus reactions in the little bulldog bat, *Noctilio albiventris*. *Biology of Reproduction* **19**:467–492.

Raud HR (1974). The regulation of ovum implantation in the rat by endogenous and exogenous FSH and prolactin: Possible role of ovarian follicles. *Biology of Reproduction* **10**:327–334.

Raychaudhuri C and Desai ID (1971). Ceroid pigment formation and irreversible sterility in vitamin E deficiency. *Science* **173**:1028–1029.

Raynaud A (1971). Foetal development of the mammary gland and hormonal effects on its morphogenesis. In *Lactation* (Ed: Falconer IR) Butterworths, London:3–29.

Raynaud JP, Mercier-Bodard C, and Baulieu EE (1971). Rat estradiol binding plasma protein. *Steroids* **18**:767–788.

Redmond RL (1986). Egg size and laying date of long-billed curlews *Numenius americanus*: Implications for female reproductive tactics. *Oikos* **46**:330–338.

Reeve JL, Kenney PA, Baxter R, and Cumming IA (1976). Effect of lupin grain, wheat and lucerne supplement on ovulation rate in maiden Border Leicester × Merino ewes. *Journal of Reproduction and Fertility* **46**:519–520.

Reid B (1965). The Adélie penguin (*Pygoscelis adeliae*) egg. *New Zealand Journal of Science* **8**:502–514.

Reilly KB and Binkley SA (1981). The menstrual rhythm. *Psychoneuroendocrinology* **6**:181–184.

Reinboth R (1967). Biandric teleost species. *General and Comparative Endocrinology* **9**:Abstract 146.

Reinboth R (Ed) (1975). *Intersexuality in the Animal Kingdom*. Springer-Verlag, Berlin.

Reiter RJ (1968). The pineal gland and gonadal development in male rats and hamsters. *Fertility and Sterility* **19**:1109–1117.

Reiter RJ (1972). Evidence for refractoriness of the pituitary-gonadal axis to the pineal gland in golden hamsters and its possible implications in annual reproductive rhythyms. *Anatomical Record* **173**:365–371.

Reiter RJ (1977). *The Pineal-1977*. Eden, Montreal.

Reiter RJ (1978). Interaction of photoperiod, pineal and seasonal reproduction as exemplified by findings in the hamster. *Progress in Reproductive Biology*, Karger, Basel **4**:169–190.

Reiter RJ (1980a). The pineal and its hormones in the control of reproduction in mammals. *Endocrine Reviews* **1**:109–131.

Reiter RJ (1980b). The pineal gland: A regulator of regulators. *Progress in Psychobiology and Physiological Psychology* **9**:323–356.

Reiter RJ (1981a). Reproductive effects of the pineal gland and pineal indoles in the Syrian hamster and the albino rat. In *The Pineal Gland. II. Reproductive Effects* (Ed: Reiter RJ) CRC, Boca Raton:45–81.

Reiter RJ (1981b). Chronobiological aspects of the mammalian pineal gland. In *Biological Rhythms in Structure and Function* (Eds: Mayersbach HV, Scheving LE, and Pauly JE) Alan R. Liss, New York:223–233.

Reiter RJ (1983). The role of light and age in determining melatonin production in the pineal gland. In *The Pineal Gland and Its Endocrine Role* (Eds: Axelrod J, Fraschini F, and Velo GP) Plenum, New York:227–241.

Reiter RJ (1986a). Pineal function in the human: Implications for reproductive physiology. *Journal of Obstetrics and Gynaecology* **6**(Supplement **2**):S77–S81.

Reiter RJ (1986b). The pineal gland and pubertal development in mammals: A state of the art assessment. In *The Pineal Gland during Development* (Eds: Gupta D and Reiter RJ) Croom-Helm, London:100–116.

Reiter RJ and Johnson LY (1974). Pineal regulation of immunoreactive luteinizing hormone and prolactin in light deprived hamsters. *Fertility and Sterility* **25**:958–969.

Reiter RJ, Blask DE, Johnson LY, Rudeen PK, Vaughan MK, and Waring PJ (1976). Melatonin inhibition of reproduction in the male hamster: Its dependency on time of day of administration and on an intact and sympathetically innervated pineal gland. *Neuroendocrinology* **22**:107–117.

Reiter RJ, Vaughan MK, Rudden PK, Vaughan GM, and Waring PJ (1984). Melatonin: Its inhibition of pineal antigonadotrophic activity in male hamsters. *Science* **185**:1169–1171.

Renfree MB (1973). Proteins in the uterine secretion of the marsupial *Macropus eugenii. Developmental Biology* **32**:41–49.

Renfree MB (1980). Embryonic diapause in the honey possum, *Tarsipes spencerae. Search (Sydney)* **11**:81.

Renfree MB (1981). Embryonic diapause in marsupials. In *Embryonic Diapause in Mammals* (Eds: Flint APF, Renfree MB, and Weir BJ) *Journal of Reproduction and Fertility*, Supplement **29**:67–78.

Reynolds HG and Turkowski F (1972). Reproductive variations in the round-tailed ground squirrel as related to winter rainfall. *Journal of Mammalogy* **53**:893–898.

Reznick D and Sexton O (1986). Annual variation of fecundity in *Sceloporus malachiticus. Journal of Herpetology* **20**:457–459.

Richardson E (1977). The biology and evolution of the reproductive cycle of *Miniopterus schreibersii* and *M. australis* (Chiroptera: Vespertilionidae). *Journal of Zoology, London* **183**:353–375.

Richter CJJ, Eding EH, Leuven FEW, and Vander Wijst GJM (1982). Effects of feeding levels and temperature on the development of the gonads in the African catfish, *Clarias lazera*. Pp 147–150. In *International Symposium on Reproductive Physiology of Fish*. Centre for Agricultural Publications and Documentations, Wageningen, Netherlands.

Richter CP (1967). Sleep and activity: Their relation to the 24-hour clock. *Proceedings of the Association for Research in Nervous and Mental Disease* **45**:8–24.

Ricklefs RE (1973). Patterns of growth in birds. II. Growth rate and mode of development. *Ibis* **115**:177–201.

Ricklefs RE (1979). Adaptive constraint and compromise in avian postnatal development. *Biological Reviews* **54**:269–290.

Ricklefs RE (1980). Geographical variation in clutch size among passerine birds: Ashmole's hypothesis. *Auk* **97**:38–49.

Ridpath MG (1972). The Tasmanian native hen, *Tribonyx mortierii*. II. The individual, the group, and the population. *CSIRO Wildlife Research* **17**:53–90.

Riedman M and Ortiz CL (1979). Changes in milk composition during lactation in the northern elephant seal. *Physiological Zoology* **52**:240–249.

Rinne JN and Wanjala T (1983). Maturity, fecundity and breeding seasons of the major catfishes (suborder: Siluroidea) in Lake Victoria, East Africa. *Journal of Fish Biology* **23**:357–363.

Roberts MG (1976). Belly soaking and chick transport in the African skimmer. *Ostrich* **47**:126.

Roberts TR (1971). Osteology of the Malaysian phallostethoid fish *Ceratostethus bicornus*, with a discussion of the evolution of the remarkable structural novelties in its jaws and esternal genitalia. *Bulletin of the Museum of Comparative Zoology* **142**:393–418.

Robertson DR and Justines G (1982). Protogynous hermaphroditism and gonochorism in four Caribbean reef gobies. *Environmental Biology of Fish* **7**:137–142.

Robertson JGM (1986). Female choice, male strategies and the role of vocalization in the Australian frog *Uperoleia rugosa. Animal Behaviour* **34**:773–784.

Robertson LAD, Chapman BM, and Chapman RF (1965). Notes on the biology of the lizards *Agama cyanogaster* and *Mabuya striata striata* collected in the Rukwa Valley, Southwest Tanganyika. *Proceedings of the Zoological Society of London* **145**:305–320.

Robinson SK (1986). Benefits, costs and determinants of dominance in a polygynous oriole. *Animal Behaviour* **34**:241–255.

Robinson TJ (1951). Reproduction in the ewe. *Biological Reviews* **26**:121–157.

Rodger JC and Young RJ (1981). Glycosidase and cumulus dispersal activities of acrosomal extracts from opossum (marsupial) and rabbit (eutherian) spermatozoa. *Gamete Research* **4**:507–514.

Rogers JG Jr and Beauchamp EK (1976). Some ecological implications of primer chemical stimuli in rodents. In *Mammalian Olfaction, Reproductive Processes and Behavior* (Ed: Doty RL) Academic, New York:181–195.

Rohwer FC (1985). The adaptive significance of clutch size in prairie ducks. *Auk* **102**:354–361.

Rokop FJ (1974). Reproductive patterns in deep sea benthos. *Science* **186**:743–745.

Romanoff AI and Romanoff AJ (1949). *The Avian Egg*. Wiley, New York.

Rood JP (1972). Ecological and behavioural comparisons of three genera of Argentine cavies. *Animal Behaviour Monograph* **5**:1–83.

Ropartz P (1976). Chemical signals in agonastic and social behavior of rodents. In *Chemical Signals in Vertebrates* (Eds: Muller-Schwarze and Mozell MM) Plenum, New York: 169–184.

Rosen DE and Gordon M (1953). Functional anatomy and evolution of male genitalia in poeciliid fishes. *Zoologica* **38**:1–47.

Rosenblatt JS (1978). Behavioral regulation of reproductive physiology: A selected review. In *Comparative Endocrinology* (Eds: Gaillard PJ and Boer HH) Elsevier/North Holland, Amsterdam:177–188.

Rosenblatt JS and Siegel HI (1980). Maternal behavior in the laboratory rat. In *Maternal Influences on Maternal Behavior* (Eds: Smotherman WP and Bell RW) Spectrum, New York:155–199.

Rosenblatt JS and Siegel HI (1981). Factors governing the onset and maintenance of maternal behavior among nonprimate mammals. In *Parental Care in Mammals* (Eds: Gubernick DJ and Klopfer PH) Plenum, New York: 13–76.

Rosenblatt JS and Siegel HI (1983). Physiological and behavioural changes during pregnancy and parturition underlying the onset of maternal behaviour in rodents. In *Parental Behaviour of Rodents* (Ed: Elwood RW) Wiley, Chichester.

Rosenzweig ML and Abramsky Z (1980). Microtine cycles: The role of habitat heterogeneity. *Oikos* **34**:141–146.

Roskaft E (1983). Male promiscuity and female adultery by the rook *Corvus frugilegus*. *Ornis Scandinavica* **14**:175–179.

Rossiter RC and Bedk AB (1966). Physiological and ecological studies on the oestrogenic isoflavones in subterranean clover (*T. subterraneum* L.). II. Effects of phosphate supply. *Australian Journal of Agricultural Research* **17**:447–456.

Roth L and Rosenblatt JS (1968). Self-licking and mammary development during pregnancy in the rat. *Journal of Endocrinology* **42**:363–378.

Rounsefell GA (1957). Fecundity of North American Salmonidae. *U.S. Fish and Wildlife Service 75 Fishery Bulletin* **122**.

Rowell TE (1974). The concept of social dominance. *Behavioral Biology* **11**:131–154.

Rüdeberg C (1968). Receptor cells in the pineal organ of the dogfish Scyliorhinus canicula Linné. *Zeitschrift fuer Zellforschung* **95**:521–526.

Rüdeberg C (1969). Light and electron microscope studies on the pineal organ of the dogfish, *Scyliorhinus canicula* L. *Zeitschrift fuer Zellforschung* **96**:548–581.

Rusak B and Zucker I (1979). Neural regulation of circadian rhythms. *Physiological Reviews* **59**:449–526.

Russell MJ (1976). Human olfactory communication. *Nature* **260**:520–522.

Russell WA, Guthrie WD, Klun JA, and Grindeland R (1975). Selection for resistance in maize to first-brood European corn borer by using leaf-feeding damage of the insect and chemical analysis for DIMBOA in the plant. *Journal of Economic Entomology* **68**:31–34.

Ryan MJ and Wagner WE (1987). Asymmetries in mating preferences between species: Female swordtails prefer heterospecific males. *Science* **236**:595–597.

Sadleir RMFS (1969). The role of nutrition in the reproduction of wild mammals. *Journal of Reproduction and Fertility*, Supplement **6**:39–48.

Sadleir RMFS (1980). Milk yield of black-tailed deer. *Journal of Wildlife Management* **44**:472–478.

Saether B-E (1985). Variation in reproductive traits in European passerines in relation to nesting site: Allometric scaling to body weights or adaptive variation. *Oecologia* **68**:7–9.

Saether B-E (1987). The influence of body weight on the covariation between reproductive traits in European birds. *Oikos* **48**:79–88.

Saint Girons H and Kramer E (1973). Le cycle sexuel chez Vipera berus (L.) en montagne. *Revue Suisse de Zoologie* **70**:191–221.

Saint Girons H and Pfeffer P (1971). Le cycle sexuel des serpentes du Cambodge. *Annales des Sciences Naturelles, Zoologie, Paris* **13**:543–572.

Sale JB (1969). Breeding season and litter size in Hyracoidea. *Journal of Reproduction and Fertility*, Supplement **6**:249–263.

Sale PF (1977). Maintenance of high diversity in coral reef fish communities. *American Naturalist* **111**:337–359.

Salthe SN (1967). Courtship patterns and phylogeny of the urodeles. *Copeia*:100–117.

Salthe SN (1969). Reproductive modes and numbers and sizes of ova in the urodeles. *American Midland Naturalist* **81**:467–490.

Salthe SN and Duellman WE (1973). Quantitative constraints associated with reproductive mode in anurans. In *Evolutionary Biology of the Anurans* (Ed: Vial JL) University of Missouri, Columbia:229–249.

Salyer A (1966). The reproductive ecology of the red-backed salamander, *Plethodon cinereus*, in Maryland. *Copeia*:183–193.

Sandell M (1984). To have or not to have delayed implantation: The example of the weasel and the stoat. *Oikos* **42**:123–126.

Sanders EH, Gardner PD, Berger PJ, and Negus NC (1981). 6-Methoxybenzoazolinone: A plant derivative that stimulates reproduction in *Microtus montanus*. *Science* **214**:67–69.

Sansum EL and King JR (1976). Long-term effects on constant photoperiods on testicular cycles of white-crowned sparrows (*Zonotrichia leucophrys gambelii*). *Physiological Zoology* **49**:407–416.

Sargent RC and Gross MR (1986). William's principle: An explanation of parental care in teleost fishes. In *The Behaviour of Teleost Fishes* (Ed: Pitcher TJ) Croom Helm, London and Sydney:275–293.

Sarnthein M (1978). Sand deserts during glacial maximum and climatic optimum. *Nature* **272**:43–46.

Sato T (1986). A brood parasite catfish of mouthbreeding cichlid fishes in Lake Tanganyika. *Nature* **323**:58–59.

Sawara Y (1974). Reproduction in the mosquito fish Gambusia affinis; a freshwater fish introduced into Japan. *Japanese Journal of Ecology* **24**:140–146.

Scaglion R (1978). Seasonal births in a western Abelam village, Papua New Guinea. *Human Biology* **50**:313–323.

Schlernitzauer DA and Gilbert PW (1966). Placentation and associated aspects of gestation in the bonnethead shark, *Sphyrna tiburo*. *Journal of Morphology* **120**:219–231.

Schlindler JF and de Vries U (1986). Ultrastructure of embryonic and processes in *Girardinichthys viviparus* (Cyprinodontiformes, Ostichthyes). *Journal of Morphology* **188**:203–224.

Schmalhausen II (1968). *The Origin of Terrestrial Vertebrates*. Academic, New York.

Schoener A (1968). Evidence for reproductive periodicity in the deep sea. *Ecology* **49**:81–87.

Schrader N (1975). Emu incubating paddy melons. *Emu* **75**:43.

Schreiber RW and Ashmole NP (1970). Sea-bird breeding seasons on Christmas Island, Pacific Ocean. *Ibis* **112**:363–394.

Schreiber RW and Schreiber EA (1984). Central pacific seabirds and the El Nĩno Southern Oscillation: 1982 to 1983 perspectives. *Science* **225**:713–716.

Schreiber RW and Schreiber EA (1986). Unusual seabird breeding parameters on Christmas Island. *Climate Diagnostics Bulletin no. 86/***4**:7–8.

Schreibman MP and Holtzman S (1975). The histophysiology of the prolactin cell in non-mammalian vertebrates. *American Zoologist* **15**:867–880.

Schroeder CR and Wesdgeforth HM (1935). The occurrence of gastric ulcers in sea mammals of the California coast, their aetiology and pathology. *Journal of the American Veterinary Association* **87**:333–342.

Schuett GW and Gillingham JC (1986). Sperm storage and multiple paternity in the copperhead, *Agkistrodon contortrix*. *Copeia*:807–811.

Schuetz AW (1972). Induction of structural alterations in the preovulatory amphibian ovarian follicle by hormones. *Biology of Reproduction* **6**:67–77.

Schultz RJ (1971). Special adaptive problems associated with unisexual fishes. *American Zoologist* **11**:351–360.

Schwassman HO (1978). Times of annual spawning and reproductive strategies in Amazonian fishes. In *Rhythmic Activity of Fishes* (Ed: Thorpe JE) Academic, London:187–200.

Schwassmann HO (1980). Biological rhythms: Their adaptive significance. In *Environmental Physiology of Fishes* (Ed: Ali MM) Plenum, New York:613–630.

Scott AP and Sumpter JP (1983). A comparison of the female reproductive cycles of autumn-spawning and winter-spawning strains of rainbow trout (*Salmo gairdneri* Richardson). *General and Comparative Endocrinology* **52**:79–85.

Scott AP, Sumpter JP, and Hardiman PA (1983). Hormone changes during ovulation in the rainbow trout (*Salmo gairdneri* R.). *General and Comparative Endocrinology* **49**:128–134.

Scott DBC (1979). Environmental timing and control of reproduction in teleost fish. In *Fish Phenology: Anabolic Adaptiveness in Teleosts* (Ed: Miller PJ) Zoological Society of London Symposium 44. Academic, London:105–132.

Scott TM (1973). Lipid metabolism of spermatozoa. In *Research Related to the Control of Fertility* (Eds: Bindon BM, Emmens CW, and Smith MSR) *Journal of Reproduction and Fertility*, Supplement **18**:65–76.

Scrimshaw NS (1945). Embryonic development in poeciliid fishes. *Biological Bulletin* **88**:233–246.

Seel DC (1970). Nestling survival and nestling weights in the house sparrow and tree sparrow *Passer* spp. at Oxford. *Ibis* **112**:1–14.

Seigel RA and Fitch HS (1985). Annual variation in reproduction in snakes in a fluctuating environment. *Journal of Animal Ecology* **54**:497–505.

Selwood L (1980). A timetable of embryonic development of the dasyurid marsupial *Antechinus stuartii* (Maclean). *Australia Journal of Zoology* **28**:650–662.

Selwood L (1981). Delayed embryonic development in the dasyurid marsupial, *Antechinus stuartii*. In *Embryonic Diapause in Mammals* (Eds: Flint APF, Renfree MB, and Weir BJ) *Journal of Reproduction and Fertility*, Supplement **29**:79–82.

Selwood L (1983). Factors influencing pre-natal fertility in the brown marsupial mouse, *Antechinus stuartii*. *Journal of Reproduction and Fertility* **68**:317–324.

Sempére A (1978). The annual cycle of plasma testosterone and territorial behavior in the roe deer. In *Environmental Physiology* (Eds: Assenmacher I and Farner DS) Springer-Verlag, Berlin:73–74.

Sempére AJ and Boissin J (1981). Relationship between antler development and plasma androgen concentrations in adult roe deer (*Capreolus capreolus*). *Journal of Reproduction and Fertility* **62**:49–53.

Sergeant DE (1973). Environment and reproduction in seals. *Journal of Reproduction and Fertility*, Supplement **19**:555–561.

Serra GB (Ed) (1983). *The Ovary. Comprehensive Endocrinology*. Raven, New York.

Serventy DL (1971). Biology of desert birds. In *Avian Biology* (Eds: Farner DS and King JR) Academic, New York:287–339.

Sessions SK (1982). Cytogenetics of diploid and triploid salamanders of the *Ambystoma jeffersonianum* complex. *Chromosoma* **84**:599–621.

Shann EW (1923). The embryonic development of the porbeagle shark, *Lamna cornubica*. *Proceedings of the Zoological Society of London* **11**:161–171.

Shapiro AM (1986). r and K selection at various taxonomic levels in the pierine butterflies of North and South America. In *Evolution of Insect Life Histories* (Eds: Taylor F and Karban R) Springer-Verlag, New York:135–152.

Shapiro DY (1984). Sex reversal and sociodemographic processes in coral reef fishes. In *Fish Reproduction: Strategies and Tactics*. (Eds: Potts GW and Wootton RJ). Academic, London:103–118.

Sharman GB (1970). Reproductive physiology of marsupials. *Science* **167**:1221–1228.

Sharp PJ (1981). Female reproduction. In *Avian Endocrinology* (Eds: Epple A and Stetson MH) Academic, New York:435–454.

Sharp PJ and Moss R (1977). The effects of castration on concentrations of luteinizing hormone in the plasma of photorefractory red grouse (*Lagopus lagopus scoticus*). *General and Comparative Endocrinology* **32**:289–293.

Sharpe RM (1982). Cellular aspects of the inhibitory actions of LH-RH on the ovary and testis. *Journal of Reproduction and Fertility* **64**:517–527.

Sharpe RM (1982). The hormonal regulation of the Leydig cell. In *Oxford Reviews of Reproductive Biology*, vol 4 (Ed: Finn CA) Oxford University, Oxford:241–312.

Sharpe RM (1984). Intratesticular factors controlling testicular function. *Biology of Reproduction* **30**:29–49.

Sharpe RM, Fraser HM, Cooper I, and Rommerts FFG (1982). Sertoli-Leydig cell communication via an LHRH-like factor. *Annals of the New York Academy of Sciences* **383**:272–294.

Shelton WL (1982). Monosex grass carp production through breeding sex-reversed broodstock. In *Proceedings of the International Symposium on Reproduction and Physiology of Fish* (Comp: Richter CJJ and Goos HJT) Centre for Agricultural Publishing Document, Wageningen:82.

Shen SC, Lin RP, and Liu CC (1979). Redescription of a protandrous hermaphroditic moray (*Rhinomuraena quaesita* Garman). *Bulletin of the Institute of Zoology, Academia Sinica* **18**:79–87.

Sherman HB (1937). Breeding habits of the free-tailed bat. *Journal of Mammalogy* **18**:176–187.

Sherry DF (1981). Parental care and the development of parental care in red junglefowl. *Behaviour* **76**:250–279.

Sherwood OD and Downing SJ (1983). The chemistry and physiology of relaxin. In *Factors Regulating Ovarian Function* (Eds: Greenwald GS and Terranova PF) Raven, New York:381–410.

Shille VM, Munro C, Farmer SW, Papkoff H, and Stabenfeldt G (1983). Ovarian and endocrine responses in the cat after coitus. *Journal of Reproduction and Fertility* **68**: 29–39.

Shimada K and Asai I (1979). Effects of prostaglandin F2a and indomethacin on uterine contraction in hens. *Biology of Reproduction* **21**:523–527.

Shine R (1980). "Costs" of reproduction in reptiles. *Oecologia* **46**:92–100.

Shine R (1983a). Reptilian reproductive modes: The oviparity-viviparity continuum. *Herpetologica* **39**:1–8.

Shine R (1983b). Reptilian viviparity in cold climates: Testing the assumptions of an evolutionary hypothesis. *Oecologia* **57**:397–405.

Shine R (1985). The evolution of viviparity in reptiles. In *Biology of the Reptilia*, vol 15 (Eds: Gans C and Billet F) Wiley, New York:604–694.

Shine R (1987a). Ecological ramifications of prey size: Food habits and reproductive biology of Australian copperhead snakes (*Austrelaps*, Elapidae). *Journal of Herpetology* **21**: 21–28.

Shine R (1987b). Reproductive mode may determine geographic distributions in Australian venomous snakes (*Pseudechis*, Elapidae). *Oecologia* **71**:608–612.

Shine R and Berry JF (1978). Climatic correlates of live-bearing in squamate reptiles. *Oecologia* **33**:261–268.

Shine R and Bull JJ (1979). The evolution of live-bearing in lizards and snakes. *American Naturalist* **113**:905–923.

Shingran VG (1976). Systems of polyculture of fishes in inland waters of India. *Journal of Fisheries Research* **33**:905–910.

Shivers CA and James JM (1971). Fertilization of antiserum-inhibited frog eggs with capacitated sperm. *Biology of Reproduction* **5**:229–235.

Short RV (1979). Sexual behavior in the red deer. In *Animal Reproduction* (Ed: Hawk HW) Beltsville Symposia in Agricultural Research 3, Allanheld, Osum, Montclair:365–372.

Short RV, Mann T, and Hay MF (1967). Male reproductive organs of the African elephant *Loxodonta africana*. *Journal of Reproduction and Fertility* **11**:517–536.

Shutt DA (1976). The effects of plant estrogens on animal reproduction. *Endeavour* **35**: 110–113.

Shutt DA and Cox RI (1972). Steroid and phyto-oestrogen binding to sheep uterine receptors in vitro. *Journal of Endocrinology* **52**:299–310.

Siegel HI and Greenwald GS (1975). Prepartum onset of maternal behavior in hamsters and the effects of estrogen and progesterone. *Hormones and Behavior* **6**:237–245.

Silver R (1983). Biparental care in birds. In *Hormones and Behaviour in Higher Vertebrates* (Eds: Balthazasrt J, Prove E, and Gilles R) Springer-Verlag, Berlin:451–462.

Silver R, Reboulleau C, Lehrman DS, and Feder AH (1974). Radioimmunoassay of plasma progesterone throughout the reproductive cycle in the male and female ring dove (*Streptopelia risoria*). *Endocrinology* **94**:437–444.

Simbotwe MP (1980). Reproductive biology of the skinks *Mabuya striata* and *Mabuya quinquetaeniata* in Zambia. *Herpetologica* **36**:99–104.

Simmonet H, Thieblot L, Melik T, and Segal V (1954). Nouvelles preuves de l'endocrinie epiphysaire. *Acta Endocrinologica* **17**:402–413.

Simon MP (1983). The ecology of parental care in a terrestrial breeding frog from New Guinea. *Behavioral Ecology and Sociobiology* **14**:61–67.

Simpson BRC (1979). The phenology of annual killifishes. In *Fish Phenology: Anabolic Adaptiveness in Teleosts* (Ed: Miller JP) Proceedings of the Zoological Society of London, no. 44. Academic, London:243–261.

Simpson E (1982). Sex reversal and sex determination. *Nature* **300**:404–406.

Simpson E, Chandler P, Goulmy E, Disteche CM, Ferguson-Smith MA, and Page DC (1987). Separation of the genetic loci for the H-Y antigen and for testis formation on the human Y chromosome. *Nature* **326**:876–878.

Simpson SM, Follett BK, and Ellis DH (1982). Modulation by photoperiod of gonadotropin secretion in intact and castrated Djungarian hamsters. *Journal of Reproduction and Fertility* **66**:243–250.

Sinclair ARE (1978). Factors affecting the food supply and breeding season of resident birds and movements of Palearctic migrants in a tropical African savannah. *Ibis* **120**:480–497.

Singh L and Jones KW (1982). Sex reversal in the mouse (*Mus musculus*) is caused by a recurrent nonreciprocal crossover involving the X and an aberrant Y chromosome. *Cell* **28**:205–216.

Sivertsen E (1941). On the biology of the harp seal (*Phoca groenlandica*). *Hvalredets Skrifter* **16**:1–166.

Skead DM (1975). Ecological studies of four estrildines in the central Transvaal. *Ostrich*, Supplement **11**:1–55.

Skinner JD, Nel JAJ, and Millar RP (1977). Evolution of time of parturition and differing litter sizes as an adaptation to changes in environmental conditions. In *Reproduction and Evolution* (Eds: Calaby JH and Tyndale-Biscoe CH) Australian Academy of Sciences, Canberra:39–44.

Skogland T (1985). The effects of density-dependent resource limitations on the demography of wild reindeer. *Journal of Animal Ecology* **54**:359–374.

Skryja DD (1978). Reproductive inhibition in female cactus mice (*Peromyscus eremicus*). *Journal of Mammalogy* **59**:543–550.

Slagsvold T (1976). Annual and geographical variation in the time of breeding of the great tit *Parus major* and the pied flycatcher *Ficedula hypoleuca* in relation to environmental phenology and spring temperature. *Ornis Scandinavica* **7**:127–145.

Slagsvold T (1981). Clutch size and population stability in birds: A test of hypotheses. *Oecologia* **49**:213–217.

Slobodkin LB (1961). *Growth and Regulation of Animal Populations*. Holt, Rinehart & Winston, New York.

Slotin CA, Heap RB, Christia JM, and Linzell JL (1970). Synthesis of progesterone by the mammary gland of the goat. *Nature* **225**:385.

Smith HD and Jørgensen CD (1975). Reproductive biology of North American desert rodents. In *Rodents in Desert Environments* (Eds: Prakash I and Ghosh PK) Junk, The Hague:305–330.

Smith HM, Sinelnik G, Fawcett JD, and Jones RE (1972). A survey of the chronology of ovulation in anoline lizard genera. *Transactions of the Kansas Academy of Science* **75**:107–120.

Smith JM (1977). Parental investment: A prospective analysis. *Animal Behaviour* **25**:1–9.

Smith KL Jr and Baldwin RJ (1984). Seasonal fluctuations in deep-sea sediment community oxygen consumption: Central and eastern North Pacific. *Nature* **307**:624–626.

Smith MJ (1981). Morphological observations on the diapausing blastocyst of some macropod marsupials. *Journal of Reproduction and Fertility* **61**:483–486.

Smith MS, Freeman ME, and Neill JD (1975). The control of progesterone secretion during the estrous cycle and early pseudopregnancy in the rat: Prolactin, gonadotropin secretion and steroid levels associated with the rescue of the corpus luteum of pseudopregnancy. *Endocrinology* **96**:219–226.

Smith RE (1968). Studies on reproduction in Costa Rican *Ameiva festiva* and *Ameiva quadrilineata* (Sauria: Teiidae). *Copeia*:236–239.

Smyth M (1966). Winter breeding in woodland mice, *Apodemus sylvaticus*, and voles, *Clethrionomys glareolus* and *Microtus agrestis*, near Oxford. *Journal of Animal Ecology* **35**:471.

Soares MJ and Talamantes F (1983). Genetic and litter size effects of serum placental lactogen in the mouse. *Biology of Reproduction* **29**:165–171.

Soloff MS, Alexandrova M, and Fernstrom MJ (1979). Oxytocin receptors: Triggers for parturition and lactation? *Science* **204**:1313–1315.

Solomon ME (1949). The natural control of animal populations. *Journal of Animal Ecology* **18**:1–35.

Soppela P and Nieminen M (1985). Thermoregulation in Cervidae. *Suomen Riista* **32**:32–42.

Southern HN (1954). Mimicry in cuckoos. In *Evolution as a Process* (Eds: Huxley J, Hardy AC, and Ford EB) George Allen and Unwin, London 219–232.

Southern HN (1970a). Prey taken by tawny owls during the breeding seasons. *Ibis* **111**:293–299.

Southern HN (1970b). The natural control of a population of tawny owls (*Strix aluco*). *Journal of Zoology* **162**:197–204.

Spears N and Clarke JR (1986). Effect of male presence and of photoperiod on the sexual maturation of the field vole (*Microtus agrestis*). *Journal of Reproduction and Fertility* **78**:231–238.

Spoor WA (1977). Oxygen requirements of embryos and larvae of the largemouth bass, *Micropterus salmoides* (Lacépède). *Journal of Fish Biology* **11**:77–86.

Spoor WA (1984). Oxygen requirement of larvae of the smallmouth bass, *Micropterus dolomieui* Lacépède. *Journal of Fish Biology* **25**:587–592.

Springer S (1948). Oviphagous embryos of the sand shark, *Carcharias taurus*. *Copeia*: 153–157.

Srebnik HH, Fletcher WH, and Campbell GA (1978). Neuroendocrine aspects of reproduction in experimental malnutrition. In *Environmental Endocrinology* (Eds:Assenmacher I and Farner DS) Springer-Verlag, New York:306–312.

Stacey N (1983a). Hormones and pheromones in fish sexual behavior. *BioScience* **33**:552–556.

Stacey NE (1983b). Hormones and reproductive behavior in teleosts. In *Control Processes in Fish Physiology* (Eds: Rankin JC, Pitcher TJ, and Duggan RT) Croom Helm, London and Canberra:117–129.

Stacey NE (1984a). Control of the timing of ovulation by exogenous and endogenous factors. In *Fish Reproduction: Strategies and Tactics* (Eds: Potts GW and Wootton RJ) Academic, London:207–222.

Stacey NE (1984b). Clonidine inhibits female spawning behavior in ovulated and prostaglandin-treated goldfish. *Pharmacology, Biochemistry and Behavior* **20**:887–891.

Stacey NE (1987). Roles of hormones and pheromones in fish reproductive behavior. In *Psychobiology of Reproductive Behavior* (Ed: Crews D) Prentice Hall, Englewood Cliffs, New Jersey:29–60.

Stacey NE, Cook AF, and Peter RE (1979). Ovulatory surge of gonadotropin in the goldfish, *Carassius auratus*. *General and Comparative Endocrinology* **37**:246–249.

Stacey NE and Goetz FW (1982). Role of prostaglandins in fish reproduction. *Canadian Journal of Fisheries and Aquatic Sciences* **39**:92–98.

Stacey NE and Kyle AL (1983). Effects of olfactory tract lesions on sexual and feeding behavior in the goldfish. *Physiology and Behavior* **30**:621–628.

Stacey NE and Liley NR (1974). Regulation of spawning behaviour in the female goldfish. *Nature* **247**:71–72.

Stamps J (1976). Egg retention, rainfall and egg laying in a tropical lizard, *Anolis aeneus*. *Copeia*:759–764.

Stamps JA and Tanaka S (1981). The relationship between food and social behavior in juvenile lizards (*Anolis aeneus*). *Copeia*:422–434.

Stearns SC (1976). Life-history tactics: A review of the ideas. *Quarterly Review of Biology* **51**:1–47.

Steel E and Hinde RA (1972). Influence of photoperiod on oestrogenic induction of nest-building in canaries. *Journal of Endocrinology* **55**:265–278.

Steele BD and Pearson WD (1981). Commensal spawning behavior of the rosefin shiner, *Notropis ardens*, in nests of the longear sunfish, *Lepomis megalotis*. *Environmental Biology of Fish* **6**:138.

Steger RW, Bartke A, Goldman BD, Soares MJ, and Talamantes F (1983). Effects of short photoperiod on the ability of golden hamster pituitaries to secrete prolactin and gonadotropins in vitro. *Biology of Reproduction* **29**:872–878.

Stein DL (1980). Aspects of reproduction of liparid fishes from the continental slope and abyssal plain off Oregon, with notes on growth. *Copeia*:687–699.

Steinberger A (1971). Inhibin production by Sertoli cells in culture. *Journal of Reproduction and Fertility*, Supplement **26**:31–45.

Steinberger A (1981). Regulation of inhibition secretion in the testis. In *Intragonadal Regulation of Reproduction* (Eds: Franchimont P and Channing CP) Academic, London:283–298.

Steinberger A (1982). Testicular inhibin: A cellular source, regulation of production, mechanism of action and physico-chemical characteristics. In *Non-steroidal Regulators in Reproductive Biology and Medicine*, vol 34:109–121, (Eds: Fujii T and Channing CP) Pergamon, New York:109–119.

Steinberger E, Browning JY, and Grotjan Jr HE (1983). Another look at steroidogenesis in testicular cells. In *Recent Advances in Male Reproduction: Molecular Basis and Clinical Implications* (Eds: D'Agata RD, Pilsett MB, Polosa P and van der Molen HJ) Raven, New York:113–120.

Steinberger A and Steinberger E (1976). Secretion of an FSH-inhibiting factor by cultured Sertoli cells. *Endocrinology* **99**:918–921.

Stephens JS Jr, Hobson ES, and Johnson RK (1966). Notes on distribution, behavior, and morphological variation in some chaenopsid fishes from the tropical eastern Pacific, with descriptions of two new species, *Acathemblemaria castroi* and *Coralliozetus springeri*. *Copeia*:424–438.

Stern JM and Lehrman DS (1969). Role of testosterone in progesterone-induced behaviour in male ring doves (*Streptopelia risoria*). *Journal of Endocrinology* **44**:13–22.

Stern JM and MacKinnon DA (1978). Sensory regulation of maternal behavior in rats: Effects of pup age. *Developmental Psychobiology* **11**:579–586.

Stewart F (1984). Mammogenesis and changing prolactin receptor concentrations in the mammary glands of the tammar wallaby (*Macropus eugenii*). *Journal of Reproduction and Fertility* **71**:141–148.

Stewart JR and Duvall D (1982). Ovariectomy fails to block egg brooding behavior in female five-lined skinks (*Eumeces fasciatus*). *Copeia*:777–779.

Stewart REA and Lavigne DM (1980). Neonatal growth of northwest Atlantic harp seals, *Pagophilus groenlandicus*. *Journal of Mammalogy* **61**:670–680.

Stewart REA and Lavigne DM (1984). Energy transfer and female condition in nursing harp seals *Phoca groenlandica. Holarctic Ecology* **7**:182–194.

Stille B, Madsen T, and Niklasson M (1986). Multiple paternity in the adder, *Vipera berus. Oikos* **47**:173–175.

Stinson CH (1977). Growth and behaviour of young ospreys *Pandion haliaetus. Oikos* **28**:299–303.

Stoddart DM (1980). *The Ecology of Vertebrate Olfaction*. Chapman and Hall, London and New York.

Stoddart DM and Braithwaite RW (1979). A strategy for utilization of regenerating heathland habitat by the brown bandicoot (*Isoodon obesulus*; Marsupialia, Peramelidae). *Journal of Animal Ecology* **48**:165–179.

Stokkan KA and Sharp PJ (1980). The roles of daylength and the testes in the regulation of plasma LH levels in photosensitive and photorefractory willow ptarmigan (*Lagopus lagopus lagopus*). *General and Comparative Endocrinology* **41**:520–526.

Stonehouse B and Gilmore D (Eds) (1977). *The Biology of Marsupials*. Macmillan, London.

Stonham MD, Everitt BJ, Hanson S, Lightman SL, and Todd K (1985). Oxytocin and sexual behaviour in the male rat and rabbit. *Journal of Endocrinology* **107**:97–106.

Stormshak F (1979). Uterine estrogen and progesterone receptors. In *Animal Reproduction* (Ed: Hawk HW) Beltsville Symposia in Agricultural Research 3 Allenheld, Osmun. Montclair:399–410.

Storrs EE and Williams RJ (1968). A study of monozygous quadrupelet armadillos in relation to inheritance. *Proceedings of the U.S. National Academy of Science* **60**:610.

Straughan IR and Main AR (1966). Speciation and polymorphism in the genus *Crinia* Tsuchdi (Anura, Leptodactylidae) in Queensland. *Proceedings of the Royal Society of Queensland* 78:11–28.

Stull JW, Brown WH, and Kooyman GL (1967). Lipids of the Weddell seal, *Leptonychotes weddelli. Journal of Mammalogy* **48**:642–645.

Sturkie PD (1976). *Avian Physiology*. Springer-Verlag, New York.

Stutterheim CJ (1982). Timing of the breeding of the redbilled oxpecker (*Buphagus erythrorhynchus*) in the Kruger National Park. *Zoologica Africana* **17**:126–129.

Sullivan TP and Sullivan DS (1982). Population dynamics and regulation of the Douglas squirrel (*Tamiasciurus douglasii*) with supplemental food. *Oecologia* **53**:264–270.

Sumpter PJ, Follett BK, and Dodd JM (1978). Studies on the purification of gonadotropin from dogfish (*Scyliorhinus canicula* L.) pituitary glands. *Annales de Biologie Animale, Biochimie et Biophysique* **18**:787–791.

Sundararaj BI (1978). Environmental regulation of annual reproductive cycles in the catfish (Heteroptneustes fossilis). In *Environmental Endocrinology* (Eds: Assenmacher I and Farner DS) Springer-Verlag, New York:26–27.

Sundararaj BI (1981). *Reproductive Physiology of Teleost Fishes*. UN Development and Progress, Rome.

Sundararaj BI, Panchanan N, and Jeet V (1978). Role of circadian and circannual rhythms in the regulation of ovarian cycles in fishes: A catfish model. In *Comparative Endocrinology* (Eds: Gaillard PJ and Boer UU) Elsevier/North Holland, Amsterdam:137–140.

Sutherland WJ, Grafen A, and Harvey P (1986). Life history correlations and demography. *Nature* **320**:88.

Svare B and Gandelman R (1976). Postpartum aggression in mice: The influence of suckling stimulation. *Hormones and Behavior* **7**:407–416.

Swanson GA, Krapus GL, and Serie JR (1979). Foods of laying female dabbling ducks on the breeding grounds. In *Waterfowl and Wetlands—An Integreted Review* (Ed: Bookout TA) The Wildlife Society, Madison, Wisconsin:47–57.

Swingland IR and Lessels CM (1979). The natural regulation of giant tortoise populations on Aldabra atoll. Movement polymorphism, reproductive success and mortality. *Journal of Animal Ecology* **48**:639–654.

Takahashi H (1975). Process of functional sex reversal of the gonad in the female guppy, *Poecilia reticulata*, treated with androgen before birth. *Development, Growth and Differentiation* **17**:167–175.

Takahashi JS, Norris C, and Menaker M (1978). Circadian periodic regulation of testis growth in the house sparrow. In *Comparative Endocrinology* (Eds: Gaillard PJ and Boer HH) Elsevier/North Holland, Amsterdam:153–160.

Talamantes F (1975). Comparative study of the occurrence of placental prolactin among mammals. *General and Comparative Endocrinology* **27**:115–121.

Talbot FH (1960). Notes on the biology of the Lutjanidae (Pisces) of the east African coast, with species reference to *L. bohar* (Forskål). *Annals of the South African Museum* **45**:549–573.

Talent LG, Krapu GL, and Jarvis RL (1982). Habitat use by mallard broods in south-central North Dakota. *Journal of Wildlife Management* **46**:629–635.

Tamarkin L, Baird CJ, and Almeida OFX (1985). Melatonin: A coordinating signal for mammalian reproduction? *Science* **227**:714–720.

Tamarkin L, Westrom WK, Hamill AI, and Goldman BD (1976). Effect of melatonin on the reproductive systems of male and female Syrian hamsters: A diurnal rhythm in sensitivity to melatonin. *Endocrinology* **99**:1534–1541.

Tamsitt JR and Valdivieso D (1963). Reproductive cycle of the big fruit-eating bat, *Artibeus lituratus* Olfers. *Nature* **198**:104.

Tamura T and Hanyu I (1980). Pineal sensitivity in fishes. In *Environmental Physiology of Fishes* (Ed: Ali MA) Plenum, New York:533–567.

Tasker CR and Mills JA (1981). A functional analysis of courtship feeding in the red-billed gull, *Larus novaehollandiae scopulinus*. *Behaviour* **77**:222–241.

Tatsukawa K and Murakami O (1976). On the food utilization of the Japanese wood mouse, *Apodemus speciosus* (Mammalia: Muridae). *Physiological Ecology of Japan* **17**:133–144.

Taubert BD (1980). Reproduction of shortnosed sturgeon (*Acipenser brevirostrum*) in Holyoke Pool, Connecticut River, Massachusetts. *Copeia*:114–117.

Taurog A (1974). The effect of TSH and long acting thyroid stimulator on the thyroid 131I metabolism and metamorphosis in the Mexican Axolotl (*Ambystoma mexicanum*). *General and Comparative Endocrinology* **24**:267–266.

Taverne MAM, Naaktgeboren C, Elsaesser F, Forsling ML, van der Weyden GC, Ellendorff F, and Smidt D (1979). Myometrial electrical activity and plasma concentrations of progesterone, estrogens and oxytocin during late pregnancy and parturition in the pigmy pig. *Biology of Reproduction* **21**:1125–1134.

Tavolga WN, Popper AN, and Fay RR (Eds) (1981). *Hearing and Sound Communication in Fishes*. Springer-Verlag, New York.

Taylor HL, Walker JM, and Medica PA (1967). Males of three normally parthenogenetic species of teiid lizards (Genus *Cnemidophorus*). *Copeia*:739–743.

Taylor MH (1986). Environmental and endocrine influences on reproduction of *Fundulus heteroclitus* (Pisces: Cyprinodontidae). *American Zoologist* **26**:159–171.

Taylor MH, DiMichelle L and Leach GJ (1977). Egg stranding in the life cycle of the mummichog, *Fundulus heteroclitus*. *Copeia*:397–399.

Taylor MH, Leach GJ, DiMichele L, Levitan WM, and Jacob WF (1979). Lunar spawning cycle in the mummichog, *Fundulus heteroclitus* (Pisces: Cyprinodontidae). *Copeia*: 291–297.

Taymor ML, Berger MJ, Thompson IE, and Karam K (1972). Hormonal factors in human ovulation. *American Journal of Obstetrics and Gynecology* **114**:445–452.

Te Winkle LE (1972). Histological and histochemical studies of post-ovulatory and preovulatory atretic follicles in Mustelus canis. *Journal of Morphology* **136**:433–458.

Teague LG and Bradley EL (1978). The existence of a puberty accelerating pheromone in the urine of the male prairie deermouse (*Peromyscus maniculatus bairdii*). *Biology of Reproduction* **19**:314–317.

Tegner HS (1964). The incidence of twins in roe deer. *Bulletin of the Mammal Society* **21**:10.

Telford SR Jr (1959). A study of the sand skink, *Neoseps reynoldsi* Stejneger. *Copeia*: 110–119.

Telford SR Jr and Campbell HW (1970). Ecological observations on an all female population of the lizard *Lepidophyma flavimaculatum* (Xantussiidae) in Panama. *Copeia*:379–381.

Temeles EJ (1985). Sexual size dimorphism of bird-eating hawks: The effect of prey vulnerability. *American Naturalist* **125**:485–499.

Temeles EJ (1986). Reversed sexual size dimorphism: Effect of resource defense and foraging behaviors on nonbreeding northern harriers. *Auk* **103**:70–78.

Temple-Smith PD and Bedford JM (1980). Sperm maturation and the formation of sperm pairs in the epididymis of the opossum, *Didelphis virginia*. *Journal of Experimental Zoology* **214**:161–171.

Temrin H (1986). Singing behaviour in relation to polyterritorial polygyny in the wood warbler (*Phylloscopus sibilatryx*). *Animal Behaviour* **34**:146–152.

Terner C (1968). Studies of metabolism in embryonic development—I. The oxidative metabolism of unfertilized and embryonated eggs of the rainbow trout. *Comparative Biochemistry and Physiology* **24**:933–940.

Terranova PF, Loker D, Garza F, and Martin N (1983). The LH surge: A signal regulating steroidogenesis of preantral follicles. In *Factors Regulating Ovarian Function* (Eds: Greenwald GS and Terranova PF) Raven, New York:39–43.

Tewksbury II HT and Conover DO (1987). Adaptive significance of intertidal egg deposition in the Atlantic silverside *Menidia menidia*. *Copeia*:76–83.

Thibault C, Ciurit M, Martinet L, Mauleon P, du Mesnil F, du Buisson P, Ortavant P, Pelletiev J, and Signoret JP (1966). Regulation of breeding season and oestrous cycles by light and external stimuli in some mammals. *Journal of Animal Science* **25** (Supplement):119–142.

Thibault RE and Schultz RJ (1978). Reproductive adaptations among viviparous fishes (Cyprinodontidae: Poeciliidae). *Evolution* **32**:320–333.

Thomas DA and Berfield RJ (1985). Ultrasonic vocalization of the female rat (*Rattus norvegicus*) during mating. *Animal Behaviour* **33**:720–725.

Thompson DBA, Thompson PS, and Nethersole-Thompson D (1986). Timing of breeding and breeding performance in a population of greenshanks (*Tringa nebularia*). *Journal of Animal Ecology* **55**:181–199.

Thompson DC (1978). Regulation of a northern grey squirrel (*Sciurus carolinensis*) population. *Ecology* **59**:708–715.

Thompson J (1977). Embryo-maternal relationship in a viviparous skink *Sphenomorphus quoyi* (Reptilia: Scincidae). In *Reproduction and Evolution* (Eds: Calaby JH and Tyndale-Biscoe CH) Australian Academy of Sciences, Canberra:279–280.

Thompson J (1982). Uptake of inorganic ions from the maternal circulation during development of the embryo of a viviparous lizard, *Sphenomorphus quoyii*. *Comparative Biochemistry and Physiology* **71A**:107–112.

Thorne JG and Foley CW (1974). Sperm metabolism: A function of the oviduct. In *The Oviduct and Its Functions* (Eds: Johnson AD and Foley CW) Academic, New York: 221–235.

Thornton VF and Evennett PJ (1969). Endocrine control of oocyte maturation and oviducal jelly release in the toad, *Bufo bufo* (L). *General and Comparative Endocrinology* **13**: 268–274.

Tilley SG (1977). Studies of life histories and reproduction in North American plethodontid salamanders. In *The Reproductive Biology of Amphibians* (Eds: Taylor DH and Guttman SI) Plenum, New York:1–41.

Timms AM and Keenleyside MHA (1975). The reproductive behaviour of *Aquidens paraguayensis* (Pisces, Cichlidae). *Zeitschrift fuer Tierpsychologie* **39**:8–23.

Tinkle DW (1962). Reproductive potential and cycles in female *Crotalus atrox* from northwestern Texas. *Copeia*:306–313.

Tinkle DW (1973). A population analysis of the sagebrush lizard, *Sceloporus graciosus*, in southern Utah. *Copeia*:284–296.

Tinkle DW and Gibbons JW (1977). The distribution and evolution of viviparity in reptiles. Miscellaneous Publications, University of Michigan, *Museum of Zoology* **154**.

Tinkle DW, Wilbur H, and Tilley SG (1970). Evolutionary strategies in lizard reproduction. *Evolution* **24**:55–74.

Tjernberg M (1983). Prey abundance and reproductive success of the golden eagle *Aquila chrysaetos* in Sweden. *Holarctic Ecology* **6**:17–23.

Toft CA, Trauger DL, and Murdy HW (1984). Seasonal decline in brood sizes of sympatric waterfowl (*Anas* and *Aythya*, Anatidae) and a proposed evolutionary explanation. *Journal of Animal Ecology* **53**:75–92.

Tokarz RR (1978). Oogonial proliferation, oogenesis and folliculogenesis in nonmammalian vertebrates. In *The Vertebrate Ovary* (Ed: Jones RE) Plenum, New York:145–179.

Toner JP and Adler NT (1986). Influence of mating and vaginocervical stimulation on rat uterine activity. *Journal of Reproduction and Fertility* **78**:239–247.

Townsend TJ and Wootton RJ (1984). Effects of food supply on the reproduction of the convict cichlid, *Cichlosoma nigrofasciatum*. *Journal of Fish Biology* **24**:91–104.

Trauth SE (1977). Winter collection of *Cnemidophorus sexlineatus* eggs from Arkansas. *Herpetological Review* **8**:33.

Travis J and Trexler JC (1986). Interactions among factors affecting growth, development, and survival in experimental populations of *Bufo terrestris* (Anura: Bufonidae). *Oecologia* **69**:110–116.

Trexler JC (1985). Variation in the degree of viviparity in the sailfin molly, *Poecilia latipinna*. *Copeia*:999–1004.

Tribble GW (1982). Social organization, patterns of sexuality, and behavior of the wrasse, *Coris dorsomaculata*, at Miyake-jima, Japan. *Environmental Biology of Fish* **7**:29–38.

Trillich F (1981). Mutual mother-pup recognition in Galapagos fur seals and sealions: Cues used and functional significance. *Behaviour* **78**:21–42.

Trivers RL (1972). Parental investment and sexual selection. In *Sexual Selection and the Descent of Man* 1871–1971 (Ed: Campbell B) Aldine, Chicago:136–179.

Troughton E (1947). Furred Animals of Australia. Scribner's, New York.

Troyer K (1987). Posthatching yolk in a lizard: Internalization and contribution to growth. *Journal of Herpetology* **21**:102–106.

Truscott B, Idler DR, Sundararaj BI, and Goswami SV (1978). Effects of gonadotropins and adrenocorticotropins on plasmatic steroids of the catfish, *Heteropneusta fossilis* (Bloch). *General and Comparative Endocrinology* **34**:149–157.

Tseng RWH (1972). Aspects of ovarian maturation in the molly, *Mollienesia latipinna*. *Biological Bulletin of Taiwan Normal University* **7**:138–146.

Tsonis CG and Sharpe RM (1986). Dual gonadal control of follicle-stimulating hormone. *Nature* **321**:724–725.

Tsui HW (1976). Stimulation of androgen production by the lizard testis: site of action of ovine FSH and LH. *General and Comparative Endocrinology* **28**:386–394.

Tucker HA (1981). Photoperiodic control of hormones, growth and lactation in cattle. In *Environmental Factors in Mammal Reproduction* (Eds: Gilmore D and Cook B) Macmillan, London.

Tucker I, Johnston PG, and Frost D (1980). Comparative physiological and biochronometric analysis of rodent seasonal reproductive cycles. In *Environmental Endocrinology* (Eds: Reiter RJ and Follet BK) *Progress in Reproductive Biology 5*. Karger, Basel:102–133.

Turkington RW, Lockwood DH, and Topper YJ (1967). The introduction of milk protein synthesis in post-mitotic mammary epithelial cells exposed to prolactin. *Biochemica et Biophysica Acta* **148**:475.

Turner BJ, Brett BH, and Miller RR (1980). Interspecific hybridization and the evolutionary origin of a gynogenetic fish, *Poecilia formosa*. *Evolution* **34**:917–922.

Turner CL (1936). The absorptive processes in the embryos of *Parabrotula dentiens*, a viviparous deep-sea brotulid fish. *Journal of Morphology* **59**:313–325.

Turner CL (1940a). Pericardial sac, trophotaeniae, and alimentary tract in embryos of goodeid fishes. *Journal of Morphology* **67**:272–289.

Turner CL (1940b). Adaptations for viviparity in jenynsiid fishes. *Journal of Morphology* **67**:219–297.

Turner CW (1973). *Harvesting Your Milk Crop*. Babson Brothers Dairy Research and Educational Service, Oak Brook, Illinois.

Turner FB (1977). The dynamics of populations of squamates, crocodilians and rhynchocephalians. In *Biology of the Reptilia*, vol 7 (Eds: Gans C and Tinkle DW) Academic, London:157–264.

Turner FB, Hayden P, Burge BL, and Robertson JB (1986). Egg production by the desert tortoise (*Gopherus agassizii*) in California. *Herpetologica* **42**:93–104.

Turner FB, Medica PA, and Lyons CL (1984). Reproduction and survival of the desert tortoise (*Scaptochelys agassizii*) in Ivanpah Valley, California. *Copeia*:811–820.

Turner G (1986). Teleost mating systems and strategies. In *The Behaviour of Teleost Fishes* (Ed: Pitcher TJ) Croom Helm, London and Sydney:253–274.

Twitty VC (1966). *Of Scientists and Salamanders*. Freeman, San Francisco.

Tyler MJ (Ed) (1983). *The Gastric Brooding Frog*. Croom Helm, Beckenham, Kent.

Tyler MJ and Carter DB (1981). Oral birth of the young of the gastric brooding frog *Rheobatrachus silus*. *Animal Behaviour* **29**:280–282.

Tyler MJ, Shearman DJC, Franco R, O'Brien P, Seamark RF, and Kelly R (1983). Inhibition of gastric acid secretion in the gastric brooding frog, *Rheobatrachus silus*. *Science* **220**: 609–910.

Tyler PA, Grant A, Pain SL, and Gage JD (1982). Is annual reproduction in deep-sea echinoderms a response to variability in their environment? *Nature* **300**:747–750.

Tyler SJ (1972). The behaviour and social organization of the New Forest ponies. *Animal Behaviour Monograph* **5**:87–196.

Tyndale-Biscoe CH (1973). *The Life of Marsupials*. Arnold, London.

Tyndale-Biscoe CH (1984). Mammals: Marsupials. In *Marshall's Physiology of Reproduction*, vol 1 (Ed: Lamming GE) Churchill Livingstone, Edinburgh:386–454.

Tyndale-Biscoe CH and Evans SM (1980). Pituitary-ovarian interactions in marsupials. In *Comparative Physiology: Primitive Mammals* (Eds: Schmidt-Nielsen K, Bolis L, and Taylor CR) Cambridge University, Cambridge:259–268.

Tyndale-Biscoe CH, Hearn JP, and Renfree MB (1974). Control of reproduction in macropod marsupials. *Journal of Endocrinology* **63**:589–614.

Tyndale-Biscoe H and Renfree M (1987). *Reproductive Physiology of Marsupials*. Cambridge University, Cambridge.

Uchida TA and Môri T (1987). Prolonged storage of spermatozoa in hibernating bats. In *Recent Advances in the Study of Bats* (Eds: Fenton EB, Racey P, and Rayner JMV) Cambridge University, Cambridge:351–365.

Uchida TA, Inoue C, and Kimura K (1984). Effects of elevated temperatures on the embryonic development and corpus luteum activity in the Japanese long-fingered bat, *Miniopterus schreibersii fuliginosus*. *Journal of Reproduction and Fertility* **71**:439–444.

Udry JR and Morris NM (1967). Seasonality of coitus and birth. *Demography* **4**:673–680.

Udvardy MDF (1969). *Dynamic Zoogeography*. Van Nostrand Reinhold, New York.

Underwood H (1979). Melatonin affects circadian rhythmicity in lizards. *Journal of Comparative Physiology* **130**:317–322.

Underwood H and Menaker M (1976). Extraretinal photoreception in lizards. *Photochemistry and Photobiology* **23**:227–243.

Urasaki H (1974). The function of the pineal gland in the reproduction of the medaka, *Oryzias latipes*. *Bulletin of Liberal Arts and Science Course, School of Medicine, Nihon University* **2**:11–17.

Urasaki H (1976). The role of pineal and eyes in the photoperiodic effect on the gonad of the medaka, *Oryzias latipes*. *Chronobiologia* **3**:228–234.

Urasaki H (1977). Response of the hypophysial-ovarian system of the teleost, *Oryzias latipes*, to administration of melatonin. *Bulletin of Liberal Arts and Science Course, School of Medicine, Nihon University* **5**:15–18.

Urasaki H, Abe T, and Mawatari SF (1981). Photoperiodicity of reproduction modulated by extraocular photoreceptive mechanism in *Oryzias latipes* I. Effect of light restriction on oocyte maturation, ovulation and spawning after removal of bilateral eyes. *Bulletin of the Liberal Arts and Science Course, School of Medicine, Nihon University* **9**:1–21.

Uzzell T (1970). Meiotic mechanisms of naturally occurring unisexual vertebrates. *American Naturalist* **104**:433–445.

Uzzell TM (1964). Relations of the diploid and triploid species of the *Ambystoma jeffersonianum* complex (Amphibia, Caudata). *Copeia*:257–300.

Uzzell TM and Goldblatt SM (1967). Serum protein of salamanders of the *Ambystoma jeffersonianum* complex. *Copeia*:602–612.

Vale W, Rivier J, Vaughan J, McClintock R, Corrigan A, Woo W, Karr D, and Spiess J (1986). Purification and characterization of an FSH releasing protein from porcine ovarian follicular fluid. *Nature* **321**:776–779.

van Balen JH (1980). Population fluctuations of the great tit and feeding conditions in winter. *Ardea* **68**:143–164.

van Beurden E (1979). Gamete development in relation to season, moisture, energy reserve and size in the Australian water-holding frog *Cyclorana platycephalus*. *Herpetologica* **35**: 370–374.

van de Graaf KM and Balda RP (1973). Importance of green vegetation for reproduction in the kangaroo rat, *Dipodomys merriami merriami*. *Journal of Mammalogy* **54**:509–512.

van den Hurk R and Lambert JGD (1983). Ovarian steroid glucuronides functions as sex hormones for male zebrafish, *Brachydanio rerio*. *Canadian Journal of Zoology* **61**: 2381–2387.

van der Horst CJ (1944). Remarks on the systematics of *Elephantulus*. *Journal of Mammalogy* **25**:77–82.

van der Lee S and Boot LM (1955). Spontaneous pseudopregnancy in mice. *Acta Physiologica et Pharmacologica Neerlandica* **4**:442–444.

van Horn RN (1975). Primate breeding season: Photoperiodic regulation in captive *Lemus catta*. *Folia Primatologica* **24**:203–220.

Van Mierop LHS and Barnard SM (1978). Further observations on the thermoregulation in the brooding female *Python molurus bivittatus* (Serpentes: Boidae). *Copeia*:615–621.

van Noordwijk AJ, van Balen JH, and Scharloo W (1980). Heritability of ecologically important traits in the great tit. *Ardea* **68**:193–203.

van Noordwijk AJ, van Balen JH, and Scharloo W (1981). Genetic variation in the timing of reproduction of the great tit. *Oecologia* **49**:158–166.

van Oordt PGWJ (1956). The role of temperature in regulating the spermatogenic cycle in the common frog (*Rana temporaria*). *Acta Endocrinologica* **23**:251–264.

van Oordt PGWJ (1982). Gonadotropic cells in the pituitary of teleosts. In *Proceedings of the International Symposium of Reproductive Physiology of Fish* (Comp: Richter CJJ and Goos HJT) Central Agricultural Publications and Documents, Wageningen:44–48.

van Wagenen G (1972). Vital statistics from a breeding colony: reproduction and pregnancy outcome in *Macaca mulatta*. *Journal of Medical Primatology* **1**:3–28.

van Wagenen G (1972). Vital statistics from a breeding colony. *Journal of Medical Primatology* **1**:3–10.

van Zinderen Bakker Sr EM (1976). The evolution of Late-Quaternary palaeoclimates of southern Africa. In *Palaeoecology of Africa and of the Surrounding Islands and of Antarctica*, vol 9 Balkema, Cape Town:160–202.

Vandenberg JG, Drickamer LC, and Colby DR (1971). Social and dietary factors in the sexual maturation of female mice. *Journal of Reproduction and Fertility* **28**:397–405.

Vandenbergh JG (1975) Hormones, pheromones and behavior. In *Hormonal Correlates of Behavior* (Eds: Sprott RL and Eleftheriou BE) Plenum, New York, **2**:551–584.

Vandenbergh JG, Finlayson JS, Dobrogosz WJ, Dills SS, and Kost TA (1976). Chromatographic separation of puberty acceleration pheromone from male mouse urine. *Biology of Reproduction* **15**:260–265.

Vandenbergh JG, Whittset JM, and Lonbardi JR (1975). Partial isolation of a pheromone accelerating puberty in female mice. *Journal of Reproduction and Fertility* **43**:515–523.

Vasal S and Sundararaj BI (1975). Response of the regressed ovary of the catfish, *Heteropneustes fossilis* (Bloch) to interrupted night periods. *Chronobiologia* **2**:224–239.

Vaughan TA and Vaughan RP (1987). Parental behavior in the African yellow-winged bat (*Lavia frons*). *Journal of Mammalogy* **68**:217–223.

Vehrencamp SL (1978). The adaptive significance of communal nesting in groove-billed anis (*Crotophaga sulcirostris*). *Behavioral Ecology and Sociobiology* **4**:1–33.

Vehrencamp SL, Bowen BS, and Koford RR (1986). Breeding roles and pairing patterns within communal groups of groove-billed anis. *Animal Behaviour* **34**:347–366.

Veith WJ (1974). Reproductive biology of *Chamaeleo pumilus pumilus* with special reference to the role of the corpus luteum and progesterone. *Zoologica Africana* **9**:161–183.

Veith WJ (1980). Viviparity and embryonic adaptations in the teleost *Clinus superciliosus*. *Canadian Journal of Zoology* **58**:1–12.

Verner J (1964). Evolution of polygyny in the long-billed marsh wren. *Evolution* **18**:252–261.

Verrell PA (1986). Male discrimination of larger, more fecund females in the smooth newt, *Triturus vulgaris*. *Journal of Herpetology* **20**:416–422.

Verrell PA (1985). Male mate choice for large, fecund females in the red spotted newt, *Notophthalmus viridescenes*: How is size assessed? *Herpetologica* **41**:382–386.

Versi E, Chiappa SA, Fink G, and Charlton HM (1982). Effect of copulation on hypothalamic content of gonadotrophic hormone-releasing hormone in the vole *Microtus agrestis*. *Journal of Reproduction and Fertility* **64**:491–494.

Vial JA and Stewart JR (1985). The reproductive cycle of *Barisia monticola*: As unique variation among viviparous lizards. *Herpetologica* **41**:51–57.

Vial JL (1968). The ecology of the tropical salamander *Bolitoglossa subpalmata* in Costa Rica. *Review of Biology in the Tropics* **15**:117–121.

Viljoen S and Du Toit SHC (1985). Postnatal development and growth of southern African tree squirrels in the genera *Funisciurus* and *Paraxerus*. *Journal of Mammalogy* **66**:119–127.

Vining DR Jr (1986). Social versus reproductive success: The central theoretical problem of human sociobiology. *Behavioral and Brain Sciences* **9**:167–216.

Visser J (1967). First report of ovoviviparity in a southern African amphisbaenid, *Monopeltis c. capensis*. *Zoologica Africana* **3**:111–113.

Visser J (1975). Oviparity in two South African skinks of the genus *Mabuya*, with notes on hatching. *Zoologica Africana* **10**:209–213.

Vitt LJ (1973). Reproductive biology of the anguid lizard, *Gerrhonotus coeruleus princeps*. *Herpetologica* **29**:176–184.

Vitt LJ (1986). Reproductive tactics of sympatric gekkonid lizards with a comment on the evolutionary and ecological consequences of invariant clutch size. *Copeia*:773–786.

Vitt LJ and Blackburn DG (1983). Reproduction in the lizard *Mabuya heathi* (Scincidae): A commentary on viviparity in new world Mabuya. *Canadian Journal of Zoology* **61**: 2798–2806.

Vitt LJ and Congdon JD (1978). Body shape, reproductive effort and relative clutch mass: Resolution of a paradox. *American Naturalist* **112**:595–608.

Vitt LJ and Cooper WE Jr (1986). Skink reproduction and sexual dimorphism: *Eumeces fasciatus* in the southeastern United States with notes on *Eumeces inexpectatus*. *Journal of Herpetology* **20**:65–76.

Vodicnik MJ, Olcese J, Delahaunty G, and de Vlaming V (1979). The effects of blinding, pinealectomy and exposure to constant dark conditions on gonadal activity in the female goldfish, *Carassius auratus*. *Environmental Biology of Fish* **4**:173–184.

vom Saal FS (1983). The interaction of circulating oestrogens and androgens in regulating mammalian sexual differentiation. In *Hormones and Behaviours in Higher Vertebrates* (Eds: Balthazart J, Prove E and Gilles R) Springer-Verlag, Berlin:159–177.

vom Saal FS, Pryor S, and Bronson FH (1981a). Effects of prior intrauterine position and housing on oestrous cycle length in adolescent mice. *Journal of Reproduction and Fertility* **62**:33–37.

vom Saal FS, Pyror S, and Bronson FH (1981b). Change in oestrous cycle length during adolescence in mice is influenced by prior intrauterine position and housing. *Journal of Reproduction and Fertility* **62**:33–37.

Vondracek B, Wurtsbaugh WA, and Cech JJ Jr (1988). Growth and reproduction of the mosquitofish, *Gambusia affinis*, in relation to temperature and ration level: Consequences for life history. *Environmental Biology of Fishes* **21**:45–57.

Vorhies CT and Taylor WP (1933). The life histories and ecology of jackrabbits. *Lepus alleni* and *Lepus californicus* ssp., in relation to grazing in Arizona. *University of Arizona Agricultural Experiment Station Bulletin* **49**:471–587.

Wachtel SS, Koo GC, and Boyse EA (1975). Evolutionary conservation of the H-Y ("male") antigen. *Nature* **254**:270–272.

Wada M (1983). Environmental cycles, circadian clock, and androgen-dependent behavior in birds. In *Avian Endocrinology* (Eds: Mikami S, Homma K and Wada M) Japan Scientific Societies, Tokyo, and Springer-Verlag, Berlin:191–200.

Wake MH (1968). Evolutionary morphology of the caecilian urogenital system I. The gonads and the fat bodies. *Journal of Morphology* **126**:291–332.

Wake MH (1970). Evolutionary morphology of the caecilian urogenital system. Part. II. The kidneys and urogenital ducts. *Acta Anatomica* **75**:321–358.

Wake MH (1976). The development and replacement of teeth in viviparous caecilians. *Journal of Morphology* **148**:33–64.

Wake MH (1977a). The reproductive biology of caecilians: An evolutionary perspective. In *The Reproductive Biology of Amphibians* (Eds: Taylor DH and Guttman SI) Plenum, New York:73–101.

Wake MH (1977b). Fetal maintenance and its evolutionary significance in the Amphibia. *Journal of Herpetology* **11**:379–386.

Wake MH (1980a). Reproduction, growth and population structure of the Central American caecilian *Dermophis mexicanus*. *Herpetologica* **36**:244–256.

Wake MH (1980b). The reproductive biology of *Nectrophrynoides malcomi* (Amphibia: Bufonidae), with comments on the evolution of reproductive modes in the genus *Nectophrynoides*. *Copeia*:193–209.

Wake MH (1981). Structure and function of the male Müllerian gland in caecilians, with comments on its evolutionary significance. *Journal of Herpetology* **15**:17–22.

Wake MH (1982). Diversity within a framework of constraints. Amphibian reproductive modes. In *Environmental Adaptation and Evolution* (Eds: Mossakowski D and Roth G) Gustav Fischer, Stuttgart:87–106.

Walker FMM and Peaker M (1978). Production of prostaglandins by goat mammary gland in vivo in relation to lactogenesis and parturition. *Journal of Endocrinology* **77**:61P–62P.

Walker FMM and Peaker M (1981). Prostaglandins and lactation. *Acta Veterinaria* **577**:299–310.

Wallace CR (1967). Observations on the reproductive behavior of the black bullhead (*lctalurus melas*). *Copeia*:852–853.

Wallace GI (1978). A histological study of early stages of pregnancy in the bent-winged bat (*Miniopterus schreibersii*) in north-eastern New South Wales. *Australia Journal of Zoology, London* **185**:519–537.

Wallace RA (1978). Oocyte growth in nonmammalian vertebrates. In *The Vertebrate Ovary* (Ed: Jones RE) Plenum, New York:469–502.

Wallace RA and Jared DW (1969). Studies on amphibian yolk. VIII. The estrogen-induced hepatic synthesis of a serum lipophosphoprotein and its selective uptake by the ovary and transformation into yolk platelet proteins in *Xenopus laevis*. *Developmental Biology* **19**:498–526.

Wallace RA, Jared DW, and Eison AZ (1966). A general method for isolation and purification of phosvitin from vertebrate eggs. *Canadian Journal of Biochemistry* **44**:1647–1655.

Walser ES (1977). Maternal behaviour in mammals. In *Comparative Aspects of Lactation* (Ed: Peaker M) Symposium of the Zoological Society of London **41**:313–331.

Walther FR (1984). *Communication and Expression in Hoofed Mammals*. Indiana University, Bloomington.

Ward JA and Samarakoon JI (1981). Reproductive tactics of the Asian cichlids of the genus *Entroplus* in Sri Lanka. *Environmental Biology of Fish* **6**:95–103.

Warham J (1975). The crested penguins. In *The Biology of Penguins* (Ed: Stonehouse B) Macmillan, London:189–369.

Warner RR (1975). The adaptive significance of sequential hermaphroditism in animals. *American Naturalist* **109**:61–82.

Warner RR (1978). The evolution of hermaphroditism and unisexuality in aquatic and terrestrial vertebrates. In *Contrasts in Behavior* (Eds: Reese ES and Lighter FJ) Wiley, New York:77–101.

Warner RR (1984). Mating behavior and hermaphroditism in coral reef fishes. *American Scientist* **72**:128–136.

Wartenberg H (1983). Structural aspects of gonadal differentiation in mammals and birds. *Differentiation* **23** (Supplement):S64–S71.

Wassarman PM (1987). The biology and chemistry of fertilization. *Science* **235**:553–560.

Wassersug RJ (1974). Evolution of anuran life cycles. *Science* **185**:377–378.

Wathes DC (1984). Possible actions of gonadal oxytocin and vasopressin. *Journal of Reproduction and Fertility* **71**:315–345.

Wathes DC and Swann RW (1982). Is oxytocin an ovarian hormone? *Nature* **297**:225–227.

Watkins D, Watson A, and Miller GR (1963). Population studies on red grouse, *Lagopus lagopus scoticus* (Lath.) in north-east Scotland. *Journal of Animal Ecology* **32**:317–377.

Watson A and Moss R (1980). Advances in our understanding of the population dynamics of red grouse from recent fluctuations in numbers. *Ardea* **68**:103–111.

Watson A and Parr R (1981). Hormone implants affecting territory size and aggressiveness and sexual behaviour in red grouse. *Ornis Scandinavica* **12**:55–61.

Watson BP (1944). The menopause patient. *Journal of Clinical Endocrinology* **4**:571–574.

Webley GE and Luck MR (1986). Melatonin directly stimulates the secretion of progesterone by human and bovine granulosa cells in vitro. *Journal of Reproduction and Fertility* **78**: 711–717.

Weekes HC (1935). On the distribution, habitat and reproductive habits of certain European and Australian snakes and lizards with particular regard to their adoption of viviparity. *Proceedings of the Zoological Society of London* **1935**:625–645.

Wehrenberg WB and Dyrenfurth I (1983). Photoperiod and ovulatory menstrual cycles in female macaque monkeys. *Journal of Reproduction and Fertility* **68**:119–122.

Weir BJ (1974). Reproductive characteristics of hystricomorph rodents. In *Biology of Hystricomorph Rodents* (Eds: Rowlands IW and Weit BJ) Zoological Society of London Symposium, no. 34. Academic, London:265–301.

Weir BJ and Rowlands IW (1977). Ovulation and atresia. In *The Ovary*, vol 1 (Eds: Zuckerman S and Weir BJ) Academic, New York:265–301.

Weir WJ (1973). The role of the male in the evocation of estrus in the cuis, *Galea musteloides*. *Journal of Reproduction and Fertility*, Supplement **19**:421–432.

Wells KD (1977). The social behaviour of anuran amphibians. *Animal Behaviour* **25**:666–693.

Wells KD (1979). Reproductive behaviour and male mating success in a neotropical toad, *Bufo typhonius*. *Biotropica* **11**:301–307.

Wells LJ (1942). The response of the testis to androgens following hypophysectomy. *Anatomical Record* **82**:565–585.

Wentworth BC, Proudman JA, Opel H, Wineland MJ, Zimmermann NG, and Lapp A (1983). Endocrine changes in the incubating and brooding turkey hen. *Biology of Reproduction* **29**:87–92.

Werner JK (1969). Temperature-photoperiod effects on spermatogenesis in the salamander *Plethodon cinereus*. *Copeia*:592–602.

Westerheim SJ (1975). Reproduction, maturation and identification of larvae of some *Sebastes* (Scorpaenidae) species in the North Pacific Ocean. *Journal of the Fisheries Research Board of Canada* **32**:2399–2411.

Westphal U, Stroupe SD, and Cheng S-L (1977). Progesterone binding to serum proteins. *Annals of the New York Academy of Sciences* **286**:10–27.

Wharton LR (1967). *The Ovarian Hormones. Safety of the Pill; Babies After Fifty*. Charles C Thomas, Springfield, Illinois.

White JB and Murphy GG (1973). The reproductive cycle and sexual dimorphism of the common snapping turtle, *Chelydra serpentina serpentina*. *Herpetologica* **29**:240–246.

White JCD (1953). Composition of whales' milk. *Nature* **171**:612.

Whitehead C, Bromage NR, Forster JRM, and Matty AJ (1978). The effects of alterations in photoperiod on ovarian development and spawning time in the rainbow trout (*Salmo gairdneri*). *Annales de Biologie Animale Biochimie Biophysique* **18**:1035–1043.

Whitten WK (1956). Modifications of the estrous cycle of the mouse by external stimuli associated with the male. *Journal of Endocrinology* **13**:399–404.

Whitworth MR (1984). Maternal care and behavioural development in pikas, *Ochotona princeps*. *Animal Behaviour* **32**:743–752.

Wiberg U, Mayerova A, Muller U, Fredga K, and Wolf U (1982). X-linked genes of the H-Y antigen system in the wood lemming (*Myopus schisticolor*). *Human Genetics* **60**:163–166.

Wickler W and Seibt U (1983). Monogamy: An ambiguous concept. In *Mate Choice* (Ed: Bateson P) Cambridge University, Cambridge:33–50.

Widdowson EW (1981). The role of nutrition in mammalian reproduction. In *Environmental Factors in Mammalian Reproduction* (Eds: Gilmore D and Cook B). Macmillan, London:145–159.

Wiebe PJ (1968a). The effects of temperature and daylength on the reproductive physiology of the viviparous seaperch, *Cymatogaster aggregata* Gibbons. *Canadian Journal of Zoology* **46**:1207–1219.

Wiebe PJ (1968b). The reproductive cycle of the viviparous sea perch, *Cymatogaster aggregata* Gibbons. *Canadian Journal of Zoology* **46**:1221–1235.

Wiegand MD (1982). Vitellogenesis in fishes. In *Proceedings of the International Symposium of Reproductive Physiology of Fish* (Eds: Richter CJJ and Goos HJT) Central Agricultural Publication Documents, Wageningen:136–146.

Wiger R (1979). Demography of a cyclic population of the bank vole *Clethrionomys glareolus*. *Oikos* **33**:373–385.

Wiger R (1982). Roles of self regulatory mechanisms in cyclic populations of *Clethrionomys* with special reference to *C. glareolus*: A hypothesis. *Oikos* **38**:60–71.

Wikman M and Tarsa V (1980). Food habits of the goshawk during the breeding season in southwestern Finland 1969–1977. *Suomen Riista* **28**:86–96.

Wilber HM and Collins JP (1973). Ecological aspects of amphibian metamorphosis. *Science* **182**:1305–1314.

Wilberg U, Mayerova A, Muller U, Fredga K, and Wolf U (1982). X-linked genes of the H-Y antigen system in the woodlemming (*Myopus schistocolor*). *Human Genetics* **60**:163–166.

Wilbrand U, Porath C, Matthaes P, and Jaster R (1959). Der Einfluss der Ovarialsteroide auf die Funktion des Atemzentrums. *Archiv für Gynakologie* **191**:507–531.

Wilbur HM, Rubenstein DI, and Fairchild L (1978). Sexual selection in toads: The roles of female choice and male body size. *Evolution* **32**:264–270.

Wildt L, Marshall G, and Knobil E (1980). Experimental induction of puberty in the infantile female rhesus monkey. *Science* **207**:1373–1375.

Wilkins NP (1967). Starvation of the herring, *Clupea harengus* L: Survival and some gross biochemical changes. *Comparative Biochemistry and Physiology* **23**:503–518.

Williams AA, Martan J, and Brandon RA (1985). Male cloacal gland complex of *Eurycea lucifuga* and *Eurycea longicauda* (Amphibia; Plethodontidae). *Herpetologica* **41**:272–281.

Williams GC (1966a). Natural selection, the cost of reproduction, and a refinement of Lack's principle. *American Naturalist* **100**:687–690.

Williams GC (1966b). *Adaptation and Natural Selection*. Princeton University, Princeton, New Jersey.

Williams EC Jr and Parker WS (1987). A long-term study of a box turtle (*Terrapene carolina*) population at Allee Memorial Woods, Indiana, with emphasis on survivorship. *Herpetologica* **43**:328–335.

Williams MJ (1974). Creching behaviour of the shelduck *Tadorna tadorna* L. *Ornis Scandinavica* **5**:131–143.

Williams RF and Hodgen GD (1983a). Mechanisms of lactational anovulation in primates: Nursing increases ovarian steroidogenesis. In *Factors Regulating Ovarian Function* (Eds: Greenwald GS and Terronova PF) Raven, New York:17–21.

Williams RF and Hodgen GD (1983b). Initiation of the primate ovarian cycle with emphasis on perimenarchial and postpartum events. In *Reproductive Physiology IV* (Ed: Greep RO) *International Review of Physiology* **27**:1–55.

Willig MR (1985a). Reproductive patterns of bats from Caatingas and Cerrado biomes in northeast Brazil. *Journal of Mammalogy* **66**:668–681.

Willig MR (1985b). Ecology, reproductive biology and systematics of *Neoplatymops mattogrossensis* (Chiroptera: Molossidae). *Journal of Mammalogy* **66**:618–628.

Wilson DE and Findley JS (1970). Reproductive cycle of a neotropical insectivorous bat, *Myotis nigracans*. *Nature* **255**:1155.

Wilson FE and Follett BK (1975). Corticosterone-induced gonadosuppression in photostimulated tree sparrows. *Life Sciences* **17**:1451–1456.

Wilson JD, George FW, and Griffin JE (1981). The hormonal control of sexual development. *Science* **211**:1278–1284.

Wilson M, Daly M, and Behrends P (1985). The estrous cycle of two species of kangaroo rats (*Dipodomys microps* and *D. merriami*). *Journal of Mammalogy* **66**:726–732.

Wilson TC and Behrens DW (1982). Concurrent sexual behavior in three groups of gray whales, *Eschrictius robustus*, during the northern migration off the central California coast. *California Fish and Game* **68**:50–53.

Wimsatt WA (1975). Some comparative aspects of implantation. *Biology of Reproduction* **12**:1–40.

Wimsatt WA and Trapido H (1952). Reproduction and the female reproductive cycle in the tropical American vampire bat, *Desmodus rotundus murinus*. *American Journal of Anatomy* **91**:415–446.

Wingfield JC, Crim JW, Mattocks PW Jr, and Farner DS (1979). Responses of photosensitive male white-crowned sparrows (*Zonotrichia leucophrys gambelii*) to synthetic luteinizing releasing hormone (Syn-LHRH). *Biology of Reproduction* **21**:801–806.

Wingfield JC and Farner DS (1980). Control of seasonal reproduction in temperate-zone birds. In *Seasonal Reproduction of Higher Vertebrates* (Eds: Reiter RJ and Follett BK) S Karger, Basel:62–201.

Winkler DW (1985). Factors determining a clutch size reduction in California gulls (*Larus californicus*): A multi-hypothesis approach. *Evolution* **39**:667–677.

Winkler DW and Walters JR (1983). The determination of clutch size in precocial birds. *Current Ornithology* **1**:33–68.

Winokur RM and Legler JM (1975). Chelonian mental glands. *Journal of Morphology* **147**:275–292.

Wise DA and Pryor TL (1977). Effects of ergocornine and prolactin on aggression in the postpartum golden hamster. *Hormones and Behavior* **8**:30–39.

Witschi E (1962). Embryology of the ovary. In *The Ovary* (Eds: Grady HG and Smith DC) Williams and Wilkins, Baltimore:1–10.

Wittenberger JF (1979). The evolution of mating systems in birds and mammals. In *Handbook of Behavioral and Neural Biology*, vol 3 (Eds: Marler P and Vandenburgh J) *Social Behavior and Communication*. Plenum, New York:271–349.

Wittenberger JF (1980). Vegetation structure, food supply and polygyny in bobolinks (*Dolichonyx oryzivorous*). *Ecology* **61**:140–150.

Wolf U, Fraccaro M, Mayerova A, Hecht T, Maraschio P, and Hameister H (1980). A gene controlling H-Y antigen on the X-chromosome. Tentative assignment by deletion mapping to Xp 223. *Human Genetics* **54**:149–154.

Wolff JO and Lidicker WZ Jr (1980). Population ecology of the taiga vole, *Microtus xanthognathus*, in interior Alaska. *Canadian Journal of Zoology* **58**:1800–1812.

Wolfson A (1960). Regulation of annual periodicity in the migration and reproduction of birds. *Cold Spring Harbor Symposia on Quantitative Biology* **25**:507–514.

Woodhead AD (1979). Senescence in fishes. In *Fish Phenology: Anabolic Adaptiveness in Fishes* (Ed: Miller PJ) *Zoological Society of London Symposium*, no. 44.179–205.

Woolfenden GE (1981). Selfish behavior in Florida scrub jay helpers. In *Natural Selection and Social Behavior: Recent Research and New Theory* (Eds: Alexander RA and Tinkle D) Chiron, New York:257–260.

Woolfenden GE and Fitzpatrick JW (1984). The Florida scrub jay. Demography of a cooperative-breeding bird. *Monography in Population Biology* 20. Princeton University, Princeton, New Jersey.

Wootton RJ (1973a). The effect of size of food ration on egg production in the female three-spined stickleback, *Gasterosteus aculeatus* L. *Journal of Fish Biology* **5**:89–96.

Wootton RJ (1973b). Fecundity of the three-spined stickleback, *Gasterosteus aculeatus* L. *Journal of Fish Biology* **5**:683–688.

Wootton RJ (1979). Energy costs of egg production and environmental determination of fecundity in teleost fishes. Fish phenology: Anabolic adaptiveness in fishes. *Zoological Society of London Symposium*, no. 44.133–159.

Wootton RJ (1982). Environmental factors in fish reproduction. In *Reproductive Physiology of Fish* (Eds: Richter CJJ and Goos HJT) Pudoc, Wageningen:210–219.

Wootton RJ (1984). Strategies and tactics in fish reproduction. In *Fish Reproduction: Strategies and Tactics* (Eds: Potts GW and Wootton RJ) Academic, London:1–12.

Wourms JP (1977). Reproduction and development in chondrichthyan fishes. *American Zoologist* **17**:379–410.

Wourms JP (1981). Viviparity: The maternal-fetal relationship in fishes. *American Zoologist* **21**:473–515.

Wourms JP and Bayne O (1973). Development of the viviparous brotulid fish, *Dinematichthys ilucoeteoides*. *Copeia*:32–40.

Wourms JP and Lombardi J (1979). Cell ultrastructure and protein absorption in the trophotaeniae epithelium, a placental analogue of viviparous fish embryos. *Journal of Cell Biology* **83**:399a.

Wourms JP and Sheldon H (1976). Annual fish oogenesis. II. Formation of the secondary egg envelopes. *Developmental Biology* **50**:338–354.

Wurtman RJ (1975). The effects of light on man and other animals. *Annual Review of Physiology* **37**:467–483.

Wurtman RJ and Axelrod J (1964). Light and melatonin synthesis in the pineal. *Federation Proceedings* **23**:206.

Wurtman RJ, Axelrod J, and Chu EW (1963). Melatonin, a pineal substance: Effect on rat ovary. *Science* **141**:277–278.

Wurtman RJ, Axelrod J, and Kelly DE (1968). *The Pineal*. Academic, New York.

Wynn AH and Zug GR (1985). Observations on the reproductive biology of *Candoia carinata* (Serpentes, Boidae). *The Snake* **17**:15–24.

Xavier F (1970). Action moderatrice de la progesterones sur la croissance des embryons chez *Nectophrynoides occidentalis* Angel. *Comptes Rendus Academie Sciences, Paris (D)* **270**:2115–2117.

Xavier F (1977). An exceptional reproductive strategy in Anura: *Nectrophrynoides occidentalis* Angel (Bufonidae), an adaptation to terrestrial life by viviparity. In *Major Patterns of Vertebrate Evolution* (Ed: Hecht MK) Plenum, New York:545–552.

Xavier F and Ozon R (1971). Recherches sur l'activité endocrine de l'ovarie de *Nectrophrynoides occidentalis* Angel (Amphibien Anoure vivipare). II. Synthesè in vitro des steroides. *General and Comparative Endocrinology* **16**:30–40.

Xavier F, Zuber-Vogeli M, and LeQuang Trong Y (1970). Recherches sur l'activité endocrine de l'ovaire de *Nectophrynoides occidentalis* Angel (Amphibian Anoure vivipare). *General and Comparative Endocrinology* **15**:425–431.

Yamamoto TO (1958). Artificial induction of functional sex-reversals in genotypic females of the medaka (*Oryzias latipes*). *Journal of Experimental Zoology* **137**:227–260.

Yanagimachi R (1973). The movement of golden hamster spermatozoa before and after capacitation. In *Sperm Capacitation* (Ed: Hamner CE) MSS Information Corporation, New York:99–102.

Yaron Z (1977). Embryo-maternal interrelations in the lizard Xantusia vigilis. In *Reproduction and Evolution* (Eds: Calaby JH and Tyndale-Biscoe CH) Australian Academy of Science, Canberra:271–277.

Yaron Z (1985). Reptilian placentation and gestation: Structure, function and endocrine control. In *Biology of the Reptilia, vol 15, Development* (Eds: Gans C and Billet F) Wiley, New York:527–603.

Yates SG (1971). Toxin-producing fungi from fescue pasture. In *Microbial Toxins, vol 7, Algal and Fungal Toxins* (Eds: Kadis S, Ciegler A, and Ajil SJ) Academic, New York:191–206.

Yogev L and Terkel J (1978). The temporal relationship between implantation and termination of the nocturnal prolactin surges in pregnant rats. *Endocrinology* **102**:160–165.

Yokoyama K and Farner DS (1976). Photoperiodic responses in bilaterally enucleated female white-crowned sparrows, *Zonotrichia leucophrys gambelli*. *General and Comparative Endocrinology* **30**:528–533.

Yom-Tov Y (1985). The reproductive rates of Australian rodents. *Oecologia* **66**:250–255.

Yoshinaga K (1978). Cyclic hormone secretion by the mammalian ovary. In *The Vertebrate Ovary* (Ed: Jones RE) Plenum, New York:691–729.

Youatt WG, Verme LJ, and Ullrey DE (1965). Composition of milk and blood in nursing white-tailed does and blood composition of their fawns. *Journal of Wildlife Management 29*:79–84.

Young IR and Renfree MB (1979). The effects of corpus luteum removal during gestation on parturition in the tammary wallaby (*Macropus eugenii*). *Journal of Reproduction and Fertility* **56**:249–254.

Young PC and Martin RB (1982). Evidence for protogynous hermaphroditism in some lethrinid fishes. *Journal of Fish Biology* **21**:475–484.

Yu JYL, Dickoff WW, Swanson P, and Gorbman A (1981). Vitellogenesis and its hormonal regulation in the Pacific hagfish, *Eptatretus stouti* L. *General and Comparative Endocrinology* **43**:492–502.

Zaborski P (1982). Expression of the H-Y antigen in nonmammalian vertebrates and its relation to sex differentiation. In *Proceedings of the International Symposium of Reproduction and Physiology of Fish* (Comp: Richter CJJ and Goos HJT) Central Agricultural Publishing and Documents, Wageningen: 64–68.

Zalisko EJ, Brandon RA, and Martan J (1984). Microstructure and histochemistry of salamander spermatophores (Ambystomidae, Salamandridae and Plethodontidae). *Copeia*:739–747.

Zarrow MX, Denenberg VH, and Anderson CO (1965). Rabbit frequency of suckling in the pup. *Science* **150**:1835–1836.

Zatz M (1981). Pharmacology of the pineal gland. In *The Pineal Gland. I. Anatomy and Biochemistry* (Ed: Reiter RJ) CRC, Boca Raton:229–242.

Zimmerman DR (1979). Selection for reproductive characteristics. In *Animal Reproduction* (Ed: Hawk HW) Beltsville Symposia in Agricultural Research 3 Allanheld, Osmun, Montclair:131–142.

Zimmerman NH and Menaker M (1975). Neural connections of sparrow pineal: Role in circadial control of activity. *Science* **190**:977–979.

Zohar Y, Breton B, and Fostier A (1982). Gonadotropic function during the reproductive cycle of the female rainbow trout, *Salmo gairdneri*, in relation to ovarian steroid secretion: In vivo and in vitro studies. In *Proceedings of the International Symposium on Reproductive Physiology of Fish* (Eds: Richter CJJ and Goos HJT) Pudoc, Wageningen:14–18.

Zoran MJ and Ward JA (1983). Parental egg care behavior and fanning activity for the orange chromid, *Entroplus maculatus*. *Environmental Biology of Fishes* **8**:301–310.

Zucker I (1985). Pineal gland influences period of circannual rhythms of ground squirrels. *American Journal of Physiology* **249**:R111–R115.

Zucker I, Johnston PG, and Frost D (1980). Comparative physiological and biochronometric analysis of rodent seasonal reproductive cycles. In *Environmental Endocrinology* (Eds: Reiter RJ and Follett BK). *Progress in Reproductive Biology 5*. Karger, Basel:102–133.

Zucker I and Licht P (1983). Circannual and seasonal variations in plasma luteinizing hormone levels on ovariectomized ground squirrels (*Spermophilus lateralis*). *Biology of Reproduction* **28**:178–185.

Zuckerman S and Baker TG (1977). The development of the ovary and the process of oogenesis. In *The Ovary*, 2nd ed, vol 1 (Eds: Zuckerman S and Weir BJ) Academic, New York:41–67.

INDEX